D0940565

UNDERSTANDING PHYSICAL CHEMISTRY

PROKOPIS SIVENAS

UNDERSTANDING PHYSICAL CHEMISTRY

Second Edition

Arthur W. Adamson

University of Southern California

W. A. BENJAMIN, INC.
Menlo Park, California • Reading, Massachusetts
London • Amsterdam • Don Mills, Ontario • Sydney

UNDERSTANDING PHYSICAL CHEMISTRY

Second Edition

Copyright © 1969 by W. A. Benjamin, Inc.
All rights reserved
SBN8053-0127-5

Library of Congress Catalog Card Number 73-80662

Manufactured in the United States of America

*The manuscript was put into production on January 15, 1969;
this volume was published on September 15, 1969*

ISBN 0-8053-0127-5
HIJKLMN-AL-7987

Preface to the First Edition

General introduction

This collection of problems and worked-out answers is offered as something different from and, I hope, more helpful than the usual study aid for introductory physical chemistry. Students who follow the procedure recommended (see below) should find that this book can function as something like a "teaching machine" in leading them in a positive way into a better understanding of physical chemistry. The process will not be effortless but, given the effort, students at any level of capability should find that their ability to do physical chemistry has been quite materially improved.

Instructors should find this a source book of reasonably stiff examination questions. The chapters on wave mechanics and group theory can be used as source material for lectures on these subjects. They provide an approach to chemical bonding that has considerable depth, yet can be handled by first-year physical-chemistry students.

As a preamble to explaining the philosophy of this book, I would like to comment briefly on the role of problems and of problem solving in the teaching of physical chemistry. It is an American tradition, and basically a correct one, I believe, that the best test of a person's understanding of a scientific subject is his ability to use it. In physical chemistry, then, examinations in this country typically tend to stress numerical problems rather than the more philosophical essay type of question.

There is a second aspect to the use of numerical problems that also is important. Such problems may be just as easy or just as hard, just as straightforward or just as subtle, as essay questions. The virtue of the numerical problem is therefore not in any intrinsic difference in the level of understanding required to answer it, as compared to essay questions. Its value is simply that, being numerical, it is concrete, and the student is thus required to so sharpen his thinking as to be able to decide exactly what the question means and exactly how he will answer it. The numerical problem is thus a pedagogical device to ensure clear and unambiguous answers and to expose sloppy thinking.

There are several distinct types of numerical problems. In the lecture course of physical chemistry as taught at the University of Southern California, homework assignments include a group of straightforward "substitution" exercises, partly assigned ahead of the lectures as to topic and, frankly, designed to direct the average student's attention to the forthcoming material. A second part of the homework will consist of relatively long and difficult problems; certain calculations are intrinsically lengthy, yet need to be appreciated at least once. In addition, on a homework assignment, students can be expected to have to wrestle a bit with more advanced subject matter.

Neither of the above two types of problems is used in our examinations, however. I am interested in how well the student understands the material; there is little value in this respect in questions asking for a recitation of memorized equations or derivations, or in mere substitution problems. The second type of homework problem is not of much use in an examination situation, either. There is no time for lengthy calculations, nor for pondering some aspect not emphasized in lecture—unless, of course, one is able to assign weekend take-home examinations.

What has evolved instead, in the case of the quizzes used here, is a type of problem that requires the use only of those definitions, derivations, and equations so central that the student can reasonably be expected to know them if he has done even the first group of homework problems. In the examination, however, a situation is described that calls for a change or adaptation of the equation or derivation. The change is simple and straightforward, but it nonetheless must be determined from an analysis of the problem.

Consider the ideal gas law—a simpler equation is difficult to imagine. Yet, until they refresh their thinking, even graduate students have trouble working the problems in Chapter 1 within the time limits. (Dare I admit this may also be true of professors?) It is not that the problems are intrinsically difficult—when you see the answers, they seem ridiculously easy and often need barely a touch of the slide rule. It is just that one has to think rather carefully to see just how the gas law must be manipulated to fit the particular situation.

Such problems are *not* "trick" problems. Each is designed to test whether the student has so well grasped a particular aspect of a principle that he can recognize its implicit presence in the question, and do so quickly.

The following analogy may help to present what I have in mind. Let us suppose that a person who has never before seen an elephant is allowed to view one, posed, a slight distance away. The various features of the animal are explained carefully, and the person feels that he knows the beast. Yet the same elephant charging toward him, screaming and with raised trunk, or running away or bathing itself or feeding or viewed from the backside or the underside or from a perch behind the ears can look very different indeed. Only after many experiences of seeing the elephant in a variety of circumstances and actions will the person finally begin to know what an elephant is like, how he will act in a given situation, what his limitations are, and how to recognize him no matter how well he is hidden in the jungle.

Physical–chemical relationships can be much like this elephant. Typically, a particular one has been derived and explained in a clear, logical, and straightforward manner and illustrated with a numerical example. It seems entirely limpid and reasonable. The student feels that he understands it perfectly and, later, cannot explain how he could possibly have had so much trouble with it on the quiz.

As with the elephant, a concept or an equation if viewed from a new reference point can suddenly look very strange. Thus the ideal gas law may seem very elementary indeed, until the student is asked to apply it to a situation involving a gas distributed between two interconnected flasks that are not at the same temperature. The freezing-point-depression effect, whereby dissolved solute B in solvent A lowers the freezing point of the latter, is given by a plain enough equation whose use is easy. It is another matter to recognize that the same equation treats the solubility of solid A in solvent B, that is, to see that this is merely a restatement of a freezing-point depression from a different point of view, and with an interchange of the labels "solvent" and "solute."

In each chapter, then, a group of centrally important equations and concepts is presented, by means of problems, in a diversity of guises. The problems are worked out in detail so that the solutions are as much a part of this study aid as are the problems themselves. Through working the problems and studying the answers, the student should gain a better perspective and a better understanding of physical chemistry. He should develop some ability to see that principles are relevant to a situation and the confidence to fashion or to modify equations as appropriate.

As the last sentence suggests, I hope that the pedagogical approach embodied in this collection of problems is one that inculcates analytical and inductive ability. Such ability is essential to the good physical chemist, and I feel that its nurture is one of the major responsibilities of the course in physical chemistry. With respect to this point, it seems to me that the teacher of a modern physical chemistry course faces a dilemma. He must acquaint the student with the more important developments in the application of wave mechanics to chemistry. The mathematical underpinning of wave mechanical calculations is generally too complicated to take up in detail;

yet if only the final results are given, the material becomes merely descriptive and not conducive to any real understanding. There is the further danger of instilling an unscientific attitude of accepting equations on the basis of authority rather than understanding. A partial resolution of this dilemma is attempted in Chapters 20 and 21 by selecting and concentrating on a group of topics in wave mechanics and chemical bonding, so chosen that it is possible to ask the student to develop a certain amount of insight into them.

In conclusion, I would greatly appreciate receiving your comments, corrections, criticisms and, perhaps above all, contributions of good problems.

To the student: a suggested procedure

First, a few points of information are in order. The problems come from actual quizzes (or, in some cases, final examinations) and the time given for their working is that which was actually allowed. You can construct a typical quiz by selecting two or three problems whose time allowance totals thirty minutes.

Beginning with Chapter 4, you will find problems requiring the use of partial derivatives and partial differential relationships. It will help, if you are shaky on this subject, to proceed to the Appendix at the end of the book. A short outline of the principal relations is given, along with a special collection of problems that you should find very enlightening.

The concepts involved and the equations needed are given at the beginning of each chapter, along with some informal comments about their use. Read these introductory sections carefully before starting on the problems. The problems may be treated as either open or closed book. In most instances it does not really matter as they are harder to do open book than closed book! (With open book, you waste too much time looking up equations you should know anyway.) As you start working a chapter you may want to use the list of equations as a kind of crib sheet, but, honestly, if you need this you are not going to be able to do the problems within the time limits.

Most of the time allowance is for thinking; the actual calculations are usually quite short. Try to train yourself not to rush into arithmetic before you know where you want to go. In grading these quizzes, we allowed 80% credit for correctly setting up the solution and 20% for correct arithmetic, which is another reason for saving the slide-rule work for the last. Keep in mind, too, that the numerical data are often carefully chosen so as to facilitate the calculation through cancellations or through opening up shortcuts or quick approximations. Approximations, however, should not impair the usual criterion of slide-rule accuracy. Thus numbers given to only one significant figure are to be understood as being accurate to 1%.

If you have trouble at first, as you try to solve these problems, give yourself about three times the indicated time, then study the answer carefully. Try to see what you missed or where you went wrong in your analysis. Try to generalize your conclusions so as to have something you can then keep in mind when going on to the next problem. Your goal is to complete the problem within the time limit and to have it substantially correct.

After completing a chapter, review all the problems in it. Make an outline of the various ways in which the subjects were approached and assign each problem to a spot in your outline. The idea is to develop a close association in your mind between each relationship or principle and the ways in which it may be used. If you become discouraged, it may help your morale to ask a graduate student to work one of the problems within the time limit!

Best wishes and good luck!

To the instructor: some further comments

Although this collection of problems is presented primarily as a new kind of study aid for students, I believe it also constitutes a good source of examination questions. The problems are largely original, although some are modifications of homework problems from various sources. The collection is large enough so that there is not much chance that a student having this book will remember a particular solution. If you want to make small changes in the numerical data given, it might be wise to check the solution, since in some cases the particular numbers used were carefully chosen to facilitate calculation.

The sequence of topics is approximately that followed in the lecture course at the University of Southern California. To an appreciable extent, then, the problems become more sophisticated as one proceeds through the collection. There is also an increasing use of actual experimental data and of situations that approximate experimental ones.

The last chapter, on group theory as applied to molecular symmetry and chemical bonding, is definitely experimental, as this topic has not yet found its way into first-year physical chemistry books. It is not particularly difficult, however, and by introducing the subject at this level students are permitted to gain a real grasp of how wave mechanics is actually used in qualitative applications to chemical bonding. The introduction to this chapter is detailed enough to serve as a text for the material, although some recommended references are listed.

Chapter 20, on wave mechanics, is likewise somewhat experimental in that the problems are restricted to a few topics, but then go into them in some detail. Thus a quite complete listing of hydrogen-like wave functions is given, and the student is expected to become really familiar with their radial and angular properties. This concentration on a few topics does imply a sacrifice of material at a more descriptive level, but I do not think it is possible otherwise to provide sufficient depth of presentation to allow the student to gain a feeling of understanding and appreciation of wave mechanics.

The paperback edition of this book is issued in two volumes (Parts One and Two) for mechanical and monetary convenience to the reader. While the division is necessarily arbitrary, the two parts do roughly correspond to the first- and second-semester portions of the physical chemistry course as taught here.

For the convenience of the reader, in the paperback editions, Part One contains the table of contents for Part Two on the back cover and Part Two contains the contents for Part One on the back cover. Both parts have the full index.

ARTHUR W. ADAMSON

Los Angeles, California
July, 1964

Preface to the Second Edition

It is a pleasant task to write a preface to the second edition of a book. After all, the first edition has then presumably met with a degree of success. I do hope that this is true in the present case, and that *Understanding Physical Chemistry* has, in fact, been of some real help both to students as a study aid and to teachers as either a short text or as a resource book.

This new edition is rather similar to the previous one: its structure and scope continue to be governed by the sentiments expressed in the preface to the first edition (which precedes). However, many new problems have been added or substituted, skimpy sections have been filled out, and a number of chapter introductions have been revised; the book is about ten percent longer. An important structural change is in the grouping of the problems of each chapter into subcategories. The intent is to allow the reader more easily to concentrate on very specific aspects of a subject.

In addition, a definite pedagogical innovation appears in Chapters 5 and 7. Group theory was introduced in the first edition (and is retained here) as a means of discussing chemical bonding that was exciting to students and that avoided the usual authoritative and descriptive approach. This experiment, although very successful in our experience at the University of Southern California, evaded a more general question on which there is currently a dichotomy of opinion. Some physical chemistry teachers believe that wave mechanics and chemical bonding should appear early in the course, because of their great importance (and perhaps also because of their popularity). Certain texts adopt this approach. The alternative argument is that it is more orderly and more profitable to the student to defer such relatively sophisticated topics until after the development of classical and macroscopic thermodynamics and chemical equilibrium. A second group of texts conforms to this position.

Both of the above points of view have merit; both encounter real objections. Chapters 5 and 7 introduce an approach newly in use in the physical chemistry course here, and one that I believe draws on the best features of the other two. Wave mechanics enters in these early chapters as a discipline that gives the energy states of translational, rotational, and vibrational motion. These formulations are then combined with the Boltzmann principle to provide the appropriate partition functions and an introduction to statistical thermodynamics. In this way, classical and quantum thermodynamics can be presented as a single, general subject without first going through all of the material on wave mechanics. Nor is statistical thermodynamics left as an appendicular subject relegated to the end of the course. A second, detailed exposure to the wave equation then occurs in Chapter 20, where the emphasis is more on the electronic energy state of molecules, and is followed in Chapter 21 by the group theoretical approach to chemical bonding.

The effects of the above sequence are to make statistical thermodynamics an integral part of the general subject of thermodynamics and to give students a double exposure to wave mechanics. However, the chapters are so structured that there should be no difficulty in using this book in conformity with some other topic sequence.

A few acknowledgements should be made. The manuscript for this edition could not have been prepared without the loyal help of my family and the able secretarial assistance of M.B. I am grateful for help in proofreading to some Very Special students.

ARTHUR W. ADAMSON

Los Angeles, California
February 1969

Contents

1

IDEAL AND NONIDEAL GASES: CONDENSATION

COMMENTS

The ideal gas law plays two different roles in physical chemistry. Both of these roles are important, but you should keep them distinct.

The first concept is that of a hypothetical state of matter—one that obeys the ideal gas law under all conditions. A number of the problems that follow are exercises in the playing of this first role. As a mild word of warning, you can expect some of the problems to be hard to do in the time limit in spite of the algebraic simplicity of the ideal gas law itself. Often, ancillary conditions are given or are inherent in the described situation; thus in Problem 10, pressure and volume are related in a special way by the nature of the experiment. These problems should help you to develop a feeling for how ideal gases behave in various simple situations and to perceive quickly any additional restrictions between the variables.

The barometric equation is included in this chapter partly as an exercise with the additional variable of height in a gravitational field. The equation also provides an easily appreciated form of the Boltzmann principle. Thus, if the temperature is constant, the pressure of an ideal gas and hence also its concentration in a gravitational field is proportional to $\exp(-mgh/kT)$. That is, the logarithm of the probability of finding a molecule at height h is proportional to the negative of its potential energy divided by kT. Much of statistical thermodynamics rests on the Boltzmann principle, as you will see in Chapters 6 and 7.

We then take a look at gases as they really are—imperfect, capable of condensation, and showing critical phenomena. The equations capable of representing such behavior can be rather complex; they must, after all, be based on more sophistical models than the one for an ideal gas. It is a wise convention in physical chemistry, however, to introduce complexities step by step, and only as needed. In the van der Waals equation, for example, the finite size of molecules is recognized, but not any details of shape; intermolecular forces are invoked, but only of the simplest form. It is no problem to elaborate on such assumptions—many theoreticians enjoy doing so. The goal at the moment, however, is for you to become familiar with the general behavior of real gases and to appreciate how amazingly successful are such relatively simple equations as that of van der Waals.

To return to the subject of the opening paragraph, the second role of the ideal gas law is that of being the limiting law or equation of state for all real gases. That is, *all* gases approach ideal behavior in the limit of zero pressure. Notice that the van der Waals and other such equations also reduce to the ideal law in this limit. As a limiting law, it is one of the fundamental laws of nature—perhaps as important as the laws of thermodynamics themselves.

EQUATIONS AND CONCEPTS

The principal equations and concepts used in the problems that follow are assembled below. This will be done at the beginning of each chapter. Primarily, these listings serve as a quick guide to the contents. Use them, if you wish, as a kind of "crib sheet."

Ideal Gas Law

$$PV = nRT, \quad \text{or} \quad PV = \frac{wRT}{M}, \quad \text{or} \quad PM = \rho RT \qquad (1\text{-}1)$$

The second two forms are merely alternative ones in which the number of moles n has been replaced by (weight)/(molecular weight) and in which w/V has been replaced by the density ρ.

Dalton's Law of Partial Pressures

$$P_{\text{tot}} = \sum P_i; \qquad P_i = N_i P_{\text{tot}} \qquad (1\text{-}2)$$

Here, P_i denotes the particle pressure of the ith species; N_i is its mole fraction. Also the ideal gas law then applies separately to each component: $P_i V = n_i RT$.

Gas Constant

You will find it convenient to keep in mind the various units in which R, the gas constant, is commonly expressed. Thus $R = 0.0821$ liter-atm/mole-deg = 82.1 cc-atm/mole-deg = 8.31 × 10^7 erg/mole-deg = 1.98 cal/mole-deg.

STP Conditions

0°C or 273°K and 1 atm pressure.

Barometric Equation

Differential form: $d \ln P = - \dfrac{Mg}{RT} dh$ (1-3)

Integrated form: $P_2 = P_1 e^{-Mgh/RT}$ (1-4)

Both forms assume the ideal gas law; in the second, it is also assumed that T and g are constant.

Van der Waals Equation

$$\left(P + \frac{a}{V^2}\right)(V - b) = RT \qquad (V = \text{molar volume}) \tag{1-5}$$

The constants a and b express the mutual attraction of molecules, and their size, respectively. A typical set of P–V isotherms is shown in Fig. 1-7. Below the critical temperature, the calculated isotherms show a maximum and a minimum, but the physical interpretation is that condensation occurs. The vapor pressure at which this happens is determined by a virtual work principle (see your text). As illustrated in the figure for the 275°C and 250°C isotherms, this principle works out to imply that the horizontal line, which gives the vapor pressure, should be located so that the *net* area between it and the dashed van der Waals curve is zero.

The maximum and minimum in the van der Waals isotherms merge to an inflection point at the critical temperature, pressure, and volume, from which one finds

$$V_c = 3b \qquad P_c = \frac{a}{27b^2} \qquad T_c = \frac{8a}{27b} \tag{1-6}$$

Finally, the minimum may occur at a negative pressure value and can then be interpreted as giving a positive tensile strength for the liquid.[1]

[1] See S. W. Benson and E. Gerjuoy, *J. Chem. Phys.* **17,** 914 (1949).

Virial Equation

$$PV/RT = 1 + B/V + C/V^2 + \cdots \tag{1-7}$$

Again, V is the molar volume; B and C depend on temperature but not on V.

Principle of Corresponding States

Plots of the compressibility factor, PV/RT, versus P/P_c are the same for all real gases at the same value of T/T_c. The behavior of real gases may be fairly well approximated by empirical plots based on this principle, as illustrated in Fig. 1-1.[2]

Also, the van der Waals equation obeys the principle of corresponding states since it can be put in the form $(\alpha + 3/\beta^2)(3\beta - 1) = 8\gamma$, where α, β, and γ are the reduced pressure, volume, and temperature, respectively.

Miscellaneous

The Boyle temperature is that temperature such that the derivative $d(PV)/dP$ becomes zero as P approaches zero. A useful rule to remember is that the normal boiling point of a liquid is usually about two-thirds of the critical temperature, expressed in absolute degrees.

PROBLEMS

1.   (13 min) A balloon contains 10 g of H_2 gas. What additional weight of argon gas (at. wt 40) must be added so that the balloon will have exactly zero buoyancy (i.e., so that its weight will be equal to that of an equal volume of surrounding air)? Assume ideal gas behavior; neglect the weight of the balloon itself and assume it to be impermeable and perfectly elastic. Average mol. wt of air is 29.

2.   (11 min) A flask filled with pure helium (at. wt 4) at 1 atm and 25°C contains 1.6 g of the gas. What weight of argon (at. wt 40) would have to be added to the bulb so that the combined weight of the two gases (at 25°C) would equal that of an equal volume of air (at 1 atm and 25°C)? (The average molecular weight of air is 29.) Give also the average molecular weight of the He–Ar mixture.

3.   (7.5 min) Air is approximately 80% nitrogen and 20% oxygen (on a mole

[2] O. A. Hougen, K. M. Watson, and R. A. Tagatz, *Chemical Process Principles, Part II* (J. Wiley & Sons, New York, 1959), Chap. 12.

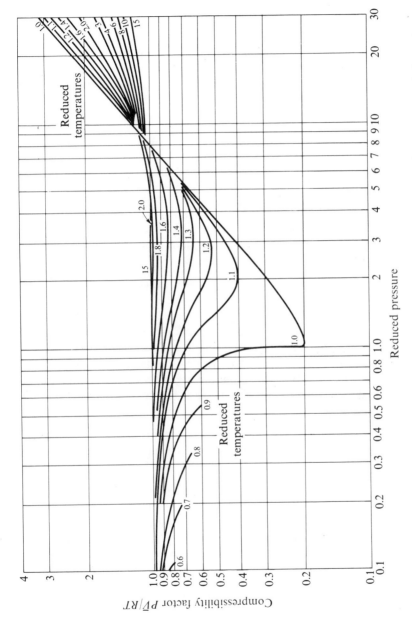

FIGURE 1-1   Hougen–Watson–Tagatz chart for calculating pressure, volume, and temperature relations at high pressure.

5

basis). If 6 g of hydrogen is added to a 22.4-liter flask maintained at 0°C and initially filled with air at 1 atm pressure, what will be the molecular weight (i.e., the average molecular weight) of the hydrogen–air mixture?

4.   (9 min) A pioneer aeronaut is planning the design of a hot-air balloon. What volume of air at 100°C should be used if the balloon is to have a gross lifting power of 200 kg (defined as the mass of displaced air minus the mass of hot air)? The ambient temperature and pressure are 25°C and 1 atm, and the average molecular weight of air is 29 g/mole, whereas that of the hot air is 32 g/mole (due to the presence of some CO_2).

5.   (12 min) An 11-liter flask contains 20 g of neon and an unknown weight of hydrogen. The gas density is found to be 0.002 g/cc at 0°C. Calculate the average molecular weight and the number of grams of hydrogen present, and also the pressure. The atomic weight of neon is 20 g/mole.

6.   (7 min) Two separate bulbs contain ideal gases A and B, respectively. The density of gas A is twice that of gas B, and the molecular weight of gas A is half that of gas B; the two gases are at the same temperature. Calculate the ratio of the pressure of gas A to that of gas B.

7.   (13 min) It is desired to prepare a gas mixture containing 5 mole % of butane and 95% of argon (a mixture of this type is used in filling Geiger–Müller counter tubes). A gas cylinder is evacuated and gaseous butane is let in until the butane pressure is 1 atm. The tank or cylinder is then weighed, and compressed argon gas forced in until a certain weight w (in grams), has been added. The volume of the cylinder is 40 liters, and the operation is carried out at 25°C. Calculate the weight of argon that gives a mixture of the desired composition, and the total pressure of the final mixture. The atomic weight of argon is 40 g/mole.

8.   (11 min) Hydrogen gas will dissociate into atoms at a high enough temperature, that is, $H_2 = 2H$. Assuming ideal gas behavior for H_2 and H, what should be the density of hydrogen at 2,000°C if it is 33% dissociated into atoms? The pressure is 1 atm.

9.   (10 min) When 2 g of gaseous substance A are introduced into an initially evacuated flask kept at 25°C, the pressure is found to be 1 atm. Three grams of gaseous substance B are then added to the 2 g of A, and the pressure is now found to be 1.5 atm. Assuming ideal gas behavior, calculate the ratio of molecular weights, that is, M_A/M_B.

10.   (9 min) An iron pipe 2 m long and closed at one end is lowered vertically into water until the closed end is flush with the water surface (see Fig. 1-2). Calculate the height h of the water level in the pipe. Miscellaneous data: 25°C, diam. of pipe is 3 in.; density of water is 1.00 g/cc; barometric pressure

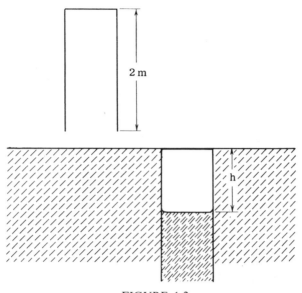

FIGURE 1-2

is 1.00 atm. Also, 1 atm $= 10^6$ dyne/cm^2 $= 10$ m hydrostatic head of water (neglect the effect of water vapor pressure).

11. (12 min) A 2-liter Dumas bulb contains n moles of nitrogen at 0.5 atm pressure and at $T°$K. On addition of 0.01 moles of oxygen, it is necessary to cool the bulb to a temperature of 10°C in order to maintain the same pressure. Calculate n and T.

12. (9 min) The two flasks shown in Fig. 1-3 are filled with nitrogen gas and, when both are immersed in boiling water, the gas pressure inside the system is 0.5 atm. One of the flasks is then immersed in an ice–water mixture, keeping the other in the boiling water. Calculate the new pressure for the system.

13. (11 min) A Dumas bulb is filled with chlorine gas at the ambient pressure and is found to contain 7.1 g of chlorine when the temperature is $T°$K. The bulb is then placed in a second thermostat bath whose temperature is 30°C hotter than the first one. The stopcock on the bulb is opened so that the chlorine-gas pressure returns to the original value. The bulb is now found to contain 6.4 g of chlorine. Calculate the value of the original temperature. If the volume of the bulb was 2.24 liters, what was the ambient pressure? The atomic weight of chlorine is 35.5.

14. (16 min) Two flasks of equal volume (see Fig. 1-4) are connected by a narrow tube (of negligible volume): Initially both flasks are at 27°C and contain 0.70 moles of hydrogen gas, the pressure being 0.50 atm. One of the

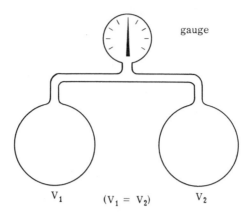

FIGURE 1-3

flasks is then immersed in a hot oil bath at 127°C, while the other is kept at 27°C. Calculate the final pressure of, and the moles of hydrogen in, each flask.

15. (12 min) Calculate the number of moles of air in a column 1 cm^2 in area and rising from sea level to an altitude of 1.5×10^6 cm (about 10 miles), allowing for the barometric decrease in pressure. Assume air to be a uniform gas of molecular weight 29, and a uniform temperature of 0°C. Note that the atmospheric pressure is equal to the total weight of gas in a column of unit area and of infinite height. Conversion factors: $R = 0.082$ liter-atm/mole-deg $= 8.31 \times 10^7$ erg/mole-deg $= 1.98$ cal/mole-deg; 1 atm $= 760$ mm Hg $= 1 \times 10^6$ dyne/cm^2.

16. (6 min) Curve 1 in the graph of Fig. 1-5 represents a plot of the variation of pressure with altitude according to the simple barometric equation which assumes T and g to be constant.

If, instead of assuming T to be constant in the derivation, one has assumed it to decrease with altitude (e.g., $T = a - bh$), would the resulting calculated

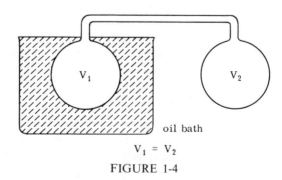

$V_1 = V_2$

FIGURE 1-4

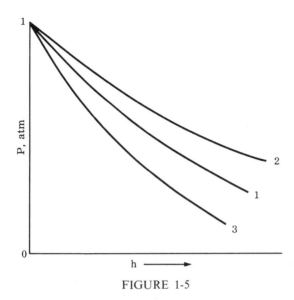

FIGURE 1-5

curve lie above or below curve 1, that is, would it resemble curve 2 or curve 3? Similarly, if instead of assuming g to be constant, one assumed it to decrease with altitude (e.g., $g = a - bh$), would the resulting curve more nearly resemble 2 or 3?

17. (15 min) The gravitational constant g decreases by $1.0 \text{ cm/sec}^2/\text{km}$ of altitude. Assuming a constant temperature of 25°C, derive a modified barometric equation that recognizes this variation in g. Calculate the pressure of nitrogen at 300 km altitude, taking sea-level pressure to be 1 atm and using this modified equation.

---

18. (11 min) Two separate bulbs contain gases A and B, respectively. The pressures and volumes are such that the PV product is the same for both gases. However, gas A is an ideal gas, and gas B is nonideal and is at a pressure and temperature less than the critical values. Explain, preferably with the aid of an appropriate graph, whether the temperature of gas B should be the same as, more than, or less than that of gas A.

19. (10 min) Doubling the temperature at constant pressure will (more than, less than) double the volume of a gas if the gas is (ideal, nonideal and below T_c, nonideal and above its Boyle temperature, nonideal and at low pressure).

20. (13 min) A nonideal gas of molecular weight 150 obeys the van der Waals equation; its critical pressure and temperature are 100 atm and 100°C, respectively.

(a) The compressibility factor PV/RT will be greater than unity at (500 atm and 80°C, 5,000 atm and 120°C, 50 atm and 60°C, 50 atm and 120°C, none of these).

(b) Calculate the value of the compressibility factor at the critical point.

21. (11 min) The curve for a reduced temperature of 0.8 is reproduced in Fig. 1-6 in terms of compressibility factor versus reduced pressure. The curve stops abruptly at point X, since condensation to liquid is supposed to occur at this point.

(a) Sketch the continuation of the line up to a reduced pressure of 1.5; that is, show how the compressibility factor should vary as one goes through and then past the condensation region.

(b) Assuming the point X represents C_2H_4 at -46°C and 17 atm, show how to calculate (or estimate) P_c, T_c, V_c, and the normal boiling point of ethylene.

22. (10 min) Make a semiquantitative plot of T versus V for water at 1 atm pressure. The range of V values should be from about 17 cc/mole to about 40 liters/mole.

23. (15 min) Some P–V plots are shown for a gas that obeys the van der Waals equation. Calculate the constants a and b for this gas. Since your calculation is necessarily approximate, it is necessary to show very clearly just how you have obtained numbers for the graph of Fig. 1-7 and how you have used them. Give the units of a and b also.

24. (12 min) Given the P versus V plot of Fig. 1-8 for a certain gas at 25°C, calculated according to the van der Waals equation using the appropriate a and b values, estimate numerical values for the following: (a) the tensile

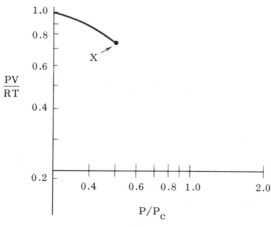

FIGURE 1-6

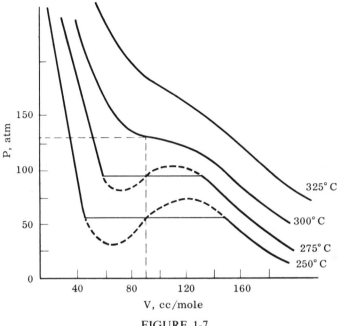

FIGURE 1-7

strength of the liquid, (b) the vapor pressure of the liquid, (c) the molar volume of the liquid, and (d) the critical volume V_c.

25. (6 min) Given the following data for a certain nonideal gas at 25°C:

ρ/P, g/liter–atm:	10	11	10	(ρ = density)
P, atm:		1	10	20

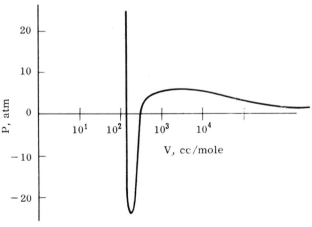

FIGURE 1-8

The critical pressure of this gas must then be (greater than 10 atm, greater than 20 atm, between 1 and 20 atm, between 1 and 10 atm, less than 20 atm, cannot tell).

26.   (5 min) The critical temperature and pressure are 32°C and 48 atm for ethane. Sketch the plot of the compressibility factor, PV/RT, versus P for the case of $t = 40°C$. Extend the plot up to 100 atm pressure.

27.   (16 min) The $P–V$ isotherm for 0°C is given in Fig. 1-9 for a gas obeying the van der Waals equation (note that *both* scales are logarithmic). Calculate (a) the value of the van der Waals constants a and b; (b) that value of P such

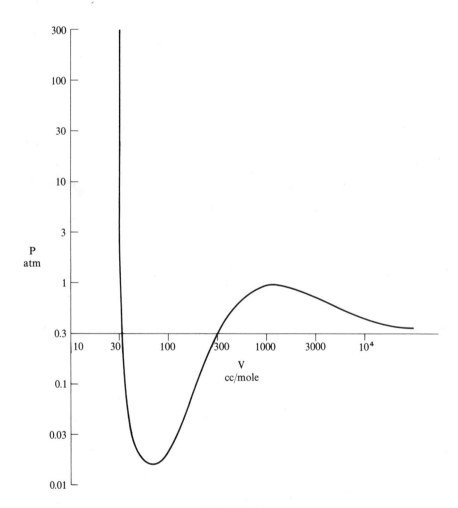

FIGURE 1-9

that the volume of the van der Waals gas is the same as for an ideal gas (the answer is *not* $P = 0$). Your answer need be accurate only to about 10%, but you should make your method of calculation clear.

28. (12 min) Fig. 1-10 shows several isotherms calculated from the van der Waals equation for a representative gas (the dashed lines locate the actual vapor pressures for the various isotherms). Locate on the graph the conditions of P and T such that (a) the b term is negligible compared to V, (b) the a/V^2 term is negligible compared to P, (c) P is negligible compared to a/V^2, and (d) V is negligible compared to b. (Answer "none" if no conditions exist under which the statement would be true.) (e) On the volume axis mark the approximate value of b.

29. (13 min) A gas obeys the van der Waals equation, with $P_c = 30$ atm and $T_c = 200°C$. The compressibility factor PV/RT will be more than one (at $P = 50$ atm, $T = 250°C$; at $P = 1$ atm, $T = 100°C$; $P = 500$ atm, $T = 500°C$; none of these). The gas will approach ideality at (low T, low density, low values of the compressibility factor, none of these). Calculate the van der Waals constant b for this gas.

30. (12 min) The critical temperature and pressure for NO gas are 177°K and 64 atm, respectively, and for CCl_4 they are 550°K and 45 atm,

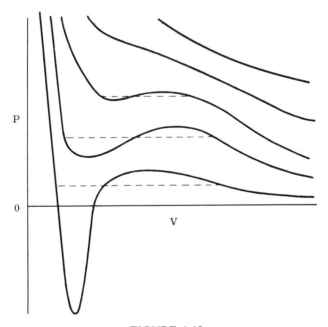

FIGURE 1-10

respectively. Which gas (a) has the smaller value of the van der Waals constant b, (b) the smaller value of the van der Waals constant a, (c) has the larger critical volume, and (d) is most nearly ideal in behavior at 300°K and 10 atm?

31.   (5 min) The compressibility plot for a certain gas is shown in Fig. 1-11 for 100°C. It follows that the critical temperature of the gas is (greater than 100°C, less than 100°C, cannot tell) and that the critical pressure of the gas is (greater than 10 atm, less than 10 atm, cannot tell).

32.   (3 min) A nonideal gas is below its critical temperature. On compression, the plot of volume versus pressure (volume as ordinate) will lie (above, below) the corresponding curve for an ideal gas.

33.   (13 min) The molecular weight of a vapor is determined by measuring the vapor density at a known P and T. Even though no chemical association or dissociation occurs, the molecular weight will be in error (in general) if it is calculated by means of the ideal gas law, since the vapor is nonideal. The critical temperature and pressure of this particular substance are 100°C and 1.0 atm, respectively.

(a) The molecular weight as calculated from the ideal gas law would be low if the vapor density were measured at (the critical point, at a sufficiently low temperature, at a sufficiently high pressure, none of these).

(b) If the vapor density were measured at 10°C (the ideal-gas-law molecular weight would be low, would be high, the measurement is impossible since the vapor would necessarily have condensed to a liquid, none of these).

(c) Explain briefly how, from measurements of vapor density, T, and P only, you would obtain an accurate molecular weight.

34.   (3 min) A certain nonideal gas is at its critical temperature and at a pressure 10% greater than its critical pressure. Doubling the pressure of the gas at constant temperature should (more than, less than) halve its volume.

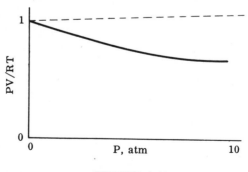

FIGURE 1-11

35. (10 min) Gases A, B, C, and D obey the van der Waals equation, with *a* and *b* values as given (liter-atm system of units):

	A	B	C	D
a	6	6	20	0.05
b	0.025	0.15	0.10	0.02

Which gas has the highest critical temperture, the largest molecules, the most nearly ideal behavior at STP? (Not necessarily the same gas!)

36. (13 min) A certain gas obeys the van der Waals equation. Its T_c is 100°C and its P_c is 90 atm. Underline the most appropriate choice to complete each statement below.

(a) The gas will approach ideality at (high pressures, low values of the PV product, low temperatures, none of these).

(b) The gas *definitely* will show a positive deviation from ideality (i.e., $PV > RT$) (at low temperatures, low PV products, around 1000°C provided the pressure is high enough, at any pressure below P_c, none of these).

(c) The gas must at least partially condense to a liquid if cooled below T_c (true, false).

37. (6 min) The experimental value of RT/V is 1.10 for 1 mole of a certain nonideal gas. The gas is at 1 atm and its temperature is below the critical temperature. If the pressure is now halved, at constant temperature, it is to be expected that the new volume will be (more than twice, less than twice) the original volume.

38. (5 min) Select the appropriate statement. The density of a nonideal gas (increases with decreasing pressure, at constant temperature; is proportional to the molecular weight of the gas; increases with increasing PV product; none of these).

39. (7 min) Given that a certain gas obeys the van der Waals equation, underline the most appropriate choice to complete each statement below.

(a) The gas approaches ideality at (high pressures, low values of the PV product, low temperatures, none of these).

(b) The equation for *n* moles of the gas is $PV = nRT$, $(P + a/V^2)(V - b) = nRT$, $(P + a/n^2 V^2)(nV - b) = nRT$, $(P + an^2/V^2)(V/n - b) = RT$, none of these.

40. (10 min) As illustrated in many texts for the CO_2 system, it takes a three-dimensional plot of *P* versus *V* versus *T* to show the complete behavior of a substance. For convenience one ordinarily deals with sections of the solid

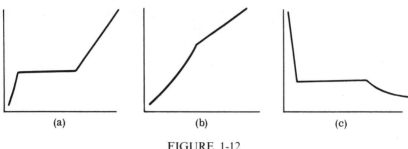

(a)   (b)   (c)

FIGURE 1-12

model so that a two-dimensional representation is possible. One thus has isotherms (P versus V or V versus P, with T constant), isobars (T versus V or V versus T, with P constant), and isochors (T versus P or P versus T, with V constant). The three diagrams of Fig. 1-12 give an isotherm, an isochor, and an isobar for water. State which diagram is which, and label each abscissa and ordinate.

41. (10 min) The values of P_c and T_c for N_2 are 126°K for T_c and 34 atm for P_c, whereas for C_2H_6 they are 48 atm and 305°K, respectively. (a) Which gas has the smaller van der Waals constant b? (b) Which has the smaller a value? (c) Which has the larger V_c value? (d) Which should show the most nearly ideal behavior at 25°C and 10 atm pressure?

42. (9 min) The van der Waals constants for gases A, B, and C are as follows.

Gas	a, liters^2–atm/mole	b, liters/mole
A	4.0	0.027
B	12.0	0.030
C	6.0	0.032

Which gas has (a) the highest critical temperature, (b) the largest molecular volume, and (c) most ideal general behavior around STP?

ANSWERS

1. For the balloon to be in equilibrium with the atmosphere, the inside and outside pressures and temperatures must be the same; also we are comparing the balloon with an equal volume of air. Since P, V, and T are the same, the balloon and the comparison sample of air must have the same number of moles; for zero buoyancy, they must also have the same weight. It follows that the average molecular weight of the mixed gases in the balloon must be the same as that of air.

$$M_{av} = \frac{\text{(total weight)}}{\text{(total moles)}} = \frac{\Sigma M_i n_i}{\Sigma n_i} = \Sigma M_i N_i$$

In the present case

$$2N_{H_2} + 40N_{Ar} = 29$$

Since $N_{H_2} + N_{Ar} = 1$, solving the above equation gives $N_{H_2} = 11/38$, and $N_{Ar} = 27/38$. The ratio n_{Ar}/n_{H_2} is then 27/11, and since there are 10/2 or 5 moles of H_2, $5 \times 27/11$ or 12.2 moles of argon must be added.

2. $W_{air} = \frac{M_{air}}{M_{He}} W_{He}$ since the comparison is for the same V, P, and T. $W_{air} = (29 \times 1.6)/4 = 11.6$ g. One therefore needs to add **10.0 g** of argon to the 1.6 of helium. $M_{av} = $ total weight/total moles, so

$$M_{av} = \frac{11.6}{\dfrac{1.6}{4} + \dfrac{10.0}{40}} = \textbf{18 g/mole}$$

3. Since we have 22.4 liters of air at STP, we have 1 mole of air.

$$M_{av} = \frac{0.80 \times 28 + 0.20 \times 32 + 6}{1 + 3} = 32.8/4 = \textbf{8.7 g/mole}$$

4. Since the mass of gas at a given P, V, T is given by PVM/RT, we have

$$w_{air} - w_{hot\ air} = 2 \times 10^5 \text{ g} = \frac{1 \times V}{0.082}(29/298 - 32/373)$$

From this, $V = \textbf{1.4} \times \textbf{10}^6$ **liters.**

5. Since the density is 2 g/liter, the 11 liters contain 22 g; hence 2 g of hydrogen in addition to the 20 of neon. This is 1 mole of each or 2 moles total. Then $M_{av} = 22/2 = \textbf{11 g/mole;}$

$$P = nRT/V = (2 \times 0.082 \times 273)/11 = \textbf{4.07 atm}$$

6. Use the gas law in the form $P = \rho RT/M$; then

$$P_A/P_B = \frac{\rho_A/\rho_B}{M_A/M_B} = 2/\tfrac{1}{2} = \textbf{4}$$

7. The moles of butane present are given by $n = PV/RT$: $n_{bu} = (1 \times 40)/0.082 \times 298 = 1.64$. One then wants 95 parts of argon to 5 of butane or 19 to 1, so

$$n_{Ar} = 19 \times 1.64 = 31.1 \text{ moles Ar} \quad \text{or} \quad 31.1 \times 40 \quad \text{or} \quad \textbf{1240 g Ar}$$

The final pressure is proportional to the change in number of moles, so $P_f = 1 \text{ atm} \times 20/1 = \textbf{20 atm.}$

8. It is convenient to think in terms of 1 mole of H_2 before dissociation, as a basis for the calculation. Then

after dissociation: 0.67 mole H_2 left
0.66 mole of H (from the dissociation
total moles: $\overline{1.33}$ of the 0.33 mole of H_2)

For a mixture of gases, the density ρ is given by $\rho = PM_{av}/RT$. In this case, M_{av} = total weight/total moles = 2 g/1.33 = 1.5 g/mole. Then

$$\rho = \frac{1 \times 1.5}{0.082 \times 2273} = \textbf{8.07} \times \textbf{10}^{-3} \textbf{ g/liter}$$

9. Since V and T are constant, P is proportional to the number of moles n. Evidently 3 g of B corresponds to half as many moles as does 2 g of A, or 6 g of B has the same number of moles as does 2 of A, and therefore $M_A/M_B = \frac{2}{6} = \frac{1}{3}$.

10. Since temperature is constant,

$$P_{fin} = P_{init}\left(\frac{\text{initial volume}}{\text{final volume}}\right)$$

$$= P_{init}\left(\frac{2 \times \text{area of pipe}}{h \times \text{area of pipe}}\right) = 2P_{init}/h$$

but $P_{fin} = h + 10$ (in meters). Thus

$h + 10 = 2 \times 10/h$ or $h^2 + 10h = 20$ and $h = \textbf{1.7 m}$

11. The condition after the addition of the oxygen is given as $(n + 0.01) = PV/RT = (1/2)(2)/0.082 \times 283 = 0.0432$. Hence

n = 0.0332

Then

$$T = PV/nR = (1/2)(2)/0.0332 \times 0.082 = \mathbf{367°K}$$

12. Here, the number of moles of gas remains constant, and we can write

$$n = n_1 + n_2 = \frac{P}{R}\left(\frac{V_1}{373} + \frac{V_2}{373}\right)$$

$$= \frac{0.5V}{R}\left(\frac{2}{373}\right) \quad \text{(initially)} \quad (V = V_1 = V_2)$$

Also

$$n = n_1' + n_2' = \frac{P'V}{R}\left(\frac{1}{273} + \frac{1}{373}\right)$$

Equating the two expressions for n, cancelling out V/R, and rearranging

$$P' = 0.5(2/373)\left(\frac{273 \times 373}{273 + 363}\right) = (0.5 \times 546)/646 = \mathbf{0.423\ atm}$$

13. The original 7.1 g corresponds to 0.1 moles, and the final 6.4 g to 10% less, hence 0.09 moles.

We are comparing two cases that involve the same volume and pressure; hence in this instance, $n_1 T_1 = n_2 T_2$ or $0.1 T_1 = 0.09(T_1 + 30)$. Or, $T_1 = \mathbf{270°K}$.

If the volume is 2.24 liters, then $P = (0.1 \times 0.082 \times 270)/2.24 = \mathbf{0.99\ atm.}$

14. Using the equation $n = PV/RT$, the initial condition is that $0.7 = P \times 2V/R \times 300$ where V is the volume of one bulb, so $V/R = 0.7 \times 300$. With one bulb heated, the same 0.7 mole is now equal to $n_1 + n_2$; that is,

$$0.7 = \frac{PV}{R}\left(\frac{1}{300} + \frac{1}{400}\right)$$

On inserting the value of V/R;

$$0.7 = P \times 0.7 \times 300(1/300 + 1/400)$$

from which $P = \mathbf{0.57\ atm}$. Also, $n_1 = 0.57\ V/300\ R = \mathbf{0.4\ mole}$; hence $n_2 = \mathbf{0.3\ mole.}$

15.   The pressure at this altitude is given by the barometric equation as

$$\log(P_2/1\text{ atm}) = -\frac{Mg}{2.3RT}h = -\frac{29 \times 980 \times 1.5 \times 10^6}{8.31 \times 10^7 \times 273 \times 2.3}$$

$$= -0.817$$

Hence $P_2 = 0.152$ atm.

Since the weight of a complete column of air extending down to sea level is 10^6 dynes/cm^2, the weight of a column extending down only to the given altitude will be 0.152×10^6 dynes/cm^2, and therefore the weight of the column between sea level and this altitude is 0.848×10^6 dyne/cm^2.

Therefore

$$0.848 \times 10^6 = nMg \quad \text{or} \quad n = 0.848 \times 10^6/29 \times 980$$

$$= \textbf{29.8 moles}$$

16.   The basic differential equation is

$$d\ln P = -\frac{Mg}{RT}\,dh$$

Hence, if T decreases with increasing h, the rate of change of $\ln P$ will increase; that is, pressure will drop more rapidly: **Curve 3.**

Correspondingly, if g decreases with increasing h, the rate of change in $\ln P$ will decrease, and pressure will drop less rapidly: **Curve 2.**

17.   The problem states that $g = 980 - 10^{-5}h$, if h is in centimeters. Inserting this into the basic differential equation gives

$$d\ln P = -\frac{M}{RT}(980 - 10^{-5}h)\,dh$$

or,

$$2.3\log(P/0.8) = -\frac{M}{RT}(980h - 5 \times 10^{-6}h^2)$$

$$= -\frac{28}{8.31 \times 10^7 \times 298}$$

$$(980 \times 3 \times 10^7 - 4.5 \times 10^9)$$

Hence

$$\log(P/0.8) = -12.1 \quad \text{or} \quad P = \textbf{6.4} \times \textbf{10}^{-13}\textbf{ atm}$$

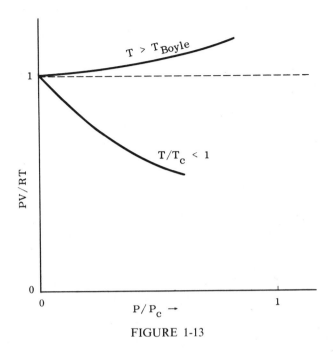

FIGURE 1-13

18. According to the principle of corresponding states, the compressibility factor PV/RT is the same for all gases at a given P/P_c and T/T_c. In this case, P and T for gas B are less than P_c and T_c, so that, as shown in Fig. 1-13, the value of PV/RT should be less than unity. Since PV is the same for both gases, it follows that T must be **greater** for gas B than for gas A.

19. If the gas is to be either more than or less than double its volume, it cannot be ideal. Referring to the diagram above, if the gas is nonideal and below T_c, doubling the temperature will increase PV/RT; hence V will more than double. If it is above its Boyle temperature, the doubling of T will decrease PV/RT and V will less than double. The third choice is contradictory, as gases approach ideality at low pressure.

20. For the compressibility factor to be greater than unity, we want a high pressure and temperature, hence **5,000 atm and 120°C.** For a van der Waals gas, $V_c = 3b$, $T_c = 8a/27Rb$, and $P_c = a/27b^2$; hence $P_cV_c/RT_c = \frac{3}{8}$.

21. At point X of Fig. 1-14 condensation begins, and V decreases with no change in pressure to the value for the liquid; there is therefore a vertical drop in PV/RT to some much lower value. When condensation is complete, V ceases to change much, so PV/RT now rises in approximate proportion to P.

If T is $-46°C$ or $227°K$ when T/T_c is 0.8, then $T_c = $ **283°K.** Further, point X occurs at 17 atm and at P/P_c of about 0.5; hence $P_c = $ (about) **34 atm.**

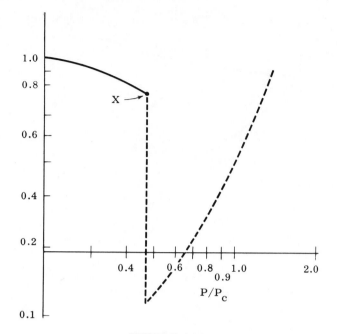

FIGURE 1-14

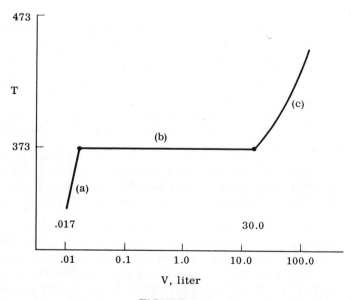

FIGURE 1-15

For a van der Waals gas, $P_c V_c / R T_c$ is $\frac{3}{8}$; hence

$$V_c = \frac{0.082 \times 283 \times 3}{34 \times 8} = \textbf{0.26 liter/mole}$$

22. In Fig. 1-15 the main points are as follow: (a) the molar volume of liquid water is 18 cc/mole, and for water at this volume to exert 1 atm pressure it must be at its boiling point; reduction in temperature leads to a slight decrease in liquid volume. (b) Any attempt to increase T simply results in vaporization and increase in V until the water is all vaporized, at which point one has 1 mole of vapor at 1 atm and 100°C, or a V value of about 30 liters. (c) The temperature may now increase, with V rising proportionally.

23. The 300°C isotherm appears to be very close to the critical one, and we can estimate P_c, V_c from the point of inflection. Then $P_c = 125$ atm, $V_c = 90$ cc/mole, and $T_c = 300$°C or 573°K. The constant b is then $V_c/3$ or **30 cc/mole,** and since $P_c = a/27b^2$, we find $a = 125 \times 27 \times 30^2 = \textbf{3} \times \textbf{10}^6 \textbf{ cc}^2$ **atm/mole**^{2}.

24. The tensile strength is given by the most negative pressure attainable, that is, about **20 atm.** The vapor pressure is estimated by locating a horizontal line in Fig. 1-16 so that the net area between it and the curve is zero, and would be about 5 atm. The molar volume of the liquid is the liquid volume at its vapor pressure, or about **110 cc/mole,** and the critical volume will be about three times larger, or about **330 cc/mole.**

25. Since density is w/V, then $\rho/P = w/PV$. Therefore, at a given temperature ρ/P varies inversely as the compressibility factor, so that the maximum

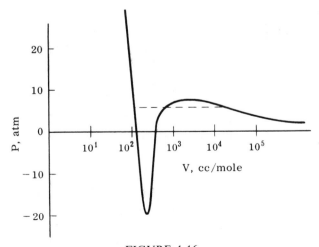

FIGURE 1-16

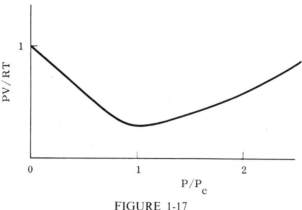

FIGURE 1-17

in ρ/P means a minimum in PV/RT. In plots of PV/RT versus P/P_c, such a minimum occurs only for $T < T_c$ and $P > P_c$ and we know that the minimum lies between 1 and 20 atm. P_c could be greater or less than 1 or 10 atm, but must be **less than 20 atm.**

26. At 40°C, $T/T_c = 1.03$. PV/RT will then vary with pressure as indicated in Fig. 1-17.

27. (a) The substance becomes essentially incompressible at a volume of about 30 cc/mole, so **b = 30** cc/mole. The second constant may be conveniently obtained by noting that at 300 cc/mole, $P \simeq 0$; hence $a/(300)^2 \times (300 - 30) = 82 \times 273$, or $a =$ **7.5 × 10⁶ cc² atm/mole².** (b) The ideal gas isotherm will cross the van der Waals one at the pressure corresponding to about 30 cc/mole volume, or at $P = 82 \times 273/30 =$ **750 atm.**

28. (a) The right-hand region, that is, where V is large. (b) The right-hand and upper region where V and/or P are large. (c) The lower left region where P and V are small, especially the region of negative pressure. (d) None. (e) Approximately the value of V where the lowest temperature isotherm cuts the $P = 0$ line.

29. PV/RT should be greater than one at the highest P and T choice, **500 atm and 500°C** as both, the first especially, are far above the critical values. Ideality will be approached at **low density.** Since for a van der Waals gas $P_c V_c/RT_c = \frac{3}{8}$, and $b = V_c/3$, then $b = 0.082 \times 473/8 \times 30 =$ **0.163 liter/ mole.**

30. For a van der Waals gas, $V_c = 3b$, $P_c = a/27b^2$, and $T_c = 8a/27bR$. Therefore b and hence V_c are proportional to T_c/P_c, and NO should have the smaller b value (this can be argued qualitatively on the grounds that NO should be the smaller molecule); similarly CCl_4 should have the largest

b and hence the largest V_c value. The constant a is proportional to $P_c V_c^2$ or to T_c^2/P_c, and will therefore be smallest for NO (again, this can be argued qualitatively on the basis that the much lower critical temperature for NO means smaller intermolecular attractive forces and hence a smaller a value). NO gas should be the more ideal since 300°K is above its critical temperature, but below that of CCl_4.

31. The critical pressure could be less than 10 atm, in which case $T_c < 100°C$; it could be greater than 10 atm, with T_c either greater or less than 100°C. Hence the answer in both cases is **cannot tell.**

32. It will lie below (except at extremely high pressures).

33. (a) Molecular weight $= w(RT/PV)$, hence is inversely proportional to the compressibility factor; the calculated value will be low under conditions such that the compressibility factor is greater than unity, that is, **at a sufficiently high pressure.**

(b) At 0°C one is below the critical temperature (and presumably at about 1 atm pressure) so PV/RT is less than one and the calculated molecular weight would be **high.**

(c) To obtain an accurate molecular weight it would be necessary to measure density at several pressures and extrapolate ρ/P versus P to zero pressure; this limiting value is now exactly equal to M/RT.

34. Under the conditions stated, the compressibility factor will be slightly past its minimum value, hence will increase when pressure is doubled. The volume will therefore be less than halved.

35. Since T_c is proportional to a/b, **gas A** has the highest T_c. Since b gives a measure of molecular volume, **gas C.** That gas having the lowest critical temperature and pressure will be most nearly ideal at STP. **Gas D** has the lowest a/b, and hence T_c, and also the lowest a/b^2, and hence P_c.

36. (a) The gas will approach ideality at low pressures; **none of these.**

(b) PV/RT would exceed unity at a high enough pressure.

(c) **False** (depends on the pressure).

37. The value of PV/RT is evidently $1/1.10 = 0.9$. The compressibility factor will increase as pressure is reduced; hence the new volume will be **more than** twice the original.

38. Strictly speaking, **none of these.** It will be approximately proportional to molecular weight, keeping P and T constant, but not exactly proportional as different gases will differently depart from ideality.

39. (a) **None of these.** (b) $(P + an^2/V^2)(V/n - b) = RT$ (replacing V/n by molar volume returns the equation to the usual form).

40. (c) is the easiest and is clearly an isotherm involving the region of condensation, with P the ordinate and V the abscissa. On reflection, (b) must be a plot of P as ordinate versus T as abscissa, hence an isochor. The first portion shows an increase in liquid vapor pressure and the subsequent linear portion, simply Charles' law for the vapor after all the liquid has evaporated. (a) Must then be an isobar, and evidently T is ordinate and V, the abscissa.

41. See Problem 30. b and thus V_c are proportional to T_c/P_c and therefore are largest for C_2H_6 (this could be argued qualitatively just on the grounds of which molecule should be largest). a is proportional to T_c^2/P_c, and hence is largest for C_2H_6. Again this could be deduced qualitatively just by noting how much higher T_c is for ethane. Since 25°C is above T_c for nitrogen but close to T_c for ethane, the nitrogen will clearly be the more ideal at this temperature. Answers are then (a) N_2; (b) N_2; (c) C_2H_6; (d) N_2.

42. T_c is proportional to a/b; hence gas B has the largest T_c. (Since the b values are not very different, this conclusion could have been reached just on the qualitative argument that the larger a value for gas B means greater intermolecular attractive forces, hence a larger T_c needed to overcome them.) b and V_c are proportional; hence gas C has the largest V_c. Since the b values are so similar, ideality around STP will largely be determined by the a values. Gas A, having the smallest a value, should be the most ideal in behavior.

2

KINETIC MOLECULAR
THEORY OF GASES

COMMENTS

The kinetic molecular theory of gases, in all its glory, can be a rather complicated affair. We restrict ourselves here to the more straightforward aspects. Equation (2-1) follows from a very simple derivation in which molecules of negligible volume are considered to be in random motion.

The Boltzmann principle [as in Eq. (1-4)] tells us that the probability of a given kinetic energy, $mc^2/2$, is proportional to $\exp(-mc^2/2kT)$, and allows the derivation of Eq. (2-2) for the probability distribution of molecular velocities. It then turns out that the average velocity in Eq. (2-1) is actually the rms velocity, while in most other situations it is the average velocity $\bar{c}$ that is needed. This is true, for example, in the case of the wall collision frequency Z [Eq. (2-3)].

The quantity Z is a very useful one. By means of it you can deal with the effusion of gases and Graham's law. Thus, if there is a small hole in a thin wall, it is safe to assume that the rate at which molecules pass through is simply the rate at which they would collide with the same area of wall. Z also gives the rate at which molecules pass through any imagined plane within a gas, and provides a means of deriving the equations for the viscosity, diffusion, and so on.

Another central equation in kinetic molecular theory is that for the frequency Z_A with which a molecule collides with others of its kind in a gas. A new concept, that of molecular diameter, is now required (after all, if molecules were points, they would never hit each other). The related quantity Z_{AA} or the frequency of bimolecular collisions is of fundamental importance to the theory of reaction kinetics (Chapter 16).

27

TABLE 2-1   SUMMARY OF GAS KINETIC THEORY QUANTITIES[a]

Quantity	Formula	Approximate value
Average velocity	$\bar{c} = \left(\dfrac{8RT}{\pi M}\right)^{1/2}$	5×10^4 cm/sec
Wall collision frequency	$Z = \frac{1}{4}\bar{c}C$	3×10^{23} coll/cm²-sec or $\frac{1}{3}$ mole/cm²-sec
Molecular collision frequency	$Z_A = \sqrt{2}\pi d^2 \bar{c}C$	5×10^9 coll/cc-sec
Bimolecular collision frequency	$Z_{AA} = \dfrac{1}{\sqrt{2}}\pi d^2 \bar{c}C^2$	1×10^5 coll/cc-sec or 6×10^{10} C^2 moles coll/liter-sec (C in moles/liter)
Mean free path	$\lambda = \dfrac{1}{\sqrt{2}\pi d^2 C}$	$1 \times 10^{-5}/P$ (P in atm)
Viscosity coefficient	$\eta = \frac{1}{2}m\lambda\bar{c}C$	3×10^{-4} poise
Self-diffusion coefficient	$\mathscr{D} = \frac{1}{3}\lambda\bar{c}$	0.2 cm²/sec
Thermal conductivity coefficient	$\kappa = \eta c_v$	4×10^{-5}

[a] Calculated for a gas at 25°C at 1 atm pressure. The molecular weight is taken to be 30 g/mole, and the area d^2 to be 10^{-15} cm² or 10 Å².

As in Chapter 1, the problems are mainly concerned with how well you can analyze a situation and how fluently you can deal with added restrictions that modify the usual relationships. You will also get some useful practice in being consistent in your choice of units for various quantities. Carelessness in this respect will lead to wildly impossible answers! You should, in fact, reach a point of knowing what the order of magnitude should be for $\bar{c}$, Z, Z_A, and so on, for a typical gas. This will help you to avoid careless errors, and, what is more important, will give you a coherent picture of what the kinetic molecular theory says a gas is like. Table 2-1 may be of help in this respect.

EQUATIONS AND CONCEPTS

Simple Kinetic Theory

$$PV = \frac{1}{3}Mc^2 \qquad c = \left(\frac{3RT}{M}\right)^{1/2} \qquad (2\text{-}1)$$

Here, c is the rms velocity; note that R should be in units of erg/mole-deg if c is to be in cm/sec.

Velocity Distribution Equations

The fraction of molecules dN/N_o having velocity between c and $c + dc$:

$$dN/N_o = \left(\frac{M}{2\pi RT}\right)^{3/2} 4\pi \, e^{-Mc^2/2RT} \, c^2 \, dc \qquad (2\text{-}2)$$

Mean velocity: $\bar{c} = (8RT/\pi M)^{1/2}$. Root-mean-square velocity: $(\bar{c}^2)^{1/2} = (3RT/M)^{1/2}$. Most probable velocity: $c_p = (2RT/M)^{1/2}$.

Wall Collision Frequency and Effusion

$$Z = \frac{1}{4}C\bar{c} \quad \text{or} \quad Z = \frac{1}{4}P\left(\frac{8}{\pi MRT}\right)^{1/2} \qquad (2\text{-}3)$$

where C is the concentration in molecules per cc, and Z is in collisions per cm^2 per second. For two gases at the same pressure and temperature, the ratio of rates of leakage through a small hole will be equal to the ratio of wall collision frequencies. Thus $v_2/v_1 = Z_2/Z_1 = (M_1/M_2)^{1/2}$. This is known as *Graham's Law*. If the velocity of effusion is expressed as weight rather than moles per unit time, then $w_2/w_1 = M_2 Z_2/M_1 Z_1 = (M_2/M_1)^{1/2}$.

Molecular Collision Frequencies

The frequency with which an individual molecule of a gas A makes collisions is

$$Z_\mathrm{A} = \sqrt{2}\pi d^2 \bar{c} C \qquad (2\text{-}4)$$

where d is the molecular diameter. The number of bimolecular collisions per cc per second, Z_AA, is

$$Z_\mathrm{AA} = \frac{1}{\sqrt{2}}\pi d^2 \bar{c} C^2 \qquad (2\text{-}5)$$

For a mixture of gases A and B,

$$Z_\mathrm{AB} = \pi d_\mathrm{AB}^2 \left(\frac{8RT}{\pi\mu}\right)^{1/2} C_\mathrm{A} C_\mathrm{B} \qquad (2\text{-}6)$$

where μ is the reduced molecular weight, $M_\mathrm{A} M_\mathrm{B}/(M_\mathrm{A} + M_\mathrm{B})$, and d_AB is the average molecular diameter.

Other Quantities

$$\text{Mean free path}: \quad \lambda = \frac{\bar{c}}{Z_A} = \frac{1}{2\pi d^2 C} \tag{2-7}$$

$$\text{Viscosity coefficient}: \quad \eta = 2Zm\lambda = \tfrac{1}{2}\rho\bar{c}\lambda \tag{2-8}$$

Here, m is the mass per molecule, and ρ is the density.

$$\text{Self-diffusion coefficient}: \quad \mathscr{D} = \tfrac{1}{3}\lambda\bar{c} \tag{2-9}$$

PROBLEMS

1. (7 min) Arrange in order of increasing value: most probable velocity of molecules of a gas, rms velocity, average velocity. Would you expect the difference between these three to increase, decrease, or remain the same with increasing temperature?

2. (5 min) Assume that for argon and krypton the vapor densities at the respective normal boiling points are the same. This means that, at their respective normal boiling points, the molecular velocity in the argon vapor is (greater than, less than, the same as, cannot tell) the molecular velocity in the krypton vapor.

3. (15 min) In the case of a two-dimensional gas, the velocity distribution equation is

$$\frac{dN}{N_o} = \left(\frac{m}{2\pi kT}\right)2\pi \, e^{-mc^2/2kT} \, c \, dc$$

Calculate the probability of a molecule having a kinetic energy $mc^2/2$ equal to or greater than a given value E. As a guide, in the derivation you are asked to make, the probability will come out as a function of E and kT only.

---

4. (12 min) Graham's law is sometimes given in terms of the rate of change dV/dt in the volume of a gas kept at constant P and T, owing to its escaping through a pinhole of area A. Derive the equation for dV/dt, that is, dV/dt as a function of P, T, mol. wt M, A, and so on.

If M and T are such that the mean velocity is 4×10^5 cm/sec, calculate dV/dt per cm^2. (Assume ideal gas behavior.)

5. (9 min) In which case, H_2 (at 1 atm and 50°K) or O_2 (at 2 atm and 200°K), would there be the greatest number of grams of gas hitting a unit area in unit time?

6. (11 min) From the kinetic theory point of view, the pressure exerted by a gas is an average resulting from many individual collisions of molecules with the wall. It is also possible to talk of the local pressure at the point of impact of a single molecule. To do this, one imagines that a molecule of radius r approaches the wall head on, flattens somewhat on impact, then rebounds elastically (see Fig. 2-1). The collision thus takes place over a time interval of about $2r/c$ sec, where c is the average velocity, and the local pressure is exerted over an area approximating the cross section of the molecule. It is thus possible to obtain an expression in which the local pressure is given as a function of m (molecular mass), c, and r. Derive this function.

7. (9 min) Two separate bulbs are filled with neon and argon gas, respectively. If conditions are such that the Ar is at twice the absolute temperature and half the density of the Ne, what is the ratio of wall collision frequencies? At. wt Ne = 20, Ar = 40.

8. (12 min) (a) The mean velocity of H_2 molecules is 2×10^5 cm/sec at $T°C$. Calculate the number of grams of hydrogen per second hitting 1 cm^2 of wall if the pressure of the gas is such that the molar volume is 1 liter.

(b) Under a new set of conditions, the hydrogen pressure is twice that in (a), but the molar volume is still 1 liter. Calculate the ratio of the new collision frequency to that in (a).

9. (6 min) What will be the ratio of final to initial Z for a given gas if its temperature is doubled at constant volume?

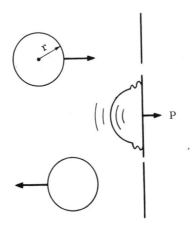

FIGURE 2-1

10. (15 min) The total volume of a vacuum line system is 22.4 liters. On pumping out the air in the system, it is found that a pinhole leak is letting in air at a rate such that the pressure in the vacuum line is increasing by 0.002 atm/sec. Air may be considered to be an ideal gas of average molecular weight 29; the temperature is 0°C (it is a cold day). Assuming that every air molecule that hits the pinhole enters the vacuum line, calculate what the area (in cm^2) of the pinhole must be. Atmospheric pressure is 1 atm. (Gas constant R : 0.082 liter-atm/mole-°K ; 82 cc-atm/mole-°K ; 1.98 cal/mole-°K ; 8.3×10^7 ergs/mole-°K. One atmosphere pressure is equivalent to 10^6 dyne/cm^2. The density of mercury is 13.6 g/cc. Acceleration due to gravity is 980 cm/sec^2.)

11. (13 min) A container is filled with an ideal gas at a certain pressure and temperature. (a) The container is cooled to one-half the original absolute temperature. (b) The amount of gas is halved, keeping the original temperature. (c) The gas is replaced by the same weight of another ideal gas, keeping the original temperature. This second gas has twice the molecular weight of the first one.

For each case, give the numerical value of the ratio of wall collision frequency Z before the change to that after the change. For which of the three cases will the average molecular velocity be the largest?

12. (10 min) Give the ratio c_1/c_2 of molecular velocities and Z_1/Z_2 of wall collision frequencies for the following changes in conditions of an ideal gas: (a) T is doubled at constant P. (b) P is doubled at constant T.

13. (11 min) A flask contains a mixture of H_2 and O_2 at 0°C and 1 atm total pressure (no sparks or catalyst present!). The mixture contains 36% by weight of H_2. Calculate the ratio Z_{H_2}/Z_{O_2} where Z denotes the wall collision frequency in moles/cm^2-sec.

14. (8.5 min) (final examination question) Gas A has twice the molecular weight of gas B. Assuming ideal behavior, if the densities and temperatures of the two gases are the same, then the grams of molecules of gas A hitting area per unit time will be _____ times the corresponding number of grams of gas B.

15. (6 min) Given the approximate wall collision frequency formula $Z = \frac{1}{6}P(3/MRT)^{1/2}$, what will be the ratio of final to initial Z values for a given gas if P is doubled at constant gas density?

16. (15 min) A 2-liter flask contains 15 g of an ideal gas at 3 atm pressure. Calculate how long it should take for 2% of the gas to escape through a pinhole 10^{-4} cm^2 in area.

17. (10 min) In the effusion situation, a complication is that rapidly moving molecules have a relatively greater chance to escape than do slowly moving ones, so that the escaping gas is hotter than the parent body of gas. As an approximation, suppose that the original gas can be considered as made up of two equal groups of molecules: group A whose velocities are all $0.5\,\bar{c}$, and group B, whose velocities are all $1.5\,\bar{c}$ (so that the overall average velocity is still $\bar{c}$).

Obtain, using simple derivations, an estimate of the ratio of numbers of group A molecules to group B molecules that effuse, and from this an estimate of the average temperature of the effusing gas, relative to the initial temperature.

---

18. (15 min) For hydrogen gas at a certain temperature and pressure, it turns out that the mean free path is 5×10^{-7} cm and the bimolecular collision frequency, Z_{AA}, is 1.6×10^{11} moles coll/liter-sec. (a) Calculate from the above data alone the viscosity of the gas. (b) Estimate the temperature of the gas if you are now also informed that the pressure is 10 atm. This last calculation is to be a quantitative one, but can involve reasonable estimates of needed quantities whose value is not given.

19. (12 min) Gases A and B (separate gases in separate flasks) are at the same pressure, and have the same molecular weight. However, gas A is at twice the absolute temperature of gas B. Calculate the ratio of the self-diffusion coefficient of gas A to that of gas B.

20. (10 min) The unit of viscosity in the cgs system is the poise. Show what the dimensions are (in g-cm-sec units).

21. (10 min) (final examination question) Show that according to the simple kinetic molecular theory of gases the viscosity is independent of pressure, and give a physical explanation of why this should be so.

22. (10 min) Show that according to the simple kinetic molecular theory of gases the thermal conductivity coefficient is independent of pressure, and give a physical explanation of why this should be so.

23. (6 min) Gas A has twice the molecular weight of gas B and is at twice the temperature; their pressure is the same. Show whether the ratio of viscosity to the self-diffusion coefficient for gas A is the same as, greater than, or less than that for gas B.

24. (12 min) For a particular gas at a certain temperature and pressure, the mean free path is 1×10^{-4} cm, and each molecule makes 2×10^{9} collisions per second with other molecules. Calculate, from this information alone, as many kinetic molecular quantities as you can.

ANSWERS

1. Order of increasing value is most probable, average, and rms. These are all proportional to $T^{1/2}$, so their ratio remains constant, but the difference between them will increase with increasing temperature.

2. The rms velocity is $c = (3RT/M)^{1/2}$, but from the ideal gas law, $RT/M = P/\rho$, then $c = (3P/\rho)^{1/2}$. Since the densities are the same and, by definition of normal boiling point, the vapor pressures are both 1 atm, it follows that the velocities are **the same**.

3. Since $dE = mc\,dc$, the equation can be written

$$\frac{dN}{N_o} = e^{-E/kT}\frac{dE}{kT}$$

We want the integral of this between the limits E and infinity; so

$$\frac{N}{N_o} = \int_E^\infty e^{-E/kT}\frac{dE}{kT} = [-e^{-E/kT}]_E^\infty = e^{-E/kT}$$

4. Since $Z = -dn/dt$, where n is the number of moles of gas remaining, one can write $Z = (P/RT)\,dV/dt$. Using the formula for Z, $dV/dt = -\frac{1}{4}(P/RT)C\bar{c}A = -\frac{1}{4}\bar{c}A$. From the value of $\bar{c}$ that is given, $dV/dt = -1 \times 10^5$ (decrease in gas volume in cc per second, for $A = 1\ \text{cm}^2$).

5. Since we want grams of gas hitting rather than moles, the value of Z must be multiplied by M, that is, $ZM = \frac{1}{4}P(8M/\pi RT)^{1/2}$. The ratio ZM for the two gases is then

$$\frac{ZM_{\text{H}_2}}{ZM_{\text{O}_2}} = \frac{1}{2}[(2 \times 200)(32 \times 50)]^{1/2} = \frac{1}{4}$$

6. The force of the impact will be $F = d(mc)/dt$ or $F = \dfrac{2mc}{2r/v} = mc^2/r$. Pressure $= F/\text{area}$, so $P = (mc^2/r)/\pi r^2 = \boldsymbol{mc^2/\pi r^3}$ (of the order of 1,000 atm for a small molecule at STP).

7. The formula $Z = \frac{1}{4}P(8/\pi MRT)^{1/2}$ is convenient to use here. First, from the ideal gas law $MP = \rho RT$ (ρ is density), so that

$$\frac{M_{\text{Ar}}P_{\text{Ar}}}{M_{\text{Ne}}P_{\text{Ne}}} = \frac{\rho_{\text{Ar}}T_{\text{Ar}}}{\rho_{\text{Ne}}T_{\text{Ne}}} \quad \text{or} \quad \frac{2P_{\text{Ar}}}{P_{\text{Ne}}} = (\tfrac{1}{2})(2)$$

and $P_{Ar} = P_{Ne}/2$. Then

$$\frac{Z_{Ar}}{Z_{Ne}} = \tfrac{1}{2}[(\tfrac{1}{2})(\tfrac{1}{2})]^{1/2} = \frac{1}{4}$$

8. (a) $Z = \tfrac{1}{4}\bar{c}C$, where C is in moles per cc if Z is in moles of coll/cm^2/sec. ZM, the grams hitting 1 cm^2/sec, is then $ZM = \tfrac{1}{4}MC\bar{c} = (1/4)(2)(0.001) \times (2 \times 10^5) =$ **100 g/cm^2-sec.** (b) For the molar volume still to be 1 liter after doubling the pressure, the temperature must have doubled; hence the new value of ZM will be $(2)^{1/2}$ times the original value since $\bar{c}$ varies as $T^{1/2}$. **141 g/cm^2-sec.**

9. The molar volume is inversely proportional to C, so Z is proportional to $(1/V)(T)^{1/2}$. The ratio of final to initial Z is then simply $2^{1/2}$ if V is constant.

10. Since the volume and temperature of the vacuum line are constant, $dn/dt = (V/RT)dP/dt = dP/dt$ (P in atm, since $22.4 = V = 0.082 \times 273$). Then

$$Z = \tfrac{1}{4}P(8/\pi MRT)^{1/2} = \tfrac{1}{4}(1 \times 10^6)\left(\frac{8}{\pi \times 29 \times 8.3 \times 10^7 \times 273}\right)^{1/2}$$

$$= 0.49 \text{ mole/cm}^2\text{-sec (note } P \text{ must be in dyne/cm}^2)$$

But $dn/dt = 0.002 = Z \times$ area; hence the area is **0.0041 cm^2.** (Perhaps the main source of trouble in this problem is with units.)

11. In each case it is necessary to make a change of variable in the wall collision formula $Z = \tfrac{1}{4}P(8/\pi MRT)^{1/2}$ so as to show explicitly the variable being held constant.

(a) Volume is constant, so replace P by nRT/V, and $Z = \tfrac{1}{4}(n/V) \times (8RT/\pi M)^{1/2}$. The ratio of old to new Z is now seen to be $2^{1/2}$ or **1.41.**

(b) Halving the amount at constant temperature (and volume) must halve the pressure. The ratio of old to new Z is then **2.**

(c) The substitution halves the number of moles (and hence the pressure) while doubling the molecular weight; hence the ratio is $(2)(2)^{1/2} =$ **2.8.**

Since $\bar{c} = (8RT/\pi M)^{1/2}$, in (a) it is reduced by $2^{1/2}$, in (b) it is **unchanged,** and in (c) it is reduced by $2^{1/2}$. Therefore of the three cases, $\bar{c}$ is the **largest for case (b).**

12. Since velocity is proportional to $(T)^{1/2}$, c_1/c_2 is $1/(2)^{1/2} =$ **0.71,** and **1,** respectively. Since Z is proportional to $P/(T)^{1/2}$, the value of Z_1/Z_2 is $(2)^{1/2}$ or **1.41** for case (a) and is $\tfrac{1}{2}$ for case (b).

13.   If the weight ratio is 36/64, the mole ratio is (36/2)/(64/32) or 9/1, thus also the ratio of partial pressures. Therefore,

$$\frac{Z_{H_2}}{Z_{O_2}} = \tfrac{9}{1}(32/2)^{1/2} = \mathbf{36}$$

14.   First, $\rho = PM/RT$ or $\rho T = PM/R$. Thus, if ρ and T are the same for the two gases, the product PM must be the same. The ratio of weights of the two gases that hit unit area per unit time will be

$$\frac{w_A}{w_B} = \frac{M_A Z_A}{M_B Z_B} = \frac{M_A P_A}{M_B P_B}\left(\frac{M_B}{M_A}\right)^{1/2}$$

But $M_A P_A = M_B P_B$ and $M_B/M_A = 1/2$, so the desired ratio $w_A/w_B = (1/2)^{1/2} = \mathbf{0.71}$.

15.   Since $M/RT = \rho/P$, if P is doubled with ρ constant, then M/T is halved and T is doubled. The $Z_2/Z_1 = (P_2/P_1)(T_1/T_2)^{1/2} = (2)(1/2)^{1/2} = \mathbf{1.41}$.

16.   The problem here is to reexpress the wall-collision-frequency equation so as to involve only those quantities that are given. We have $Z = \tfrac{1}{4}P(8/\pi MRT)^{1/2}$. First convert to dw/dt or grams hitting unit area per second by multiplying by $M: dw/dt = \tfrac{1}{4}P(8M/\pi RT)^{1/2}$. Next, from the ideal gas law, $PV = wRT/M$, it follows that $M/RT = w/PV$, so that

$$dw/dt = \tfrac{1}{4}P\left(\frac{8w}{\pi PV}\right)^{1/2} = \frac{1}{4}\left(\frac{3wP}{V}\right)^{1/2}$$
$$= \frac{1}{4}\left(\frac{3 \times 15 \times 3 \times 10^6}{2{,}000}\right)^{1/2} = 58 \text{ g/cm}^2\text{-sec}$$

For the pinhole in question, dw/dt is the 0.0058 g/sec. The time for 2% or 0.3 g of gas to escape is then $0.3/0.0058 = \mathbf{52}$ **sec.**

17.   The ratio of effusion rates for the two groups will be that of their mean velocity values, since the concentrations are equal, or $n_A/n_B = 0.5/1.5$. The respective mole fractions of cold and hot molecules escaping is thus 0.25 and 0.75. Since kinetic energy E is proportional to $\bar{c}^2$, it follows that the average E of the escaped molecules is proportional to $(0.25)(0.5\bar{c})^2 + (0.75)(1.5\bar{c})^2$ or $1.75\bar{c}^2$. Since T and E are proportional, the average temperature of the escaping gas is thus **1.75 times** that of the rest of the gas.

18.   (a) It is necessary to do some eliminating of variables. First, $Z_{AA} = \tfrac{1}{2}Z_A C$ so $C = 2Z_{AA}/Z_A$. Also, $\lambda = \bar{c}/Z_A$, so $\bar{c} = \lambda Z_A$. On substituting into the equation for η,

$$\eta = \frac{1}{2}m\lambda\bar{c}C = \frac{1}{2}m\lambda(\lambda Z_A)\left(\frac{2Z_{AA}}{Z_A}\right) = m\lambda^2 Z_{AA}$$

Then

$$\eta = \left(\frac{2}{6 \times 10^{23}}\right)(5 \times 10^{-7})^2\left(\frac{1.6 \times 10^{11} \times 6 \times 10^{23}}{1,000}\right)$$

$$\eta = 8 \times 10^{-5} \text{ poise}$$

(b) From the equation for λ, $C = 1/(2\pi d^2\lambda)$, and if d^2 is estimated to be, for instance, $5 \times 10^{-16}\,\text{cm}^2$, then C is 9.1×10^{20} molecules per cc or 1.5×10^{-3} moles/cc. From the ideal gas law, T is then $P/RC = 10/82 \times 1.5 \times 10^{-3} = \mathbf{81°K}$.

19. Again it is necessary to change some variables. The expression for $\mathscr{D}$ is $\frac{1}{3}\lambda\bar{c}$, and λ is proportional to $1/d^2 C$ or to $T/d^2 P$ while $\bar{c}$ is proportional to $(T/M)^{1/2}$. It then follows that $\mathscr{D}$ itself is proportional to $T^{3/2}/d^2 P M^{1/2}$. The pressure and molecular weight are the same for the two gases, and since the molecular weights are the same, the d^2 values should not be much different. Thus $\mathscr{D}_A/\mathscr{D}_B \approx (T_A/T_B)^{3/2} = 2^{3/2} = \mathbf{2.8}$.

20. Since $\eta = \frac{1}{2}m\lambda\bar{c}C$, the dimensions are (g) (cm) (cm sec^{-1}) (cm^{-3}) or $\text{g cm}^{-1}\text{sec}^{-1}$.

21. The conclusion that viscosity is independent of pressure follows from the equation

$$\eta = \frac{1}{2}m\lambda\bar{c}C = \frac{m\bar{c}}{2\sqrt{2\pi d^2}}$$

which contains no quantities that depend on pressure. The qualitative explanation is that while the number of molecules that cross between adjacent shear planes decreases with decreasing pressure, the mean free path increases, so that the amount of momentum transfer is unchanged.

22. The equation $\kappa = \eta c_v$ contains no pressure dependent terms. The qualitative explanation is similar to that of Problem 21. The flux of heat carrying molecules across unit area in the gas decreases with decreasing pressure, but elementary length of heat transport, the mean free path, increases; the two effects just cancel.

23. From the equation of Table 2-1, the ratio $\eta/\mathscr{D}$ is proportional to mC and hence to the density ρ. By the ideal gas law, $\rho = PM/RT$; thus if molecular weight and temperature are doubled at constant pressure, ρ and therefore $\eta/\mathscr{D}$ are **unchanged**.

24.   The product $Z_A \lambda$ equals $\bar{c}$; $\mathscr{D} = \frac{1}{3}\lambda\bar{c}$. Then $\bar{c} = 1 \times 10^{-4} \times 2 \times 10^9 = 2 \times 10^{-5}$ cm/sec, and $\mathscr{D} = 1 \times 10^{-4} \times 2 \times 10^{-5}/3 = \mathbf{6.7\,cm^2/sec}$.

3

SOME PHYSICAL PROPERTIES OF MOLECULES

COMMENTS

We concentrate, in this section, on three useful and widely used physical properties that reflect molecular structure: absorption of light, index of refraction, and dielectric constant. As with an iceberg, there is a lot more to these topics than the simple equations below indicate. Light absorption lifts a molecule to a higher energy state—electronic, vibrational, or rotational, or some combination of these—so that an absorption spectrum tells us something about the electronic and geometric organization of a molecule. Roughly speaking, electronic transitions require light of an energy corresponding to the visible or ultraviolet region of the spectrum; a change in vibrational state requires energy corresponding to light in the deep red or ir region; and radiation in the microwave region suffices to bring about changes in the rotational state of a molecule.

Index of refraction depends on the polarizability, or ease of electrical distortion, and is largely an atomic property, only moderately dependent on the presence of chemical bonding. Polarizabilities or molar refractions are therefore nearly additive properties. Dielectric constants reflect both polarizability and the presence of dipoles. Knowledge of the dipole moment of a molecule can be very useful, as it may settle questions about molecular geometry as well as indicate how polar the bonds are.

The main concern in this section is to acquaint you with the Beer–Lambert law of light absorption in a variety of situations, with the additivity principle that has more applications than just to molar refraction, and with the workings of the interrelations between index of refraction, dielectric constant, dipole moment, and molecular structure. As you gain appreciation of how these properties behave in various circumstances, hopefully the theoretical background of them will become more real and more interesting.

EQUATIONS AND CONCEPTS

Beer–Lambert Law

$$I/I_0 = e^{-klC} \tag{3-1}$$

where $l =$ path length and C denotes concentration. Alternative form: $D = \varepsilon lC$ where D is the optical density (absorbancy) defined as $D = \log I_0/I$, ε is the extinction coefficient (or absorbancy index), and l and C are in units of centimeters and moles/liter, respectively.

for a mixture of absorbing species: $\quad D = \Sigma D_i = l\Sigma \varepsilon_i C_i \tag{3-2}$

Molar Refraction

$$M_r = \frac{M}{\rho} \frac{n^2 - 1}{n^2 + 2} \tag{3-3}$$

where M/ρ is the volume in which 1 mole is present, and n is the index of refraction. To a fair degree of approximation, the molar refraction of a molecule can be expressed as the sum of molar refractions for the atoms in the molecule.

Molar Polarization

$$\mathbf{P} = \frac{M}{\rho} \frac{D - 1}{D + 2} \tag{3-4}$$

where M/ρ again denotes the volume in which 1 mole is present and D is the dielectric constant. Also $\mathbf{P} = \frac{4}{3}\pi N\alpha + \frac{4}{3}\pi N(\mu^2/3kT)$, where α is the polarizability and the term $\frac{4}{3}\pi N\alpha$ may be approximated, if so desired, by $\mathbf{M}_r$, and μ is the dipole moment. Dipole moment is defined as $\mu = ed$, where e is the value in electrostatic units (esu) of equal and opposite charges, which are separated by distance d. A customary unit is the debye: 1 debye $= 10^{-18}$ esu-cm. The net dipole moment of a molecule is the vector sum of the contributions from each bond.

PROBLEMS

1. (10 min) Cell A is 1 cm deep and, when filled with a certain solution, the percent transmission for light of a given wavelength is 70%. Cell B is 2 cm deep and, when filled with another certain solution, transmits 60% of light of this same wavelength. The percent transmission for light passing through both cells (of this wavelength) will then be 10%, 24%, 70%, 60%, none of these within 1%.

2. (9 min) The light absorption is measured vertically through a cuvette (of square cross section) containing a 1-cm depth of a 0.02-M aqueous solution of a chromic salt (see Fig. 3-1a). At 350 mμ the optical density is found to be 0.5. If pure water is added so that the solution is diluted to the point where the depth is 3 cm, as in Fig. 3-1b, the optical density becomes 0.6. Neglecting any absorption by the cuvette itself, calculate the optical density of 1 cm thickness of pure water and the extinction coefficient at this wavelength for the chromic salt.

3. (12 min) Given the following molar refractions: CH_3I, 19.5; CH_3Br, 14.5; HBr, 9.9; and CH_4, 6.8; calculate the value for CH_2BrI.

4. (6 min) The molar polarization of $NH_3(g)$ obeys the equation $P = A + B/T$ where A and B have the values 5.6 and 12,000 in cc/mole, respectively. Calculate the dielectric constant of $NH_3(g)$ at STP.

5. (6 min) The extinction coefficient for $Coen_2Br_2^+$ is 40 at 650 mμ. Calculate the percent transmission for a 5-cm cell filled with 0.01 m solution. Neglect solvent absorption.

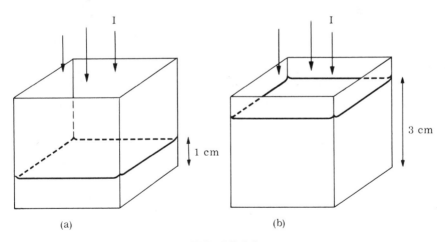

FIGURE 3-1

6. (10 min) The index of refraction of a gaseous normal paraffin (i.e., of formula C_nH_{2n+2}) is found to be 1.00139 when the gas is at STP. Given that the atomic refractions are 1.1 and 2.42 (cc/mole) for H and C, respectively, determine the formula for the hydrocarbon, that is, the value of n in the formula.

7. (10 min) A 0.003-M solution of $Co(NH_3)_6^{3+}$ transmits 75% of incident light of 500 mμ if the path length is 1 cm. Calculate the extinction coefficient and the percent absorption for a 0.001-M solution (other factors remaining constant).

8. (10 min) If Δ is defined as the decimal part of the index of refraction ($\Delta = n - 1$), show that for an ideal gas Δ should be proportional to pressure at constant temperature, at low pressures.

9. (8 min) The total optical density D of a 0.05-M solution of $Cr(H_2O)_6(ClO_4)_3$ is 0.60 at 310 mμ, and is 1.10 for a 0.10-M solution. Assuming Beer's law to hold, the optical density of the solvent is most nearly (0, 0.1, 0.2, 0.4, 0.6, 1.1), and the molar extinction coefficient of the salt is most nearly (0, 1, 2, 3, 10, 12, 20). The cell depth is 1 cm.

10. (12 min) Ninety percent of incident light of wavelength 570 mμ is absorbed by a solution that is 1 M in $Cr(H_2O)_6^{3+}$ (species A) and 1.5 M in SCN^-, using a 1-cm cell. On standing, a reaction takes place whereby all of A is converted into $Cr(H_2O)_5SCN^{2+}$ (species B). The extinction coefficient for B is three times that for A at this wavelength.
   (a) Calculate the optical density D for the original solution.
   (b) Assuming that A and B are the only absorbing species involved, calculate D and the percent transmission for the solution after standing.
   (c) The answer to (b) is obtained assuming no absorption by solvent. If it is recognized that the optical density of the solvent at 370 mμ is actually 0.1, then show what the value of D should be for the final solution.

11. (12 min) A pure liquid hydrocarbon is known to belong to the series C_nH_{2n+2}, but its actual molecular weight is not known. Its density is 0.66 and its index of refraction is 1.38. In addition, one knows that the atomic refraction is 1.10 for H and 2.42 for C. Determine from the above information the value of n in the formula C_nH_{2n+2} and then the molecular weight of the compound.

12. (11 min) A 0.03-M solution of $Co(C_2O_4)_3^{3-}$ has an optical density of 2.0 at 660 mμ, using a 1-cm cell. Calculate (a) the value of the extinction coefficient ε, (b) the value of I/I_0, and (c) the percent absorption for a 0.015-M solution in the same cell. Neglect solvent absorption.

13. (7 min) Arrange in order of decreasing dipole moment (group together compounds having the same value): Cl_2, SO_2, CO_2, H_2O, o-dinitrobenzene, HI, CH_4.

14. (4 min) A 2-cm cell filled with a 0.005-M solution of bromine transmits 1% of light of 436 mμ. Calculate the extinction coefficient ε for bromine at this wavelength.

15. (7 min) Sugden's parachor **P** is another type of molar volume that is handled in the same way as is the molar refraction. Calculate **P** for C_2H_6 given that **P** = 110, 73, and 71 for CH_3Cl, CH_4, and HCl, respectively.

16. (7 min) The index of refraction for liquid water at 25°C is 1.33; calculate n for water vapor at a pressure and temperature such that the density is 1 g/liter. No other information is needed, except that the density of liquid water is 1 g/cc.

17. (15 min) The optical density of 2×10^{-4} M K_2PtCl_6 is 0.50 at 264 mμ, using a 1-cm cell.

(a) To what percent transmission does this correspond?

(b) What is the extinction coefficient ε for K_2PtCl_6 at 264 mμ?

(c) For what concentration of K_2PtCl_6 would the percent absorption be 90% (1-cm cell, 264 mμ)?

(d) If the same 1-cm cell when filled with water gives an optical density of 0.15 with light of 264 mμ, what is the correct value of ε for K_2PtCl_6?

18. (12 min) Given the data of Table 3-1 as an aid, for each substance plot the molar polarization **P** against $1/T$ on Fig. 3-2. The plots need be

TABLE 3-1

Substance	M_r, cc/mole	μ, debyes[a]
(a) benzene	26	0
(b) CH_4	6.8	
(c) C_6H_5Cl	31	1.7
(d) *o*-dinitrobenzene		

[a] You are expected to make your own estimates of missing information.

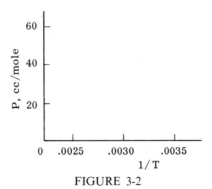

FIGURE 3-2

semiquantitative only, but close attention should be paid to the **relative** positions and slopes of the lines.

19.   (4 min) A 0.001-M solution of $PtBr_6^{2-}$ absorbs 99% of the incident light of wavelength 400 mμ, using a 1-cm cell. For what length cell would 90% of the incident light be absorbed, other aspects remaining the same?

20.   (9 min) Calculate the index of refraction of chlorine gas at STP. The molar refraction for chlorine is 11.2 cc/mole.

21.   (9 min) List in order of increasing dipole moment (put those of equal dipole moment in a common category): H_2, o-dinitrobenzene, He, Br_2, H_2O, CCl_4, HI, HCl.

22.   (9 min) Given the following molar refractions: $CH_3CH_2CH_2CH_3$, 20.6; CH_3CH_2OH, 12.9; and CH_3OH, 8.3; calculate that for

$$CH_3CH_2CH_2CH_2OH$$

23.   (12 min) The extinction coefficient for cis-$Coen_2Cl_2^+Cl^-$ is 50 at 387 mμ (in aqueous solution). The percent transmission for a 1-cm cell was determined for a certain solution, and it was concluded that the concentration was 0.006 M.

(a) Calculate the percent transmission that was measured.

(b) It was then discovered that an impurity was present, which contributed an optical density of 0.15 to the solution. What was the correct concentration of the complex?

24.   (5 min) A 2-cm cell filled with a certain gas at STP absorbs 60% of incident light of wavelength 580 mμ. Give the percent transmission for a 2-cm cell with the gas at 2 atm (same wavelength and T).

25.   (15 min) Given that

$$P \text{ (cc/mole)} = \frac{M}{\rho}\frac{D-1}{D+2} = \frac{4\pi N}{3}[\alpha + \mu^2/3kT]$$

(D = dielectric constant, α = polarizability, μ = dipole moment) and that for $CH_3Cl(g)$ **P** is 90 cc/mole at $1/T = 0.004$, and 50 cc/mole extrapolated to $1/T = 0$, (a) calculate α and μ for CH_3Cl and (b) calculate the dielectric constant for $CH_3Cl(g)$ at 500°K and 20 atm pressure.

26.   (15 min) The extinction coefficients for ferrocyanide and ferricyanide ions are 250 and 1000, respectively, at 320 mμ (in $M^{-1} \times cm^{-1}$). (a) What length of cell, when filled with 1×10^{-3} M ferricyanide solution, would absorb 99% incident light of 320 mμ? (b) How many moles per liter of ferricyanide must be added to a 10^{-3}-M solution of ferrocyanide so that the resulting solution will absorb 90% of the incident light of 320 mμ when placed in a 1-cm cell?

27. (6 min) Given that the molar refractions for CH_4 and C_2H_6 are 6.8 and 11.4, respectively, calculate the atomic refractions for C and H.

28. (18 min) A 0.01-M solution of $KCr(C_2O_4)_2(H_2O)_2$ in a 1-cm cell absorbs half as much light of wavelength 250 mμ as does a 0.001-M solution of Br_2 in a 2-cm cell. Calculate ε for the oxalate complex if ε for Br_2 is 150.

29. (10 min) The molar refraction M_r is 1.643 for oxygen in an ether group, and is 6.818 for methane and 13.279 for dimethyl ether (CH_3-O-CH_3). Calculate the value for diethyl ether.

30. (10 min) The extinction coefficient for Fe^{3+} is $250 \, M^{-1} \, cm^{-1}$ at 290 mμ. Calculate the percent of light absorbed by a 2-cm cell filled with $2 \times 10^{-3} \, M$ solution of ferric nitrate. Calculate also the percent of light transmitted if the concentration is doubled. Neglect absorption by solvent.

31. (10 min) Given the plot of Fig. 3-3 of molar polarization versus $1/T$, list the four substances in order of increasing dipole moment and also in order of increasing polarizability α.

32. (10 min) The concentration of ferrous iron in an unknown is to be determined by measuring the absorption of light (of 540 mμ) by the solution (with added dioxime to give the proper colored complex). It is found that a 3-cm-length cell absorbs 64% of the incident light when filled with the unknown solution. From previous work, it is known that a 0.003-M solution of ferrous iron, similarly treated, will absorb 40% of the incident light when placed in a 1-cm cell. Calculate the concentration of ferrous iron in the unknown solution.

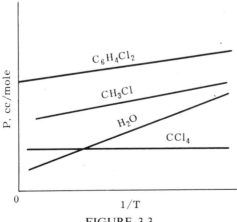

FIGURE 3-3

33. (10 min) Given the following molar refractions: $CH_4 = 6.82$, $C_2H_6 = 11.44$, $C_3H_8 = 16.06$, $CH_3CH_2CH_2OH = 17.58$, calculate the molar refraction of glycol, $HO-CH_2-CH_2-OH$, and the index of refraction (density is 1.115 g/cc).

34. (12 min) Given the partially completed table of information (Table 3-2), make semiquantitative plots (use Fig. 3-4) of the expected variation of the molar polarization of each substance with $1/T$. Pay special attention to having correct relative slopes and intercepts.

35. (10 min) Assume for the purpose of this problem that the bond angle in H_2S is exactly 90° and that the $H-S$ bond length is 1.8 Å. If the dipole moment of each $H-S$ bond is 0.78 debyes, calculate (a) the dipole moment for the molecule as a whole and (b) the effective charge on each hydrogen atom. (One electronic charge equals 4.8×10^{-10} esu.)

36. (9 min) A spectrophotometer cell, when filled with liquid A, transmits 50% of incident light of a certain wavelength (corresponding to an optical density of 0.3) and, when filled with liquid B, transmits only 25% of light of this wavelength. What would be the optical density at this wavelength if the cell were filled with a mixture of equal volumes of the two liquids? Show your work.

TABLE 3-2

Compound	Molar refraction, cc/mole	Dipole moment, debyes
Benzene	26	
o-Dichlorobenzene		
Water	3.8	1.8
Ethanol	12.8	1.7

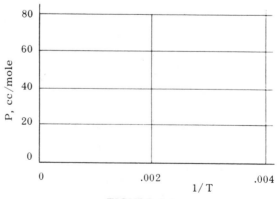

FIGURE 3-4

ANSWERS

1. The answer is just the product of the fractions transmitted or 42%.

2. Since the cross section is constant, that part of the absorption resulting from the chromic ion is unchanged by the increase in depth of solution; hence the difference $0.6 - 0.5$ is due to the extra 2 cm of water. The optical density for 1 cm of water is then **0.05.** That caused by chromic ion is then $0.5 - 0.05$ or 0.45, and its extinction coefficient is $\varepsilon = 0.45/1 \times 0.02 = $ **22.5.**

3. Molar refractions are treated as additive, and there are several ways of combining the data to get the desired answer; here is one.

$$\mathbf{M}_r(CH_2) \quad = \mathbf{M}_r(CH_3Br) - \mathbf{M}_r(HBr) = 14.5 - 9.9 = 4.6$$

$$\mathbf{M}_r(H) \quad = [\mathbf{M}_r(CH_4) - \mathbf{M}_r(CH_2)]/2 = (6.8 - 4.6)/2 = 1.1$$

$$\mathbf{M}_r(I) \quad = \mathbf{M}_r(CH_3I) - \mathbf{M}_r(CH_2) - \mathbf{M}_r(H) = 13.8$$

$$\mathbf{M}_r(CH_2BrI) = \mathbf{M}_r(CH_3Br) - \mathbf{M}_r(H) + \mathbf{M}_r(I) = \mathbf{27.2}$$

4. **P** at 273°K is then $\mathbf{P} = (5.6 + 12{,}000)/273 = 49.4$ cc/mole, and this divided by 22,400 cc/mole molar volume at STP gives 0.00221 for $(D - 1)/(D + 2)$. Since D the dielectric constant will be nearly unity, as a good approximation, $D = 1 + 3 \times 0.00221 = $ **1.00663.**

5. Using the relationship $D = \varepsilon Cl$, $D = 40 \times 5 \times 0.01 = 2$. Hence $-\log I_0/I = 2$ and $I/I_0 = 0.01$, so 1% transmission.

6. Since the molar volume at STP is 22,400 cc/mole,

$$\mathbf{M}_r = 22{,}400 \frac{(1.00139)^2 - 1}{(1.00139)^2 + 2}$$

As a good approximation, $(1.00139)^2 = 1.00278$, so $\mathbf{M}_r = 22.4 \times 2.78/3 = 20.7$. Then $20.7 = 2.42n + 1.1(2n + 2) = 4.62n + 2.2$, whence $\boldsymbol{n = 4}$.

7. $D = \log I_0/I = \log 1/0.75 = 0.126$; hence $\varepsilon = 0.126/0.003 = $ **42.** For the 0.001-M solution, D will then be 0.042, and $I_0/I = 1.10$, which corresponds to 91% transmission or 9% absorption. This last answer can be obtained by the shortcut of noting that, if the concentration is reduced by a factor of three, then I/I_0 goes to its cube root: $(I/I_0)^{1/3} = (0.75)^{1/3} = 0.91$.

8. The molar volume of an ideal gas is RT/P; hence $\mathbf{M}_r = (RT/P) \times (n^2 - 1)/(n^2 + 2)$. At low pressure n will be close to unity, so $n^2 = $ about $1 + 2\Delta$ and $n^2 + 2 = $ about 3. The equation then becomes $\Delta = (3\mathbf{M}_r/2RT)P = $ const $\times P$.

9.   The situation is that the observed optical density is the sum of that resulting from the dissolved salt and that caused by solvent water; that is, $D = D_{salt} + D_w$. For the first case, $0.60 = D_{0.05\,salt} + D_w$, whereas in the second, the optical density due to the salt should be doubled, so $1.10 = 2D_{0.05\,salt} + D_w$. Subtracting the two equations gives $D_{0.05\,salt} = 0.50$ and $D_w = \textbf{0.1}$. Hence $\varepsilon_{salt} = 0.5/0.05 = \textbf{10}$.

10.   (a) $D = \log I_0/I = \log 1/0.1 = \textbf{1}$.

(b) The extinction coefficient of the absorbing species is tripled and so is D; that is, $D = \textbf{3}$.

(c) D, owing to the salt, will now be $1 - 0.1 = 0.9$, and this is tripled in the final solution. The final total D is then $2.7 + 0.1 = \textbf{2.8}$.

11.   We proceed as follows:

$$\textbf{M}_r = \frac{M}{0.66}\frac{n^2 - 1}{n^2 + 2} = \frac{M}{0.66}\frac{0.902}{3.9} = 0.35\,M$$

Also $\textbf{M}_r = 2.42n + 1.1(2n + 2) = 4.62n + 2.2$, and $M = 12n + 2n + 2 = 14n + 2$. Therefore $4.62n + 2.2 = 0.35(14n + 2) = 4.90n + 0.7$ or $n =$ about **5** and $M = \textbf{72}$.

12.   (a) $\varepsilon = D/C = 2.0/0.03 = \textbf{66.7}$. (b) $\log I_0/I = 2$, so $I/I_0 = \textbf{0.01}$. (c) D for a 0.015-M solution will be 1; hence $I/I_0 = 0.1$, so **90%** absorption.

13.   Molecules with a center of symmetry will have zero net dipole moment. This applies to Cl_2 and CO_2, similarly to CH_4 as a regular tetrahedron. If a net moment is expected, it should be larger the more polar the bonds, and the less similar moments should oppose each other. The sequence is then o-dinitrobenzene, H_2O, SO_2, HI, (Cl_2, CO_2, CH_4).

14. $\varepsilon = D/2C$; $D = \log I_0/(0.01 \times I_0) = 2$. Hence $\varepsilon = 2/(0.005 \times 2) = \textbf{200}$.

15.   Treating **P** as an additive quantity, $\textbf{P}_{CH_2} = \textbf{P}_{CH_3Cl} - \textbf{P}_{HCl} = 110 - 71 = 39$. Then $\textbf{P}_{C_2H_6} = \textbf{P}_{CH_4} + \textbf{P}_{CH_2} = 73 + 39 = \textbf{112}$.

16.   From the data for liquid water:

$$\textbf{M}_r = 18\frac{1.33^2 - 1}{1.33^2 + 2} = 3.67$$

Then

$$3.67 = \frac{18}{0.001}\frac{n^2 - 1}{n^2 + 2}$$

Since n for the vapor will be close to unity, a good approximation will be

$n^2 - 1 = 3 \times 3.67 \times 10^{-3}/18 = 6.12 \times 10^{-4}$. Again, since n is close to unity, if $n = 1 + \Delta$, then $n^2 - 1 = 2\Delta$. Hence $\Delta = 3.06 \times 10^{-4}$ and $n = \mathbf{1.000306}$.

17. (a) $\text{Log } I_0/I = D = 0.5$; hence $I/I_0 = 0.316$, so **31.6%** transmission.
(b) $\varepsilon = D/Cl$, so $\varepsilon = 0.50/(2 \times 10^{-4}) = \mathbf{2,500}$.
(c) 90% absorption of 10% transmission corresponds to $D = 1$; hence $C = 1/\varepsilon = \mathbf{4 \times 10^{-4}} \, \mathbf{M}$.
(d) The net optical density resulting from the salt is now $0.5 - 0.15$ or 0.35; hence $\varepsilon = 0.35/(2 \times 10^{-4}) = \mathbf{1,750}$.

18. First approximate the missing items in the table. The dipole moment for CH_4 should be zero on grounds of symmetry, and both M_r and μ should be the largest for o-dinitrobenzene as it is a large molecule with two polar groups acting somewhat in parallel.

The relationship involved is $P = M_r + \frac{4}{3}\pi N\mu^2/3kT$; the plots of Fig. 3-5 should then consist of straight lines, of intercepts M_r, and slopes given by $4\pi N\mu^2/9k$. An order of magnitude calculation gives $(12/9)(36 \times 10^{46}) \times (10^{-36})/8 \times 10^7$ or about 6,000. For a μ of one, P should increase by 6 cc/mole for each increment of 0.001 in $1/T$. With this in mind, one can now sketch qualitative answers as indicated in the graph.

19. 99% absorption corresponds to $I_0/I = 100$ and hence $D = 2$, whereas 90% absorption corresponds to $D = 1$. Therefore the cell length must be **halved**.

20. Since the molar volume of a gas at STP is 22,400 cc/mole, $11.2 = 22,400[(n^2 - 1)/(n^2 + 2)]$. Since n is close to unity, it follows that $n^2 - 1 = (3 \times 11.2)/22,400 = 0.0015$. Hence $n = \mathbf{1.00075} \, [(1.00075)^2 \simeq 1.0015]$.

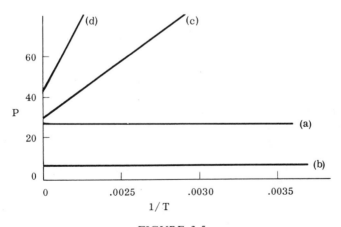

FIGURE 3-5

21. Molecules with a center of symmetry have zero dipole moment (H_2, He, Br_2), as does a regular tetrahedron (CCl_4). Otherwise, the dipole moment increases the more polar the bond and, if more than one like bond, the less they tend to cancel. **Answer**: (H_2, He, Br_2, CCl_4), HI, HCl, H_2O, o-dinitrobenzene.

22. Since molar refractions are treated as additive:

$$M_r(CH_2) = M_r(CH_3CH_2OH) - M_r(CH_3OH) = 12.9 - 8.3 = 4.6$$

Hence

$$M_r(CH_3CH_2CH_2CH_2OH) = M_r(CH_3CH_2OH) + 2M_r(CH_2)$$
$$= 12.9 + 9.2 = \mathbf{22.1}$$

23. (a) Since $D = \varepsilon Cl$, the observed D was $D = 50 \times 0.006 = 0.3$. Since $D = \log I_0/I$, the observed I/I_0 must have been 0.5, that is, 50% transmission.
   (b) Allowing for the impurity, the net D was then $0.3 - 0.15$ or 0.15; hence the correct concentration was one-half 0.006 or **0.003 M.**

24. Initially, $I/I_0 = 0.4 = \varepsilon^{-kCl}$. Doubling the pressure doubles the concentration, hence squares I/I_0, which now has the value 0.16. **Answer**: 16%.

25. At $1/T = 0$, $\mathbf{P} = 4\pi N/3\alpha$; hence $\alpha = (50 \times 3)/(4\pi \times 6 \times 10^{23}) = \mathbf{1.98}$ $\times$ $\mathbf{10^{-23}}$ **cc/molecule.** (a) At $1/T = 0.004$, the second term in the equation is $90 - 50 = 40$ cc/mole; hence

$$40 = \frac{4\pi N\mu^2}{9kT}$$

or

$$\mu^2 = (40 \times 9 \times 8.3 \times 10^7)/(4\pi \times 36 \times 10^{46} \times 0.004)$$
$$= 8.3 \times 10^8/16\pi \times 10^{43}$$
$$\mu^2 = 1.64 \times 10^{-36} \quad \text{and} \quad \mu = \mathbf{1.28 \text{ debyes}}$$

   (b) At 500°K, $1/T = 0.002$; hence $\mathbf{P} = 50 + 40/2 = 70$. M/ρ, the molar volume will be $22,400 (500/273)(1/20) = 2,050$ cc/mole. Then $(D - 1)/(D + 2) = 70/2050 = 0.0326$ and $D = \mathbf{1.096.}$

26. (a) 99% absorption means $I/I_0 = 0.01$, and $D = \log I_0/I = 2$. Since $D = \varepsilon Cl$, $l = 2/1000 \times 10^{-3} = \mathbf{2 \text{ cm.}}$
   (b) 90% absorption means a D of 1. Since D's are additive,

$$1 = \varepsilon_{ferro}C_{ferro} + \varepsilon_{ferri}C_{ferri} = 10^{-3} \times 250 + 1000 C_{ferri}$$

Hence

$$1000\, C_{\text{ferri}} = 1 - 0.25 = 0.75 \quad \text{and} \quad C_{\text{ferri}} = \textbf{7.5} \times \textbf{10}^{-4}\, \textbf{\textit{M}}$$

27.   Molar refractions are taken to be additive, so $2M_r\,(\text{H}) = 2M_r\,(\text{CH}_4) - M_r\,(\text{C}_2\text{H}_6) = 13.6 - 11.4 = 2.2$; hence for H, $M_r = \textbf{1.1}$. Then $M_r\,(\text{C}) = M_r\,(\text{CH}_4) - 4 \times 1.1 = \textbf{2.4}$.

28.   Let A denote the complex and B denote Br_2, then the statement is $[(I_0 - I)/I_0]_A = \frac{1}{2}\,[(I_0 - I)/I_0]_B$ or $2(I/I_0)_A = 1 + (I/I_0)_B$. The Br_2 solution has an optical density, $D = 150 \times 0.001 \times 2 = 0.3$; hence $(I/I_0)_B = 0.5$. Therefore $(I/I_0)_A = 0.75$ and $D_A = 0.126$; therefore $\varepsilon_A = 0.126/0.01 = \textbf{12.6}$.

29.   Using the additivity principle, we find $M_r\,(\text{C}_2\text{H}_6) = M_r$ (dimethyl ether) $- M_r\,(\text{O}) = 13.279 - 1.643 = 11.636$. Then $2M\,(\text{H}) = 2M_r\,(\text{CH}_4) - M_r\,(\text{C}_2\text{H}_6) = 2$ and $M_r\,(\text{CH}_2) = M_r\,(\text{CH}_4) - 2M_r\,(\text{H}) = 4.818$. Finally, M_r (dimethyl ether) $= M_r$ (dimethyl ether) $+ 2M_r\,(\text{CH}_2) = \textbf{22.915}$.

30. $D = \varepsilon Cl = 250 \times 2 \times 10^{-3} \times 2 = 1$; hence $I/I_0 = 0.1$ or **90%** absorbed. If C is doubled, $D = 2$; hence $I/I_0 = 0.01$ or **1%** transmitted.

31.   This is merely a matter of intercepts and slopes. The former are a measure of the polarizability α and the latter, of the dipole moment. The two sequences are

dipole moment: $\text{CCl}_4, \text{C}_6\text{H}_4\text{Cl}_2, \text{CH}_3\text{Cl}, \text{H}_2\text{O}$

$\quad\quad\quad\alpha$: $\text{H}_2\text{O}, \text{CCl}_4, \text{CH}_3\text{Cl}, \text{C}_6\text{H}_4\text{Cl}_2$

32.   The two equations are

$$0.36 = \varepsilon^{-klC} = \varepsilon^{-k3C}$$
$$0.60 = \varepsilon^{-kl0.003}$$

Since $0.36 = (0.60)^2$, then $(3kC) = 2(0.003k)$ or $C = \textbf{0.002}$.

33.   (a) Using the additivity principle, $2M_r(\text{H}) = 2M_r(\text{CH}_4) - M_r(\text{C}_2\text{H}_6) = 13.64 - 11.44 = 2.2$ or $M_r(\text{H}) = 1.1$. $M_r(\text{CH}_3-\text{CH}_2-\text{CH}_2-\text{OH}) - M_r(\text{CH}_3-\text{CH}_2-\text{CH}_3) = M_r(\text{OH}) - M_r(\text{H}) = 1.52$; hence $M_r(\text{OH}) = 2.62$. $M_r(\text{CH}_2) = M_r(\text{CH}_4) - 2M_r(\text{H}) = 6.82 - 2.2 = 4.62$. Then $M_r(\text{glycol}) = 2M_r(\text{OH}) + 2M_r(\text{CH}_2) = 5.24 + 9.24 = \textbf{14.48}$.

   (b) The molar volume of glycol $= M/\rho = 62/1.115 = 55$ cc/mole. Hence $14.48 = 55[(n^2 - 1)/(n^2 + 2)]$ or $n^2 + 2 = 3.85(n^2 - 1)$ or $n^2 = 2.05$, $n = \textbf{1.43}$.

34.   First, we can complete the table with rough guesses as to the missing values. M_r for o-dichlorobenzene should definitely be larger than that for

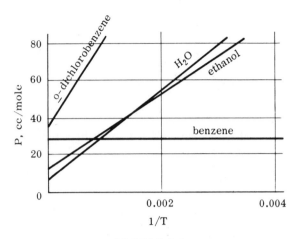

FIGURE 3-6

benzene, about 35 cc/mole. The dipole moment of benzene should be zero since the molecule has a center of symmetry, whereas that for *o*-dichlorobenzene should be quite large, as there are two nearly parallel polar bonds— let us guess a value of about 3. The plots in Fig. 3-6 of **P** versus $1/T$ should be linear, with intercept $\mathbf{M}_r$ and slope proportional to μ^2. Very roughly, the term in μ^2 contributes about 20 cc/mole to **P** at 25°C and $\mu = 1$. With these points in mind, the required straight-line plots are those shown in the figure.

35. (a) The force triangle is evidently that shown in Fig. 3-7, so that the net dipole moment is given by the hypotenuse of a right-angle triangle. Thus $\mu = (2 \times 0.78^2)^{1/2} = \mathbf{1.10\ debyes.}$

(b) Since $\mu = ed$, we have $0.78 \times 10^{-18} = 1.8 \times 10^{-8} \times Z \times 4.8 \times 10^{-10}$, whence $Z = \mathbf{0.09}$ units of electronic charge.

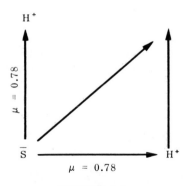

FIGURE 3-7

36. The optical density for pure liquid B is log 4 or 0.62. On mixing equal volumes, each liquid is diluted to half of its original concentration, so contributes only half of its original optical density. The total optical density of the mixture must then be $D = 0.15 + 0.31 = $ **0.46.**

4

FIRST LAW OF THERMODYNAMICS

COMMENTS

We come now to the first law of thermodynamics—a most serious topic indeed! It and the other two laws (Chapter 7) form the foundation of classical physical chemistry. Along with the first law itself comes the idea of variables of state—quantities that depend only on the state of the system and not on how it got there. Pressure, volume, temperature, and, now, energy and enthalpy, are such quantities. Mathematically, this means making the acquaintance with the total (or exact) differential and with partial differentials and the relations between them.

In contrast to state variables are those quantities that *do* depend on path, such as q the heat absorbed and w the work done by the system as the change occurs. One learns that there are reversible paths where, at each stage along the change, the system is essentially at equilibrium. For our purposes in this chapter, this means that temperature and pressure are uniform; later we shall add the additional meaning that the chemical potential of each substance is uniform throughout the system. These reversible paths come in a variety of common kinds, and you will need to become familiar with them. The concept of heat capacity as a function of type of path enters the picture, too.

Then there are irreversible paths or processes. These are much harder to deal with generally, because in a system of nonuniform pressure or temperature it is hard to calculate q and w. For example, in a reversible isothermal expansion, the pressure is everywhere uniform, and hence the pressure inside the system equals the external pressure, and the work done is $\int P_{ext}\, dV = \int P_{int}\, dV$. For an ideal gas, we know that $P_{int} = RT/V$ and if the temperature is constant, the integral becomes $RT \int dV/V = RT \ln V_2/V_1$. If, however, the isothermal expansion is irreversible, then P_{int} is not uniform, and therefore $P_{int} \neq P_{ext}$. The work is still $\int P_{ext} \, dV$, but now we cannot replace P_{ext} by P_{int}, and so cannot use the ideal gas law to relate pressure and volume.

There are a few special cases in which w can be obtained easily for an irreversible process, however. If, say, the gas expands isothermally into an evacuated flask, we know that no external work is done, so $w_{irrev} = 0$. Or suppose that the gas is confined in a piston and cylinder, with the piston held down by a catch. On releasing the catch, the piston flies up, and the work done is just the pressure resulting from the piston (i.e., the weight of the piston divided by its area) times the volume it sweeps out. So again, w_{irrev} can be calculated.

Many of the problems that follow begin with: "A mole of an ideal, monatomic gas . . . ." This repetition is not due to any lack of imagination; it is no trouble to take several moles of a nonideal, polyatomic gas and thereby add a lot of interest (and time) to the problem. After all, these were examination questions, and the quoted phrase tells you that you can use the ideal gas law, and that C_v and C_p are $3R/2$ and $5R/2$, respectively. You can then proceed with maximum efficiency to show how well you understand the workings of the first law.

EQUATIONS AND CONCEPTS

First Law of Thermodynamics

$$dE = \delta q - \delta w \tag{4-1}$$

(hereafter written dq and dw, although they are not exact differentials).

Definitions

Enthalpy H: $H = E + PV$. Work done by the system: $dw = P \, dV$ or $W = \int P \, dV$. Heat capacity $C : C = dq/dT$. C_p, C_v, and so on, are so written to indicate that P, V, or some other variable is held constant during the temperature change. Joule–Thompson coefficient μ: $\mu = (\partial T/\partial P)_H$.

Special Relations for an Ideal Gas

For any change, reversible or not: $dE = C_v \, dT$ and $dH = C_p \, dT$. Also, $dE = dH - R \, dT$.

The relationships for the following *reversible* processes involving an ideal gas are useful:

$$\text{Isothermal:} \quad dE = 0 \qquad dH = 0 \tag{4-2}$$

$$w = RT \ln \frac{V_2}{V_1} = RT \ln \frac{P_1}{P_2} \tag{4-3}$$

$$q = \Delta E + w \tag{4-4}$$

Isobaric: $dE = C_v \, dT \qquad dH = C_p \, dT \qquad q = \Delta H = \int C_p \, dT$ (4-5)

$$w = P\Delta V = R\Delta T \tag{4-6}$$

Isochoric: $dE = C_v \, dT \qquad dH = C_p \, dT \qquad q = \Delta E = \int C_v \, dT$ (4-7)

$$w = 0 \tag{4-8}$$

Adiabatic: $q = 0 \quad$ so $\quad dE = -dw \quad$ or $\quad C_v \, dT = -P \, dV$ (4-9)

and $$C_v \ln \frac{T_2}{T_1} = -R \ln \frac{V_2}{V_1} \tag{4-10}$$

or $$C_p \ln \frac{T_2}{T_1} = R \ln \frac{P_2}{P_1} \tag{4-11}$$

Total and Partial Differentials

Total differential of $z = f(x, y)$: $dz = (\partial z / \partial x)_y \, dx + (\partial z / \partial y)_x \, dy = M \, dx + N \, dy$. Test that an equation $dz = M \, dx + N \, dy$ is a total differential:

$$\left(\frac{\partial M}{\partial y}\right)_x = \left(\frac{\partial N}{\partial x}\right)_y \tag{4-12}$$

Other relationships:

$$0 = \left(\frac{\partial z}{\partial x}\right)_y + \left(\frac{\partial z}{\partial y}\right)_x \left(\frac{\partial y}{\partial x}\right)_z$$

Also, if w is another function of x and y, then $(\partial z / \partial x)_w = (\partial z / \partial x)_y + (\partial z / \partial y)_x (\partial y / \partial x)_w$. (See Appendix.)

PROBLEMS

1. (18 min) One mole of an ideal monatomic gas initially at STP experiences a reversible process in which the volume is doubled. The nature of the process is unspecified, but ΔH is 500 cal and g is 400 cal. (a) Calculate the final T and P, and ΔE and w for the process. (b) Suppose that the gas were taken to the same final conditions by a process involving an isochoric change and an isothermal one, both reversible. Calculate ΔH, ΔE, q, and w for this sequence.

2. (12 min) Suppose that the quantity D is defined as $D = H + RT$. D is thus a state function. (a) Show that $C_p = (\partial D/\partial T)_p - R$. (b) Relate the coefficient $(\partial T/\partial P)_D$ to other partial differential quantities.

3. (7 min) Many texts give the general equation:

$$C_p - C_v = [P + (\partial E/\partial V)_T](\partial V/\partial T)_P$$

Show how from this equation one obtains the conclusion that for an ideal gas $C_P - C_v = R$, using the first law of thermodynamics, results of fundamental experiments, and the like.

4. (6 min) Derive from the first law of thermodynamics and related definitions:

$$C_v = - \left(\frac{\partial E}{\partial V}\right)_T \left(\frac{\partial V}{\partial T}\right)_E$$

---

5. (10 min) A cyclic process involving 1 mole of an ideal monatomic gas has a w of 100 cal/cycle. Per cycle, q is then (zero, 100 cal, -100 cal, cannot tell since the process is not stated to be reversible).

6. (12 min) An ideal gas undergoes a reversible isothermal expansion from an initial volume of V_1 to a final volume $10V_1$ and thereby does 10,000 cal of work. The initial pressure was 100 atm. (a) Calculate V_1. (b) If there were 2 moles of gas, what must its temperature have been?

7. (18 min) One mole of an ideal gas of $C_v = 5.0$ cal/mole-deg initially at STP is put through the following reversible cycle. A: State 1 to state 2, heated at constant volume to twice the initial temperature. B: State 2 to state 3, expanded adiabatically until it is back to the initial temperature. C: State 3 to state 1, compressed isothermally back to state 1. Calculate q, w, ΔE, and ΔH for steps A and B and for the cycle.

8. (18 min) The volume of 1 mole of an ideal monatomic gas initially at 2 atm and 25°C (and 12.2 liters volume) is doubled by (a) isothermal expansion, (b) adiabatic expansion, and (c) expansion along the path $P = 0.1V + b$, where P is in atm and V in liters/mole. All paths are reversible. Calculate the final P for each case. Sketch each path on a plot of P versus V. Arrange the ΔE's in order of decreasing magnitude. Do likewise for the w's. Calculate q, w, and ΔE for one path.

9. (18 min) One mole of an ideal monatomic gas is carried through the cycle of Fig. 4-1, consisting of steps A, B, and C, and involving states 1, 2, and 3. Fill in Tables 4-1 and 4-2. Assume reversible steps.

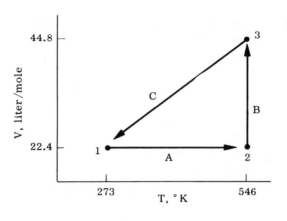

FIGURE 4-1

TABLE 4-1

State	P, atm	V, liter	T,°K
1	_____	22.4	273
2	_____	22.4	546
3	_____	44.8	546

TABLE 4-2

Step	Name of process	q, cal	w, cal	ΔE, cal
A	_____	_____	_____	_____
B	_____	_____	_____	_____
C	_____	_____	_____	_____
	cycle	_____	_____	_____

10.   (21 min) One mole of an ideal monatomic gas initially at STP is taken through the reversible sequence of steps shown in Fig. 4-2. Fill in the information called for in Tables 4-3 and 4-4.

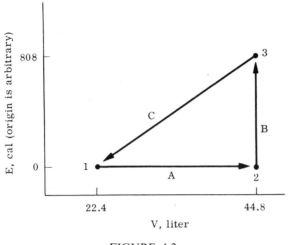

FIGURE 4-2

TABLE 4-3

State	P, atm	V, liter	T, °K
1	1	22.4	273
2	_____	44.8	_____
3	_____	44.8	_____

TABLE 4-4

Step	Type of process	q, cal	w, cal	ΔE, cal
A	_____	_____	_____	_____
B	_____	_____	_____	_____
C	_____	_____	_____	_____
	cycle	_____	_____	_____

11. (18 min) One mole of an ideal monatomic gas is put through the three-step cycle shown on the graph of Fig. 4-3. Complete the information called for in Tables 4-5 and 4-6.

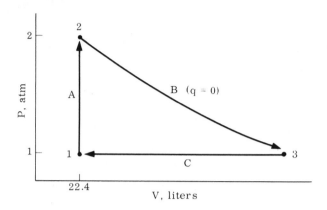

FIGURE 4-3

TABLE 4-5

State	P, atm	V, liter	T, °K
1	1	22.4	_____
2	2	22.4	_____
3	1	_____	_____

TABLE 4-6

Step	Type of process	q, cal	w, cal	ΔE, cal
A	_____	_____	_____	_____
B	_____	_____	_____	_____
C	cycle	_____	_____	_____

12. (3 min) Derive the equation $(\partial H/\partial P)_T = -\mu C_p$, where μ is the Joule–Thompson coefficient.

13. (21 min) One mole of an ideal monatomic gas is taken through the three steps shown on the P versus T plot of Fig. 4-4.
    (a) Sketch and similarly label the same three steps on a plot of P versus V.
    (b) Complete the following table:

	Step A	Step B	Step C	Cycle
ΔE	_____	_____	-810 cal	_____
q	_____	1356 cal	_____	_____
w	-373 cal	_____	_____	_____

(c) The above steps are all reversible. The numerical magnitude of w for the cycle is therefore a (maximum, minimum).

14. (5 min) One mole of an ideal monatomic gas is to be taken from $P_1 = 1$ atm and $T_1 = 300°$K to $P_2 = 10$ atm and $T_2 = 600°$K by some combination of isobaric, isothermal, adiabatic, and isochoric steps. (a) Give that path, involving only the above types of processes, which will require the least reversible work for the over-all change (you may use some or all of the types of steps). (b) Calculate this minimum work.

15. (15 min) One mole of an ideal monatomic gas initially at $P_1 = 2$ atm, $T_1 = 273°$K is taken to a pressure of $P_2 = 4$ atm by the reversible path defined by $P/V = $ constant. Calculate V_1, V_2, and T_2. Calculate ΔE, ΔH, q, and w (in cal).

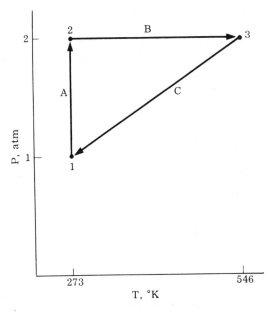

FIGURE 4-4

16. (24 min) One of an ideal monatomic gas is reversibly (1) expanded from 10 atm and 2 liters to 5 atm, isothermally, and (2) expanded from 10 atm and 2 liters to 5 atm, adiabatically. (a) Calculate q, w, ΔE, and ΔH (in cal) for processes (1) and (2). (b) Sketch, in a P versus V diagram, the path taken by the gas in each process. (c) There is a third process (3) which would show as a straight line on the P versus V diagram, and is such that process (2) plus process (3) gives the same final state as process (1). What is the nature of (3) (isothermal, isochoric, isobaric, etc.)?

17. (6 min) Referring to Problem 15, calculate the heat capacity of the gas along this path, that is, the heat absorbed per degree rise in temperature under the restriction $P/V = $ constant.

18. (19 min) The heat capacity ratio γ for a gas is determined in the following simple experiment: A carboy is filled with the gas to a pressure of 1.10 atm (laboratory pressure is 1.00 atm). The stopper to the carboy is then suddenly removed, so that the gas expands adiabatically; after a few seconds, the stopper is replaced and the gas is allowed to warm up to room temperature, and the pressure is now found to be 1.03 atm. Assuming the gas to be ideal, calculate C_p and C_v.

19. (24 min) One mole of an ideal monatomic gas is put through the indicated cycle (see Fig. 4-5). Step A: isochoric reduction in pressure; step B: isobaric increase in volume; step C: return to initial state by straight-line path (this is *not* isothermal). Assuming the steps to be reversible, calculate q, w, and ΔE for each step and for the entire cycle, in calories.
    Summary: State 1: 4 atm, 11.2 liters, 546°K; state 2: 2 atm, 11.2 liters, 273°K; state 3: 2 atm, 22.4 liters, 546°K.

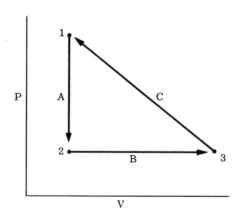

FIGURE 4-5

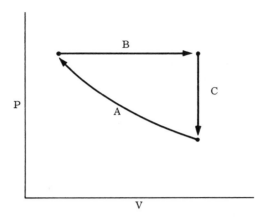

FIGURE 4-6

20. The equation $dE = (\partial E/\partial V)_T\, dV + (\partial E/\partial T)_V\, dT$ may be written $dE = C_v\, dT + (\partial E/\partial V)_T\, dV$ since $(\partial E/\partial T)_V = C_v$. One can also write $dE = dq + (\partial E/\partial V)_T\, dV$, since $C_v\, dT = dq$. By comparison with the first law statement $(dE = dq - P\, dV)$, it appears that $(\partial E/\partial V)_T = -P$. This conclusion is *not* correct; explain the error in the derivation.

21. (21 min) One mole of a perfect monatomic gas is put through a cycle consisting of the following three reversible steps: (A) isothermal compression from 2 atm and 10 liters to 20 atm and 1 liter; (B) isobaric expansion to return the gas to the original volume of 10 liters with T going from T_1 to T_2; (C) cooling at constant volume to bring the gas to the original pressure and temperature. The steps are shown schematically in Fig. 4-6.

(a) Calculate T_1 and T_2.
(b) Calculate ΔE, q, and w, in calories, for each step and for the cycle.

22. (30 min) One mole of an ideal monatomic gas experiences the reversible steps shown in Fig. 4-7. Fill in the values called for by the blanks in Tables 4-7 and 4-8. If the above cycle were carried out irreversibly and in a manner such that the net work were zero, calculate or explain what can be said about the values of ΔE and q for the cycle.

TABLE 4-7

State	P, atm	V, liters	T, °K
1	2	_____	_____
2	4	_____	_____
3	4	_____	_____

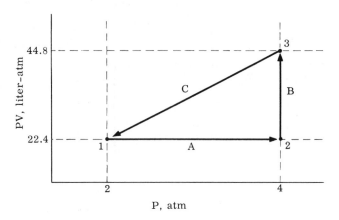

FIGURE 4-7

TABLE 4-8

Path	Type (isothermal, etc.)	q, cal	w, cal	ΔE, cal
A				
B				
C				
	Cycle			

23. (18 min) One mole of an ideal monatomic gas initially at 2 atm and 11.2 liters is taken to a final pressure of 4 atm along the reversible path defined by $PT = $ constant. Calculate (a) the final volume and temperature, (b) ΔE and ΔH, and (c) the work done w.

24. (22 min) One mole of an ideal monatomic gas initially at STP is compressed reversibly to two atm pressure along the path: $PT = $ constant. (a) Calculate the final T and V for the gas. (b) Sketch qualitatively (but with reasonable attention to detail) the appearance of this path on a plot of P versus V. (c) Calculate q, w, ΔH, and ΔE for the process.

ANSWERS

1. (a) Since $\Delta H = C_p \Delta T$ and $C_p = 5R/2$ or 5, it follows that $\Delta T = 100°$. The new temperature is then **373°K**, and $\Delta E = C_v \Delta T = $ **300 cal.** $w = q - \Delta E = 400 - 300 = $ **100 cal.** The final pressure is $(373/273)(1/2) = $ **0.69 atm.**

(b) Various two step sequences are possible and a moment's reflection can save much time at this point. A very convenient sequence would be the following: (1) isothermal expansion to twice the volume. For this $w = RT \times \ln(V_2/V_1) = 1.98 \times 273 \times 2.3 \times 0.3 = $ **373 cal.** Since ΔE is zero, $q = $ **373 cal** also. (2) It then remains to heat at constant volume from 273° to 373°. For this $q = 100 \times 3\,R/2 = $ **300 cal.** w is zero. For the over-all sequence, ΔH and ΔE must be the same as in (a), and $w = $ **373 cal** and $g = $ **673 cal.**

2. (a) Since $C_p = (\partial H/\partial T)_P$, it follows that $C_p = (\partial D/\partial T)_P - R$. (b) Take $D = f(T, P)$; then

$$dD = \left(\frac{\partial D}{\partial T}\right)_P dT + \left(\frac{\partial D}{\partial P}\right)_T dP$$

from which

$$\left(\frac{\partial T}{\partial P}\right)_D = -\frac{(\partial D/\partial P)_T}{(\partial D/\partial T)_P}$$

3. A fundamental property of an ideal gas is that $(\partial E/\partial V)_T = 0$. From the ideal gas law, $(\partial V/\partial T)_P = R/P$. Insertion of these results in the equation gives $C_p - C_v = (P + O)(R/P) = R$.

4. This looks like one of the relations from a total differential expression, so try writing $C_v = (\partial E/\partial T)_V$. It is now seen that the desired equation follows from

$$dE = \left(\frac{\partial E}{\partial V}\right)_T dV + \left(\frac{\partial E}{\partial T}\right)_V dT$$

by dividing through by dT, keeping E constant:

$$0 = \left(\frac{\partial E}{\partial V}\right)_T \left(\frac{\partial V}{\partial T}\right)_E + \left(\frac{\partial E}{\partial T}\right)_V$$

5. This one is easy. For any cycle ΔE is zero and hence $q = w = $ **100 cal.**

6. (a) $w = nRT \ln V_2/V_1 = 2.3\ nRT$. But $P_1 V_1 = nRT = w/2.3 = 10{,}000/2.3 = 4{,}340$ cal. Hence $V_1 = (4{,}340/100)(0.041 \text{ liter-atm/cal}) = $ **1.78 liters.** (b) If $n = 2$, then $T = 4{,}340/2 \times 1.98 = $ **1,100°K.**

7. In problems of this kind do the easiest things first:
(i) For the cycle, ΔH and ΔE are zero.
(ii) Step A is isochoric, hence $w = 0$; $\Delta E = C_v \Delta T = 5 \times 273 = 1365$ cal; $\Delta H = C_p \Delta T = $ **1,911 cal;** $q = C_v \Delta T = $ **1,365 cal.**

(iii) Step B is adiabatic, so $q = 0$; $\Delta E = C_v \Delta T = -1,365\,\text{cal}$; $\Delta H = -1,911\,\text{cal}$; $w = -\Delta E$.

(iv) Step C is isothermal, hence ΔE and ΔH are zero. $q = w = -RT \ln V_3/V_1$. Now $V_2 = V_1$ and, for step 2, $C_v \ln T_2/T_3 = -R \ln V_2/V_3 = R \ln V_3/V_1$. But $T_2/T_3 = 2$, so $R \ln V_3/V_1 = 5 \times 2.3 \log 2 = 3.45$. Therefore w for step C is $-3.45 \times 273 = -940\,\text{cal}$.

(v) For the cycle, then, $w = 0 + 1,365 - 940 = -425\,\text{cal}$. $q = w$.

8. (a) If volume is doubled isothermally, the $P_{\text{final}} = 1$ atm.

(b) If volume is doubled adiabatically, then

$$\log \frac{T_2}{T_1} = -\left(\frac{R}{C_v}\right) \log \frac{V_2}{V_1} = \left(\frac{R}{C_p}\right) \log \frac{P_2}{P_1}$$

Hence

$$\log \frac{P_2}{P_1} = -\left(\frac{C_p}{C_v}\right) \log 2 = -0.5 \qquad \frac{P_2}{P_1} = 0.316 \qquad P_2 = 0.73\,\text{atm}$$

(c) If expansion is along the given path, $P_2 = 0.1 \times 24.4 + b$ but $b = P - 0.1 V = 2 - 0.1 \times 12.2 = 0.78$ or $P_2 = 3.22\,\text{atm}$. The paths are sketched in Fig. 4-8.

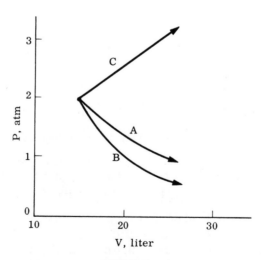

FIGURE 4-8

ΔE's: The ΔE's will be in the same relation as the ΔT's. ΔT_a is zero, ΔT_b may be obtained from the equation for an adiabatic expansion:

$$\log \frac{T_2}{T_1} = -\left(\frac{R}{C_v}\right) \log \frac{V_2}{V_1} = -\frac{0.3}{1.5} = -0.2$$

so

$$T_2 = 0.632 \times 298 = 188°\text{K} \qquad \Delta T_b = -110°$$

T_2 for step C is $3.22 \times 24.4/R = 958°\text{K}$, $\Delta T_c = 660$. Thus $\Delta T_b < \Delta T_a < \Delta T_c$ and likewise for the ΔE's. (The relative positions could have been guessed: ΔT_b must be negative as one had an adiabatic expansion; ΔT_a is zero; ΔT_c must be positive as the process called for an increase in both P and V.)

The relative order of the w's is given most quickly simply by noting the relative areas under the three paths in the P versus V plot: $w_b < w_a < w_c$. Step A is probably the easiest way by which to get q, w, and ΔE. $\Delta E = 0$, $q = w = RT \ln V_2/V_1 = 1.98 \times 298 \times 2.3 \log 2 = $ **408 cal.**

9.   Table 4-1: for state 1, P is obviously 1 atm (22.4 liters is the STP volume); hence for state 2, P is 2 atm as T is doubled. For state 3, P must be back to 1 atm, as V is doubled.

Table 4-2: Processes A, B, and C must be isochoric, isothermal, and isobaric, respectively; you need only notice which quantity is constant during them.

ΔE: these values can be written down immediately as equal to $C_v \Delta T$ or $3 \Delta T$: A: $3 \times 273 = $ **819 cal;** B: $3 \times 0/2 = $ **0 cal;** C: $-$**819 cal;** cycle: **0 cal.**

q: these values likewise can be written down immediately as $q = C \Delta T$. A: $q = 3 \times 273 = $ **819 cal;** B: $q = w$; C: $q = -5 \times 273 = -$**1,365 cal.**

w: A: $w = 0$ (V constant); B: $w = RT \ln V_2/V_1 = 1.98 \times 546 \times 2.3 \log 2 = $ **748 cal;** C: $w = q - \Delta E = -1,365 + 819 = -$**546 cal.** Then for the cycle, $q = 819 + 748 - 1,365 = $ **202 cal** and $w = q$.

10.   Table 4-3: E and hence T must be constant for step A, so $T_2 = T_1 = 273$, and P_2 is therefore 0.5 atm. E increases by 808 cal in step B, so $\Delta T = 808/(3R/2) = 273$. T_3 is therefore 543°K, and P_3 is 1 atm.

Table 4-4. The steps are isothermal, isochoric, and isobaric, respectively. Step A: $\Delta E = 0$, so $q = w = RT \ln V_2/V_1 = 1.98 \times 273 \times 2.3 \log 2 = $ **377 cal.**

Step B: $\Delta E = $ **808 cal** (from graph). Since $\Delta V = 0$, $w = 0$, and $q = \Delta E$.

Step C: The endpoints of step C are at the same pressure, but, technically, it should be shown that P is constant at all points on the path. This can be done by noting that, for the path, $dE = (C_v \times 273/22.4) \, dV$ and that $dE = C_v \, dT$, hence $dT = (273/22.4) \, dV$, true only if P is constant at 1 atm. The process is then isobaric, $q = C_p \Delta T = (5R/2) \, \Delta T = -\mathbf{1{,}350 \, cal}$; $\Delta E = -\mathbf{808}$, hence $w = -1{,}350 + 808 = -\mathbf{542 \, cal}$. Cycle: $\Delta E = \mathbf{0}$, $w = 377 - 542 = -\mathbf{165}$ cal, $q = w$.

11. Table 4-5: T_1 is evidently 273° K, and T_2 is 546°K. State 3 is determined by noting that step B is adiabatic, and since P_3/P_2 is given, the appropriate equation is $C_p \ln T_3/T_2 = R \ln P_3/P_2$ or $\log T_3/T_2 = (2/5)$ $\log (1/2)$ or $T_3/T_2 = 0.758$ and $T_3 = 414$°K. V_3 is then $22.4 \times 414/273 = 34$ liters.

Table 4-6: The processes are isochoric, adiabatic, and isobaric, respectively.
Step A: $w = 0$, $\Delta E = q = C_v \Delta T = (3R/2)273 = \mathbf{808 \, cal.}$
Step B: $q = 0$, $-w = \Delta E = C_v \, \Delta T = (3R/2)(-132) = -\mathbf{392 \, cal.}$
Step C: $\Delta E = C_v \Delta T = -\mathbf{416 \, cal}$; $q = C_p \Delta T = -\mathbf{695 \, cal}$; $w = q - \Delta E = -\mathbf{279 \, cal.}$
Cycle: $\Delta E = \mathbf{0}$, $w = q = 808 + 0 - 695 = \mathbf{113 \, cal.}$

12. The Joule–Thompson coefficient is defined as $\mu = (\partial T/\partial P)_H$. On inserting this in the equation, inspection of the arrangement of variables suggests that the equation might come from the total differential for H:

$$dH = (\partial H/\partial P)_T \, dP + (\partial H/\partial T)_P \, dT$$

If we divide through by dP at constant H, then $0 = (\partial H/\partial P)_T + (\partial H/\partial T)_P \times (\partial T/\partial P)_H$. On inserting the definition of μ, and remembering that $C_p = (\partial H/\partial T)_P$, the above rearranges to the desired equation.

13. The required P versus V plot is shown in Fig. 4-9. To construct it, V_1 must be 22.4 liters and V_2 11.2 liters. Since T is doubled in B, V_3 is back to 22.4 liters.
Step A: isothermal, so $\Delta E = 0$, $q = w = -\mathbf{373 \, cal.}$
Step B: isobaric; $\Delta E = C_v \Delta T = (3R/2)273 = \mathbf{808 \, cal.}$ $w = q - \Delta E = 1356 - 808 = \mathbf{548 \, cal.}$
Step C: isochoric; $w = 0$, so $q = \Delta E = -\mathbf{810 \, cal.}$
Cycle: $\Delta E = 0$, $q = w = -373 + 1356 - 810 = \mathbf{173 \, cal.}$
Work is done by the cycle, so w is a maximum.

14. The situation is shown schematically in Fig. 4-10. The idea is to get from state 1 to state 2 by a route having a minimum area. Since we have a hypothetical ideal gas, we cool at constant volume to 0°K, reducing P to zero, then compress isothermally to the final volume ($V_{int} = 24.5$ liters, $V_{final} = 4.9$ liters), and heat isochorically to 600°K. $w = 0$, $q = \Delta E = C_v \Delta T = (3R/2)300 = \mathbf{890 \, cal.}$

15. V_1 must be 11.2 liters; since P/V is constant, V_2 must be 22.4 liters. Combining $P/V = $ constant with $PV = RT$: $T/V^2 = $ constant, so $T_2 = $

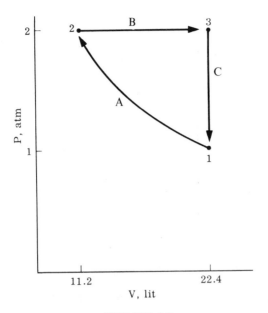

FIGURE 4-9

$4T_1 = 1{,}092°\text{K}$. $\Delta E = C_v\Delta T = (3R/2)819 = 2{,}420\,\text{cal}$; $\Delta H = C_p\Delta T = 4{,}040$ cal. To get w, we must evaluate $w = \int P\,dV$. From the initial conditions,

$$P/V = \text{constant} = 2/11.2 = 0.178$$

then

$$w = 0.178\!\int V\,dV = 0.089(V_2^2 - V_1^2) = 0.089 \times 375$$

$$= 33.3\,\text{liter-atm}$$

or

$$w = 823\,\text{cal}; q = \Delta E + w = \textbf{3,230 cal.}$$

16. (a) The initial T_1 is 244°K. For step A, $T_2 = 244°\text{K}$; V_2 then is 4 liters. $\Delta E, \Delta H = \mathbf{0}; q = w = RT\ln V_2/V_1 = 1.98 \times 244 \times 2.3 \log 2 = \textbf{334 cal.}$ For process B, $q = \mathbf{0}$; T_2 is given by $C_p \log T_2/T_1 = R \log P_2/P_1$ or $\log T_2/T_1 = 0.4$ $\log 1/2 = \mathbf{-0.12}$; $T_2 = 185°\text{K}$. Hence $-w = \Delta E = (3R/2)59$ or **175 cal.** $\Delta H = (5R/2)59 = \textbf{298 cal.}$

(b) See Fig. 4-11.

(c) The adiabatic processes gives a lower final temperature and therefore volume than the isothermal one. Step C is then an isobaric heating back to 244°K.

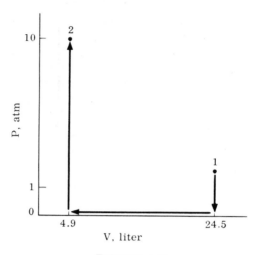

FIGURE 4-10

17.   From the first law, $dE = dq - dw$, and for an ideal gas

$$C_v \, dT = dq - P \, dV$$

where $P \, dV$ is governed by $P/V = b$ (a constant). We want $(dq/dT)_{\text{path}} = C_v + P(dV/dT)_{\text{path}}$. To get the last term, combine $P/V = b$ with $PV = RT$ and get $V^2 = RT/b$. Then $2V \, dV = (R/b) \times dT$ and $dV/dT = R/2bV$. Then

$$\left(\frac{dq}{dT}\right)_{\text{path}} = C_v + P\left(\frac{R}{2bV}\right) = C_v + \frac{R}{2}$$

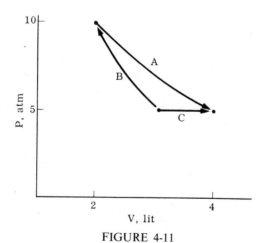

FIGURE 4-11

18. Combining the equations $C_v \log T_2/T_1 = -R \log V_2/V_1$ and $C_p \log T_2/T_1 = R \log P_2/P_1$ for an adiabatic process, one obtains

$$-\left(\frac{C_p}{C_v}\right) \log \frac{V_2}{V_1} = \log \frac{P_2}{P_1}$$

We know P_2/P_1; it is $1.00/1.10$. The final volume is that of the carboy V_i plus the gas that escaped. We know, however, that gas remaining was $1.03/1.10$ of V_1; hence V_2 was $1.10 V_1/1.03$.

Then $-(C_p/C_v) \log (1.10/1.03) = \log (1.00/1.10)$, from which $C_p/C_v = 1.38$. But $C_p = C_v + R$, whence $C_v = $ **5.2**, $C_p = $ **7.2**.

19. Step A: since the step is isochoric, $w = $ **0**. $q = \Delta E = C_v \Delta T = -(3R/2)273 = $ **−808 cal.**

Step B: $\Delta E = C_v \Delta T = 808$ cal. $q = C_p \Delta T = (5R/2)273 = $ **1,350 cal**; $w = q - \Delta E = $ **542 cal.**

Step C: $\Delta E = 0$, since there is no over-all change in temperature. T varies along the path, however, so w is not given by the isothermal-work formula. It can be obtained, however, as the area under the path $\int P \, dV$. From the geometry of the straight-line path, this area is $2 \times 11.2 + (1/2)2 \times 11.2$ or -33.6 liter-atm (negative as the direction is from left to right) or -820 cal. Then $q = w$. For the cycle: $\Delta E = 0$, $q = w = $ **−278 cal.**

20. The difficulty is that the first law statement is applicable to any path; q and w are not defined; only their difference equals ΔE. The first equation says that dE is the sum of dq_V (i.e., absorption of heat on a constant volume heating) and a second term corresponding to a **second** step, one at constant temperature.

21. $T_1 = P_1 V_1/R = $ **244°K**; $T_2 - 20 \times 10/0.082 = $ **2,440°K.**

Step A: $\Delta E = $ **0**, $w = RT \ln V_2/V_1 = 1.98 \times 244 \times 2.3 \log 1/10 = $ **−1,110 cal**; $q = w$.

Step B: $\Delta E = C_v \Delta T = (3R/2)2,200 = $ **6,520 cal**; $q = C_p = (5R/2)220 = 10,900$ cal; $w = q - \Delta E = $ **4,380 cal.**

Step C: $w = 0$; $q = \Delta E = C_v \Delta T = $ **− 6,520 cal.**

Cycle: $\Delta E = 0$; $q = w = -1110 + 4380 = $ **3,270 cal.**

22. Table 4-7: State (1): $V_1 = 22.4/2 = $ **11.2 liters**. Hence $T_1 = $ **273°K** ($PV = 22.4$). State (2): $V_2 = 22.4/4 = $ **5.6 liters**. Also $T_2 = $ **273°K** (same PV value). State (3): $V_3 = 44.8/4 = $ **11.2 liters**; $Y_3 = 2 \times 273 = $ **546°K.**

Table 4-8: The processes are evidently isothermal, isobaric, and isochoric, respectively.

For A: $\Delta E = 0$, $q = w = RT \ln V_2/V_1 = 1.98 \times 273 \times 2.31 \log 1/2 = $ **− 373 cal.**

For B: $\Delta E = C_v \Delta T = (3R/2)273 = $ **808 cal;** $q = C_P \Delta T = (5R/2)273 = $ **1,350 cal;** $w = q - \Delta E = $ **542 cal.**

For C: $w = 0$; $\Delta E = q = C_v \Delta T = -808$ cal.

Cycle: $\Delta E = 0$, $q = w = -372 + 542 = $ **169 cal.** For the irreversible cycle, ΔE is still zero; if $w = 0$, then $q = $ **0.**

23. By comparison with the molar volume at STP, the initial temperature is evidently 273°K. Then $PT = 2 \times 273 = 546$.

(a) The final temperature must then be $T = 546/4 = $ **136°K,** so the final volume must be $V = 0.082 \times 138/4 = $ **2.82 liters.**

(b) $\Delta E = C_v \Delta T = (3R/2)(136 - 273) = $ **−402 cal.** $\Delta H = C_p \Delta T = (5R/2)(136 - 273) = $ **−672 cal.**

(c) To get w, we must evaluate $\int P \, dV$. To begin, we must express P in terms of V. Thus $P = 546/T = 546R/PV$, so $P^2 = 546R/V$. The integral then becomes $w = (546R)^{1/2} \times \int dV/V^{1/2} = 2(546R)^{1/2}(V_2^{1/2} - V_1^{1/2}) = 2(P^2V)^{1/2}(V_2^{1/2} - V_1^{1/2}) = 2(P_2V_2 - P_1V_1) = 2(4 \times 2.82 - 2 \times 11.2) = -22.4$ liter-atm or **−547 cal.**

24. (a) The first condition is that $PT = k$, from which $k = 1 \times 273 = 273$ atm-deg. After compression to 2 atm, T_2 is then $273/2 = $ **136°K** and V_2 is therefore $0.082 \times 136/2 = $ **5.58 liters,** that is, one-quarter of the original volume since pressure is doubled and temperature is halved.

(b) The initial and final states are shown in Fig. 4-12. It remains to determine if the connecting line is straight or bowed. One way is to calculate a

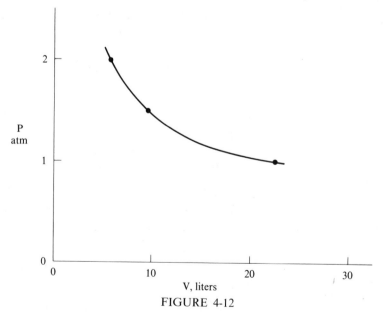

FIGURE 4-12

midway point. Thus at $P = 1.5$, $T = 273/1.5 = 182°K$ and $V = 0.082/1.5 = 9.95$ liters. Were the line straight, the volume should be halfway between 22.4 and 5.58 or at 14.0 liters. The actual volume is less than this, so the line is concave, as shown.

(c) First, $\Delta E = C_v \Delta T = -\frac{3}{2}R \times 136 = $ **−408 cal**, and $\Delta H = C_p \Delta T = -\frac{5}{2}R \times \Delta T = $ **−680 cal**. Getting w is a little more trouble; it is necessary to evaluate $\int P \, dV$ for the path. To do this P is expressed in terms of V as follows:

$$PV = RT = Rk/P \quad \text{or} \quad P^2 V = Rk \quad \text{and} \quad P = (Rk/V)^{1/2}$$

Then

$$w = P \, dV = (Rk)^{1/2} \int (1/V^{1/2}) \, dV = 2(Rk)^{1/2} |V^{1/2}|_{22.4}^{5.58}$$

$$= 2(0.082 \times 273)^{1/2}(2.36 - 4.72) = -22.4 \text{ liter-atm}$$

$$= -\textbf{546 cal}$$

Finally, $q = w$.

5

HEAT CAPACITY OF GASES; SOME STATISTICAL THERMODYNAMICS

COMMENTS

You learned in the chapter on the first law of thermodynamics that for an ideal gas $C_p - C_v = R$. There is a good deal more than this to the subject of heat capacity. One aspect is the explanation and prediction of the actual C_v values for a gas.

For an ideal monatomic gas, kinetic molecular theory tells us that the kinetic energy is $(3/2)RT$ per mole [Eq. (2-1)] and the Boltzmann analysis, that each of the three velocity components are independent. The conclusion is that a total heat capacity of $3R/2$ is associated with translation, as a result of the contribution $R/2$ from each degree of translational freedom.

In the case of a diatomic molecule, it would take six velocity components to describe the motions of the atoms in a molecule, and, in general, $3n$ components for a molecule with n atoms. It makes sense in a number of ways, however, to regroup these degrees of freedom into three for translation of the molecule as a whole, two or three for its rotation (two if the molecule is linear, and three otherwise), and then to assign the rest to vibrational motions. The simple conclusion is still that the total C_v should be given by a contribution of $R/2$ from each degree of freedom, but with the important addendum that since vibration involves both kinetic and potential energy, it should contribute doubly to C_v. This is, in brief, the *Equipartition Principle*, and a number of the problems are designed to test your ability to use it.

In the case of a crystalline solid, there is no translation, and, if the lattice positions are occupied by atoms, no rotation. The degrees of freedom are thus all vibrational. The equipartition principle becomes the law of Dulong and Petit, which states that heat capacity of such a solid is 6 cal/deg-mole.

It turns out, unfortunately, that the equipartition principle at best gives a maximum or limiting C_v value; the problem is the following. The principle tacitly implies that molecules smoothly gain or lose translational, rotational, and vibrational energy as the temperature changes, but quantum theory denies this. We accept that all energy is quantized, and may be gained or lost by a molecule only in discrete amounts. For a small, tight molecule like HCl, the first excited vibrational level is about 8.1 kcal/mole above the ground level. This is 13.7 times the value of RT at 25°C, and the Boltzmann principle assigns a probability weight of only $\exp(-13.7)$ or 1.13×10^{-6} to such an excited state. At 35°C, the energy of the state is 13.3 times RT, and the probability weight inches up to 1.68×10^{-6}. The net effect is that a 10°C rise in temperature has transferred approximately the fraction 0.55×10^{-6} of HCl molecules from the ground to the excited vibrational state. Per mole, the energy gained would be $8{,}100 \times 0.55 \times 10^{-6}$ or about 4.5×10^{-3} cal, corresponding to a vibrational heat capacity contribution of 4.5×10^{-4} cal/deg as compared to the equipartition value of 2 cal/deg.

The quantum effect diminishes in importance if the spacing between energy levels is decreased. Thus, for I_2, the first excited vibrational level is 820 cal/mole above the ground state, and the calculation corresponding to the above one now gives 0.65 cal/deg for the vibrational contribution to the molar heat capacity. Rotational energy levels are spaced only a calorie per mole or so apart, and the quantum effect is essentially gone by around 25°C. Translational energy levels are so closely spaced that one generally ignores quantum effects completely, in the above sense.

The first, approximate summary of the above situation is that for gases around 25°C, C_v will contain the full equipartition contributions from translation and rotation, but only a fraction of that from vibration. For polyatomic molecules, a rough rule to use is that only about 20% of the equipartition value for the vibrational part of C_v is developed at 25°C.

The calculation for HCl and I_2 made use only of the first excited vibrational state; in principle the population of yet higher states should have been considered. The correction would have been negligible for HCl, but not entirely so for I_2. In general, then, one needs the sum of Boltzmann terms for the entire series of states involved. This type of summation turns out to be an extremely useful one; it is called the *Partition Function Q*. The average energy of a system can be expressed easily in terms of Q and hence also its heat capacity. The quantum effect then reduces to one of evaluating Q and some of its derivatives.

At this point quantum mechanics becomes very useful since it provides fairly good approximate theoretical expressions for the spacings of translational, rotational, and vibrational levels. From these the corresponding Q's can be derived, and, in turn, simple closed expressions for the heat capacity contributions.

Rigorous quantum statistical thermodynamics is quite a glorious thing. All kinds of theoretical niceties and details have been skipped in the outline that follows, as, for example, why it is a good approximation (ordinarily) to write the total partition function as a product of the separate ones for translation, rotation, and vibration. The main purposes of the problems in this chapter are simply to test your appreciation of the physical picture involved, and your ability to handle the principal relationships.

EQUATIONS AND CONCEPTS

Equipartition Principle

For a molecule with n atoms, the heat capacity per mole of gas, C_v, is made up as follows.

translation : $3R/2$

rotation : $2R/2$ (linear molecules)

$3R/2$ (nonlinear molecules)

vibration : $(3n - 5)R$ (linear molecules)

$(3n - 6)R$ (nonlinear molecules)

An approximate rule is that at 25°C the actual value of C_v contains about 20% of the equipartition vibrational contribution, but the full equipartition values for translation and rotation.

Statistical Thermodynamics

The partition function Q is

$$Q = \sum g_i e^{-\varepsilon_i/kT} \tag{5-1}$$

where g_i is the number of states happening to have an identical energy ε_i. The average energy $\bar{\varepsilon}$ is the energy of each state weighted by its probability

$$\bar{\varepsilon} = \frac{\sum \varepsilon_i g_i e^{-\varepsilon_i/kT}}{\sum g_i e^{-\varepsilon_i/kT}} \tag{5-2}$$

It is an assumption of statistical mechanics that the time average energy of a molecule $\bar{\varepsilon}$ is the same as the average energy of a large collection of molecules at any instant, or, per molecule that $\bar{\varepsilon} = E$. With this substitution, and using Q and the expression for dQ/dT, Eq. (5-2) becomes

$$E = \frac{kT^2}{Q}\frac{dQ}{dT} = kT^2\frac{d\ln Q}{dT} \tag{5-3}$$

or, per mole

$$E = RT^2\frac{d\ln Q}{dT} \tag{5-4}$$

It is usually assumed that a molecule can have translational, rotational, and vibrational energy states independent of each other, that is, without any interaction. One then finds that

$$Q_{\text{tot}} = Q_{\text{trans}}Q_{\text{rot}}Q_{\text{vib}} \tag{5-5}$$

Since E involves a logarithm of Q, it follows that

$$E_{\text{tot}} = E_{\text{trans}} + E_{\text{rot}} + E_{\text{vib}} \tag{5-6}$$

Finally, for the moment, heat capacity C_v is given by $(\partial E/\partial T)_V$, and when this operation is applied to Eq. (5-3) and (5-4) one obtains

$$C_v \text{ (per molecule)} = \frac{k}{T^2}\frac{\partial^2 \ln Q}{\partial(1/T)^2} \tag{5-7}$$

$$C_v \text{ (per mole)} = \frac{R}{T^2}\frac{\partial^2 \ln Q}{\partial(1/T)^2} \tag{5-8}$$

Equations (5-7) and (5-8) can be applied to translational, rotational, and vibrational heat capacity separately, in view of Eq. (5-6).

Some Wave Mechanical Results—Vibrational States

In the case of a simple harmonic oscillator (a fair approximation to a single vibrating bond) wave mechanics comes up with the result

$$\text{vib} = (n + \tfrac{1}{2})h\nu_0 \tag{5-9}$$

where h is Planck's constant and n is a quantum number that can take on the values 0, 1, 2, and so on. The frequency v_0 is the characteristic frequency of the oscillator, and $(1/2)hv_0$ is the zero point or minimum possible vibrational energy; hv_0 corresponds to the energy difference between the ground and first excited vibrational states.

The vibrational partition function is then

$$Q_{vib} = \sum_n \exp(-\varepsilon_n/kT) = [\exp(-hv_0/kT) + \exp(-3hv_0/2kT) + \ldots] \tag{5-10}$$

or

$$Q_{vib} = x(1 + x^2 + x^4 + \cdots) \quad \text{where} \quad x = \exp(-hv_0/2kT)$$

Use of the binomial theorem gives

$$Q = \frac{x}{(1 - x^2)} = \frac{\exp(hv_0/2kT)}{\exp(hv_0/kT) - 1} \tag{5-11}$$

Substitution of this result into Eq. (5-8) leads (eventually) to

$$C_v \text{ (per mole)} = R\left(\frac{hv_0}{kT}\right)^2 \frac{\exp(hv_0/kT)}{[\exp(hv_0/kT) - 1]^2} \tag{5-12}$$

Translational Partition Function

The ideal gas law can be derived on a classical basis by considering a molecule bouncing back and forth in a box, and the wave mechanical picture is similar. Again, a box is present, but now only those energies which correspond to standing waves are allowed. The three directions in space can be treated independently, so that one can, in this sense, talk about a one-dimensional, a two-dimensional, and a three-dimensional box and hence gas.

In one dimension, the set of energy states is found to be

$$\varepsilon = \frac{n^2 h^2}{8a^2 m} \tag{5-13}$$

where a is the length of the one-dimensional box, m is the mass of the atom or molecule, and n is again a quantum number which takes on integral values. The partition function is then

$$Q_{trans} = \sum_n \exp(-n^2 h^2/8a^2 mkT)$$

$$= \int_0^\infty \exp(-n^2 h^2/8a^2 mkT)\, dn \tag{5-14}$$

The summation can be replaced by an integration because of the extremely close spacing of the energy levels. On evaluating the definite integral,

$$Q_{\text{trans}} = \left(\frac{2\pi mkT}{h^2}\right)^{1/2} a \tag{5-15}$$

In the case of a two-dimensional box, the quantum numbers are given by $(n_x^2 + n_y^2)$, where n_x and n_y are the quantum numbers for the x and y directions, and separately take on all integral values. Similarly, for a three-dimensional box, one has $(n_x^2 + n_y^2 + n_z^2)$. The partition functions are

$$Q_{\text{2-dim}} = \sum_{n_x} \sum_{n_y} \exp[-(n_x^2 + n_y^2)h^2/8a^2mkT] = Q_x Q_y = Q^2 \tag{5-16}$$

$$Q_{\text{3-dim}} = \sum_{n_x} \sum_{n_y} \sum_{n_z} \exp[-(n_x^2 + n_y^2 + n_z^2)h^2/8a^2mkT]$$

$$= Q_x Q_y Q_z = Q^3 \tag{5-17}$$

On again replacing summations by integrations,

$$Q_{\text{2-dim}} = (2\pi mkT/h^2)a^2 = (2\pi mkT/h^2)A \tag{5-18}$$

$$Q_{\text{3-dim}} = (2\pi mkT/h^2)^{3/2} a^2 = (2\pi mkT/h^2)^{3/2} V \tag{5-19}$$

where A and V are the area and volume, respectively.

The most common and important case is that for three dimensions. Equation (5-19) may be simplified to the form $Q = bT^{3/2}$ where b is the appropriate collection of constants. Insertion of this into Eq. (5-4) gives

$$E = RT^2 \frac{d\ln Q}{dT} = \frac{RT^2}{Q}\frac{dQ}{dT} = \frac{RT^2}{Q}\frac{3}{2}(bT^{1/2})$$

$$= 3RT/2 \tag{5-20}$$

The result is thus the same as the classical one, namely that of the energy of translation in three dimensions as given by the equipartition principle.

The heat capacity C_v is now trivial to obtain; it is simply the equipartition value $3R/2$.

Rotational Partition Function

The set of rotational energy levels for a (rigid) molecule is given by wave mechanics as

$$\varepsilon_{\text{rot}} = J(J + 1)h^2/8\pi^2 I \tag{5-21}$$

where I is the moment of inertia, and J is again a quantum number taking on the values 0, 1, 2, and so on. Molecular moments of inertia are around 10^{-39} g cm^2—in a diatomic molecule I involves the product of the mass of an atom times the square of the interatomic distance. The product $h^2/8\pi^2 I$ is then about 5×10^{-16} erg per molecule, or about 1 cal/mole. The spacing of rotational levels is thus much smaller than RT at room temperature, and again the summation is usually replaced by an integration.

In deriving the partition function, it is necessary to consider whether there are two or three possible axes of rotation, and the statistical weights of various states; the resulting expression is

$$Q_{\text{rot}} = \left(\frac{1}{\pi\sigma}\right)\left(\frac{8\pi^3 I k T}{h^2}\right)^{n/2} \qquad (5\text{-}22)$$

where n is the number of independent rotation axes (2 or 3) and σ is a symmetry number given by the number of indistinguishable positions into which the molecule can be turned by rotations. For an asymmetric molecule, I is now the geometric mean of the three principal moments of inertia.

Equation (5-22) can be written as $Q = b'T^{n/2}$, where b' stands for the collection of constants. Then on applying the same procedure as in obtaining Eq. (5-20), we find

$$E_{\text{rot}} = \left(\frac{n}{2}\right)RT \qquad (5\text{-}23)$$

The rotational energy is thus either $3RT/2$ or RT, depending on whether the molecule is linear or nonlinear, and C_v correspondingly is either $3R/2$ or R. Thus again the equipartition value of C_v comes out of the statistical mechanical treatment. As with the translational case, the equipartition result is obtained only with the assumption that the level spacings are much less than kT.

PROBLEMS

1. (9 min) Estimate the value of C_p for each substance in the following reaction, and hence ΔC_p for the reaction:

$$C(s) + 2H_2O(g) = CO_2(g) + 2H_2(g)$$

2. (12 min) Locate the following (a–f) in the correct range of C_p values:

$$C_p, \text{cal/deg-mole}$$

(a) $H_2(g)$ at $-150°C$	20–25
	15–20
(b) $N_2(g)$ at 25°C	12–15
	10–12
(c) $He(g)$ at 1,000°C	8–10
	7
(d) $NaBr(s)$ at 25°C	6
	5
(e) Pb at 25°C	4
	3
(f) $C_2H_5OH(g)$ at 25°C	2
	1

3. (9 min) Estimate C_p (in cal/deg-mole) for each of the following, to about 1 cal/deg-mole:

(a) argon gas at 1,000°C   (d) $C_2H_6(g)$ at 25°C
(b) Be(metal) at $-100°C$ (e) Au(metal) at 500°C
(c) $H_2(g)$ at 300°C   (f) $H_2S(g)$ at 25°C

4. (12 min) Estimate the following C_v values to the nearest 2 cal/deg-mole (all at 25°C): $Ar(g)$, $H_2(g)$, $CO_2(g)$, $N_2(g)$, $Ag(s)$, C(diamond), $C_6H_6(g)$, $H_2O(g)$, $H_2O(l)$.

5. (9 min) Substance X is an ideal gas containing n atoms per molecule. (a) C_p for $X(g)$ and for $N_2(g)$ is the same at 0°C (consider the contribution of vibrational degrees of freedom to the heat capacity to be negligible at 0°C). (b) The difference between the equipartition values of C_p for $X(g)$ and $N_2(g)$ is approximately 6 cal/deg-mole. Show from the above information what can be concluded about n and any other aspects concerning the structure of X.

6. (12 min) (a) The equipartition value of γ ($= C_p/C_v$) for ozone is 1.15; assuming ideal gas behavior, is the molecule linear or nonlinear?

(b) A lead–silver alloy has a C_v of 0.0383 cal/deg-g. What is the composition of the alloy? (Atomic weights: Pb, 207; Ag, 107.)

7. (9 min) Calculate or estimate (to the nearest whole number) the following heat capacities in cal/deg-mole: (a) C_p for H_2 at 25°C, (b) C_v for $NaCl(s)$ at 300°C, (c) C_p for $Ag(s)$ at 25°C, (d) γ (equipartition) for CH_4, (e) γ for argon at 1,000°C.

8. (10 min) Estimate the equipartition values of C_p for each substance in the reaction:

$$C_2H_2(g) \text{ (acetylene)} + H_2O(g) = C_2H_4O(g) \text{ (ethylene oxide)}$$

9. (12 min) The value C_p for gaseous benzene at 50°C is 14 cal/deg-mole. Calculate the number of vibrational degrees of freedom this corresponds to, assuming equipartition. To what fraction of the total number of vibrational degrees of freedom for benzene does the above number correspond? Assume ideal gas behavior.

10. (10 min) When water molecules are adsorbed on a metal surface, they apparently lose two of their three degrees of rotational freedom, but retain two out of the three degrees of translational freedom. (a) Suggest what happens to the missing degrees of freedom. (b) Calculate the equipartition C_v for the adsorbed water; be consistent with your answer in (a).

11. (6 min) What weight of NaCl should have an equipartition heat capacity of 6 cal/deg-mole?

---

12. (12 min) The rotational partition function for a certain gas is

$$\cdot Q = (1/3\pi)(8\pi^3 I^k T/h^2)^{3/2}$$

so that Q is proportional to $T^{3/2}$. Show that the rotational heat capacity is therefore $3R/2$ per mole.

13. (12 min) The value of C_v for vibration as given by Eq. (5-12) approaches a limiting value as T approaches infinity. Show what this limiting value is.

14. (12 min) In the case of I_2, hv_0/kT is 0.68 at 25°C; substitution of this result into Eq. (5-12) gives a vibrational heat capacity contribution of $0.85R$. In the case of HCl, hv_0/kT is 6.8 at 25°C. Explain at what temperature the vibrational heat capacity of HCl should be the same as that of iodine gas. Comment on your result.

15. (15 min) Estimate, with explanation, the temperature at which the quantum mechanical translational heat capacity of an ideal gas might begin to deviate from the equipartition value.

16. (12 min) Calculate the energy per mole of a two-dimensional monatomic gas at 25°C. (Such a calculation is not a pure fantasy; the answer here might be applied, for example, to argon adsorbed on the surface of an adsorbent, over which it is supposed to move freely.)

17. (15 min) The value of hv_0/kT is 0.68, and the vibrational heat capacity contribution is $0.97R$ for I_2 at 25°C. (a) Show what this contribution should be at 0°K. (b) HCl must be at about 3,000°K to have the same vibrational heat capacity as does I_2 at 25°C. Calculate the zero-point energy of HCl.

ANSWERS

1. C(s): assume Dulong and Petit's law, $C_p = $ **6 cal/deg-mole.**[1] H_2O(g): assuming 20% of vibrational heat capacity, $C_p = R + (3R/2)$(trans) + $(3R/2)$(rot) + $0.2(3R)$(vib) or $C_p = 4.6R = $ **9.1 cal/deg-mole.** CO_2(g): (linear molecule) $C_p = R + (3R/2)$(trans) + $(2R/2)$(rot) + $0.2(4R)$(vib) = $4.3R = $ **8.5 cal/deg-mole.** H_2(g): $R + (3R/2)$(trans) + $(2R/2)$(rot) + $0.2R$ (vib) = $3.7R = $ **7.3 cal/deg-mole.** Then $\Delta C_p = $ **−1.1 cal/deg.**

2. (a) H_2: no vib at -150°C, so $C_p = R + (3R/2)$(trans) + $(2R/2)$(rot) = $3.5R \cong $ **7.**

(b) N_2: at 25°C take 20% of equipartition vibrational heat capacity, so $C_p = R + (3R/2)$(trans) + $(2R/2)$(rot) + $0.2R$(vib) = $3.7R \cong $ **7.4 ≅ 7.**

(c) He: $C_p = R + (3R/2)$(trans) = $2.5R \cong $ **5.**

(d) NaBr: assume Dulong and Petit's law, but remember there are 2 moles of atoms so $C_p \cong C_v \cong $ **12.**

(e) Pb: assume Dulong and Petit's law, $C_p \cong C_v \cong $ **6.**

(f) C_2H_5OH: $C_p = R + (3R/2)$(trans) + $(3R/2)$(rot) + $0.2(21R)$(vib) = $8.2R \cong $ **16.**

3. (a) $C_p = R + (3R/2)$(trans) = $2.5R \cong $ **5.**

(b) Probably only about 20% of the Dulong and Petit maximum of 6, since Be atoms are light and the temperature is small. $C_p \cong C_v \cong 0.2 \times 6 \cong $ **1.2.**

(c) Allow maybe 30% of equipartition vibrational heat capacity, $C_p = R + (3R/2)$(trans) + $(2R/2)$(rot) + $0.3R$ (vib) = $3.8R \cong $ **7.**

(d) At 25°C allow about 20% of equipartition vibrational heat capacity, $C_p = R + (3R/2)$(trans) + $(3R/2)$(rot) + $0.2(18R)$(vib) = $7.6R \cong $ **15.**

(e) Allow the full Dulong and Petit value (heavy atoms, high temperature), so $C_p \cong C_v = $ **6.**

(f) Allow 20% of vibrational heat capacity, $C_p = R + (3R/2)$(trans) + $(3R/2)$(rot) + $0.2(3R)$(vib) = $4.6R \cong $ **9.**

4. Ar(g): only translation so $3R/2$ or **3.**

H_2(g): allow 20% of equipartition vibrational heat capacity, so $C_v = (3R/2)$(trans) + $(2R/2)$(rot) + $0.2(R)$(vib) = $2.7R \cong $ **5.**

CO_2(g): $C_v = (3R/2)$(trans) + $(2R/2)$(rot) + $0.2(4R)$(vib) = $3.3R \cong $ **6.**

N_2(g): same as for H_2.

[1] Actually, one should recognize that for C(s) the value at 25°C will be about 20% of 6 or about 1.5 cal/deg-mole.

Ag(s):   heavy atoms, soft, so assume equipartition Dulong and Petit value: **6.**

C(s):   light atoms, hard, so guess that about 20% of D and P value: **1.**

C_6H_6(g): $C_v = (3R/2)$(trans) $+ (3R/2)$(rot) $+ 0.2(30R)$(vib) $= 9R = $ **18.**

H_2O(g): $C_v = (3R/2)$(trans) $+ (3R/2)$(rot) $+ 0.2(3R)$(vib) $= 3.6R \cong $ **7.**

H_2O(l): $C_v = $ **18.** (You should know that the specific heat of water is about 1 cal/deg-g !)

5.   The heat capacity of N_2 is made up of R (for C_p) $+ (3R/2)$(trans) $+ (2R/2)$ (trans) $+ (2R/2)$(rot) $+ R$(vib). At 0°C only translation and rotation are to be considered, so X must also have only two degrees of rotational freedom; that is, X must be linear. However, the extra 6 cal/deg-mole at equipartition means that X has three more vibrational degrees of freedom than N_2, which means one more atom. X is therefore a linear triatomic molecule.

6.   (a) If linear, the equipartition C_v would be $(3R/2)$(trans) $+ (2R/2)$(rot) $+ (4R)$(vib) or $6.5R$, and γ would then be $7.5/6.5 = 1.15$. If nonlinear, C_v would be $(3R/2)$(trans) $+ (3R/2)$(rot) $+ (3R)$(vib) or $6R$ and γ is then $7/6 = 1.17$. The molecule is therefore **linear.**

   (b) Assuming 6 cal/deg-mole as the Dulong and Petit value, the average molecular weight of the alloy must be $6/0.0383 = 157$. In terms of mole fractions N, then $157 = N_{Ag} 107 + N_{Pb} 207 = 107 + 100 N_{Pb}$ or $N_{Pb} = $ **0.50.**

7.   (a) Assume 20% of equipartition vibrational heat capacity: $C_p = R + (3R/2)$(trans) $+ (2R/2)$(rot) $+ (0.2R)$(vib) $= 3.7R \cong $ **7.**

   (b) Take Dulong and Petit value (but for 2 moles of atoms): **12.**

   (c) Take Dulong and Petit value: **6.**

   (d) $C_v = (3R/2)$(trans) $+ (3R/2)$(rot) $+ (9R)$(vib) $= 12R$; $\gamma = 13/12 = $ **1.08** $\cong$ **1.**

   (e) $C_v = 3R/2$; $\gamma = 5/3 = $ **1.66** $\cong$ **2.**

8.   Acetylene is a rare four-atom molecule that is linear. $C_p = R + (3R/2)$ (trans) $+ (2R/2)$(rot) $+ 0.2(7R)$(vib) $= 4.9R = $ **9.6 cal/deg-mole.** H_2O(g): $C_p = 9.1$ cal/deg-mole (see Problem 1). C_2H_4O: $C_p = R + (3R/2)$(trans) $+ (3R/2)$(rot) $+ 0.2(15R)$(vib) $= 7.0R = $ **13.8 cal/deg-mole.**

9. C_p for translation and rotation only should be $R + 3R/2 + 3R/2$ or $4R = 7.93$ cal/deg-mole. This leaves 6.07 for vibration, corresponding to about 3.1 vibrational modes at equipartition. There should be $3 \times 12 - 3 - 3$ or 30 vibrational modes in all, so the above number corresponds to about 10% of this total (exactly, if R is taken to be 2 cal/deg-mole).

10.   (a) Evidently the water is essentially bonded to the surface, and the missing three degrees of freedom must reappear as vibrations against the adsorbent. (b) On this basis, $C_v = (2R/2)$(trans) $+ (R/2)$(rot) $+ (9 - 3)R$(vib) $= $ **15 cal/deg-mole.**

11. The answer should be that weight which contains one mole of atoms, or half a formula weight of NaCl, that is, **27.2 g.**

12. For the present purpose, Q may be written simply as $Q = aT^{3/2}$. Equation (5-4) then reads

$$E = (RT^2/aT^{3/2})\, d(aT^{3/2})/dT = \frac{3RT}{2}$$

and C_v is dE/dT or $3R/2$.

13. At large T, the exponent hv_0/kT becomes small, and the exponential in the numerator of Eq. (5-12) can be replaced by unity, and, in the denominator, by $1 - hv_0/kT$. The equation then reduces to $C_v = R$.

14. Equation (5-12) involves only the ratio hv_0/kT. We thus want to make this ratio equal to 0.68 for HCl, instead of 6.8. Evidently, then, the temperature must be $10 \times 298°$ or **2,980°K.** The comment is that by this temperature it is possible that HCl has become somewhat unstable, so that the actual gas may contain appreciable amounts of H_2 and Cl_2.

15. Deviations from the wave mechanical result should begin to be important when the integration is no longer a good approximation for the summation [Eq. (5-14)], or when the energy level spacing is comparable to kT. This temperature would then be about $T = h^2/8a^2mk$. Assuming, for example, 1 cm^3 volume and a molecular weight of 30, $T = (6.5 \times 10^{-27})^2 6 \times 10^{23}/8 \times 1^3 \times 30 \times 1.36 \times 10^{-16} = 8 \times 10^{-16}$ °K !

16. From Eq. (5-18), we can write $Q = aT$. Insertion in Eq. (5-4) then gives $E = (RT^2/aT)\, d(aT)/dT = RT = 1.98 \times 298 = $ **590 cal/mole.**

17. (a) The exponentials of Eq. (5-12) become infinite at $T = 0°K$, and since there is a square term in the denominator, C_v must go to **zero.** (b) hv_0/kT must be 0.68 for HCl at 3,000°K. The zero-point energy hv_0 is then $1.98 \times 3000 \times 0.68$ or **4,000 cal/mole.**

6
THERMOCHEMISTRY

COMMENTS

The topic of this section should constitute a bit of a breather between the first and second laws of thermodynamics. Be careful with your handling of signs, though!

Thermochemistry *is* straightforward; you add or subtract ΔH's or ΔE's as you do the equations to which they apply. Since we have no idea of what absolute energies are for things, we always deal with Δ quantities. Often, for convenience, these are heats of formation. They may, however, be heats of combustion, and note Problem 4 for another type.

The Δ quantities change with temperature according to the integral of ΔC_p or ΔC_v. In an adiabatic reaction, the products warm or cool as they are formed, but ΔH and ΔE are state functions, independent of path, so we choose a path of convenience. That is, we imagine that the reaction takes place isothermally, then use the heat produced or absorbed to see how much the products can be warmed or cooled.

EQUATIONS AND CONCEPTS

"Δ" Operator

The sum of values for the products minus the sum of values for the reactants. Thus ΔH, ΔE, ΔC_p, Δn, and so on.

Hess's Law

ΔH's and ΔE's are added and subtracted as are the equations, and the value for a particular reaction is independent of the choice of reactions combined in order to obtain it. That is, ΔH and ΔE are independent of path.

86

Heats of Formation and Combustion

Heat of formation is the ΔH per mole required to form the species from the elements in their standard states. Heat of combustion is the heat of reaction of the compound with oxygen to give CO_2 and H_2O, usually $H_2O(l)$.

Change of ΔH and ΔE with Temperature

$$\Delta H_2 = \Delta H_1 + \int_{T_1}^{T_2} \Delta C_p \, dT \tag{6-1}$$

and

$$\Delta E_2 = \Delta E_1 + \int_{T_1}^{T_2} \Delta C_v \, dT \tag{6-2}$$

Miscellaneous Topics and Relations

Variation of heat capacity with temperature, usually expressed as a power series in T. Relation between ΔH and ΔE: $\Delta H = \Delta E + \Delta n_g RT$ where n_g is the mole number of gaseous species and Δn_g is the difference between moles of gaseous products and gaseous reactants.

Heat of solution. Maximum temperature of an explosion.

PROBLEMS

1. (13 min) Given: heat of formation at 25°: $H_2O(g)$: $- 57.8$ kcal; $CH_4(g)$: -17.9 kcal.

Heat of combustion at 25°C to CO_2 and $H_2O(g)$: $CH_4(g)$:$- 192.0$ kcal. Calculate ΔH at 25°C for $C(s) + 2H_2O(g) = CO_2(g) + 2H_2(g)$. Calculate ΔE at 25°C for this same reaction.

2. (21 min) Given the following information:
Reaction: $4C_2H_5Cl(g) + 13O_2(g) = 2Cl_2(g) + 8CO_2(g) + 10H_2O(g)$.
$\Delta H_{298°K} = - 1,229.6$ kcal.
heat of combustion per mole of ethane $[C_2H_6(g)]$ to $CO_2(g)$ and $H_2O(g) = - 341$ kcal.
heat of formation per mole of $H_2O(g)$: -57.8 kcal.
heat of formation per mole of $HCl(g)$: -21 kcal.
  (a) Calculate $\Delta H_{298°K}$ for the reaction $C_2H_6(g) + Cl_2(g) = C_2H_5Cl(g) + HCl(g)$.

(b) Assuming ΔC_p for the first reaction to be $-10\,cal/deg$, calculate $\Delta H_{398°K}$.

(c) Calculate $\Delta E_{298°K}$ for the first reaction.

3. (30 min) Given the following information:

heats of formation at 298°K: $CO_2(g)$ $-94.0\,kcal$

                            $C_2H_4O_2(l)$(acetic acid) $-116.4\,kcal$

                            $H_2O(g)$ $-57.8\,kcal$

heat of combustion of $CH_4(g)$ to give $CO_2(g)$ and $H_2O(g)$: $-192.7\,kcal$

heat of vaporization of water at 100°C: $9.4\,kcal/mole$

C_p values (in cal/deg-mole): $C_2H_4O(g)$(acetaldehyde):   12.5

                            $CO(g)$:   7.5 $CH_4(g)$:   9.0

                            $H_2O(g)$: 7.3 $H_2O(l)$: 18.0

(a) Calculate the heat of formation of $H_2O(l)$ at 298°K.

(b) Calculate $\Delta H_{298°K}$ for the reaction: $C_2H_4O_2(l) = CH_4(g) + CO_2(g)$.

(c) Calculate the temperature at which ΔH for the reaction $C_2H_4O(g) = CH_4(g) + CO(g)$ should be zero. $\Delta H_{298°K}$ is $-4.0\,kcal$.

4. (9 min) The "heat of total cracking" H_{TC} for a hydrocarbon is defined, for the purpose of this question, as the $\Delta H_{298°K}$ for a reaction of the type

$$C_nH_m(g) + (2n - m/2)H_2(g) = nCH_4(g)$$

Given that H_{TC} is $-15.6\,kcal$ for $C_2H_6(g)$ and $-20.9\,kcal$ for $C_3H_8(g)$, and that the heat of formation of $CH_4(g)$ at 25°C is $-17.9\,kcal$, calculate $\Delta H_{298°K}$ for

$$CH_4(g) + C_3H_8(g) = 2C_2H_6(g)$$

5. (21 min) Given the reaction below, whereby H_2 is burned in excess O_2, $H_2(g, 25°C) + 10O_2(g, 25°C) = H_2O(g, 25°C) +$ excess $O_2(g, 25°C)$ and also that the heat of formation of $H_2O(g, 25°C)$ is $-58\,kcal$ and that C_p is $6.5\,cal/deg$-mole for H_2 and for O_2 and $7.5\,cal/deg$-mole for $H_2O(g)$; (a) calculate ΔE at 25°C for reaction; (b) calculate $\Delta H_{498°K}$ for the reaction; and (c) calculate the maximum temperature in an adiabatic explosion of the mixture in a sealed bomb, if the reactants are initially at 25°C.

6. (30 min) Given the following information:

heats of formation at 298°K: $CO_2(g)$ $-94\,kcal$

                            $H_2O(g)$ $-58\,kcal$

heats of reaction at 298°K:

$H_2O(g) + C(s) = H_2(g) + CO(g)$(water gas reaction) $\Delta H = 32\,kcal$

$H_2(g) = 2H(g)$ $\Delta H = 103\,kcal$

$O_2(g) = 2O(g)$ $\Delta H = 34\,kcal$

heat capacities: Take C_p to be 7.00 cal/deg-mole for all gases and 2.5 cal/deg-mole for C(s).

Calculate (a) the heat of combustion of graphite at 298°K, in kcal/g; (b) the heat of combustion, at 298°K, of the H_2 and CO formed by the complete reaction of water with 1 g of carbon; (c) ΔE for the water–gas reaction at 298°K; (d) ΔH at 600°K for the water–gas reaction; (e) the H—O bond energy.

7. (18 min) The heat of formation of $C_2H_5OH(l)$ is -66 kcal/mole, while the heat of combustion to $CO_2(g)$ and $H_2O(l)$ of the isomeric CH_3—O—$CH_3(g)$ is -348 kcal/mole. The heat of formation of $H_2O(l)$ is -68 kcal/mole, and the heat of combustion of carbon to $CO_2(g)$ is -94 kcal/mole (all data for 25°C). (a) Calculate $\Delta H_{298°K}$ for the isomerization reaction

$$C_2H_5OH(l) = CH_3\text{—}O\text{—}CH_3(g)$$

(b) Assuming the answer to part (a) to be -10 kcal, what would $\Delta E_{298°K}$ be?

8. (12 min) The heat of formation of HBr(g) from $H_2(g)$ and $Br_2(g)$ at 25°C is -9 kcal/mole. Assuming that all diatomic gases have a constant C_p of 7 cal/deg-mole, (a) show what the heat of formation of HBr would be at 125°C. (b) If 1 mole of H_2 and 99 moles of Br_2 are exploded in an insulated bomb, then, neglecting the heat capacity of the bomb itself, calculate the final temperature if the initial temperature was 25°C.

9. (15 min) The heats of combustion of $(CH_2)_3$, carbon, and H_2 are -500.0, -94.0, and -68.0 kcal/mole, respectively (where burned to carbon dioxide and liquid water). The heat of formation of CH_3CH=CH_2 is 4.9 kcal/mole. (a) Calculate the heat of formation of cyclopropane $(CH_2)_3$. (b) Calculate the heat of isomerization of cyclopropane to propylene.

10. (12 min) One mole of NaCl is dissolved in sufficient water to give a solution containing 12 percent NaCl by weight. The ΔH for this reaction is 774.6 cal at 20°C and 700.8 cal at 25°C. The heat capacity of solid NaCl is 12 cal/deg-mole and, for water, 18 cal/deg-mole. Calculate the heat capacity of the solution in cal/deg-g.

11. (18 min) Given the following data:

(ethyl benzene) + $3H_2(g)$ = (ethylcyclohexane)   (1)

$$\Delta H_{298°K} = -48.3 \text{ kcal}$$

$$CH{=}CH_2(l) + 4H_2(g) =$$

(styrene)

(I)

$$\Delta H_{298^\circ K} = -74.65 \text{ kcal}$$

The heat of combustion of ethylcyclohexane to water vapor and CO_2 is -1238.23 kcal/mole at $298^\circ K$, and the heats of formation of water vapor and of CO_2 are -58.32 and -94.05 kcal/mole, respectively, at $298^\circ K$. Calculate (a) the heat of hydrogenation of styrene to ethylbenzene, and (b) the heat of formation of ethylbenzene.

12. (12 min) For the process $Na_2CO_3 \cdot 10H_2O + \text{water} = \text{solution}$, $\Delta H = 16.2$ kcal. We propose to utilize this effect to cool beverage cans (e.g., beer) by equipping the cans with an outer jacket containing this salt, and by adding water to this reservoir when cooling is desired (see Fig. 6-1). As a typical case, 0.2 mole (60 g) of salt are used, the jacket holds 200 g of water, and the inner container holds 200 g of beverage (99.99 % water); the initial temperature is $20^\circ C$. heat capacities are: water = 1 cal/deg-g, solution = 0.8 cal/deg-g, solid salt = 0.20 cal/deg-g, and container = 10 cal/deg. Calculate the temperature to which the beverage would be cooled.

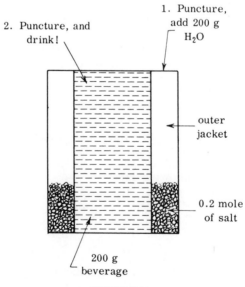

FIGURE 6-1

13. (19 min) Given that

heats of combustion at 298°K : CH₂=CHCN(g) −488.67 kcal
                                   C(graphite)   −94.05
                                   H₂(g)   −68.33
heats of formation at 298°K :   HCN(g)   31.00
                                   C₂H₂   54.19

[In the combustion of acrylonitrile, nitrogen ends up as $N_2(g)$.]
    (a) Calculate $\Delta H_{298°K}$ for the reaction HCN(g) + C₂H₂(g) = CH₂=CHCN(g), and (b) calculate the difference between ΔH and ΔE for this reaction.

14. (11 min) For the process $(NH_4)_2C_2O_4(s) + H_2O(l)$ = solution of 1 mole of each $\Delta H_{298°K} = 0.8$ kcal. C_p values are $(NH_4)_2C_2O_4(s)$, 20 cal/deg-mole ; H₂O(l), 1 cal/deg-g ; the solution of 1 mole of each, 40 cal/deg. If 1 mole of ammonium oxalate and 1 mole of water, both at 25°C, are mixed in an insulated vessel, what will be the temperature of the resulting solution?

15. (18 min) Given the following data : Heat of combustion of ethyl acetate $(CH_3COOC_2H_5)(l)$ to give $CO_2(g)$ and H₂O(l): − 536.9 kcal/mole at 25°C. Heats of formation, per mole at 298°K : $CH_3COOH(l)$: − 116.7 ; $C_2H_5OH(l)$: − 66.3 ; $CO_2(g)$: −94.0 ; H₂O(g): −57.8. Heat of vaporization of water at 298°K is 10.5 kcal/mole. Calculate ΔH and ΔE at 298°K for the reaction

$$CH_3COOH(l) + C_2H_5OH(l) = CH_3COOC_2H_5(l) + H_2O(g)$$

16. (12 min) Given, for the reaction $CO(g) + H_2O(g) = CO_2(g) + H_2(g)$, that $\Delta H_{298°K} = -10$ kcal, and the following heat capacities (C_p) per mole :

CO: $6.60 + 1.0 \times 10^{-3}T$ H₂: $6.6 + 1.0 \times 10^{-3}T$
H₂O: $7.3 + 2.0 \times 10^{-3}T$ CO₂: $7.3 + 3.0 \times 10^{-3}T$

Calculate ΔH for the above reaction at 1,000°K.

17. (30 min) Given the following information :

heats of formation per mole at 298°K : C₂H₅OH(l)   −66.0.kcal
                                              CO₂(g)   −94.0
                                              H₂O(l)   −68.3
heats of combustion, at 298°K, to give
  H₂O(l) and/or CO₂(g):   CO   −68 kcal
                                              CH₄(g)   −212
heat capacities C_p in cal/deg-mole :   CH₄ = 5.0, CO₂ = 7.0,
                                              C₂H₅OH(l) = 32.

(a) Calculate $\Delta H_{298°K}$ for the reaction $3CH_4(g) + CO_2(g) = 2C_2H_5OH(l)$.

(b) Calculate $\Delta E_{298°K}$ for the reaction in (a).

(c) Calculate $\Delta H_{-100°C} - \Delta H_{298°K}$ for reaction in (a).

18.   (16 min) Given the values of ΔH at 25°C and 1 atm for the reactions:

(1) $C_2H_6(g) + \frac{7}{2}O_2(g) = 2CO_2(g) + 3H_2O(l)$ $\Delta H = -372.8$ kcal
(2) $H_2(g) + \frac{1}{2}O_2(g) = H_2O(l)$ $\Delta H = -68.3$
(3) $2C(\text{graphite}) + 3H_2(g) = C_2H_6(g)$ $\Delta H = -20.2$
(4) $2C(\text{graphite}) + 2H_2(g) = C_2H_4(g)$ $\Delta H = 12.6$

(a) Calculate ΔH for the hydrogenation of ethylene at 25°C:

$$C_2H_4(g) + H_2(g) = C_2H_6(g)$$

(b) Calculate the heat of combustion of ethylene at 25°C.

(c) Calculate ΔE for the reaction in (a), making and stating any reasonable assumptions that are necessary.

19.   (5 min) Given the following data: $C_2H_6(g, 25°C) + \frac{7}{2}O_2(g, 25°C) = 2CO_2(g, 25°C) + 3H_2O(g, 25°C)$, $\Delta E = 262.6$ kcal.

C_v in cal/deg-mole: C_2H_6 8.0 H_2O 6.2
                                        O_2 4.8 CO_2 5.7

Calculate the maximum temperature achieved when 0.1 mole of ethane is exploded with 1 mole of oxygen in a bomb calorimeter. The reactants are initially at 25°C (make any reasonable assumptions necessary to complete the calculation).

20.   (6 min) Explain in a qualitative (but definite) manner whether the enthalpy of formation of CCl_4 should increase (become more positive) or decrease with increasing temperature. A possibly useful datum is that the heat capacity of graphite is about 2 cal/deg-mole.

21.   (24 min) The heat of total hydrogenation (H_{TH}) is defined as the ΔH for the reaction of an organic compound with hydrogen to convert the carbon to methane and any oxygen to water (gas). The heats of total hydrogenation at 25°C are -89.9 kcal/mole for acetylene, C_2H_2 and -84.1 kcal/mole for ethylene oxide, C_2H_4O. The heat of formation of $H_2O(g)$ is -57.8 kcal/mole at this temperature.

(a) Calculate ΔH at 25°C for $C_2H_2 + H_2O(g) = C_2H_4O$. (b) Calculate ΔE at 25°C for the above reaction. (c) Calculate ΔH at 598°K for the above reaction. Assume C_v values are 7, 5, and 12 cal/deg-mole for C_2H_2, $H_2O(g)$, and C_2H_4O, respectively.

22. (18 min) Carbon monoxide is burned in an adiabatic, constant volume calorimeter according to the reaction $CO + \frac{1}{2}O_2 = CO_2$. There is, however, a deficiency of oxygen present so that instead of $\frac{1}{2}$ mole per mole of CO, there is only x mole present. The mixture of CO and O_2 initially is at 25°C and, after the reaction, the product gases are at 125°C. The heats of formation per mole at 25°C are -26 kcal and -94 kcal for CO and CO_2, respectively, and the respective C_p values are 7 and 8 cal/deg-mole. C_p for oxygen is 6.8 cal/deg-mole.

(a) Calculate ΔH for the reaction $CO + \frac{1}{2}O_2 = CO_2$. (b) Calculate x, the moles of O_2 per mole of CO present in the combustion experiment.

ANSWERS

1. The only process that involves C(s) is the heat of formation of CH_4, so start with it:

$$C(s) + 2H_2 = CH_4 \qquad \Delta H_1 = -17.9$$
now add $\qquad CH_4 + 2O_2 = CO_2 + 2H_2O \qquad \Delta H_2 = -192.2$
then add $\qquad 4H_2O = 4H_2 + 2O_2 \qquad \Delta H_3 = 4 \times 57.8$

This gives the desired equation so $\Delta H = -17.9 - 192.2 + 231.2 = $ **21.3 kcal.** Within slide rule accuracy, $\Delta E = \Delta H - \Delta n_g RT = 21.3 - 1.98 \times 298/1000 = $ **20.7 kcal.**

2. (a) Start with the first reaction written backwards:

$$10H_2O + 8CO_2 + 2Cl_2 = 13O_2 + 4C_2H_5Cl \quad \Delta H_1 = 1229.6 \text{ kcal}$$

Then get in the 4 moles of C_2H_6 needed by adding the combustion of 4 moles:

$$4C_2H_6 + 14O_2 = 8CO_2 + 12H_2O \qquad \Delta H_2 = -4 \times 341 \text{ kcal}$$

Now get in the desired 4 moles of HCl:

$$2Cl_2 + 2H_2 = 4HCl \qquad \Delta H_3 = -4 \times 21 \text{ kcal}$$

To make things come out, add

$$2H_2O = O_2 + 2H_2 \qquad \Delta H_4 = 2 \times 57.8 \text{ kcal}$$

$$4C_2H_6 + 4Cl_2 = 4C_2H_5Cl + 4HCl \qquad \Delta H = 1229.6 - 1364 - 84 + 115.6 = -103 \text{ kcal}$$

(b) $\Delta H_{398°K} = \Delta H_{298°K} + \Delta C_p \Delta T = -1229.6 + (10 \times 100)/1000 = -1230.6 \text{ kcal.}$

(c) $\Delta E = \Delta H - \Delta n_g RT = -1229.6 - 3RT = -1231.4 \text{ kcal.}$

3. (a) Heat of formation of $H_2O(l)$: $-57.8 - 9.4 = -67.2 \text{ kcal.}$

(b) We want acetic acid on the left, so begin with the reverse of its formation:

$$C_2H_4O_2 = 2C(s) + 2H_2 + O_2 \qquad \Delta H_1 = 116.4 \text{ kcal}$$

We want CH_4 on the right, so add the reverse of its combustion:

$$CO_2 + 2H_2O = CH_4 + 2O_2 \qquad \Delta H_2 = 192.7 \text{ kcal}$$

Now, eliminate $C(s)$ using the formation of CO_2:

$$2C(s) + 2O_2 = 2CO_2 \qquad \Delta H_3 = -2 \times 94.0 \text{ kcal}$$

To make things come out, add the formation of H_2O:

$$2H_2 + O_2 = 2H_2O \qquad \Delta H_4 = -2 \times 57.8 \text{ kcal}$$

$$C_2H_4O_2 = CH_4 + CO_2 \qquad \Delta H = 116.4 + 192.7 - 188$$
$$-115.6 = 5.5 \text{ kcal}$$

(c) $\Delta H_T = \Delta H_{298°K} + \Delta C_p \Delta T.$ $\Delta C_p = 9.0 + 7.5 - 12.5 = 4.0.$ Hence $\Delta H_T = 0 = -4,000 + 4\Delta T$ or $\Delta T = 1,000,$ and $T = 1298°K.$

4. There is a real shortcut here. Any standard reaction in which all species except the one of interest are the same behaves like a heat of formation or combustion. A ΔH of reaction involving such species is simply given by the appropriate sum or difference of the ΔH's of the standard reactions. Here, H_{TC} has this property; moreover H_{TC} is zero for CH_4. The desired ΔH is then simply $2H_{TC}(C_2H_6) + H_{TC}(C_3H_8) = 2 \times 15.6 + (-20.9) = 10.3 \text{ cal.}$ (Check the answer the long way, if you don't believe me!)

5. (a) $\Delta H_{298°K}$ is simply the heat of formation of $H_2O(g)$ or -58 kcal. $\Delta E = \Delta H - \Delta n_g RT = -58 - (-\frac{1}{2})RT/1,000 = -57.7 \text{ kcal.}$ (Δn_g is determined by the stoichiometry of the reaction that takes place.)

(b) $\Delta H_{598°K} = \Delta H_{298°K} + \Delta C_p \Delta T = -58 + (-2.25)200/1,000 = -58.45 \text{ kcal.}$

(c) The maximum temperature is that obtained by applying the ΔE of reaction to heating up the products; here $H_2O(g) + 9.5O_2(g)$; so $57,700 = C_v \Delta T = (5.5 + 9.5 \times 4.5)\Delta T = 69.2\Delta T$ or $\Delta T = 1,200°,$ $T = 1,225°C.$

6. (a) The molar heat of combustion of graphite is the same as the heat of formation of CO_2; per gram it would be $-94/12 = -7.82 \text{ kcal.}$

(b) The answer must be the same as for (a), plus the energy added by the water–gas reaction, so $-7.82 - 32/12 = $ **−10.5 kcal.**

(c) $\Delta E = \Delta H - \Delta n_g RT = 32 - RT/1{,}000 = $ **31.43 kcal.**

(d) $\Delta H_{600°K} = \Delta H_{298°K} + \Delta C_p \times 302 = 32 + 4.5 \times 302/1{,}000 = $ **33.4 kcal.**

(e) We take the bond energy for the O—H bond to be half the ΔH for the reaction $H_2O = 2H + O$. This reaction is the sum of

$$H_2O = H_2 + \tfrac{1}{2}O_2 \quad \Delta H_1 = 58 \text{ kcal}$$
$$H_2 = 2H \quad \Delta H_2 = 103 \text{ kcal}$$
$$\tfrac{1}{2}O_2 = O \quad \Delta H_3 = \tfrac{1}{2}34 = 17 \text{ kcal}$$

or $178/2 = 89$ kcal. Strictly, we want one-half of the ΔE, so subtract 0.5 kcal to get **88.5 kcal.**

7. We want C_2H_5OH on the left, so write the formation in reverse:

$$C_2H_5OH = 2C + 3H_2 + \tfrac{1}{2}O_2 \quad \Delta H_1 = 66 \text{ kcal}$$

Bring in CH_3—O—CH_3 by adding its combustion, also in reverse:

$$2CO_2 + 3H_2O = CH_3—O—CH_3 + 3O_2 \quad \Delta H_2 = 348 \text{ kcal}$$

Now add the formation of H_2O and the combustion of C:

$$2C + 2O_2 = 2CO_2 \quad \Delta H_3 = -94 \times 2 \text{ or } -188 \text{ kcal}$$
$$3H_2 + \tfrac{3}{2}O_2 = 3H_2O \quad \Delta H_4 = -3 \times 68 = -204 \text{ kcal}$$

The desired ΔH is then $66 + 348 - 188 - 204 = $ **22 kcal.**

(b) $\Delta E = \Delta H - \Delta n_g RT = -10 - RT/1{,}000 = $ **−10.6 kcal.**

8. (a) The formation reaction is $\tfrac{1}{2}H_2(g) + \tfrac{1}{2}Br_2(g) = HBr(g)$. The molecules are all diatomic and there are the same number on both sides, so $\Delta C_p = 0$. ΔH at 125°C is therefore also **−9 kcal/mole.**

(b) After the reaction, there are 2 moles of HBr and 98 of Br_2 or 100 moles of diatomic gas—the C_v of the mixture is then $100(7 - 2) = 500$. The increase in temperature is then $18{,}000 = 500\,\Delta T$ or $\Delta T = 36$ (ΔE and ΔH are both $-18{,}000$); T is $25 + 36 = $ **61°C.**

9. The heat of combustion of $(CH_2)_3$: $(CH_2)_3 + \tfrac{9}{2}O_2 = 3CO_2 + 3H_2O$ is related to its heat of formation: $\Delta H_c = 3H_{CO_2} + 3H_{H_2O} - H_{(CH_2)_3}$ (H denotes heat of formation), so $H_{(CH_2)_3} = 500.0 - 3 \times 94 - 3 \times 68 = $ **14 kcal.** (Note that the heat of combustion of carbon is also the heat of formation of CO_2; likewise, the heat of combustion of H_2 is the same as the heat of formation of water.)

For the isomerization, $\Delta H_1 = H_{CH_3CH=CH_2} - H_{(CH_2)_3} = 4.9 - 14.0 = -9.1\,\text{kcal}.$

10.   Take 100 g of solution as a basis—the process is then

$12\,\text{g NaCl} + 88\,\text{g H}_2\text{O} = 100\,\text{g solution}$
$(12/58.2 = 0.205\,\text{mole})$

Per mole of NaCl, ΔC_p is given by the change in ΔH per degree, that is, $(700.8 - 774.6)/5$ or -14.7. Per 12 g or 0.205 moles, $\Delta C_p = -3.01$. Then $-3.01 = C_p$ (solution) $- 2.5$ (C_p of NaCl) $- 88$ (C_p of H$_2$O) or, C_p (solution) $= 87$ or, per gram, **0.87 cal/deg-g.**

11.   (a) Subtracting the two reactions gives

$C_8H_8(\text{styrene}) + H_2 = C_8H_{10}\ (\text{ethylbenzene})$
$\Delta H = -74.65 - (-48.3) = -\textbf{26.4 kcal}$

(b) For the combustion of ethylcyclohexane,

$C_8H_{16} + 12O_2 = 8CO_2 + 8H_2O \qquad \Delta H_c = -1238.23\,\text{kcal}$
$\Delta H_c = 8H_{CO_2} + 8H_{H_2O} - H_{C_8H_{16}}$
$\qquad = -8 \times 94.05 - 8 \times 58.32 - H_{C_8H_{16}}$

from which $H_{C_8H_{16}} = 1238.23 - 8 \times 152.37 = 19.27\,\text{kcal}$. From the first reaction given,

$-74.65 - 48.3 = H_{C_8H_{16}} - H_{C_8H_{10}}$

or

$H_{C_8H_{10}} = 48.3 + 19.3 = \textbf{67.6 kcal}$

12.   For 0.2 mole of salt, q_p will be $0.2 \times 16{,}200$ or 3,240 cal. This heat is then abstracted from the final system, whose heat capacity will be 0.8×260 (solution) $+ 1 \times 200$ (beverage) $+ 10$ (container) $= 418\,\text{cal/deg}$. The ΔT will then be $3240/418 = 7.8$ so the final temperature is **12.2°C.**

13.   (a) We want acrylonitrile on the right, so write its combustion in reverse:

$3CO_2 + \tfrac{3}{2}H_2O + \tfrac{1}{2}N_2 = CH_2{=}CHCN + 3\tfrac{3}{4}O_2$
$\Delta H_1 = 488.67\,\text{kcal}$

C_2H_2 must be on the left, so write its formation in reverse:

$$C_2H_2 = 2C + H_2 \qquad \Delta H_2 = -54.19$$

Also,

$$HCN = \tfrac{1}{2}H_2 + C + \tfrac{1}{2}N_2 \qquad \Delta H = -31.00$$

To tidy up, add the combustion of the proper number of moles of C and H_2:

$$3C + 3O_2 = 3CO_2 \qquad \Delta H_3 = -3 \times 94.05$$
$$\tfrac{3}{2}H_2 + \tfrac{3}{4}O_2 = \tfrac{3}{2}H_2O \qquad \Delta H_4 = -\tfrac{3}{2}68.33$$

The sum gives the desired reaction, for which ΔH is then

$$488.67 - 54.19 - 282.15 - 102.50 - 31.00 = \textbf{18.83 kcal}$$

(b) Since $\Delta H = \Delta E + \Delta n_g RT$, then in this case $\Delta H - \Delta E = (-1)RT = -0.59\,\textbf{kcal.}$

14. When the dissolution reaction occurs, 800 cal are absorbed. The total heat capacity of the solution is 40 cal/deg, so there will be a 20-deg drop; that is, the final temperature will be 5°C. (The data for the salt and for water were superfluous.)

15. The combustion reaction for $CH_3COOC_2H_5$ is

$$CH_3COOC_2H_5 + 5O_2 = 4CO_2 + 4H_2O \qquad \Delta H = -536.9 \text{ kcal}$$

Then $-536.9 = 4H_{CO_2} + 4H_{H_2O} - H_{ester}$ or $H_{ester} = 536.9 - 649.2 = -112.3$ kcal. [Note that H for $H_2O(l)$ is required here.] Then ΔH for the desired reaction is

$$\Delta H = H_{ester} + H_{H_2O} - H_{CH_3COOH} - H_{C_2H_5OH}$$
$$= -112.3 - 57.8 - (-116.7) - (-66.3) = \textbf{12.9 kcal}$$
$$\Delta E = \Delta H - \Delta n_g RT = 12.9 - RT = \textbf{12.3 kcal}$$

16. The basic equation is

$$\Delta H_{1000°K} = \Delta H_{298°K} + \int_{298}^{1000} \Delta C_p \, dT$$

On combining the equations for C_p, $\Delta C = 0 + 10^{-3}\,T$. The integral then becomes $(10^{-3}/2)(1{,}000^2 - 298^2) = (10^{-3} \times 9.1 \times 10^5)/2 = 460$, so $\Delta H_{1{,}000\,°K} = -10 + 0.46 = -9.54\,\textbf{kcal.}$

17. (a) It is probably quickest to get the heat of formation of CH_4 first. The combustion reaction is $CH_4 + 2O_2 = CO_2 + 2H_2O$ and $\Delta H_c = H_{CO_2} + 2H_{H_2O} - H_{CH_4}$, whence $H_{CH_4} = 212 - 94.0 - 2 \times 68.3 = 212 - 230.6 = -18.6\,\text{kcal.}$ Then ΔH for the desired reaction is $\Delta H = 2H_{C_2H_5OH} - 3H_{CH_4} - H_{CO_2}$ or

$$\Delta H = -2 \times 66.0 - (-3 \times 18.6) - (-94.0) = \textbf{17.8 kcal}$$

(b) $\Delta E = \Delta H - \Delta n_g RT = 17.8 - (-4)RT/1{,}000 = \textbf{20.2 kcal.}$
(c) $\Delta H_{173\,°K} = \Delta H_{298\,°K} + \Delta C_p \Delta T.$ $\Delta C_p = 2 \times 32 - 3 \times 5 - 7 = 42$ cal/deg, so the desired difference is $42(-125) = \textbf{−5,250 cal.}$

18. (a) The simplest way is just to take the difference between (3) and (4); that is, the desired ΔH is $\Delta H_3 - \Delta H_4 = -20.2 - 12.6 = \textbf{−32.8 kcal.}$ (b) Reaction (1) plus reaction (a) plus the reverse of reaction (2) gives the desired heat of combustion, so $\Delta H_c = -372.8 + (-32.8) - (-68.3) = \textbf{−337.3 kcal.}$ (c) $\Delta E = \Delta H - \Delta n_g RT$ if the gases are ideal. Then $\Delta E = -32.8 - (-1)RT/1000 = \textbf{−32.2 kcal.}$

19. The reaction will use 0.35 mole O_2, leaving 0.65, and produces 0.2 mole CO_2, 0.3 mole H_2O. Neglecting the heat capacity of the calorimeter and assuming the C_v values do not change with temperature and that the reaction does go to completion, the total C_v for the mixture after reaction will be $0.65 \times 4.8 + 0.2 \times 5.7 + 0.3 \times 6.2 = 6.13$. ΔT is then given by

$$0.1 \times 262{,}000 = 6.13\Delta T \quad \text{or} \quad \Delta T = 4{,}280 \quad \text{or} \quad T = \textbf{4,300°C.}$$

20. The heat capacity, C_p, for Cl_2 should be about $(3R/2 + 2R/2 + 0.2R + 20) = 7.4$ cal/deg-mole, and for CCl_4, about $(3R/2 + 3R/2 + 0.2 \times 9 \times 2) = 11.6$ cal/deg-mole. ΔC_p for the reaction is then about $(11.6 - 2 \times 7.4 - 2) = -5.2$ cal/deg. This in turn means that ΔH should become more negative with increasing temperature, that is, **decrease.**

21. (a) Note Problem 4. $\Delta H = H_{TH}(C_2H_4) - H_{TH}(C_2H_4O) = -89.8 - (-84.1) = \textbf{−5.7 kcal.}$ (b) $\Delta E = \Delta H - \Delta n_g RT$, where $\Delta n = -1$. Then $\Delta E_{298°} = -5.7 - (-1.98 \times 298/1{,}000) = \textbf{−5.1 kcal.}$ (c) $\Delta C_p = (12 - 7 - 9) = -2$ cal/deg. Then $\Delta H_{598} = -5.7 - 2 \times 300/1{,}000 = \textbf{−6.3 kcal.}$

22. (a) $\Delta H = -94 - (-26) = -68\,\text{kcal}$. (b) Let y denote the moles of O_2 per mole of CO; this is a deficiency, so all the O_2 reacts, to produce $2y$ moles of CO_2 and leaving unreacted $1 - 2y$ moles of CO. The heat generated by this much reaction at 25°C, q_p, would be $2y(68,000)$ or $136,000y$ cal. The heat capacity of the reacted mixtures would be $2y(8) + (1 - 2y)(7)$ or $2y + 7\,\text{cal/deg}$, and the heat q_p must then also equal $(2y + 7)(100)$ or $2,000y + 700$. On solving the two statements for q_p, $y = 700/134,000 = \mathbf{5.2 \times 10^{-3}}$.

7

SECOND LAW OF THERMODYNAMICS; SOME MORE STATISTICAL MECHANICS

COMMENTS

The combined first and second law statement and the use of H, A, G, and S are examined in this chapter. In many cases, ideal gases are involved, so that the relationships you have to use are not very complicated. The main difficulty will be in seeing what to do rather than how to do it. A good deal of attention is paid to irreversible processes, and it is important to remember that S is a state function, so that ΔS does not depend on path. On the other hand, to calculate ΔS from the data, it may be necessary to make use of the relationship $\Delta S = q_{rev}/T$, which means that you will want to formulate a set of reversible steps that take you from the given initial to the given final condition.

The criteria for equilibrium must be kept in mind. Two of the more useful ones are that dS is zero for a reversible process occurring in an isolated system and that dG is zero for a reversible process at constant T and P. Also, the Carnot cycle is used a good deal, as are applications of it to heat engines and heat pumps.

Since E, A, H, G, and S are state functions, they each have a total differential, usually expressed in terms of those variables most convenient for the particular quantity. The total differentials, in turn, give rise to a number of cross-differential or Euler relationships (see Appendix), and there is now considerable scope for derivations of partial differential equations. A number of problems are of this type.

Turning next to statistical mechanics, the second law relationship between entropy and heat capacity can be combined with Eq. (5-7) to express entropy in terms of Q, the partition function. If the third law of thermodynamics holds, that is, if $S_{0°K} = 0$, then absolute entropy values can be obtained. The free energy can also be evaluated from Q, so that the entire set of thermodynamic quantities can be obtained by means of the statistical mechanical approach. Remember, though, that the applications here are of the simplest type. The ability to obtain thermodynamic quantities from Q is no panacea; the analytical evaluation of Q is usually a very tricky matter, and is often accomplished only by slipping in various distinctly simplifying assumptions.

EQUATIONS AND CONCEPTS

Definitions

$$H = E + PV \qquad A = E - TS \qquad G = H - TS \tag{7-1}$$

Combined First and Second Law Statements

$$dE = T\,dS - P\,dV \quad \text{so} \quad \left(\frac{\partial E}{\partial S}\right)_V = T \qquad \left(\frac{\partial E}{\partial V}\right)_S = -P \tag{7-2}$$

$$dH = T\,dS + V\,dP \qquad \left(\frac{\partial H}{\partial S}\right)_P = T \qquad \left(\frac{\partial H}{\partial P}\right)_S = V \tag{7-3}$$

$$dA = -S\,dT = P\,dV \qquad \left(\frac{\partial A}{\partial T}\right)_V = -S \qquad \left(\frac{\partial A}{\partial V}\right)_T = -P \tag{7-4}$$

$$dG = -S\,dT + V\,dP \qquad \left(\frac{\partial G}{\partial T}\right)_P = -S \qquad \left(\frac{\partial G}{\partial P}\right)_T = V \tag{7-5}$$

Entropy in Terms of q

$dS = dq_{rev}/T$ or $dS = C\,d\ln T$ where $C =$ heat capacity. Thus $\Delta S = \int C_p\,d\ln T$ for a constant-pressure process and $\Delta S = \int C_v\,\ln T$ for a constant volume process (both should be reversible).

Ideal Gas

$$dS = C_v\,d\ln T + R\,d\ln V \tag{7-6}$$

$$\Delta S = \int C_v\,d\ln T \quad \text{for a constant-volume process} \tag{7-7}$$

$$\Delta S = \int R\, d \ln V = R \ln \frac{V_2}{V_1} \quad \text{for a constant-temperature process} \quad (7\text{-}8)$$

$$\Delta S = \int C_v\, d \ln T + R \ln \frac{V_2}{V_1} \quad \text{for any process} \quad (7\text{-}9)$$

Note that the change need not be reversible here, since an irreversible change from V_1 and T_1 to V_2 and T_2 can always be replaced by two reversible steps consisting of an isothermal change from V_1 to V_2 followed by an isochoric change from T_1 to T_2, and the over-all ΔS is again given by the above equation. An alternative form is $\Delta S = \int C_p\, d \ln T - R \ln P_2/P_1$. If the ideal gas is monatomic, we have

$$\Delta S = \tfrac{3}{2}R \ln \frac{T_2}{T_1} + R \ln \frac{V_2}{V_1} = \tfrac{5}{2}R \ln \frac{T_2}{T_1} - R \ln \frac{P_2}{P_1} \quad (7\text{-}10)$$

Reversible Adiabat

ΔS is zero for a reversible adiabatic change. For an ideal gas it follows that

$$C_v \ln \frac{T_2}{T_1} = -R \ln \frac{V_2}{V_1} \quad (7\text{-}11)$$

$$C_p \ln \frac{T_2}{T_1} = R \ln \frac{P_2}{P_1} \quad (7\text{-}12)$$

Carnot Cycle, Heat Engines, and Heat Pumps

The two basic equations for a Carnot cycle operating between an upper temperature T_2 and a lower temperature T_1 are -

$$w = q_1 + q_2 \quad (7\text{-}13)$$

$$\frac{q_1}{T_1} + \frac{q_2}{T_2} = 0 \quad (\text{i.e.,}\ \Delta S_1 + \Delta S_2 = 0) \quad (7\text{-}14)$$

In the case of a heat engine, we want w in terms of q_2, so eliminate q_1 :

$$w = -\left(\frac{T_1}{T_2}\right)q_2 + q_2 = q_2\left(\frac{T_2 - T_1}{T_2}\right) \quad (7\text{-}15)$$

In the case of a refrigerator, we want q_1 in terms of w, so eliminate q_2 :

$$w = q_1 - \left(\frac{T_2}{T_1}\right)q_1 \quad \text{or} \quad q_1 = w\left(\frac{T_1}{T_1 - T_2}\right) \quad (7\text{-}16)$$

The quantity in parentheses is negative, but w is negative and q_1 is positive, so the signs come out all right.

Euler Relationships

If $dz = M\,dx + N\,dy$ is a total differential, then:

$$\left(\frac{\partial M}{\partial y}\right)_x = \left(\frac{\partial N}{\partial x}\right)_y \qquad (7\text{-}17)$$

Thus

$$\left(\frac{\partial T}{\partial V}\right)_S = -\left(\frac{\partial P}{\partial S}\right)_V \qquad \text{etc.} \qquad (7\text{-}18)$$

Statistical Thermodynamics

Since $dS = dq_{\text{rev}}/T$ it follows that at constant volume

$$S = S_0 + \int_0^T C_v\, d\ln T \qquad (7\text{-}19)$$

where S_0 is the entropy at $0°\text{K}$. On replacing C_v by the statistical mechanical expression of Eq. (5-7), we obtain

$$S - S_0 = R \int_0^T \frac{1}{T^3} \frac{\partial^2 \ln Q}{\partial (1/T)^2}\, dT = -R \int_0^T \frac{\partial^2 \ln Q}{\partial (1/T)^2}\, d\!\left(\frac{1}{T}\right) \qquad (7\text{-}20)$$

and integration by parts leads to

$$S - S_0 = RT\frac{\partial \ln Q}{\partial T} + R \ln Q - R \ln Q_0 \qquad (7\text{-}21)$$

or

$$S - S_0 = \frac{E}{T} + R \ln Q - R \ln Q_0 \qquad (7\text{-}22)$$

where Q_0 is the partition function at $0°\text{K}$. One concludes that S_0 and $R \ln Q_0$ are not unrelated constants, but that as a separate condition $S_0 = R \ln Q_0$, that is, that the entire entropy at $0°\text{K}$ is given by the probability of the lowest energy state. It then follows that

$$S = \frac{E}{T} + R \ln Q \tag{7-23}$$

At $0°K$, all the molecules are in the lowest possible energy state, and Q_0 is just g_0, the statistical weight of the ground state. In the case of a perfect crystalline solid, there are no alternative structures or arrangements possible at $0°K$, so that in this case $g_0 = 1$ and S_0 is therefore zero. This last amounts to a statement of the *Third Law of Thermodynamics*.

Combination of Eq. (7-23) with the definitions of A and G [Eq. (7-1)] gives

$$A = R \ln Q \tag{7-24}$$
$$G = R \ln Q + PV \tag{7-25}$$

Sackur–Tetrode Equation

A very important application of wave mechanics to statistical mechanics is the calculation of the absolute entropy of an ideal gas. Equation (5-19) gives Q for a monatomic ideal gas (i.e., translation only), but it is now necessary to add a slight complexity to make the treatment adequately rigorous. It is assumed that translational energies are independent, that is, that the particular energy that one molecule has does not affect the chance of some other molecule having some other particular energy; then for one mole of molecules

$$Q_{\text{tot}} = Q_{\text{trans}}^N \tag{7-26}$$

The idea is essentially the same as that involved in obtaining Eq. (5-5). The energy per molecule is

$$E_{\text{molec}} = kT^2 \frac{\partial \ln Q}{\partial T} \tag{7-27}$$

and the energy per mole is then

$$E_{\text{mole}} = kT^2 \frac{\partial \ln Q_{\text{tot}}}{\partial T} \tag{7-28}$$

but $\partial \ln Q_{\text{tot}} = \partial \ln Q^N = N \, \partial \ln Q$, and since $kN = R$, the final result for E is the same as given in Eq. (5-4).

In the case of translation, there is an added aspect which, while it cancels out and does not affect the evaluation of E, does affect that of S. In considering a set of N molecules, Eq. (7-26) implies N distinguishable molecules, and consequently Q_{tot} as given by Eq. (7-26) overcounts the probabilities by the number of ways of permuting N molecules—molecules which are, in fact, not distinguishable. As a consequence, the correct expression for Q_{tot} is

$$Q_{tot} = \frac{1}{N!} Q^N \tag{7-29}$$

For translation, then

$$Q_{tot} = \frac{1}{N!} \left[\left(\frac{2\pi mkT}{h^2} \right)^{3/2} V \right]^N \tag{7-30}$$

Since $N!$ is a very large number, it can be approximated by Sterling's formula, $\ln N! = N \ln N - N$, so that Eqs. (7-30) and (7-23) become

$$S_{trans} = \frac{E}{T} - k(N \ln N - N) + Nk \ln \left[\left(\frac{2\pi mkT}{h^2} \right)^{3/2} V \right] \tag{7-31}$$

As in the case of Eq. (7-28), Eq. (7-23) is used in the more correct form

$$S = \frac{E}{T} + k \ln Q_{tot} \tag{7-32}$$

In the case of an ideal gas, $E/T = 3R/2$, and Eq. (7-31) can be put in the form

$$S_{trans} = \frac{3R}{2} + R + R \ln \left[\left(\frac{2\pi mkT}{h^2} \right)^{3/2} \left(\frac{1}{N} \right) \left(\frac{RT}{P} \right) \right] \tag{7-33}$$

Further manipulation gives

$$S_{trans} = \frac{5R}{2} + R \ln \left[\left(\frac{2\pi}{h^2} \right)^{3/2} \left(\frac{1}{N^4} \right) (R^{5/2}) \right] + R \ln \left(\frac{T^{5/2} M^{3/2}}{P} \right) \tag{7-34}$$

where M is now the molecular weight. For the ideal gas at one atmosphere pressure the final result is

$$S_{trans} \text{(cal/deg-mole)} = R \ln(T^{5/2} M^{3/2}) - 2.31 \tag{7-35}$$

which is the famous Sackur–Tetrode equation [units must be watched carefully in deriving (7-35) from (7-34)].

The same derivation as the above can be applied to the case of a two-dimensional gas. However, Q is now given by Eq. (5-18), so that the expression corresponding to Eq. (7-33) is

$$S_{\text{trans}} \text{ (2-dim)} = 2R + R \ln \left[\left(\frac{2\pi mkT}{h^2} \right) \left(\frac{A}{N} \right) \right] \qquad (7\text{-}36)$$

On inserting numerical values for the various constants,

$$S_{\text{trans}} \text{ (2-dim)} = R \ln(MT\sigma) + 65.8 \text{ (cal/deg-mole)} \qquad (7\text{-}37)$$

where σ denotes the area available per molecule.

The complication of distinguishability, that is, with $N!$, appears only in the case of the translational entropy; rotational and vibrational entropies per mole are just N times those per molecule. Thus insertion of expression (5-22) for Q_{rot} and Eqs. (5-23) and (7-23) gives

$$S_{\text{rot}} = R(\ln Q_{\text{rot}} + 1) \qquad (7\text{-}38)$$

PROBLEMS

1. (10 min) A refrigerator operates at 50% of ideal efficiency, that is, the ideal work is 50% of the actual work. If it operates between 0°C and 25°C, calculate the work to freeze 1 kg of ice (heat of fusion 80 cal/g) and the heat discharged at 25°C.

2. (6 min) Show what percent T_1 is of T_2 for a heat engine whose ideal efficiency is 10%.

3. (10 min) Figure 7-1 shows a carnot cycle as often depicted in texts. The same cycle may, alternatively, be given as a plot of S versus T, as is done in Fig. 7-2.

(a) Indicate which are the corresponding steps in the two figures. Do this by labeling the arrows in Fig. 7-2 with A, B, C, and D.

(b) An ideal heat engine operating between $T_1 = 273°K$ and T_2 produces 1,000 cal of work per cycle. The entropy changes that the working fluid goes through are shown in Fig. 7-2. Calculate q_1 and q_2 for one cycle, and T_2.

4. (15 min) A homeowner is fortunate enough to possess an ideally operating refrigerator; nonetheless he feels it is using too much power and calls in a repair man, who moves the machine back from the wall to have better air circulation around the hot coils. The homeowner now finds his power consumption to be halved. Assuming that the machine operated between

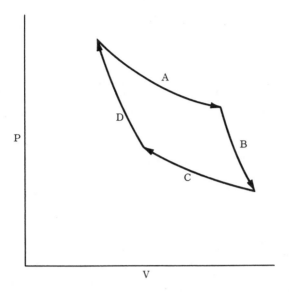

FIGURE 7-1

0°C and an upper temperature T_2, and after adjustment, operated between 0°C and 30°C, calculate (a) the watts/hour needed (after adjustment) to convert 1 kg water at 0°C to 1 kg ice at 0°C and (b) the original upper temperature T_2. (1 cal = 4.2 J or W/sec, heat of fusion of water is 80 cal/g.)

5. (13 min) A reversible heat engine absorbs heat q_2 at 900°K, per cycle, and evolves heat q_1 at 300°K. Its work output is used to run a hoist, and, owing to friction in the pulleys, 10% of w is converted into heat at 300°K.

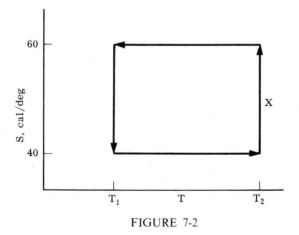

FIGURE 7-2

For the system engine plus pulleys, the *total* heat evolved is 12,000 cal per cycle. (a) Calculate q_1, q_2, and w. (b) Calculate ΔS per cycle, for the system engine plus pulleys.

6.   (12 min) A homeowner has the idea of using an extra refrigerator to cool his living room during the summer. He therefore sets up the machine in the middle of the room, leaving the refrigerator door open to get the benefit of its cooling coils. Room temperature is 25°C, and it may be assumed that the refrigerator is operating between 25°C and 0°C. The machine ordinarily is capable of freezing 1 kg of ice/h with these operating temperatures. Calculate the temperature change in the living room, that is, the new temperature, after 1 h of operation of the refrigerator. Assume ideal operation, and that the heat capacity of the room is 100 kcal/deg. The heat of fusion of water is 80 cal/g.

7.   (14 min) A farmhouse has a heat pump for heating purposes; on a particular day it is operating between 25°C and 0°C. Since no electricity is available, the heat pump is operated by a heat engine (e.g., a gasoline motor) operating between 1,000°C and 25°C. Both machines operate ideally (fortunate farmer!).

Calculate the performance factor for the system, that is, the ratio of the amount of heat delivered to the house at 25°C to the amount of heat produced at 1,000°C by the burning gasoline. The motor is located outside the house.

8.   (12 min) A heat engine operating between 1,000°C and 25°C produces work that is entirely used to run a refrigerating machine, operating between 0°C and 25°C. Calculate the ratio of the heat absorbed by the engine to that absorbed by the refrigerator, assuming ideal operation for both.

9.   (12 min) An ideally operating refrigerator works between 0°C and T°C. It will freeze 1 kg of ice per hour (that is, remove 80 kcal of heat per hour from water at 0°C). At the same time, the total heat output of the refrigerator to the room is 100 kcal/h. Calculate the value of T°C.

---

10.   (12 min) Calculate (or give with explanation) ΔH and ΔS when a 1-kg bar of copper at 100°C is placed in 2 kg of water at 0°C in an insulated container maintained at 1 atm pressure. Heat capacities: Cu, 0.1 cal/deg-g; $H_2O(l)$, 1 cal/deg-g.

11.   (7 min) Calculate ΔG for the process of Problem 2. Actually there is insufficient information, so explain what additional information is needed, and set up the equations to show clearly how you would make the calculations.

12.   (12 min) One mole of an ideal monatomic gas is taken from the state (22.4 liter, 273°K, $S = 20$ cal/deg) to the state (2 atm, 303°K). Calculate ΔE, ΔH, ΔS, and ΔG for this change.

13. (10 min) Give a process for which (a) $\Delta E = 0$, (b) $\Delta H = 0$, (c) $\Delta A = 0$, (d) $\Delta G = 0$, and (e) $\Delta S = 0$. State all necessary conditions or restrictions clearly.

14. (10 min) One mole of an ideal, monatomic gas initially at STP expands isothermally and *irreversibly* to 44.8 liter, under conditions such that $w = 100$ cal. Calculate ΔS and ΔG.

15. (20 min) One mole of a perfect monatomic gas initially at volume $V_1 = 5$ liter, pressure P_1, and temperature $T_1 = 298°$K experiences the following reversible changes: (A) Isothermal compression to one half the volume, the new volume and pressure being $V_2 = \frac{1}{2}V_1$, and P_2. (B) Cooling at constant volume, until the pressure is returned to the original value of P_1, the final temperature being T_2. These changes are shown schematically in Fig. 7-3. Notice that process C, reduction in volume at constant pressure P_1, is equivalent to the sum of steps A and B.

(a) Calculate P_1, P_2, and T_2. Also q, w, ΔE, ΔH, ΔS, and ΔG for A and B separately.

(b) Are the magnitudes (without regard for sign) of ΔE, q, and w for step C greater than, less than, or equal to the values of these quantities for the sum of steps A and B?

16. (12 min) One mole of an ideal monatomic gas initially at 10 atm pressure and 0°C is allowed to expand against a constant external pressure of 1.0 atm. Conditions are such that the final volume is 10 times the initial volume; the final gas pressure equals the external pressure.

(a) Calculate the initial and final volume and the final temperature.

(b) Calculate q, w, ΔE, ΔH, ΔS, and ΔG for the process.

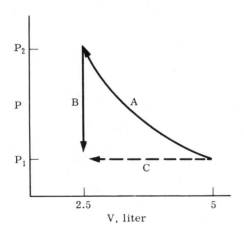

FIGURE 7-3

17. (19 min) One mole of an ideal monatomic gas is contained in a piston and cylinder at 300° K and 10 atm. The pressure is suddenly released to 2 atm, so that the gas expands adiabatically at a constant pressure of 2 atm. (a) Explain whether ΔS for this expansion should be positive, negative, or zero. (b) Calculate ΔE and ΔS. To speed your calculation, 288 cal of work are done by the gas in this expansion.

18. (3 min) An *isolated* system comprises 1 mole of ideal gas, a heat reservoir at T_1, a machine, and a source of energy for the machine. The gas, at T_1, first expands freely to twice its initial volume; it is then compressed by the machine (at T_1) back to the initial volume; the gas again is allowed to expand freely to twice the initial volume. What is the *minimum* ΔS for the system at the end of these three steps?

19. (18 min) One mole of He is heated from 200°C to 400°C at a constant pressure of 1 atm. Given that the absolute entropy of He at 200°C is 32.3 cal/deg-mole, and assuming that He is a perfect gas, calculate ΔG, ΔH, and ΔS for this process: ΔG comes out negative; does this mean that the process is spontaneous? Explain.

20. (16 min) One mole of an ideal monatomic gas undergoes an irreversible adiabatic process in which the gas ends up at STP and for which ΔS is 5 cal/deg and w is 300 cal. The third law entropy for the gas at STP is 45 cal/deg-mole. Calculate ΔE and ΔG for the process and also the initial state of the gas (its initial P, V, and T).

21. (12 min) An ideal gas expands isothermally at 25°C, but somewhat irreversibly, producing 1,000 cal of work. The entropy change is 10 cal/deg. Calculate the degree of irreversibility, that is, the ratio w_{actual}/w_{rev} where w_{rev} is the reversible work for the isothermal expansion of the same gas to the same final volume.

22. (12 min) One mole of an ideal monatomic gas initially at STP undergoes an irreversible isothermal expansion to 0.5 atm pressure, thereby doing 100 cal of work. Calculate or explain the value of ΔE, ΔH, ΔG, q, and ΔS. Calculate also the reversible work for this isothermal change of state.

23. (12 min) It has been possible to supercool small drops of water to -40°C. Such drops are unstable, of course, and eventually nucleation occurs and ice crystals form. Assume that such a drop is thermally isolated so that, when spontaneous ice formation occurs, it does so adiabatically (and isobarically). C_p is 1 cal/deg for water (l) and 0.5 cal/deg-g for water (s), and the heat of fusion of water is 80 cal/g at 0°C. Calculate the final temperature of the drop after the spontaneous process has occurred, and ΔH and ΔS for the process, in cal/g.

24. (9 min) Calculate ΔH and ΔG for the process

$$CH_3OH(l, 64°C, 1\ atm) = CH_3OH(g, 64°C, 0.5\ atm)$$

(The normal boiling point of CH_3OH is 64°C, and the heat of vaporization is 260 cal/g.)

25. (18 min) One gram of water enclosed in a vial is placed in an evacuated flask maintained at 25°C. By means of a lever, the vial is broken so that the water is free to vaporize and, when equilibrium is reached, one-half the water has vaporized. The vapor pressure of water at 25°C is 24 mm Hg, and the enthalpy of vaporization is 590 cal/g. Calculate q, w, ΔH, ΔG, and ΔS for this process.

26. (15 min) One mole of liquid water is allowed to expand into an evacuated flask of volume such that the final pressure is 0.3 atm. The bulb containing the liquid and the flask are thermostated so that a constant temperature of 100°C is maintained. It is found that 11,000 cal of heat are absorbed when this process occurs. Calculate or explain what the values are for w, ΔE, ΔH, ΔS, and ΔG. Neglect PV quantities for liquid water.

27. (19 min) One mole of water vapor at 100°C is compressed to 2 atm. It is sufficiently dust free so that it supersaturates; after a while, however, condensation occurs and goes to completion, since T and P are kept constant. The process is thus $H_2O(g, 100°C, 2\ atm) = H_2O(l, 100°C, 2\ atm)$. Data: heat capacities: C_p for the vapor is 7 cal/deg-mole, and is 18 cal/deg-mole for the liquid. Heat of vaporization under these conditions: 11 kcal/mole. It is permissible to assume the vapor to be ideal and liquid water to be incompressible at a molar volume of 18 cc. Calculate ΔH, ΔG, and ΔS for the process.

28. (18 min) 100 g of ice at 0°C is dropped into an insulated beaker containing 150 g of water at 100°C. Calculate ΔS for the process that then occurs. (You may take the heat of fusion of ice to be 80 cal/g, and the heat capacities of water and of ice to be 1 and 0.5 cal/deg-g, respectively.)

---

29. (12 min) (a) Derive $(\partial S/\partial V)_E = P/T$. (*Hint:* Make use of the total differential for E expressed as a function of S and V.) (b) Verify the above equation for an ideal gas; that is, evaluate $(\partial S/\partial V)_E$ directly.

30. (9 min) Derive from the first and second laws of thermodynamics and related definitions: $(\partial T/\partial P)_S = (\partial V/\partial S)_P$.

31. (7 min) Show that $(\partial H/\partial P)_S = V$, using the first and second laws of thermodynamics and related definitions.

32. (10 min) Derive from the first and second laws of thermodynamics and related definitions: $\partial(A/T)/\partial T = -E/T^2$.

33. (7 min) Derive from the first and second laws and related definitions: $(\partial A/\partial V)_T = -P$.

34. (15 min) Derive from the first and second laws and related definitions: $(\partial V/\partial T)_P = -(\partial S/\partial P)_T$.

---

35. (12 min) Given the process:

$$0.20 \text{ mole } O_2 \text{ (g, 0.2 atm)} + 0.8 \text{ mole } N_2 \text{ (g, 0.8 atm)} = \text{mixture (g)}$$

all at 25°C, which is carried out by having the oxygen and nitrogen initially in separate flasks, and then by opening the stopcock connecting the two flasks. (a) Calculate the final P. (b) Calculate q, w, ΔE, ΔS, and ΔG for the process. (c) Calculate the **reversible** q and w for the isothermal process returning the mixture to its initial state.

36. (12 min) The element krypton exists in nature as a mixture of about 10% Kr^{82}, 70% Kr^{84}, and 20% Kr^{86}. Calculate ΔS and ΔG at 25°C for the process:

$$1 \text{ mole Kr} \begin{bmatrix} \text{natural mixture,} \\ 1 \text{ atm, 25°C} \end{bmatrix}$$

$$= 0.1 \text{ mole } Kr^{82}(1 \text{ atm, 25°C}) + \text{mixture of} \begin{pmatrix} 0.7 \text{ mole } Kr^{84} \\ 0.2 \text{ mole } Kr^{86} \\ \text{at 1 atm, 25°C} \end{pmatrix}$$

---

37. (12 min) (a) Explain why, according to the Sackur–Tetrode equation, the entropy of an ideal gas depends on the molecular weight of the gas. It is not sufficient to copy a derivation! What is wanted is an explanation of what is going on physically that brings about this effect. (b) Explain whether the Sackur–Tetrode equation does or does not predict that the heat capacity of a gas should depend on its molecular weight. Here, the best approach will be through a derivation rather than physical arguments.

38. (10 min) The rotational partition function for a diatomic molecule (ignoring a nuclear spin factor) is $8\pi^2 IkT/h^2$, where I is the moment of inertia. Calculate the difference in rotational entropies per mole for diatomic molecules AA and BB. Both gases are at 25°C, but AA has twice the molecular weight and twice the moment of inertia of BB.

39. (10 min) Show how the entropy of an ideal monatomic gas should vary with temperature, according to the Sackur–Tetrode equation. Obtain the same result by another means.

40. (10 min) Give a qualitative, but definite explanation of whether the rotational entropy for CH_4 gas at STP should be greater, the same as, or less than the translational entropy.

41. (13 min) For a diatomic molecule, $Q_{rot} = 8\pi^2 IkT/h^2$ (as noted in Problem 38). (a) Estimate (to within about 30%) the value of Q_{rot}/T for H_2. (b) Estimate the temperature at which S_{rot} might come out negative on application of Eq. (7-23). (c) Why is the situation described in (b) nonsense and how does such a situation appear to be possible?

42. (14 min) A vapor adsorbed on a liquid surface can behave very much like an ideal two-dimensional gas. Equation (7-37) gives the entropy for such a gas in terms of σ, the area available per molecule. Derive the expression for the translational entropy of a two-dimensional gas in a standard state defined as one in which the molecules are the same average distance apart as for an ideal three-dimensional gas at STP.

43. (18 min) An alternative and somewhat more general form of Eq. (5-4) is

$$E^0 = E_0^0 + \tfrac{3}{2}RT + RT^2\, d\ln Q_{int}/dT$$

where E^0 is the energy in a chosen standard state, E_0^0, the energy of the lowest state (zero-point energy). The equipartition value for translational energy is assumed, and Q_{rot} and Q_{vib} have been lumped together as $Q_{int} = Q_{rot}Q_{vib}$. Given the above approach, derive with the use of other equations of this chapter the expression

$$(G^0 - E_0^0)/T = -\tfrac{3}{2}\ln M - \tfrac{5}{2}\ln T - R\ln Q_{int} + 7.27$$

for an ideal gas at 1 atm pressure. [The quantity $(G^0 - E_0^0)/T$ is known as the free energy function.] (This may look a bit forbidding. As a hint, a good place to start is with the defining equations for G and H.)

44. (6 min) Q_{vib} and its contribution to the heat capacity of gases was fairly well emphasized in Chapter 5, yet has hardly been mentioned in the present chapter. Why might this be? (A serious answer, please.)

45. (10 min) Derive Eq. (7-38). (*Hint:* focus on the temperature dependence of Q_{rot}.)

46. Suppose that the molecules of a certain substance can exist either in a ground energy state (of energy taken to be zero) or an excited state of energy

ε; there are no other excited states. Write the expression for the partition function. Calculate the average energy per mole and the entropy for the case of infinite temperature.

ANSWERS

1. Here, we want w in terms of q_1: $w = q_1 - (T_2/T_1)q_1 = -80,000 \times 25/273 = -7,350$ cal. The actual work will be twice this: $-$**14,700 cal.** The heat discharged at 25°C will be 80,000 plus 14,700 = **94,700.**

2. We want the equation $w = q_2 - q_2T_1/T_2$. If $w/q_2 = 0.1$, then $(1 - T_1/T_2) = 0.1$, or $T_1/T_2 = 0.9$ so T_1 is 90% of T_2.

3. Step (a) is isothermal and at the higher temperature, so it must correspond to the step marked X in Fig. 7-2; the other labels follow in counterclockwise order.

(b) For the cycle $q_1/T_1 + q_2/T_2 = 0 = \Delta S_1 + \Delta S_2$. From Fig. 7-2, $\Delta S_1 = -20$ and $\Delta S_2 = 20$, so $q_1 = -20\,T_1$ and $q_2 = 20\,T_2$. The work, $w = q_1 + q_2 = 1,000$ cal, so $-20 \times 273 + 20T_2 = 1,000$ and $T_2 = $ **323°K;** $q_2 = -$**6,460 cal.**

4. (a) Use the equation $w = q_1 - (T_2/T_1)q_1 = 80,000 \times (1 - 303/273) = -8,800$ cal (per h). In J/sec, $8,800 \times 4.2/3,600 = $ **10.1 W.**

(b) The original work must have been $-17,600$ cal for the same job; hence $1 - T_2/T_1 = -17,600/80,000 = -0.22$ so $T_2/T_1 = 1.22$ and $T_2 = 333$°K or 60°C (inspection of the formula for w would have told you that to double w, the temperature difference must be doubled, and hence be 60° instead of 30°).

5. (a) First, $0.1w + |q_1| = 12,000$ cal. Also, $w = q_1 - (T_2/T_1)q_1 = (600/300)|q_1| = 2|q_1|$. (It is easier to use the magnitude of q_1 and argue intuitively that it is then to be added to $0.1w$.) Then $0.2|q_1| + |q_1| = 12,000$ and $|q_1| = 10,000$ or $q_1 = -$**10,000 cal.** Next, $q_2 = -(T_2/T_1)q_1 = -(900/300)q_1 = $ **30,000 cal.** Finally, $w = q_1 + q_2 = $ **20,000 cal.**

(b) Since the engine is reversible, $\Delta S = 0$, per cycle, for it. Its surroundings also receive the frictional heat, which is 2,000 cal from part (a), at 300°K, which produces an entropy of $2,000/300 = 6.67$ cal/deg per cycle.

6. Here $w = q_1 - (T_2/T_1)q_1 = -80,000 \times 25/273 = 7,350$ cal (per h). ($q_1 = 80,000$ and $q_2 = -87,350$ cal.) The homeowner is evidently in for a surprise, as the net effect is that the work of running the refrigerator appears as heat! The room heats up by 7,350 cal/h or $7,350/10^5 = 0.0735$ deg/h.

7. For the heat pump, $w = q_1 + q_2 = q_2(1 - T_1/T_2)$, while for the heat engine, $w' = q_1' + q_2' = q_2'(1 - T_1'/T_2')$. Since $w' = -w$, it follows that $-q_2(1 - T_1/T_2) = q_2'(1 - T_1'/T_2')$, and hence that the desired ratio of $-q_2/q_2' = (T_2' - T_1')T_2/(T_2 - T_1)/T_2' = 975 \times 298/25 \times 1273 = $ **9.1.**

8. For the engine, $w = q_1 + q_2 = q_2(T_2 - T_1)/T_2$. For the refrigerator, $w' = q_1' + q_2' = q_1'(T_1' - T_2')/T_1'$. Since $w = w'$, it follows that $q_2(1,000 - 25)/1,273 = q_1'(25 - 0)/273$ and hence that $q_2/q_1' = (25/273)(1,273/975) = $ **0.12.**

9. Since $q_1/T_1 + q_2/T_2 = 0$, it follows that $T_2 = -(q_2/q_1)T_1 = (100/80)273 = 341°K$ or **68°C.**

10. Since the system is adiabatic at constant pressure, $\Delta H = 0$. ΔS, however, is the sum of the entropy changes in the copper and the water. First, the final temperature is given by $1,000 \times 0.1(100 - t) = 2,000 \times 1 \times t$, or $t = 4.77°C$. For the copper,

$$\Delta S = C_p \ln T_2/T_1 = 1,000 \times 0.1 \times 2.3 \log 277.9/373$$
$$= -28.7 \, \text{cal/deg}.$$

For the water,

$$\Delta S = 2,000 \ln(277.9/273.1) = 2,000 \ln(1 + 0.018)$$
$$\cong 2,000 \times 0.018 = 36 \, \text{cal/deg}.$$

The net ΔS is then $-28.7 + 36 = $ **7 cal/deg.**

11. Since the initial and final states are not at the same temperature, ΔG must be assembled as follows. For the copper, $\Delta G = \Delta H_{Cu} - (T_2 S_2 - T_1 S_1)_{Cu}$; for the water, $\Delta G = \Delta H_{H_2O} - (T_2 S_2 - T_1 S_1)_{H_2O}$. ΔG net is the sum (the ΔH's will cancel). The additional information needed is then the absolute entropy of Cu and H_2O at some particular temperature, from which the values at the desired temperatures can be calculated using the given heat capacities, that is,

$$S_2 = S_1 + \int_{T_1}^{T_2} C_p d \ln T$$

12. First, complete the identification of initial and final states:

Initial: $V = 22.4$ liter   Final: $V = 12.4$ liter
          $T = 273°K$ $T = 303°K$
So $P = 1$ atm $P = 2$ atm
          $S = 20$ cal/deg-mole

Since the gas is ideal, $\Delta E = C_v \Delta T = (3R(2) \times 30 = \mathbf{90\,cal};$ $\Delta H = C_p \Delta T = (5R/2) \times 30 = \mathbf{150\,cal}.$

For ΔS, the general equation could be used, but since ΔG also is wanted, it is better to write out explicitly the two steps:

(1) 22.4 liter, 273°K, 1 atm = 24.8 liter, 303°K, 1 atm

$$\Delta S_1 = C_p \ln T_2/T_1 = (5R/2) \ln 1.11 = 0.53\,\text{cal/deg (so } S_{\text{final}} = 20.53)$$

$$\Delta G_1 = \Delta H_1 - (T_2 S_2 - T_1 S_1) = 150 - (303 \times 20.53 - 273 \times 20)$$
$$= -600\,\text{cal}$$

(2) 24.8 liter, 303°K, 1 atm = 12.4 liter, 303°K, 2 atm

$$\Delta S_2 = R \ln P_1/P_2 = 1.98 \times 2.3 \log \tfrac{1}{2} = -1.37\,\text{cal/deg}$$
$$\Delta G_2 = RT \ln P_2/P_1 = 417\,\text{cal}$$

Over-all, then, $\Delta S = 0.53 + (-1.37) = \mathbf{-0.84\,cal/deg},$ $\Delta G = -600 + 417 = \mathbf{-183\,cal}.$

13. (a) $\Delta E = 0$ for any isochoric adiabatic process $(q_v = 0)$. (b) Similarly, $\Delta H = 0$ for any isobaric adiabatic process $(q_p = 0)$. (c) Since $dA = -S\,dT - P\,dV$, it will be zero for any reversible constant T and V process. (d) Since $dG = -S\,dT + V\,dP$, ΔG will be zero for any reversible constant T and P process. (e) Since $dS = q_{\text{rev}}/T$, ΔS will be zero for any reversible adiabatic process.

14. (Since the gas is ideal and the process is isothermal, $\Delta E = 0$, so $q = w = 100$ cal.) Actually, all that is needed is the knowledge that the final state is 273°K and 44.8 liter (so $P = 0.5$ atm). Hence $\Delta S = R \ln V_2/V_1 = 1.98 \times 2.3 \log 2 = \mathbf{1.37\,cal/deg}$, and $\Delta G = RT \times \ln P_2/P_1 = 1.98 \times 273 \times 2.3 \log 0.5 = \mathbf{-374\,cal}.$

15. (a) From the ideal gas law, $P_1 = RT_1/V_1 = 4.89$ atm. After the isothermal compression to half the volume, pressure must be doubled, so $P_2 = 9.78$ atm. To halve this pressure by cooling, $T_2 = T_1/2 = 149°K$.

Step A: Since it is isothermal, $\Delta E = \Delta H = \mathbf{0}.$ $q = w = RT \times \ln V_2/V_1 = 1.98 \times 298 \times 2.3 \log \tfrac{1}{2} = \mathbf{-408\,cal}.$ $\Delta S = R \ln V_2/V_1 = \mathbf{-1.36\,cal/deg}$ and $\Delta G = RT \ln P_2/P_1 = \mathbf{408\,cal}.$

Step B: Since it is isochoric, $w = \mathbf{0}.$ $q = \Delta E = C_v \Delta T = (3R/2) \times (-149) = \mathbf{-442\,cal}.$ $\Delta H = C_p \Delta T = (5R/2)(-149) = \mathbf{-740\,cal}.$ $\Delta S = C_v \ln T_2/T_1 = (3R/2)2.31 \log \tfrac{1}{2} = \mathbf{-2.06\,cal/deg}.$ Since $\Delta G = \Delta H - (T_2 S_2 - T_1 S_1)$, it cannot be obtained without a knowledge of the absolute entropies and these are not given.

(b) ΔE is independent of path and must be the same. Since the area under step C in the P–V plot of Fig. 7-3 is less than under steps A plus B, w must be smaller in magnitude for step C. Since $\Delta E = q - w$, q must likewise be smaller.

16. (a) V_1 is 2.24 liters; hence V_2 is 22.4 liters. Since the final pressure equals the external pressure of 1 atm, the final temperature must be 0°C. The over-all process is then isothermal; hence $\Delta E = \Delta H = 0$.

(b) The work done is $P \, \Delta V = 1 \text{ atm} \times (22.4 - 2.24) = 20.2$ liter-atm or **493 cal.** Since ΔE is zero, $q = w = $ **493 cal.** $\Delta S = R \ln V_2/V_1 = 1.98 \times 2.3 \log 10 = $ **4.55 cal/deg** and $\Delta G = RT \ln P_2/P_1 \, (P = P_{int}) = -1.98 \times 273 \times 2.3 \log 0.1 = -$**1,240 cal.**

17. (a) This is an irreversible process for which $q = 0$; therefore $\Delta S > 0$.

(b) Since $q = 0$, $\Delta E = C_v \Delta T = -w$ or $(3R/2) \times \Delta T = -288$ and $\Delta T = -97°$, so $T_2 = 203°$K, and therefore $V_2 = 8.32$ liters. Since $\Delta S = C_p \ln T_2/T_1 - R \ln P_2/P_1$, $\Delta S = (5R/2)2.3 \times \log(203/300) - 1.98 \times 2.3 \log 2/10 = -1.92 + 3.19 = $ **1.27 cal/deg.**

18. Consider first the gas only: step 1 plus step 2 returns it to the initial state, so $\Delta S_1 = 0$. Step 3 (as does step 1) involves $\Delta S = R \ln 2 = 1.36$ cal/deg. For the rest of the system, in step 2, $w = RT \ln P_2/P_1 = 1.98 \times T_1 \times 2.3 \log 2 = 1.36 \, T_1 = q$, since the gas suffers no energy change. Therefore $1.36 T_1$ cal are delivered to the heat reservoir, and it gains $1.36(T_1/T_1)$ entropy or 1.36 cal/deg. The minimum total ΔS is then 2×1.36 cal/deg.

19. $\Delta S = C_p \ln T_2/T_1 = (5R/2) \times 2.3 \log(673/473) = 1.75$ cal/deg. Then $\Delta S_2 = 32.3 + 0.8 = 34.1$ cal/deg; $\Delta H = C_p \Delta T = (5R/2)200 = 990$ cal. Then $\Delta G = \Delta H - (T_2 S_2 - T_1 S_1) = 990 - (673 \times 34.1 - 473 \times 32.3) = -6,560$ cal. The process need not be spontaneous; if ΔG is negative for a process at constant T and P, then it is spontaneous.

20. By the first law, $\Delta E = 0 - w = -300$; hence $\Delta T = -300/(3R/2) = -101°$, and $T_1 = 273 + 101 = $ **374°K.** By the second law equation $\Delta S = C_v \ln(T_2/T_1) + R \ln(V_2/V_1)$, which applies to any change of state, reversible or not, $R \ln(V_2/V_1) = 5 - (3R/2) \ln(273/374) = 5.94$. Then $\log(V_2/V_1) = 1.30$, and $V_1 = 22.4/20 = $ **1.12 liter.** P_1 is then $0.082 \times 374/1.12 = $ **27.3 atm.**

$\Delta G = \Delta H - \Delta(TS)$, with $\Delta H = -101(5R/2) = -500$. Also, $S_1 = 45 - 5 = 40$, so $\Delta G = -500 - (273 \times 45 - 374 \times 40) = $ **2,170 cal.**

21. In this case, the amount of gas is not specified, so $\Delta S = nR \ln(V_2/V_1)$, where n is the number of moles. w_{rev} is just $nRT \ln(V_2/V_1)$ or $w_{rev} = T\Delta S = 2,980$. $w_{actual}/w_{rev} = 1,000/2,980 = $ **0.34.**

22. The process is isothermal, so $\Delta E = 0$ and $\Delta H = 0$. Also, then $q = w = $ **100 cal.** ΔS is given by $-R \ln P_2/P_1 = -R \ln 0.5/1 = $ **1.37 cal/deg;** this is also q_{rev}/T and w_{rev} is therefore $273 \times 1.37 = $ **373 cal.** $\Delta G = \Delta H - T\Delta S = 0 - 373 = -$**373 cal.**

23. It is first necessary to decide whether the drop ends up entirely as ice (at $T \leq 0°C$) or whether it only partly freezes (and so is at $0°C$). A little reflection suggests that the 80 cal/g liberated if it all froze should be more than enough to raise the temperature to $0°C$, and that the drop will therefore not entirely solidify. To check this, set up the two reversible steps and work out the heat balance (clearly q_p and also ΔH are zero here):

1 g (1) $H_2O(l, -40°C) = H_2O(l, 0°C)$ $\qquad \Delta H_1 = 40 \text{ cal/g}$

(2) $xH_2O(l, 0°C) = xH_2O(s, 0°C)$ $\qquad \Delta H_2 = -80x$

Since $\Delta H_1 + \Delta H_2 = 0$, x must be 0.5 g.

The fact that x is a physically possible answer confirms the guess as to which happens. Then $\Delta S_1 = C_p \ln T_2/T_1 = 1 \times 2.3 \log(273/233) = 0.159$ cal/deg. $\Delta S_2 = q/T = -40/273 = -0.146$. ΔS net is then **0.013 cal/deg.**

24. Set up a reversible path as follows:

(1) $CH_3OH(l, 64°C, 1 \text{ atm}) = CH_3OH(g, 64°C, 1 \text{ atm})$

$\Delta H_1 = 260 \times 32 = 8,310 \text{ cal/mole}$

$\Delta G_1 = 0$

(2) $CH_3OH(g, 64°C, 1 \text{ atm}) = CH_3OH(g, 64°C, 0.5 \text{ atm})$

$\Delta H_2 = 0$

$\Delta G_2 = RT \ln P_2/P_1 = 1.98 \times 337 \times 2.3 \log \frac{1}{2} = -460 \text{ cal}$

Over-all then, $\Delta H = \mathbf{8,310 \text{ cal/mole}}$, $\Delta G = \mathbf{-460 \text{ cal/mole}}$.

25. The process is evidently

$$\tfrac{1}{2} g \ H_2O[l, 25°C \text{ (and 24 mm Hg)}] = \tfrac{1}{2} g \ H_2O(g, 25°C, 24 \text{ mm Hg})$$

Then $q = 590/2 = 295$ cal, $w = 0$ (no external work is done). We assume q is q_v, so $\Delta E = 295$ cal. The corresponding reversible process would simply involve placing the vial in a piston and cylinder at $25°C$, and slowly expanding until half the liquid vaporized, so that for this process $q = q_p = q_v + P\,\Delta V = 295 + (1/36)$ moles $\times RT = 312$ cal (neglecting the volume of liquid). ΔH is then 312 cal. 312 is the q_{rev} so $\Delta S = 312/298 = \mathbf{1.05 \text{ cal/deg.}}$ ΔG is zero since we have a reversible constant T and P process. Answers based on $\Delta H = 295$ cal were equally accepted, in which case q would be 278 cal.

26. No external work is done, so $w = 0$ and $q = q_v = \Delta E = \mathbf{11,000 \text{ cal.}}$ $\Delta H = \Delta E + \Delta(PV) = 11,000 + P_g V_g = 11,000 + RT = \mathbf{11,740 \text{ cal.}}$ To get the other quantities, set up the reversible steps:

(1) $H_2O(l, 100°C) = H_2O(g, 100°C, 1 \text{ atm})$

$\Delta S_1 = q/T = 11,740/373 = 31.5 \text{ cal/deg}$

$\Delta G_1 = 0$

(2) $H_2O(g, 100°C, 1 \text{ atm}) = H_2O(g, 100°C, 0.3 \text{ atm})$

$\Delta S_2 = R \ln V_2/V_1 = 1.98 \times 2.31 \log(1/0.3) = 2.37 \text{ cal/deg}$

$\Delta G_2 = RT \ln P_2/P_1 = -886 \text{ cal}$

Then ΔS total is **33.9 cal/deg** and ΔG total is **−886 cal.**

27.   The process is at constant P, so $q = q_p = \Delta H = -11,000$ **cal.** Now set up equivalent reversible steps:

(1) $H_2O(g, 100°C, 2 \text{ atm}) = H_2O(g, 100°C, 1 \text{ atm})$

$\Delta H_1 = 0$

$\Delta S_1 = R \ln V_2/V_1 = 1.37 \text{ cal/deg}$

$\Delta G_1 = RT \ln P_2/P_1 = -509 \text{ cal}$

(2) $H_2O(g, 100°C, 1 \text{ atm}) = H_2O(l, 100°C, 1 \text{ atm})$

(3) $H_2O(l, 100°C, 1 \text{ atm}) = H_2O(l, 100°C, 2 \text{ atm})$

Since the liquid is incompressible and V_L is small, ΔH_3, ΔS_3, and ΔG_3 are small and may be neglected.

For process (2), then, ΔH_2 must be $-11,000 = q_2$; hence $\Delta S_2 = -11,000/373 = -29.5 \text{ cal/deg}$. ΔG_2 is zero, as we have a constant T and P reversible process. Over-all, then, $\Delta H = -11,000$ **cal,** $\Delta S = -28.1$ **cal/deg,** and $\Delta G = -509$ **cal.**

28.   It is first necessary to set up a heat balance to determine the final temperature. Preliminary inspection suggests that all of the ice will melt, so the over-all process can be written as the sum of the following three:

$$100 \text{ g } (s, 0°C = 100 \text{ g } (l, 0°C)   \qquad q_1 = 100 \times 80$$
$$100 \text{ g } (l, 0°C) = 100 \text{ g } (l, t°C)   \qquad q_2 = 100t$$
$$150 \text{ g } (l, 100°C) = 150 \text{ g } (l, t°C)   \qquad q_3 = -150(100 - t)$$

Since $q_{net} = 0$, it then follows that $8,000 + 100t = 15,000 - 150t$, or $t = 28°C$ or $301°K$. The corresponding entropy changes are $\Delta S_1 = 8,000/273 = 29.3$, $\Delta S_2 = C_v \ln(T_2/T_1) = 100 \ln(301/273) = 9.7$, $\Delta S_3 = 150 \ln(301/373) = -32.1$. The over-all ΔS is then $29.3 + 9.7 - 32.1 = $ **6.9 cal/deg.**

29.   (a) Following instructions (i.e., the hint), write $dE = (\partial E/\partial S)_V \, ds +$

$(\partial E/\partial V)_S \, dV$. The desired partial derivative follows, on dividing by dV at constant E:

$$0 = \left(\frac{\partial E}{\partial S}\right)_V \left(\frac{\partial S}{\partial V}\right)_E + \left(\frac{\partial E}{\partial V}\right)_S$$

or

$$\left(\frac{\partial S}{\partial V}\right)_E = -\frac{(\partial E/\partial V)_S}{(\partial E/\partial S)_V}$$

The two partial derivatives can be evaluated by comparing with the first law equation $dE = T \, dS - P \, dV$, that is, $(\partial E/\partial S)_V = T$ and $(\partial E/\partial V)_S = -P$. Substitution then yields the equation to be derived.

(b) For an ideal gas, E is a function of T only, so the partial derivative becomes $(\partial S/\partial V)_T$. Since $dS = R \, d \ln V = R \, dV/V$ for an isothermal process with an ideal gas, $dS/dV = (\partial S/\partial V)_T = R/V = P/T$.   Q.E.D.

30.   We have a derivative by P with S constant equal to one by S with P constant, so this is clearly a Euler or cross-differentiation relationship. We look for something equal to $T \, dS + V \, dP$, and a little reflection yields $dH = T \, dS + V \, dP$ (from $dE = T \, dS - P \, dV$ and $H = E + PV$).

31.   These types of relationships generally come from one of the versions of the combined first and second laws. Try $dH = T \, dS + V \, dP$; dividing by dP with S constant yields the desired partial derivative.

32.   It is necessary to recognize that the partial derivative must be at constant V: $\partial(A/T)/\partial T = (1/T)(\partial A/\partial T)_V - A/T^2$ but $dA = -S \, dT - P \, dV$ so $(\partial A/\partial T)_V = -S$ and $\partial(A/T)/\partial T = -S/T - A/T^2 = -(1/T^2)(TS + A) = -E/T^2$.

33.   Try various versions of the combined first and second laws. In fact, the one for dA suggests itself: $dA = -S \, dT, - P \, dV$ (from $dE = T \, dS - P \, dV$ and $A = E - TS$). Division by dV with T constant yields the desired partial differential.

34.   We have a differential by T with P constant equal to one by P with T constant, which suggests a Euler or cross-differential relationship. We therefore look for a relationship involving $V \, dP - S \, dT$ and a little reflection yields $dG = -S \, dT + V \, dP$ (from $dE = T \, dS - P \, dV$, and $G = H - TS$, $H = E + PV$).

35.   (a) The volume of 0.2 mole at 0.2 atm is the same as that of 0.8 mole at 0.8 atm, so the two flasks must have been equal in volume. We end up with 1 mole of gas in twice the volume of one flask, so the final pressure must be 0.5 atm.

(b) The final partial pressures must be 0.4 for N_2 and 0.1 for O_2. Then for the O_2, $\Delta S = 0.2R \ln V_2/V_1$, and for the N_2, $\Delta S = 0.8R \ln V_2/V_1$ or ΔS total is $R \ln V_2/V_1 = 1.98 \times 2.3 \log 2 = \textbf{1.37 cal/deg.}$ Similarly, ΔG total is $RT \ln P_2/P_1 = 1.98 \times 298 \times 2.3 \log \frac{1}{2}$ $(P_2/P_1 = \frac{1}{2}$ for each gas) or $\Delta G = \textbf{−409 cal.}$ From the nature of the operation no work is performed, so $w = \textbf{0.}$ The gases are ideal so their energies do not change with volume; hence $\Delta E = \textbf{0.}$ Finally, it follows that $q = \textbf{0.}$

(c) Since ΔE is zero (the gases being ideal), $q = w$. Also, $q_{rev} = T \Delta S$ or $q = w = 298(-1.37) = \textbf{−409 cal.}$

36. Since the Kr^{84} and Kr^{86} are not separated, their mixture may be treated as a single species, and the process can be rewritten as

$$1 \text{ mole } (10\% \text{ A}, 90\% \text{ B}) = 0.1A + 0.9B$$

ΔS is then $-\Delta S = -R(0.1 \ln 0.1 + 0.9 \ln 0.9) = 0.642$ or $\Delta S = \textbf{−0.642 cal/deg.}$ Since ΔH is zero, $\Delta G = -T \Delta S = \textbf{192 cal.}$

37. (a) Referring back to Eq. (5-13) the wave mechanical expression for the separation of translational energy states contains the mass m of the particle in the denominator. This means that the larger the molecular weight, the more closely spaced are these states, and the lower the temperature at which the system can gain energy. (b) The Sackur–Tetrode equation predicts that the translational heat capacity should not depend on the molecular weight of the gas. The equation can be written $S = (\text{constant}) + (5R/2) \ln T$, and since $C_p = (\partial S/\partial \ln T)_P$, it then follows directly that $C_P = 5R/2$.

38. Since $S = E/T + R \ln Q$, $S_{AA} - S_{BB}$ for rotation will reduce to $R \ln Q_{AA}/Q_{BB}$, the rotational energy being expected to be at the same equipartition value in both cases. The ratio of Q's further reduces to I_{AA}/I_{BB} or 2, so the answer is $R \ln 2$ or **1.38 cal/deg-mole.**

39. According to the Sackur–Tetrode equation $S_{trans} = R \ln T^{5/2}$ plus temperature-independent quantities. Then $dS/dT = (5R/2)(1/T)$. As the alternative route, C_p is $5R/2$ by the equipartition principle, and $dS = C_p d \ln T$, or $S_{trans} = C_p \ln T$ plus constants; again, $dS/dT = (5R/2)(1/T)$.

40. The entropies are obtained through the equation $dS = C_v d \ln T$, so the lower the temperature at which equipartition heat capacity is reached, the larger will be this integral. In view of the very close spacing of translational energy levels as compared with rotational, the translational entropy should be (and is) much larger than the rotational.

41. (a) It will be about $(1/6 \times 10^{23})(1 \times 10^{-8})^2$, and inserting values for the other constants, Q_{rot}/T is about 0.04. (b) From Eq. (7-23), $S = E/T +$

$R \ln Q$; the maximum possible E/T is the equipartition value of $2R/2$ or R. S must be negative, then, if $\ln Q < 1$, or $Q < 1/e$, or if T is less than about $10°K$. (c) The situation in (b) is nonsensical since S as an absolute entropy must be greater than zero. The difficulty is evidently that the given expression for Q, being based on an integration, is not accurate at such a low temperature. From Eq. (5-21) $\Delta\varepsilon_{1,0}$ between the states with $J = 1$ and $J = 0$ is $h^2/8\pi^2 I$, so the expression for Q itself (by the integration procedure) turns out to be just $kT/\Delta\varepsilon_{1,0}$. At $10°K$, kT is thus about one-third of $\Delta\varepsilon_{1,0}$, and it is to be expected that the integration approximation would be poor.

42. For a gas at STP, the volume per molecule is $22,400/6 \times 10^{23}$ or 3.7×10^{-20} cc/mole, corresponding to an average distance apart of 3.3×10^{-7} cm, and an area $\sigma \simeq 1.1 \times 10^{-13}$ cm^2. S_{trans} (two-dimensional) then becomes $R \ln(1.1 \times 10^{-13}) + R \ln(MT) + 65.8 = R \ln(MT) + 6.6$.

43. First, by definition $G^0 = H^0 - TS^0 = E^0 + PV - TS^0 = E^0 + RT - TS^0$. The translational part of S^0 is given by the Sackur–Tetrode equation, and the rest by $S_{\text{int}}^0 = E_{\text{int}}^0/T + R \ln Q_{\text{int}} = RT\, d \ln Q_{\text{int}}/dT + R \ln Q_{\text{int}}$. On substitution into the defining equation for G^0,

$$G^0 = E_0^0 + \tfrac{3}{2}RT + RT^2\, d \ln Q_{\text{int}}/dT + RT - \tfrac{3}{2}RT \ln M - \tfrac{5}{2}RT \ln T$$
$$+ 2.31T - RT^2\, d \ln Q_{\text{int}}/dT - RT \ln Q_{\text{int}}$$

On cancelling terms, and rearranging,

$$(G^0 - E_0^0)/T = -\tfrac{3}{2}R \ln M - \tfrac{5}{2}R \ln T - R \ln Q_{\text{int}} + (2.31 + \tfrac{3}{2}R + R)$$

The constants total to 7.27, to give the desired result.

44. Q_{vib} can easily make an appreciable contribution to the heat capacity of a gas around room temperature, but does not add much to the entropy of the gas. As a reason for the latter, entropy development can be regarded in terms of the integral $C_v\, d \ln T$, and the vibrational contribution begins to appear only at a relatively high temperature.

45. From Eqs. (7-23) and (5-4)

$$S_{\text{rot}} = E_{\text{rot}}/T + R \ln Q_{\text{rot}} = RT\, d \ln Q_{\text{rot}}/dT + R \ln Q_{\text{rot}}$$

Since Q_{rot} may be written as equal to aT, it follows that $d \ln Q_{\text{rot}}/dT = 1/T$, so that S_{rot} is simply $R + R \ln Q_{\text{rot}}$.

46. Q in this case is simply $1 + \exp(-\bar{\varepsilon}/kT)$ and $\bar{\varepsilon}$ is

$$\bar{\varepsilon} = \frac{0 + e^{-\varepsilon/kT}}{1 + e^{-\varepsilon/kT}}$$

As infinite temperature $\exp(-\varepsilon/kT)$ approaches unity, so $\bar{\varepsilon}$ becomes $\varepsilon/2$ or $E = N\varepsilon/2$, and $S = E/T + R \ln Q = \boldsymbol{R \ln 2}$.

8

LIQUIDS AND THEIR SIMPLE PHASE EQUILIBRIA

COMMENTS

We now consider the application of the combined first and second laws of thermodynamics to simple phase equilibria. The general cross-differentiation equation $(\partial P/\partial T)_V = (\partial S/\partial V)_T$ becomes the Clapeyron equation $dP/dT = \Delta H/T\,\Delta V$ when applied to the equilibrium between two phases, since ΔS can now be written as $\Delta H/T$, and it is not necessary to retain the partial differential form. If one of the two phases is an ideal gas, then further approximations lead to the very useful Clausius–Clapeyron equation. These two relationships, plus some semiempirical rules such as Trouton's rule, are presented to you in this chapter in quite a variety of disguises.

The present chapter is an appropriate one in which to introduce surface tension, its manifestations, and its determination. The Laplace equation is fundamental to this topic of capillarity, and its use will be required over and over again in the problems that follow. Watch for variations such as the case of maximum bubble pressure, and the rise of a liquid between parallel plates.

Be careful in your choice of units. ΔH will sometimes be needed in cc-atm/mole units and pressure differences in capillarity situations will generally be in dyne/cm², not atm.

EQUATIONS AND CONCEPTS

Clapeyron Equation

$$\frac{dP}{dT} = \frac{\Delta H}{T\,\Delta V} \tag{8-1}$$

Clausius–Clapeyron Equation

$$d \ln \frac{P}{dT} = \frac{\Delta H}{RT^2} \tag{8-2}$$

$$\ln \frac{P_2}{P_1} = \frac{\Delta H}{R} \left(\frac{1}{T_1} - \frac{1}{T_2} \right) \tag{8-3}$$

$$\ln P = \text{constant} - \frac{\Delta H}{RT}, \quad \text{or} \quad P = P^0 \exp \left[-\frac{\Delta H}{R} \left(\frac{1}{T} - \frac{1}{T^0} \right) \right] \tag{8-4}$$

Effect of Mechanical Pressure on Vapor Pressure

$$RT \ln \frac{P'}{P} = \int V \, dP_{\text{mech}} \tag{8-5}$$

Semiempirical Rules and Laws

Trouton's rule. $\Delta H_v/T_b = 21$ cal/deg-mole where T_b denotes the normal boiling point and ΔH_v, the heat of vaporization per mole. Another observation is that the normal boiling point of a liquid is often about two-thirds of its critical temperature.

Law of Rectilinear Diameters

The sum of the densities of a liquid and its equilibrium vapor is a constant, independent of temperature. A more realistic version is that the sum varies linearly with temperature.

Capillarity

Laplace equation. $\Delta P = \gamma(1/R_1 + 1/R_2)$ where ΔP is the pressure difference across a curved surface, γ is the surface tension, and R_1 and R_2 are the two radii of curvature. The signs are such that the surface is convex toward the low-pressure side; thus the pressure inside a spherical drop or soap bubble is greater inside than outside. For surfaces that are sections of a sphere, the Laplace equation becomes

$$\Delta P = \frac{2\gamma}{r} \tag{8-6}$$

where r is the radius of the sphere.

Capillary rise. $\Delta P = \rho g h = 2\gamma/r$ where r is the radius of the capillary and h, the height of rise. For a nonzero contact angle, the equation becomes

$$\Delta P = \rho g h = \frac{2\gamma \cos \theta}{r} \tag{8-7}$$

Drop-weight method. $W_{ideal} = 2\pi r \gamma$. This is known as Tate's law, and actual drop weights will differ from the ideal weight by a substantial correction factor f, so that

$$W_{actual} = 2\pi r \gamma f \tag{8-8}$$

This correction factor can be expressed as a graphical function of $r/V^{1/3}$ where V is the drop volume. Note that in the case of a liquid that wets the tube, r is the outside radius.

Maximum bubble-pressure method.

$$P = P_{hyd} + \frac{2\gamma}{r} \tag{8-9}$$

P_{hyd} is the hydrostatic pressure as determined by the depth of immersion of the tube out of which bubbles are formed.

Pendant drop method. For a pendant drop, the basic equation of Laplace can be maneuvered into the form

$$\gamma = \frac{\rho g d_e^2}{H} \tag{8-10}$$

where d_e is the equatorial or maximum diameter of the drop, and H is a function of a shape factor S. S is defined as d_s/d_e, where d_s is the diameter of the drop measured a distance d_e up from the bottom. The following is an abbreviated $1/H$ versus S table:

S	0.70	0.75	0.80	0.85	0.90	0.95	1.00
$1/H$	0.80	0.67	0.57	0.48	0.41	0.36	0.31

PROBLEMS

1. (21 min) The normal boiling point of pyridine is 114°C. At this temperature, its vapor density is 2.5 g/liter, and the density of the liquid is 0.8000 g/cc. At a certain higher temperature T', the liquid has expanded to where its density is 0.7900 g/cc. Calculate, or estimate with explanation, (a) the heat of vaporization of pyridine, (b) the boiling point of pyridine on a mountain top where the atmospheric pressure is 740 mm Hg instead of 760 mm Hg, and (c) the vapor density of pyridine at the temperature T'.

2. (14 min) The semilog plots for the vapor pressures of liquids A and B are shown in Fig. 8-1. (a) Calculate ΔH_v for liquid A. (b) Liquids A and B

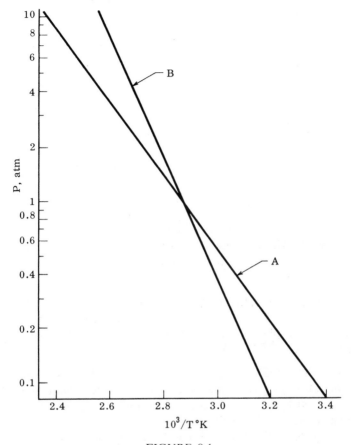

FIGURE 8-1

cannot both have the same Trouton constant. Explain how this can be concluded from the data. Give a brief argument as to which liquid should be considered the more associated.

3. (4.5 min) The law of rectilinear diameters (in its simplest form as illustrated in Fig. 8-2) is not compatible with the van der Waals statement that $V_c = 3b$. Show why this statement is true.

4. (18 min) The vapor pressure of liquid A is 50 mm Hg at 46°C; this vapor pressure is 0.50 mm greater than that of solid A at the same temperature. At 45°C, the vapor pressure of the liquid is 1.00 mm Hg greater than that of the solid. ΔH_v is 9.0 kcal. (a) Estimate the melting point of A. (b) Calculate the heat of fusion of A, and its heat of sublimation.

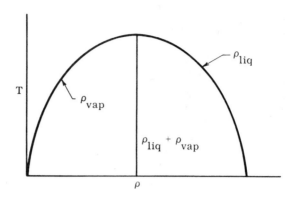

FIGURE 8-2

5. (18 min) The vapor pressure of CCl_4 increases by 4% per degree around 25°C. Calculate ΔH_v and the normal boiling point of CCl_4. List all the assumptions and approximations involved in the derivation of the equation used to obtain ΔH_v.

6. (12 min) A particular liquid obeys both the Clausius–Clapeyron equation and Trouton's rule. With this information calculate the vapor pressure of the liquid, in atmospheres, at a temperature equal to one-third of the normal boiling point (expressed in °K).

7. (15 min) The heat of vaporization of water is 9.7 kcal/mole, whereas that for liquid A is 7.0 kcal/mole. The vapor pressure of water and of A are the same at 150°C. (a) Plot P versus $1/T$ on a semilog plot, for water and for A. (The plots may be approximate, but it must be clear how they were obtained.) (b) Determine from the plot the normal boiling point of liquid A. (c) Which of the two liquids more nearly obeys Trouton's rule? (On a problem such as this, the appropriate graph paper would be available.)

8. (30 min) The following data are given for *p*-chloroaniline:

mol. wt: 127   vp at the mp: 5.0 mm Hg
normal mp: 70°C   vp at 100°C: 20.0 mm Hg
ΔH_f = 4,700 cal/mole   densities at the mp: solid 1.45 g/cc
                                  liquid 1.15 g/cc

(a) Calculate the heat of vaporization. (b) State the assumptions and approximations involved in the equation(s) used in (a). (c) Estimate the normal boiling point and the heat of sublimation. State whether the mp under 100 atm pressure would be greater or less than 70°C (explain). (d) Air at 1 atm and 100°C is bubbled through the liquid (also at 100°C) at the rate of 3.8 mole/h. Assuming the effluent air to be saturated with vapor, how long will it take for 12.7 g of the liquid to be evaporated? (Effluent gas also at 1 atm.)

9. (8 min) The normal boiling point of a liquid obeying Trouton's rule is 120°C. Calculate the vapor pressure of the liquid at 121°C (or, better, the increase in vapor pressure from 120°C to 121° C).

10. (12 min) The vapor pressure of acetonitrile is changing at the rate of 0.030 atm/deg in the vicinity of its normal boiling point, which is 80°C. Calculate the heat of vaporization.

11. (12 min) The straight-line plots of $\ln P$ versus $1/T$ are sketched in Fig. 8-3 for various liquids obeying the simple Clausius–Clapeyron equation. These lines all meet at $1/T$ equal to zero; show that this behavior is required by Trouton's rule.

12. (12 min) The straight-line plot of $\ln P$ versus $1/T$ for the vapor pressure of a certain liquid whose normal boiling point is 27°C extrapolates to $P = 10^5$ atm at infinite temperature. Using the Clausius–Clapeyron equation, calculate from these data the heat of vaporization of the liquid.

---

13. (18 min) A certain liquid of molecular weight 60 has a critical temperature of 400°C. Its melting point is 15.000°C as normally measured and is 14.980°C at its triple point (where the system is subjected only to its own low vapor pressure). The solid and liquid densities are 0.85 and 0.80 g/cc, respectively. Calculate, using empirical or semiempirical relationships where necessary, ΔH_v, ΔH_s, ΔH_f, and the vapor pressure at the triple point.

14. (9 min) Explain whether the melting point of a solid substance A will be raised or lowered by pressure, given that solid A does *not* float on liquid A. Illustrate your explanation by suitable equations.

15. (18 min) The melting point of glacial acetic acid is 16°C at 1 atm pressure. Calculate the melting point under its own vapor pressure (essentially

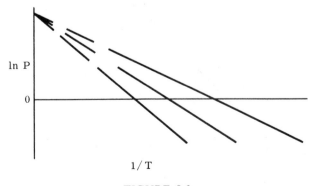

FIGURE 8-3

zero pressure). The heat of fusion is 2,700 cal/mole, densities for liquid and solid acetic acid are 1.05 and 1.10 g/cc, respectively, and the molecular weight is 60. Set up the appropriate equation and insert the proper data so as to obtain an equation whose only unknown is the desired quantity.

The normal boiling point of acetic acid is 118°C. Show how to estimate the heat of sublimation of solid acetic acid from this and the other data given (obtain a numerical answer).

16.   (15 min) It is desired to calculate the vapor pressure of an equilibrium mixture of solid and liquid benzene which is under 100 atm of inert gas pressure. The information available includes ΔH_f and ΔH_v, the molar heats of fusion and vaporization at the melting point (under normal pressure), which is T_1; the densities of the solid and liquid ρ_1 and ρ_2; the molecular weight M; the normal boiling point of the liquid T_2; and general constants such as R. Set up equations for calculating this vapor pressure. Make clear what equations you would use, the sequence of their use, and the actual data needed.

17.   (24 min) For a certain substance the change in entropy on melting is 3 cal/deg per cc of solid which melts. The melting point under 1 atm pressure is 6°C, and the densities of the solid and liquid are 0.90 and 0.85 g/cc, respectively. Calculate the melting point under 10^4 atm pressure.

Also, the vapor pressure versus temperature curve for the solid and the liquid, at 1 atm pressure, are sketched in Fig. 8-4. Show qualitatively how

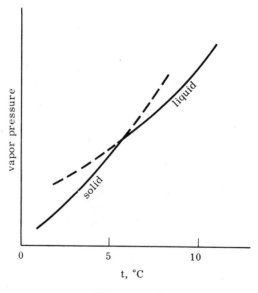

FIGURE 8-4

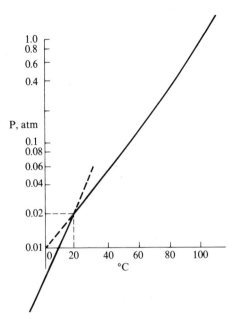

FIGURE 8-5

these two curves should look for the solid and liquid under 100 atm pressure (the pressurization is done by means of inert gas pressure, so these vapor pressure curves can be obtained experimentally). (The curves should be of the correct shape, relative position, etc.)

18. (24 min) A substance A of mol. wt 44 melts at 20°C under its own vapor pressure of 0.02 atm. The low pressure region of the phase diagram is shown in Fig. 8-5. In addition the melting point changes by 2% (using °K) per 100 atm applied pressure. The densities of the solid and liquid forms are about 1.6 and 1.1 g/cc, respectively. Calculate, or if semiempirical rules are applicable, estimate (a) the heat of fusion, (b) the heats of vaporization and sublimation, and (c) the value of ΔG for the process A (s, 10°C, 1 atm) = A (l, 10°C, 1 atm).

19. (6 min) ΔH is 0.07 kcal for the transition S (rhombic) = S (monoclinic). Monoclinic sulfur is in equilibrium with rhombic at 1 atm pressure and 115°C, and at 100 atm pressure the two are in equilibrium at 120°C. Show which of the two forms is the more dense.

20. (7.5 min) Water, which wets glass, rises in a given capillary to a height h. If, as shown in Fig. 8-6, the capillary is broken off, so that its length above the surface is only $h/2$, will the water then flow over the edge? Explain your

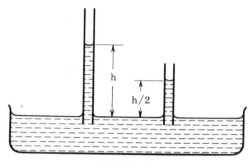

FIGURE 8-6

answer in terms of specific detailed analysis of just what happens. Use sketches.

21. (12 min) A capillary tube of radius 0.1 mm is inserted through the two-layer system shown in Fig. 8-7. The capillary rise of the water–benzene meniscus is 4.0 cm. The glass–water–benzene contact angle is 40° (cos θ = 0.76), and the densities of water and of benzene are 1.00 and 0.80, respectively. Calculate the interfacial tension between water and benzene.

22. (12 min) A length of uniform-bore capillary tubing is bent in an S shape, and one end is immersed in a liquid of surface tension 25 dyne/cm and density 0.80 g/cc. The radius of the capillary is 0.050 cm; the liquid wets it.

After immersing one end of the capillary, some of the same liquid is added to the other end until the pressure of the trapped air is sufficient to force the meniscus back down to the level of the liquid in the container. This final situation is illustrated in Fig. 8-8. If the external air pressure is 1 atm (10^6 dyne/cm^2), calculate or explain what the values of P_1, P_2, P_3, P_4, and P_5 must be.

23. (6 min) (a) Liquid A has half the surface tension and twice the density of liquid B, at 25°C. If the capillary rise is 1.0 cm for liquid A, then, in the

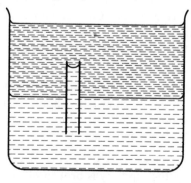

FIGURE 8-7

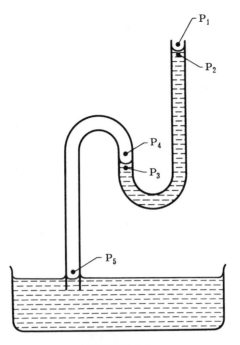

FIGURE 8-8

same capillary, the rise will be (0.25, 0.5, 1.0, 2, 4, 10 cm, none of these). (b) Given the same data as in (a), the maximum bubble pressure for liquid A will be (0.25, 0.5, 1.0, 2, 4, 10, none of these) times that for liquid B, using the same tube for both liquids.

24. (9 min) A thin-walled tube 0.10 cm in diameter is lowered into a dilute aqueous detergent solution until its open end is 10 cm below the surface. The maximum air pressure just insufficient for bubbles to grow and break away is found to be 11.6 cm, as read on a *water* manometer. Calculate the surface tension of this solution.

25. (15 min) In the drop-weight method for surface tension determination, the actual weight W of a drop is equal to the ideal weight W_i multiplied by a correction factor Ψ where Ψ is a function of $r/V^{1/3}$. Here, r is the radius of the tip, and V, the volume of the actual drop. For $r/V^{1/3} = 0.5$, it has been found that $\Psi = 0.65$.

(a) Give the equation relating W_i to the surface tension of the liquid. (b) Calculate what size of tip (i.e., r value) should be used for a liquid of $\gamma = 26$, and density equal to 0.8 g/cc, in order that $r/V^{1/3}$ will in fact be 0.5. Using this tip size and the above liquid, what will be the actual weight of the drop?

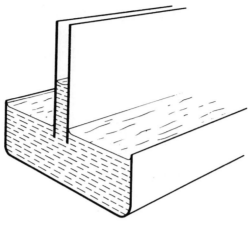

FIGURE 8-9

26. (10 min) Two vertical, parallel plates, 0.1 cm apart are partly lowered into a liquid density 1.10, which wets the plates (see Fig. 8-9). Derive the formula for the capillary rise of the liquid. If the rise is 1.30 cm, what is the surface tension? (Assume that the plates are wide enough that end effects can be neglected.)

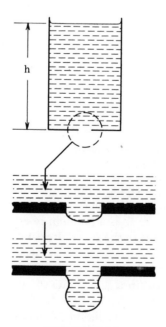

FIGURE 8-10

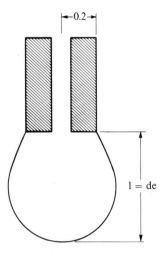

FIGURE 8-11

27. (12 min) A metal cylinder has a small pin hole in the bottom. The hole is smooth and circular, and 0.04 mm in diameter. Calculate the depth to which the container can be filled with water before the water will start dripping out through the hole. $\gamma = 72$, density = 1.0; assume that water fails to wet the metal so that the sequence of drop shapes is that shown in Fig. 8-10. Sketch the shape of the meniscus or nascent drop when the container is filled to the maximum depth possible before dripping occurs.

28. (10 min) A drop of water at 25°C is formed, hanging from a tip with 0.2-in. radius as shown in Fig. 8-11. The drop is such that its length l is equal to d_e. Calculate an approximate value for l.

29. (15 min) One commonly used method of measuring surface tension consists of placing a cylindrical tube so that its opening is just below the

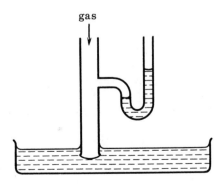

FIGURE 8-12

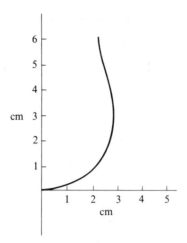

FIGURE 8-13

surface of the liquid in question, and slowly bubbling an inert gas through the liquid (see Fig. 8-12). A manometer connected to the tube permits the measurement of the difference in pressure between the gas in the tube and atmospheric pressure. Show that the gas pressure is a maximum when the radius R of the bubble is equal to the radius r of the tube. (A brief logical argument is wanted, based on the laws governing surface tension effects.) Assuming the proposition to be correct, calculate the surface tension of the liquid if the maximum pressure difference is 0.30 mm Hg, given that the temperature is 25°C, the density of the liquid is 1.5 g/cc, that of mercury is 13.6 g/cc, and the radius of the tube is 0.10 cm.

30. (12 min) The profile of an enlarged photograph of a pendant drop of benzene, surface tension 30 dyne/cm and density 0.9 g/cc, is shown in Fig. 8-13. Calculate the magnification factor.

ANSWERS

1. (a) Using Trouton's rule, $\Delta H_v = 21 \times 387°K = $ **8.13 kcal.**

(b) From the Clausius–Clapeyron equation, $d \ln P/dT = \Delta H/RT^2$ or $dP/dT = \Delta H \times P/RT^2$. As an approximation, take P to be 750 mm Hg, so $dP/dT = 8{,}130 \times 750/1.98 \times 387^2 = 20.5$ mm Hg/deg. The temperature at which P is 740 mm Hg is then reduced by 20/20.5 or by **1.0°.** More exactly,

$$\log 760/740 = \frac{\Delta H}{2.3R}\left(\frac{1}{T_b'} - \frac{1}{T_b}\right)$$

which gives **0.98°.**

(c) From the law of rectilinear diameters, the sum of the liquid and vapor densities should be invariant. At 114°C it is $800 + 2.5$ g/liter. The vapor density at T' is then $802.5 - 790$ or **12.5 g/liter.**

2. (a) From the Clausius–Clapeyron equation, $d \ln P/d(1/T) = -\Delta H/R$, so the slope of the line for liquid A can be used directly. It is $[\log 1 - \log 0.1]/[2.9 - 3.4] \times 10^{-3}$ or about $-2,000$. ΔH_v is then $2,000 \times 2.3 \times 1.98 = $ **9.2 kcal.**

(b) The simplest way of showing this is by writing the integrated form of the Clausius–Clapeyron equation: $\ln P = \text{const} - \Delta H/RT$. By Trouton's rule, when $P = 1$ atm, $\Delta H/T = 21$, so the constant of integration must be $21/R$. It is thus the same for all liquids, or in other words, all liquids should show the same intercept at $1/T = 0$. This is clearly not the case here, so Trouton's rule cannot be valid in this instance.

(c) Liquid B has the higher heat of vaporization (steeper slope) and hence is probably the more associated.

3. In its simplest form, the law of rectilinear diameters states that the sum of liquid and equilibrium vapor densities is a constant. For a liquid around its boiling point, the vapor density is so small that this sum is essentially just the liquid density. Then, according to the LRD, at the critical temperature, the liquid and vapor densities are equal, and each must then be half the density of the liquid around its boiling point. In other words, the critical volume is predicted to be twice (not three times) that of the liquid and hence only twice that of the volume of 1 mole of molecules (to which b is supposed to correspond).

4. (a) At the melting point, liquid and solid must have the same vapor pressure. If the difference is 1 mm at 45°C and 0.5 mm at 46°C, it should be close to zero at **47°C.**

(b) The most efficient way to get at the heat of fusion is probably from the Clausius–Clapeyron equation: $dP/dT = \Delta H \times P/RT^2$. For the vapor at 46°C, then $[dP/dT]_v = 9,000 \times 50/1.98 \times 319^2 = 2.2$ mm/deg. Since the vapor pressure of the solid is increasing 0.5 mm/deg faster than is that of the liquid, $(dP/dT)_s = 2.7$ mm/deg. The heat of sublimation of the solid is then

$$\Delta H_s = \frac{2.7 \times 1.98 \times 320^2}{52.7} = \textbf{10.35 kcal}$$

The heat of fusion is then $\Delta H_s - \Delta H_v = $ **1.35 kcal.**

5. We use the Clausius–Clapeyron equation in the form

$$dP/dT = P \times \Delta H_v/RT^2$$

If P increases by 4% per degree, then $(dP/P)/dT$ is 0.04; so $\Delta H_v = 0.04 \times 1.98 \times 298^2 = \textbf{7.04 kcal}$.

The normal boiling point can be estimated from Trouton's rule: $T_b = 7{,}040/21 = \textbf{335°K}$. The assumptions and approximations involved in the derivation of the Clausius–Clapeyron equation are as follow. (1) First and second laws of thermodynamics; (2) equilibrium between liquid and vapor; (3) neglect molar volume of liquid in comparison with that of vapor; (4) assume ideal behavior for the vapor.

6. Using the integrated form, $\ln P_2/P_1 = (\Delta H/R)(1/T_1 - 1/T_2)$, let $P_1 = 1$ atm, and T_1 the normal boiling point; ΔH is then $21 T_1$. With these substitutions, $\ln P_2 = (21T_1/R)(1/T_1 - 3/T_1)$, since T_2 is to be one-third of T_1. Simplification then gives $\ln P_2 = -42/R$, from which $P_2 = \textbf{6.4} \times \textbf{10}^{-10}$ **atm.**

7. (a) We know P_w is 1 atm at 373°K; by the Clausius–Clapeyron equation, $\log P_2/P_1 = (\Delta H/2.3R)(1/T_1 - 1/T_2)$, the vapor pressure at 150°C will be $\log P_2 = (9{,}700/2.3 \times 1.98)(2.69 - 2.63) \times 10^{-3} = 0.70$ or $P_2 = 5$ atm.

(b) Since the plot will be a straight line (see Fig. 8-14), we now draw it between the two points. For liquid A, the point at 150°C is the same, and at 100°C its vapor pressure is given by

$$-\log\frac{P_1}{5} = \frac{7{,}000}{2.3 \times 198} \times 0.33 \times 10^{-3} = \frac{7}{9.7} \times 0.7 = 0.5$$

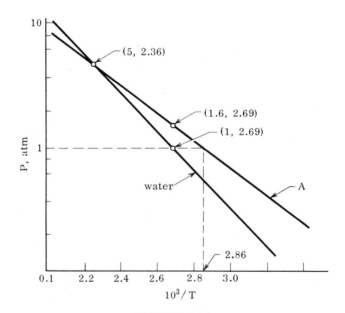

FIGURE 8-14

or

$$P_1 = 0.32 \times 5 = 1.6\,\text{atm}$$

The line for liquid A can then be located. Extension of the line gives a normal boiling point of $10^3/T = 2.86$ or $T = 350°\text{K}$.

(c) $\Delta H/T_b$ is $9,700/373 = 26$ for water and $7,000/350 = 20$ for liquid A, so **liquid A** is the closer to obeying the Trouton rule value of 21.

8. (a) The heat of vaporization is given by the Clausius–Clapeyron equation: $\log 20/5 = (\Delta H/2.3 \times 1.98)(1/343 - 1/373)$ or $\Delta H = 0.60 \times 2.3 \times 1.98/0.24 \times 10^{-3} = \textbf{11.4 kcal.}$

(b) The assumptions and approximations involved in the derivation of the Clausius–Clapeyron equation are (1) first and second laws of thermodynamics, (2) molar volume of liquid negligible compared with that of the vapor, (3) vapor behaves ideally, (4) ΔH does not vary with temperature, and (5) liquid and vapor are in equilibrium.

(c) From Trouton's rule, the normal boiling point would be $T_b = 11,400/21 = 543°\text{K}$. The heat of sublimation is the sum of ΔH_f and ΔH_v, or $4.7 + 11.4 = \textbf{16.1 kcal.}$ From the Clapeyron equation, $dP/dT = \Delta H_f/T\,\Delta V$. Since the solid is more dense, its molar volume is less than that for the liquid and ΔV is positive for the process; solid $\rightleftharpoons$ liquid. Since ΔH is also positive for this process, dP/dT must be positive, and the melting point under 100 atm pressure will be **greater** than 70°C.

(d) The mole fraction of vapor in the effluent gas will be $20/760$ or 0.0263, or the ratio of moles of air to moles of vapor will be 37.4. Now 12.7 g of liquid corresponds to 0.1 mole; hence 3.74 moles of air are needed, and air must be bubbled through for **1 h.**

9. We write $d\ln P/dT = \Delta H/RT^2$ or $dP/dT = P\,\Delta H/RT^2$. At the boiling point, $P = 1$ atm, and $\Delta H/T = 21$, so $dP/dT = 21/RT = 0.027$. The vapor pressure at 121°C is then **1.027 atm.**

10. The Clausius–Clapeyron equation is best used here in the form

$$(dP/P)/dT = \frac{\Delta H}{RT^2}$$

where $(dP/P)/dT$ is given a 0.030 (P is 1 atm at the normal boiling point). Then $\Delta H = 0.030 \times 1.98 \times 353^2 = \textbf{7.4 kcal/mole.}$

11. The integrated form of the Clausius–Clapeyron equation may be written $\ln P = \text{const} - \Delta H/RT$. According to Trouton's rule, when $P = 1$ atm, $\Delta H/T = 21$; hence the constant equals $21/R$, that is, $\ln P = 21/R$

$- \Delta H/RT$. Thus all liquids are supposed to have a common intercept at $1/T = 0$.

12.   The integrated form of the Clausius–Clapeyron equation is $\ln P = A - \Delta H/RT$, and, evidently, $A = \ln 10^5$, so the equation becomes

$$\log 1/10^5 = -\Delta H/R \times 300 \times 2.3$$

Then $\Delta H = $ **6,820 cal.**

13.   (a) The heat of vaporization can be estimated as follows: the normal boiling point is about two-thirds of the critical temperature, and thus is $2 \times 637/3$ or about 450°K. By Trouton's rule, ΔH_v is then 21×450 or **9.4 kcal.**

(b) The heat of fusion is given by the Clapeyron equation: $dP/dT = \Delta H_f/T_f \, \Delta V$. The molar volumes of the liquid and solid are $60/0.8$ and $60/0.85$ or 75 and 70.7 cc/mole, respectively, so ΔV is 4.3 cc/mole. ΔH_f is then $288 \times 4.3 \times (1 \text{ atm} - 0)/(15.000 - 14.980)$ or $\Delta H_f = 6.2 \times 10^4$ (in cc-atm units!) or $6.2 \times 10^4/41.3 = $ **1.5 kcal.** ΔH_s is the sum of ΔH_v and ΔH_f or **10.9 kcal.** Finally, the vapor pressure at the triple point is given by the Clausius–Clapeyron equation:

$$\log 760/P = \frac{9,400}{2.3 \times 1.98}(1/288 - 1/450)$$

or

$$\log 760/P = 2.57; \quad P = \textbf{2.05 mm Hg}$$

14.   According to the Clapeyron equation, $dP/dT = \Delta H/T \, \Delta V$. If the solid is more dense than the liquid, its molar volume is less, so ΔV must be positive for the process: solid $\rightleftharpoons$ liquid. Since ΔH_f is always positive, the right-hand side of the Clapeyron equation must be positive and hence also dP/dT. The melting point is therefore raised on application of pressure.

15.   The Clapeyron equation applies here:

$$dP/dT = \Delta H/T \, \Delta V$$

Solving for dT, or ΔT, $\Delta T = \Delta P \times T \times \Delta V/\Delta H$, where ΔP is -1 atm, T is 289°K, ΔH is $2,700 \times 41.5 = 1.13 \times 10^5$ cc-atm and $\Delta V = V_l - V_s = 60(1/1.05 - 1/1.10)$. From Trouton's rule, the normal boiling point is related to ΔH_v by $\Delta H_v/T_b = 21$, hence $\Delta H_v = 21 \times 391 = 8,200$ cal/mole. $\Delta H_s = \Delta H_f + \Delta H_v = 8,200 + 2,700 = $ **10,900 cal/mole.**

16. First, we determine the temperature of the freezing mixture under 100 atm pressure, using the Clapeyron equation $\Delta P/\Delta T = \Delta H/T\,\Delta V$, where ΔH is the heat of fusion, and ΔV is given by $\Delta V = M(1/\rho_2 - 1/\rho_1)$; T is approximately the melting point T_1. T is then $T_1 + \Delta T$. Next, use the Clausius–Clapeyron equation to calculate the vapor pressure of the liquid at this temperature.

$$\ln P/1 = \frac{\Delta H_v}{R}\left(\frac{1}{T_2} - \frac{1}{T}\right)$$

where T is the above temperature. Finally, it is necessary to recognize that the liquid at T is under 100 atm pressure so that its free energy is increased by $\int V\,dP$. That is, $RT \ln P'/P = \int V\,dP = (M/\rho_2) \times 100\,\text{atm}$, where P is the vapor pressure obtained from the Clausius–Clapeyron equation and P' is the desired final answer.

17. For 1 g solid, the volume changes from 1.11 cc to 1.18 cc, so ΔV is 0.07 cc. Per cc of solid, ΔV is then $0.9(1/0.85 - 0.9) = 0.06$ cc. From the Clapeyron equation, $\Delta P/\Delta T = \Delta S/\Delta V = (3/0.06) \times (41) = 2{,}050\,\text{atm/deg}$, so for $\Delta P = 10^4\,\text{atm}$, $\Delta T = 4.9°$.

The effect of mechanical pressure will be to raise both vapor-pressure curves ($RT \ln P'/P = \int V\,dP_{\text{mech}}$), but evidently the crossing point must shift to the right (compare Figs. 8-4 and 8-15), so the vapor-pressure curve for the liquid must be raised more than that for the solid (as the $\int V\,dP_{\text{mech}}$ predicts).

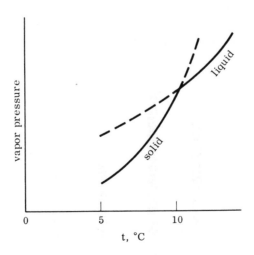

FIGURE 8-15

18.   (a) ΔV for the fusion process is $44(1/1.1 - 1/1.6)$ or $12.5\,\text{cc/mole}$, so the Clapeyron equation becomes

$$\frac{dP}{dT} = \frac{\Delta H_f}{12.5T}$$

or

$$\Delta H_f = \frac{12.5\, dP}{dT/T}$$

Take dP to be $100\,\text{atm}$, and dT/T to be 0.02; this gives $\Delta H_f = 100 \times 12.5/0.02 = 6.25 \times 10^4\,\text{cc-atm}$ or **1,520 cal/mole.** (b) The heat of vaporization could be obtained from the Clausius–Clapeyron equation, using points read off the figure; the wording of the question allows a simpler approach. The normal boiling point is about $100°\text{C}$, so by Trouton's rule, $\Delta H_v = 21 \times 373 =$ **7,820 cal/mole.** $\Delta H_s = \Delta H_f + \Delta H_v = 1,520 + 7,820 =$ **9,340 cal/mole.** (c) The effect of 1 atm pressure will be negligible, and ΔG is given by $RT \ln P/P_s = 1.98 \times 283 \ln 0.03/0.01$, using values of P and P_s read off the graph; ΔG is then about **650 cal/mole.**

19.   By the Clapeyron equation, $dP/dT = \Delta H/T\,\Delta V$. Since dP/dT is positive, in this case, ΔV must have the same sign as ΔH, that is, is positive for $V_m - V_r$. Therefore rhombic sulfur has the smaller molar volume and is the more dense.

20.   The water will not flow over the edge! As shown in Fig. 8-16, the meniscus will rise to the end of the capillary, then flatten to a point where its radius of curvature is just half of what it would normally be; the pressure drop across the meniscus is then in hydrostatic balance with the pressure drop along the column of liquid.

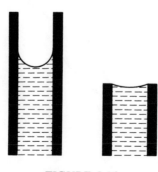

FIGURE 8-16

21. The relevant equation is $\Delta\rho gh = 2\gamma\cos\theta/r$; so the interfacial tension γ is

$$\gamma = \frac{(1.00 - 0.80) \times 980 \times 4 \times 0.01}{2 \times 0.76} = \textbf{5.1 dyne/cm}$$

22. We need the value of ΔP, the pressure drop across the meniscus; this is given by $\Delta P = 2\gamma/r = 2 \times 25/0.05 = 1{,}000 \text{ dyne/cm}^2$. We can now reason as follows: P_1 is just atmospheric pressure, 10^6 dyne/cm^2. P_2, just under the meniscus, must be 1,000 dyne/cm^2 less or $\textbf{0.999} \times \textbf{10}^6$. P_5, just above the bottom meniscus, must be $\textbf{1.001} \times \textbf{10}^6$ and, neglecting the hydrostatic drop in air pressure, will be equal to P_4. P_3 must be 1,000 less, or equal to $\textbf{1.000} \times \textbf{10}^6 \textbf{ dyne/cm}^2$.

23. For capillary rise, $\rho gh = 2\gamma/r$ or h is proportional to γ/ρ. If liquid B has twice the surface tension and half the density of liquid A, the two ratios will combine to give **four** times the height of capillary rise for liquid B. (b) The maximum bubble pressure, ΔP, is given by $\Delta P = 2\gamma/r$. It will, therefore, be **twice** as much for liquid B as for liquid A.

24. In the bubble-pressure method, the maximum pressure due to meniscus curvature occurs when the nascent bubble has a minimum radius of curvature, which, for small tubes, is that of the tube itself. ΔP is then $2\gamma/r$, where r is 0.05 cm in this case. The height h of water corresponding to this pressure is ρgh, so $h = 2 \times \gamma/1.0 \times 980 \times 0.05 = 0.0408\gamma$. The total pressure is the above, plus the hydrostatic head at the level of immersion; that is, $11.6 = 10 + 0.0408\gamma$ or $\gamma = 1.6/0.0408 = \textbf{39.3 dyne/cm.}$

25. (a) By Tate's law, $W_i = 2\pi r\gamma$.
  (b) If $r/V^{1/3}$ is to be 0.5, then $V = r^3/0.125$ and $W = \rho Vg = 0.8\,gr^3/0.125$. But $W = 2\pi r\gamma\Psi = 2\pi r\gamma \times 26 \times 0.65$, so we have the equation $2\pi r\gamma \times 26 \times 0.65 = 0.8\,r^3g/0.125$, whence $r^2 = 0.017$ and $r = \textbf{0.148 cm.}$ $W = 2\pi r\gamma = 2\pi \times 0.148 \times 26 = 24.1$ or, in grams, $W = \textbf{0.0246.}$

26. The only difference between this case and that for capillary rise in a cylindrical capillary is that one of the radii of curvature is infinite; that is, $\Delta P = \gamma(1/R_1 + 1/R_2)$ where $1/R_2$ is now zero. Hence $\rho gh = \gamma/r$ (assuming $R_1 = r$, the half-distance between plates). In this present case, $\gamma = r\rho gh = 0.05 \times 1.10 \times 980 \times 1.30 = \textbf{70 dyne/cm.}$

27. The analysis is similar to that for the maximum bubble-pressure method. For any height less than the maximum, the hydrostatic head ρgh must be just balanced by an equal ΔP across the meniscus, where ΔP will be given by the Laplace equation, $\Delta P = 2\gamma/R$ (R is the radius of curvature of the drop, assumed to be a section of a sphere). h will be a maximum when

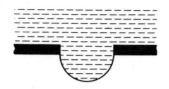

FIGURE 8-17

R is a minimum, and this will occur when the nascent drop is just hemispherical, with radius equal to that of the hole, as shown in Fig. 8-17. Then $\rho g h_{\max} = 2\gamma/r$, or $h_{\max} = 2 \times 72/1 \times 980 \times 0.002 = $ **73 cm.**

28.   Since length of the drop is just d_e, the diameter d_s is also the diameter of the tip, or 0.4 cm. S is then $0.4/d_e$ or $l = 0.4/S$. From Eq. (8-10) $d_e^2/H = \gamma/\rho g = 72/980$ or $l^2/H = 0.073$. It follows that $(0.4)^2/S^2 H = 0.073$ or $S^2 H = 0.16/0.073 = 2.2$. By trial and error, this requirement is satisfied if S is about 0.92 (and $1/H$ is 0.39), and therefore $l = 0.4/0.92 = $ **0.44 cm.**

29.   The pressure is given by the Laplace equation $\Delta P = 2\gamma/R$, where R is the radius of the bubble (if a section of a sphere); for ΔP to be a maximum, R must be at a minimum, and this will indeed be true when the bubble is just hemispherical, as illustrated in Fig. 8-18. At this point, $R = r$, so the proposition is correct. For the particular case, then, $\Delta P = \rho_{Hg}gh = 13.6 \times 980 \times 0.03 = 400$ dyne/cm^2. Then $\gamma = 400 \times 0.1/2 = $ **20 dyne/cm.**

30.   The maximum diameter, d_e, is 2×2.8 or 5.6 cm, by measuring on Fig. 8-12. The diameter 5.6 units up is 2×2.0 or 4.0 cm, so $d_s/d_e = 4.0/5.6 = 0.71$; S is then 0.71, and from the table given following Eq. (8-10), $1/H$ is about 0.77. By Eq. (8-10), d_e^2 is $30/0.9 \times 980 \times 0.77 = 0.044$, and $d_e = 0.21$ cm. The magnification factor is then $5.6/0.21$ or about 27 fold. As reproduced here, however, 1 cm on the scale of Fig. 8-13 measures 0.7 cm actual distance. The magnification factor of Fig. 8-13 is thus 27×0.7 or about **19 fold.**

FIGURE 8-18

9

SOLUTIONS

COMMENTS

This chapter is concerned primarily with solutions of nonelectrolytes. You will become acquainted with Raoult's law, both as a limiting law and as an ideal law. In this second capacity, the Raoult's law statement that $P_i = N_i P_i^0$ is taken to apply to all components of a solution, at all compositions. As a limiting law, the statement: $P_A = N_A P_A^0$, approaches validity only in the limit of N_A approaching unity, but now applies to nonideal solutions. For such solutions a second limiting law, Henry's law, applies in the limit of N_A approaching zero; this second limiting behavior may be written $P_A = k_A N_A$. You are expected to keep these two limiting laws in mind in drawing partial pressure versus composition diagrams.

Such diagrams may show just the plot of total vapor pressure versus solution composition; this will be a straight line connecting the two P^0 values if the solution is ideal, and will otherwise be curved and may have a minimum or a maximum. A second curve may now be included, which gives the plot of vapor pressure versus equilibrium vapor composition. The combined plot takes on the character of a phase diagram in that it shows a liquid, a vapor, and a two-phase region; a system whose gross composition and vapor pressure is such as to locate a point in the two-phase region will consist of some vapor and some liquid. Furthermore, the relative amounts of the two phases can be obtained by a simple graphical method of making a material balance calculation, known as the *lever principle*.

Vapor-pressure diagrams may be converted to boiling-point ones, and vice versa. Many of the problems deal with these interconversions, and it is helpful to remember that the two types of diagrams for a given system resemble each other if one is turned upside down. Thus a system showing a maximum in its vapor-pressure diagram will show a minimum in its boiling-point diagram. Qualitatively, positive deviation from ideality is interpreted on a molecular basis as being due to a tendency toward immiscibility of the two components. In terms of intermolecular forces, this amounts to saying that A–A and B–B type forces are greater than A–B type forces. The increasing degree of manifestation of this tendency leads to maximum vapor-pressure (and minimum boiling-point) diagrams and, finally, to the limiting case of complete immiscibility. At this extreme, we get into steam-distillation problems.

Conversely, negative deviation from ideality is interpreted qualitatively as due to a tendency toward association or compound formation. Increasing degree of this tendency leads to minimum vapor-pressure (and maximum boiling-point) diagrams.

Finally, you will find a small introduction to partial molar quantities via a few problems involving partial molar volumes.

EQUATIONS AND CONCEPTS

Ideal Solutions

Some Raoult's law equations are the following:

$$P_i = N_i P_i^0 \quad \text{(Raoult's law)} \tag{9-1}$$

For two components,

$$P_{tot} = P_A + P_B = N_A P_A^0 + N_B P_B^0 = P_B^0 + N_A(P_A^0 - P_B^0) \tag{9-2}$$

The vapor composition is $Y_i + P_i/P_{tot}$, and for two components

$$Y_A = \frac{N_A P_A^0}{P_B^0 + N_A(P_A^0 - P_B^0)}, \qquad Y_B = 1 - Y_A \tag{9-3}$$

There is no volume change and no heat of mixing in the case of ideal solutions.

Nonideal Solutions

Raoult's law and Henry's law are now limiting laws. Thus

$$P_A \rightarrow N_A P_A^0 \quad \text{as} \quad N_A \rightarrow 1 \tag{9-4}$$

and

$$P_A \rightarrow N_A k_A \quad \text{as} \quad N_A \rightarrow 0 \tag{9-5}$$

where k_A is the Henry's law constant. Deviation from ideality may be expressed by the use of activities or activity coefficients based on either Raoult's law or Henry's law:

$$P_A = a_A P_A^0 = f_A N_A P_A^0 \qquad \text{(Raoult's law)} \qquad (9\text{-}6)$$

$$P_A = a'_A k_A = y_A N_A k_A \qquad \text{(Henry's law)} \qquad (9\text{-}7)$$

Remember that, if k_A/P_A^0 is greater than unity, then for a two component system, k_B/P_B^0 must also be greater than unity, and similarly, if k_A/P_A^0 is less than unity, k_B/P_B^0 will be less than unity. That is, both components will show the same direction of deviation from ideality.

The total volume of a two-component solution is given by

$$V = n_A \overline{V}_A + n_B \overline{V}_B \qquad (9\text{-}8)$$

where $\overline{V}_A$ and $\overline{V}_B$ are the partial molar volumes. If V per mole is plotted against mole fraction, the two $\overline{V}$'s are given by the intercepts of the tangent to the plot at the given composition.

Regular Solutions

The effective mole fraction of a component may be written as $a_i = P_i/P_{tot}$ where a denotes activity or effective composition. In the case of a nonideal solution, a_i will not be equal to N_i and we may write $a_i = f_i N_i$ where f_i is called the activity coefficient.

So-called regular solutions show a symmetric behavior of the activity coefficients, and for a two-component solution this is given by:

$$\log f_A = -\alpha N_B^2 \quad \text{and} \quad \log f_B = -\alpha N_A^2 \qquad (9\text{-}9)$$

where α has the same value in both equations.

Immiscible Liquids

Each component contributes its full vapor pressure to the total vapor pressure. For two-component systems, it then follows that $Y_A = P_A^0/(P_A^0 + P_B^0)$. In the case of steam distillation, $P_w^0 + P_x^0 = 1$ atm, and P_w^0 is known from the boiling temperature, and hence so is P_x^0, and therefore Y_x for the vapor.

Lever Principle

If a two-component system of over-all composition N_A is partitioned into n_l moles of composition N'_A and n_v moles of vapor composition Y_A, then

$$\frac{n_l}{n_l + n_v} = \frac{Y_A - N_A}{Y_A - N'_A} \qquad (9\text{-}10)$$

In the case of diagrams showing both the liquid- and vapor-composition lines, the differences $(Y_A - N_A)$ and $(Y_A - N'_A)$ may be read off the graph directly and may in fact be measured in arbitrary units by means of a ruler since their ratio is dimensionless. If composition is in weight fraction, the lever principle gives the weight fraction of the system present as liquid; that is,

$$\frac{w_l}{w_l + w_v} = \frac{W_{A(vapor)} - W_{A(over\,all)}}{W_{A(vapor)} - W_{A(liquid)}} \tag{9-11}$$

where W denotes weight-fraction composition.

PROBLEMS

1. (26 min) Liquids A and B form ideal solutions. A mixture of the vapors which is 40 mole % in A is contained in a piston and cylinder arrangement which is kept at a certain constant temperature T. The system is then slowly compressed. Given that P_A^0 and P_B^0 are 0.4 and 1.2 atm, respectively, at T, calculate the total pressure at which liquid first begins to condense out and also the composition of this liquid. Calculate the composition of that solution whose normal boiling point is T.

2. (10 min) Using the vapor-pressure plots in Fig. 9-1, calculate the normal boiling point of (a) pure liquid A and B, (b) a solution of $N_A = 0.25$ and (c) a

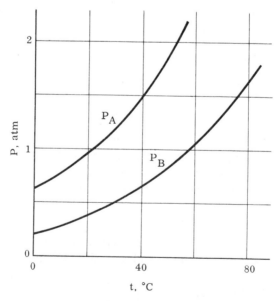

FIGURE 9-1

solution of $N_A = 0.75$. [Your answers to (b) and (c) need be correct only to a few degrees, provided the procedure is clear.] (Assume ideal solutions.)

(d) Calculate the vapor compositions for the solutions in (b) and (c) when at their normal boiling points, and make a semiquantitative plot of the boiling-point diagram. Label the phases present in each phase region. (Assume ideal solutions.)

(e) Calculate the normal boiling points of mixtures (b) and (c), assuming now that A and B are completely immiscible.

3. (10 min) Liquids A and B form ideal solutions. They are isomeric so that their molecular weights are the same (and hence weight and mole fractions). A solution of composition N_A is found to have a vapor pressure of 650 mm Hg at 50°C. It is then distilled (with no reflux) until half has been collected as condensate. The condensate has a composition $N'_A = 0.60$, and the residual liquid a composition $N''_A = 0.40$ and a vapor pressure of 600 mm Hg at 50°C. Calculate N_A, P_A^0, and P_B^0.

4. (9 min) Liquids A and B form an ideal solution. At 50°C, the total vapor pressure of a solution consisting of 1 mole of A and 2 moles of B is 250 mm Hg; on addition of 1 more mole of A to the solution, the vapor pressure increases to 300 mm Hg. Calculate P_A^0 and P_B^0.

5. (15 min) Toluene and xylene form an ideal solution. At 20°C the vapor pressures of pure toluene and xylene are 22 and 5 mm Hg, respectively. Make a graph showing how the vapor pressure of a solution should vary with composition. Show, on the same graph, the approximate appearance of the plot of equilibrium vapor compositions and explain how one particular point on this plot is calculated.

If the force of attraction between a toluene and a xylene molecule were less than in the ideal case, that is, the toluene–xylene pair were less strongly attracted than the toluene–toluene and xylene–xylene pairs, show graphically the changes qualitatively to be expected in the above plot.

6. (10 min) (Final examination question) Benzene and toluene form essentially ideal solutions, and a particular mixture which consists of two moles of benzene and three moles of toluene has a total vapor pressure of 280 mm Hg at 60°C. If one additional mole of benzene is added to the solution, the new total vapor pressure is now 300 mm Hg. Calculate the vapor pressures of pure benzene and toluene at 60°C.

7. (7 min) A and B form ideal solutions, and the vapor pressures of pure A and B are 300 mm Hg and 100 mm Hg, respectively, at 50°C. The vapor in equilibrium with a certain solution at 50°C is 0.5 mole fraction in A. Calculate the composition of this solution, and its total vapor pressure. This may be done graphically, using Fig. 9-2.

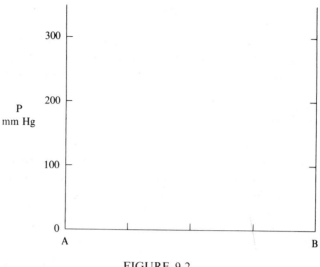

FIGURE 9-2

8. (18 min) Liquids A and B form an ideal solution. A certain solution contains 25 mole % of A, whereas the vapor in equilibrium with the solution at 25°C contains 50 mole % of A. The heat of vaporization of A is 5 kcal/mole; that of B is 7 kcal/mole.

(a) Calculate the ratio of the vapor pressure of pure A to that of pure B at 25°C.

(b) Calculate the value for this same ratio at 100°C. (It is not necessary to obtain a final numerical answer to (b); set up the equations and substitute in numbers so that the desired ratio is the only unknown.)

9. (15 min) Two solutions of A and B are available. The first is known to contain 1 mole of A and 3 moles of B, and its total vapor pressure is 1.0 atm. The second is known to contain 2 moles of A and 2 moles of B; its vapor pressure is greater than 1 atm, but it is found that this total vapor pressure may be reduced to 1 atm by the addition of 6 moles of C. The vapor pressure of pure C is 0.80 atm. Assuming ideal solutions, and that all these data refer to 25°C, calculate the vapor pressures of pure A and of pure B.

10. (21 min) Given that liquids A and B form ideal solutions and that the vapor pressures of the pure liquids at 80°C are 100 and 600 mm Hg respectively, plot the total vapor pressure of solutions at 80°C versus their mole fraction. Calculate several vapor compositions and, by means of these points, draw on your diagram a semiquantitative vapor-composition curve.

A solution containing 40 mole % of B is placed in a previously evacuated container of such size that, at 80°C, one-third of the liquid vaporizes. Calculate the compositions of the final liquid and vapor phases.

Solution of $N_B = 0.6$ is evaporated in an open container at a constant temperature of 80°C until the total vapor pressure falls to 80 % of its initial value (i.e., by 20%). What is the final liquid composition, and roughly how many moles of liquid remain after the evaporation to this point, assuming that initially there was 1 mole of liquid?

11. (12 min) Water and CCl_4 are completely immiscible; their vapor pressures are shown in Fig. 9-3 as a function of temperature. A *gaseous* mixture of water and CCl_4 (i.e., a mixture of the vapors) at 1 atm constant total pressure is cooled from 100°C. At 80°C pure liquid water begins to condense out.

Estimate, by means of the graph, the mole fraction of water in the vapor mixture and the temperature to which the mixed vapor must be cooled before liquid CCL_4 begins to condense out.

12. (10 min) When a mixture of water and chlorobenzene (mutually immiscible) is distilled at an external pressure of 740.2 mm Hg, the mixture is found to boil at 90.3°C, at which temperature the vapor pressure of pure water is 530.1 mm Hg. Calculate the percent by weight of chlorobenzene in the distillate. (Molecular weight is 112.)

13. (12 min) In a steam distillation of an organic oil, the mixture of oil and water (mutually immiscible) boils at 99°C, under 1 atm pressure. The vapor

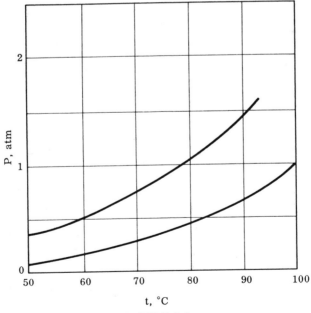

t, °C

FIGURE 9-3

pressure of water is 730 mm Hg at the temperature. The collected condensate is 10% by weight organic compound; calculate the molecular weight of the compound.

14. (15 min) Liquid water and benzene may be considered as virtually immiscible; information on their vapor pressure is given below. A closed container holds 0.6 mole of water and 0.4 mole of benzene and initially is at 100°C. Under this condition, only the mixed vapors are present, and the total pressure is 1 atm. The container is then cooled to 60°C. Calculate the number of moles of water (or benzene) now present as liquid, and the new total pressure of the vapor phase. The vapors are ideal gases.

The vapor pressure of water is 0.2 atm at 60°C, and the vapor pressure of benzene is 0.5 atm at 60°C and 2 atm at 100°C.

---

15. (15 min) Liquids A and B form nonideal solutions, and the total vapor pressure versus composition (for a given temperature) is plotted in Fig. 9-4.

(a) Draw on the graph qualitatively (but carefully) the separate vapor pressure versus composition plots for each component, that is, P_A and P_B.

(b) Add also to the graph the probable appearance of the vapor-composition line.

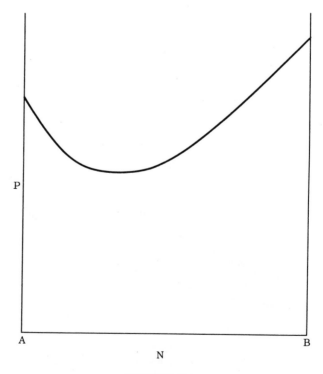

FIGURE 9-4

(c) Sketch in a separate diagram the probable appearance of the boiling-point diagram for this system.

16. (4 min) Give a brief chemical or molecular explanation of why, if one component of a nonideal binary solution shows negative deviation from ideality in its variation of partial pressure with composition, the other component will do likewise.

17. (9 min) The vapor-pressure diagram for the chloroform–isopropyl ether system is such as to suggest that there is a tendency toward compound formation. Sketch the probable appearance of the two partial-pressure and the total vapor-pressure curves at 60°C. At this temperature, the vapor pressures of chloroform and of isopropyl ether are 700 mm Hg and 800 mm Hg, respectively. Although the curves need only be qualitative, pay attention to their general shape.

18. (18 min) Figure 9-5 shows the ethyl acetate–water system. A solution of mole fraction 50% is boiled in an open beaker until the boiling point rises by

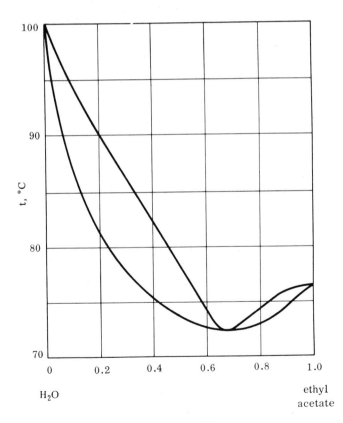

FIGURE 9-5

5°C. Calculate or make clear how you obtain graphically (a) the composition of the solution when the boiling was stopped and the initial and final boiling points, (b) the average composition of the distillate (if the vapors had been collected and condensed), (c) the moles of ethyl acetate remaining in the liquid when the boiling was stopped, assuming that the initial mixture contained 1 mole, (d) the composition of the last drop of liquid had the boiling been continued, and (e) the probable appearance of the vapor-pressure diagram for this system.

19. (9 min) The boiling-point diagram for two miscible liquids is shown in Fig. 9-6. (a) On boiling a 50 mole % solution in an open container, what will be the composition of the first vapor and of the last vapor? (b) If the solution is heated in a closed system at 1 atm constant pressure, what will be the liquid composition when half the solution is vaporized? What will be the temperature?

20. (15 min) Acetone and ether form a nonideal solution. At 30°C, the vapor pressure of pure acetone is 280 mg Hg, and that of pure ether is 650 mm Hg. The total vapor pressure of a 0.5 mole fraction solution is 600 mm Hg at 30°C, and the Raoult's law activity coefficient of the ether in this solution is 1.30.

(a) Calculate the activity coefficient for the acetone in the above solution. (b) Sketch the appearance of the vapor-pressure and boiling-point diagrams

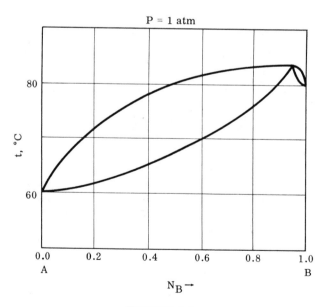

FIGURE 9-6

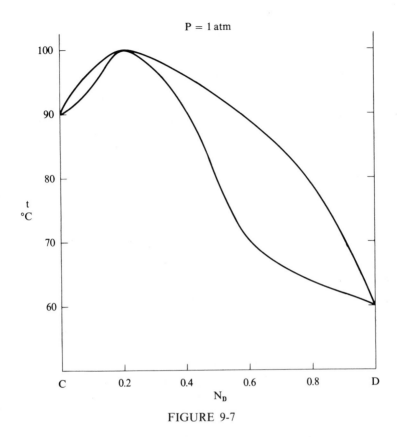

P = 1 atm

t
°C

C   0.2   0.4   0.6   0.8   D

N_D

FIGURE 9-7

for this system. Show the partial vapor-pressure curves on the vapor-pressure diagram, and the liquid- and vapor-composition curves on both diagrams.

21. (12 min) The normal boiling-point diagram is shown for liquids C and D in Fig. 9-7. One mole of solution of N_D = 0.6 is boiled in an open (non-refluxing) distillation setup until the boiling point rises 10°C, the vapor being condensed and collected in another container. Calculate the moles of liquid remaining and the normal boiling point (n.b.p.) of the condensate. Give also the partial pressure of C in equilibrium with a maximum boiling solution.

22. (15 min) A particular nonideal solution of A and B is normal boiling at 60°C, and for this solution, the Raoult's law activity coefficients of A and B and 1.3 and 1.6, respectively. The activity of A is 0.6 and P_A^0 is 400 mm Hg. Calculate (a) the mole fraction of A in the vapor phase which is in equilibrium with the solution and (b) the value of P_B^0.

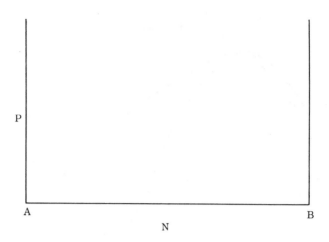

FIGURE 9-8

23. (10 min) Liquids A and B form nonideal solutions. The Henry's law constant for liquid A is 0.2 atm ($P_A = k_A N_A$), and the vapor pressures of the respective pure liquids are 1 atm and 0.5 atm, all at 25°C. Draw qualitative but careful sketches of the plots of P_A and P_B versus composition, using Fig. 9-8.

24. (12 min) Liquids A and B form nonideal solutions, with $P_A^0 = 700$ mm Hg and $P_B^0 = 400$ mm Hg. The Raoult's law activity coefficients are given in the table below. (a) Calculate k_A, the Henry's law constant for A. (b) Calculate also the Henry's law activity coefficient for A at $N_A = 0.6$.

N_A	1	0.9	0.8	0.7	0.6	0.5	0.4	0.3	0.2	0.1
y_A	1.00	0.99	0.905	0.845	0.750	0.645	0.570	0.535	0.440	0.430

25. (18 min) Two-tenths mole of a solution of composition N_B is placed in a 2-liter vessel at 25°C. The vessel initially is evacuated so that, on introduction of the solution, partial evaporation occurs yielding residual solution of composition N_B' and equilibrium vapor of composition Y_B (still at 25°C). The equilibrium vapor is withdrawn and condensed. The normal boiling point of this condensate is 45°C.

By means of the graphs in Fig. 9-9, calculate (a) the composition of the final condensate, (b) the value of N_B', (c) the total vapor pressure of this residual solution at 25°C, (d) the total moles of equilibrium vapor, and (e) the value of N_B.

26. (16.5 min) One mole of a 50 mole % solution of propyl alcohol in ethyl alcohol is distilled until the boiling point of the solution rises to 90°C. The condensate is allowed to accumulate in a cooled receiver and, after being mixed to insure uniformity, its vapor pressure is found to be 1,066 mm Hg

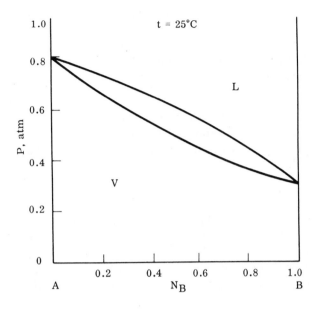

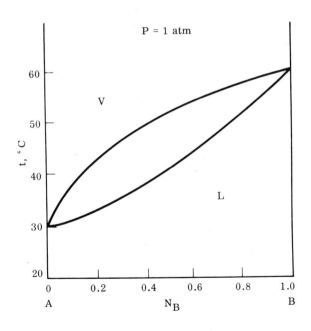

FIGURE 9-9

when measured .at 90°C. The vapor pressures of pure ethyl and propyl alcohols are 1,190 and 574 mm Hg at 90°C, respectively. Assuming that the solutions and vapors are ideal, calculate (a) the mole fraction of ethyl alcohol in the 90°C boiling liquid in the distilling flask, (b) the mole fraction of ethyl alcohol in the distillate, and (c) the number of moles of ethyl alcohol that were distilled.

27. (12 min) Liquids A and B form nonideal, but regular, solutions; P_A^0 and P_B^0 are 0.75 atm and 1.50 atm at 25°C, respectively. If for a solution of $N_A = 0.50$, P_A at 25°C is 0.25 atm, calculate the activity coefficient of species A and the value of P_B.

---

28. (10 min) The graph of Fig. 9-10 gives the average molar volume V for the system ethyl iodide–ethyl acetate, as a function of N, the mole fraction of ethyl iodide. Calculate or obtain graphically the partial molar volumes $\overline{V}_1$ and $\overline{V}_2$ for ethyl iodide and ethyl acetate, respectively, for a solution of composition $N_1 = 0.75$. Calculate ΔV for the process: 3 ethyl iodide + ethyl acetate = solution.

29. (7.5 min) The molar volume of pure methanol is 40 cc/mole. Also, the volume of a solution containing 1,000 g of water and n moles of methanol is given by $V = 1,000 + 35n + 0.5n^2$. Calculate the partial molar volume $\overline{V}$ for methanol for m (molality) = 0 and for $m = 1$. Calculate also ΔV for the process: 55.5 H_2O + CH_3OH = solution.

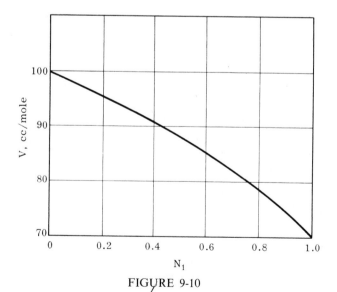

FIGURE 9-10

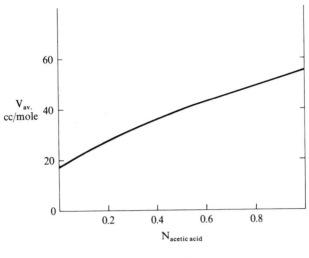

FIGURE 9-11

30. (12 min) Figure 9-11 gives the average molar volume of solutions of acetic acid (HAc) in water. Calculate, showing clearly your procedure, (a) the partial molar volumes for water and for acetic acid for a solution which is 0.2 mole fraction in acetic acid, (b) the total volume of a solution containing 0.2 mole of acetic acid and 0.8 mole of water, and (c) the volume change when 0.2 mole of acetic acid is mixed with 0.8 mole of water.

ANSWERS

1. At the beginning of condensation, the vapor will still be essentially 0.4 mole fraction in A; hence $P_A/P_{tot} = 0.4$. But $P_{tot} = P_A^0 N_A + P_B^0 N_B = 0.4 N_A + 1.2 N_B = 1.2 - 0.8 N_A$. The first equation then becomes: $0.4 N_A/(1.2 - 0.8 N_A) = 0.4$, from which $N_A = $ **0.67**. P_{tot} is then $1.2 - 0.8 \times 0.67 = $ **0.67 atm.** If the solution is boiling at T, then $P_{tot} = 1$ atm. Hence $N_A = (1.2 - 1)/0.8$ or $N_A = $ **0.25.**

2. (a) The normal boiling points are simply those temperatures at which the vapor pressures are 1 atm, namely, about **22°C** for A and **58°C** for B.

(b) The equation 1 (atm) $= P_A^0 \times 0.25 + P_B^0 \times 0.75$ must be satisfied; with a few trials, an approximate solution is found at about **45°C**, with $P_A^0 = 1.7$ and $P_B^0 = 0.75$.

(c) The equation is now $1 = 0.75 P_A^0 + 0.25 P_B^0$, for which a solution is found at about **27°C**, with $P_A^0 = 1.2$ and $P_B^0 = 0.5$.

(d) The mole fractions of A in the vapor are given by $Y_A = P_A/1$ or about **0.42** and **0.9** for solutions (b) and (c), respectively. These various values are located in Fig. 9-12 to provide an approximate boiling-point plot.

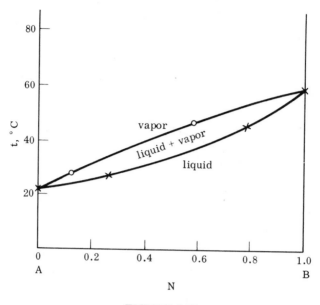

FIGURE 9-12

(e) If A and B are immiscible, one now seeks a solution to: 1 (atm) = $P_A^0 + P_B^0$, which is found at approximately **10°C.**

3. By material balance in A, we can write $N_A = (N_A' + N_A'')/2$ since if there were 1 mole of solution, after distillation there would be 0.5 mole of condensate and 0.5 mole of residual liquid. Then $N_A = (0.6 + 0.4)/2 = $ **0.5.**

We have the total vapor pressure of the original and residual solutions; hence

$$650 = 0.5P_A^0 + 0.5P_B^0 \quad \text{and} \quad 600 = 0.4P_A^0 + 0.6P_B^0$$

On solving these equations simultaneously, $P_A^0 = $ **900** mm Hg and $P_B^0 = $ **400** mm Hg.

4. The two conditions yield two equations:

$$250 = \tfrac{1}{3}P_A^0 + \tfrac{2}{3}P_B^0 \quad \text{and} \quad 300 = \tfrac{1}{2}P_A^0 + \tfrac{1}{2}P_B^0$$

Simultaneous solution yields $P_B^0 = $ **150 mm Hg** and $P_A^0 = $ **450 mm Hg.**

5. As shown in Fig. 9-13, the liquid-composition line is simply the straight line connecting the two P^0 values. The vapor-composition line is curved, and must lie below the liquid line. As an illustration, the vapor composition in equilibrium with liquid of toluene mole fraction 0.2 will be $Y_T = 0.2 \times 22/$ $(0.2 \times 22 + 0.8 \times 5) = 0.53$. In this case a positive deviation from ideality is expected, and the qualitative appearance of the vapor-pressure diagram is as shown in Fig. 9-13.

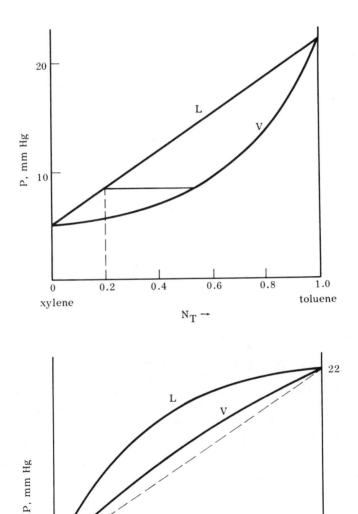

FIGURE 9-13

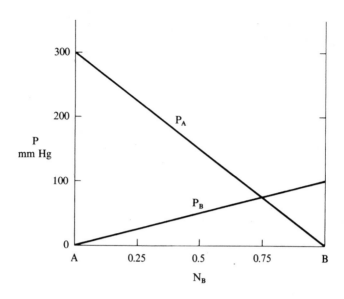

FIGURE 9-14

6. The two solutions are 0.4 and 0.5 mole fraction in benzene, so the simultaneous equations are

$$280 = 0.4(P_B^0 - P_T^0) + P_T^0$$
$$300 = 0.5(P_B^0 - P_T^0) + P_T^0$$

On solving these, $P_T^0 = $ **200 mm Hg**, and $P_B^0 = $ **400 mm Hg.**

7. As shown in Fig. 9-14, the point of crossing of the two partial pressure lines gives the composition of solution for which the vapor will be 0.5 mole fraction; this is a solution **0.25 mole fraction** in A. The total vapor pressure is then 2×80 or about **160 mm Hg.**

8. (a) We can write $Y_A/Y_B = P_A/P_B$ where Y denotes mole fraction in the vapor phase; in this case $Y_A = Y_B$. Then $1 = P_A/P_B = 0.25P_A^0/0.75 P_B^0$ or $P_A^0/P_B^0 = $ **3.**

(b) For a solution of the same composition, but at 100°C, we can write

$$\ln P_A'/P_A = \ln P_A'^0/P_A^0 = \frac{\Delta H_A}{R}(1/298 - 1/373)$$

and similarly for component B. If the two equations are then subtracted, we obtain

$$\ln R'/R = \frac{(\Delta H_A - \Delta H_B)}{R}(1/298 - 1/373)$$

where R denotes the ratio P_A/P_B. Insertion of $R = 3$, and of values for R (gas constant) and the ΔH's completes the solution.

9. The data provide the following two expressions for total pressure:

$$1 = \tfrac{1}{4}P_A^0 + \tfrac{3}{4}P_B^0 \quad \text{or} \quad 4 = P_A^0 + 3P_B^0$$

and

$$1 = \tfrac{1}{5}P_A^0 + \tfrac{1}{5}P_B^0 + \tfrac{3}{5}P_C^0 \quad \text{or} \quad 5 = P_A^0 + P_B^0 + 2.4$$

On solving these two equations simultaneously, one obtains $P_B^0 = \textbf{0.7 atm}$ and $P_A^0 = \textbf{1.9 atm.}$

10. The liquid line in Fig. 9-15 is obtained simply by drawing a straight line connecting the two P^0 values. A few vapor compositions may be calculated from the equation $Y_B = P_B/P_{tot}$. Thus for $N_B = 0.2$, $P_B = 120$ and from the

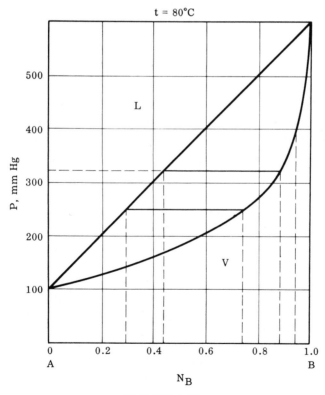

t = 80°C

P, mm Hg

500

400

300

200

100

L

V

0   0.2   0.4   0.6   0.8   1.0
A   B

N_B

FIGURE 9-15

graph, $P_{tot} = 200$, hence $Y_B = 0.6$. Similarly, for $N_B = 0.4$, one finds $Y_B = 240/300 = 0.80$, and for $N_B = 0.6$, $Y_B = 360/400 = 0.90$. These points then allow a rough curve of vapor composition to be drawn in.

Since $\frac{1}{3}$ of the $N_B = 0.4$ solution evaporates, one looks for that tie line which will, by the lever principle, predict this proportioning. As shown on the graph, the ends of this line give liquid and vapor compositions of approximately 0.3 and 0.75 mole fraction in B, respectively.

The initial vapor pressure at 80°C is read off the graph as 400 mm Hg, and at 80% of this value, or 320 mm Hg, the liquid composition is read off the graph to be about $N_B = \mathbf{0.45}$. The initial and final vapor compositions similarly can be read off the graph as approximately 0.92 and 0.90, giving an average value of about 0.91. By the lever principle, that is, by material balance, the moles of liquid remaining must then be about

$$n_l = (0.91 - 0.6)/(0.91 - 0.45) = \mathbf{0.67}$$

11.   From the graph, the vapor pressure of water at 80°C is 0.45 atm; since the total pressure of the vapor mixture is 1 atm, the partial pressure of the CCl_4 must then also be 0.55 atm. The vapor is thus **45 mole %** in water. At the point where CCl_4 begins to condense out, we have both liquid phases present, and must now look for a temperature at which the sum of the liquid vapor pressures is 1 atm. Inspection of the graph gives about **70°C.**

12.   The chlorobenzene must be supplying $740.2 - 530.1$ or 210.1 mm Hg vapor pressure and the ratio of moles of it to moles of water in the distillate must then be 210.1/530.1 or 0.40. Per 100 g of water, or 5.55 mole, there will then be 0.4×5.55 or 2.2 moles of chlorobenzene and hence 2.2×112 or 246 g. The weight % chlorobenzene will then be 246/346 or **0.71 (71%).**

13.   The compound evidently exerts 760–730 or 30 mm Hg vapor pressure, and the ratio of moles of compound to water in the condensate must be 30/730 or 0.041. 100 g of condensate would have 90 g of water or 5 mole, and hence should have 0.041×5 or 0.205 mole of compound. Since 10 g of compound are present, the molecular weight is 10/0.205 or **49.**

14.   The vapors at 100°C evidently consist of 0.6 atm of water vapor and 0.4 atm of benzene vapor. On cooling to 60°C, if no condensation were to occur, the total pressure should drop to $1 \times 333/373 = 0.89$, and the two partial pressures, to 0.53 atm and 0.36 atm, respectively. However, 0.53 atm exceeds the vapor pressure of liquid water, so condensation of water will take place until the water-vapor pressure drops to 0.2 atm. At this point the total pressure will be $0.2 + 0.36$ or 0.56 atm (there being no condensation of the benzene). Mole H_2O condensed: $0.6(0.53 - 0.2)/0.53 = \mathbf{0.37}$.

15.   The curves for P_A and P_B are shown on Fig. 9-16. The main points are that they should show a negative deviation from ideality, and should

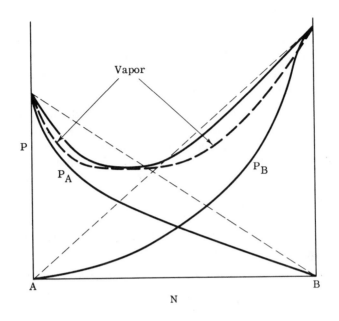

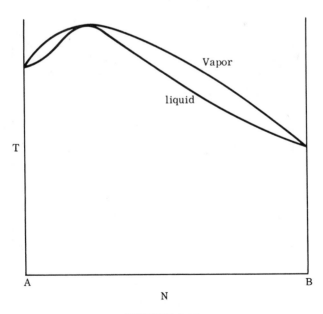

FIGURE 9-16

approach the Raoult's law line at one end and a Henry's law line at the other. The vapor-composition line should touch the liquid-composition line at the point of minimum pressure. The boiling-point diagram is expected to be a maximum boiling-point type, with the vapor line now above the liquid-composition line, and touching it at the maximum. Also, the boiling point of A should appear higher than that of B.

16.   Negative deviation is considered to be due to the presence of some degree of association, that is, to a situation where the forces between A molecules or between B molecules are less than those between A and B molecules. Clearly, the situation is symmetric; if A is partially involved in association with B, B is similarly involved. Both partial pressures should then show a negative deviation from ideality.

17.   Negative deviation is expected if there is a tendency toward compound formation. The partial-pressure plots of Fig. 9-17 should show an approach to Raoult's law at one end and to Henry's law at the other.

18.   (See Fig. 9-18.) (a) The initial boiling point is given by the liquid line at 50%, that is, **74°C**; the final boiling point, of course, is then **79°C**. At 79°C, the liquid line gives a composition of about $N_2 = 0.25$.

(b) The initial vapor composition is given by the vapor line as 0.6, and the final one as about 0.50. The average value is then about **0.55.**

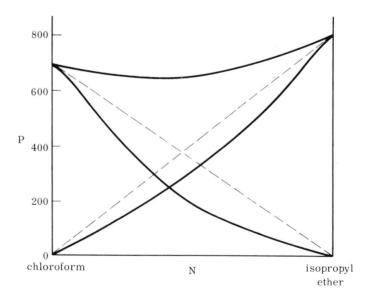

FIGURE 9-17

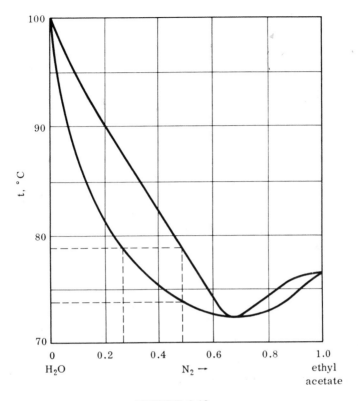

FIGURE 9-18

(c) By material balance in ethyl acetate:

initial moles = moles left + moles vapor
$$0.5 \times 1 = 0.25 \times n + 0.55(1 - n)$$

from which n, the moles of liquid left, is about 0.1. The moles of ethyl acetate remaining is then about **0.042.**

(d) The composition of the last drop of liquid is pure water (we are boiling away in an open system).

(e) The general appearance of the vapor-pressure diagram is sketched in Fig. 9-19.

19. (a) The 50% composition line intercepts the liquidus line at about 68°C, and the horizontal line drawn over to the vapor-composition curve intersects it at about $N_B = $ **0.1.** The last vapor composition will be that of the maximum boiling point, or about $N_B = $ **0.95.**

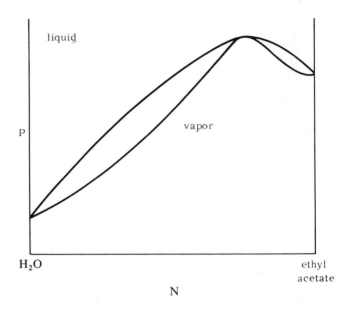

FIGURE 9-19

(b) On half vaporization in a closed system, the temperature must be such that the tie line is bisected by the system composition line, and, as indicated in Fig. 9-20, this occurs at about **75°C**. The liquid composition is $N_B = 0.75$.

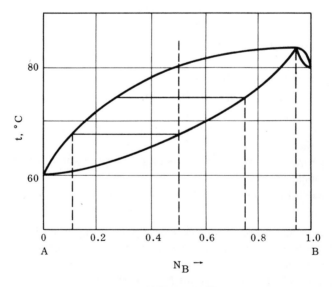

FIGURE 9-20

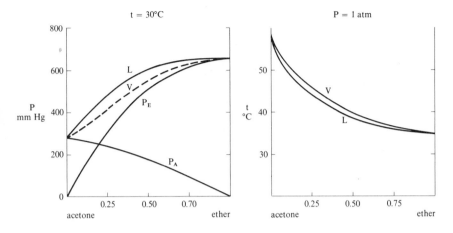

FIGURE 9-21

20. (a) P_E, the partial pressure of the ether, is evidently $1.30 \times 0.5 \times 650 = 420$ mm Hg, so P_A must be 600–420 or 180 mm Hg. The activity coefficient for acetone is then $150/0.5 \times 280 = $ **1.07.** (b) Briefly, both partial pressure curves should show positive deviation from ideality, should approach ideal behavior as the pure component is approached in composition, and should approach a Henry's law line in the limit of high dilution. The boiling-point diagram could, but need not, show a minimum boiling. Qualitative curves are shown in Fig. 9-21.

21. The first vapor to come off has the composition $Y_D = 0.9$, as indicated in Fig. 9-22, while a solution boiling 10° higher, at 80°C, would have $N_D = 0.5$ and would produce vapor of about $0.78 Y_D$. The average vapor composition is then about 0.84, and by the material balance or level principle

$$\frac{n_l}{n} = \frac{(0.84 - 0.6)}{(0.84 - 0.5)} = 0.71$$

The amount of liquid remaining is then **0.71 mole.** In answer to the second question, at the maximum boiling composition, liquid and vapor compositions are the same, namely $N_D = Y_D = 0.2$. Since the total pressure is 1 atm, P_D would then be **0.2 atm.**

22. The actual vapor pressure of A above the solution must be 0.6×400 or 240 mm Hg. Since the solution is normal boiling, P_B must then be $760 - 240$ or 520 mm Hg. Also the mole fraction of A in the vapor must be just $240/760$ or **1.31 (a).** Since activity is equal to activity coefficient times mole fraction, N_A must be $0.6/1.3$ or 0.46, and N_B is then 0.54. The activity of B is $0.54 \times 1.6 = 0.86$, and P_B^0 is then $520/0.86$ or **610 mm Hg (b).**

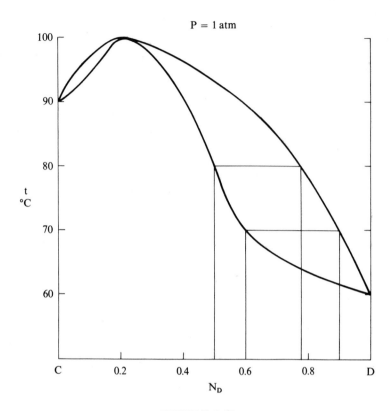

FIGURE 9-22

23. The required plots are shown in Fig. 9-23. Since k_A is less than P_A^0, there is negative deviation from ideality, and B must likewise show a negative deviation.

24. (a) $P_A = y_A N_A P_A^0$, and as N_A approaches zero, it is evident that y_A approaches about 0.43, so that P_A is approaching the straight line $P_A = (0.43 \times P_A^0) N_A = 300\, N_A$. The Henry's law constant is then **300 mm** Hg. (b) P_A should then be 300×0.6 at $N_A = 0.3$, or 180 mm Hg, if Henry's law were being obeyed. The actual P_A is $0.75 \times 0.6 \times 700$ or 314 mm Hg, so the Henry's law activity coefficient is $314/180$ or **1.75**. (*Note:* it will probably help to calculate and plot a full P_A versus N_A graph, although it was not necessary to do so to work the problem.)

25. (a) By marking off the 45°C line in Fig. 9-24 one reads off **0.6** as the mole fraction in B of the condensate. (b) As vapor, before condensation, the tie line shows it was in equilibrium at 25°C with liquid of composition $N_B' = $ **0.8**. (c) The tie line in (b) is located at $P = $ **0.45 atm.** (d) The moles of this vapor that were present must be $n = PV/RT = 0.45 \times 2/0.082 \times 298 =$

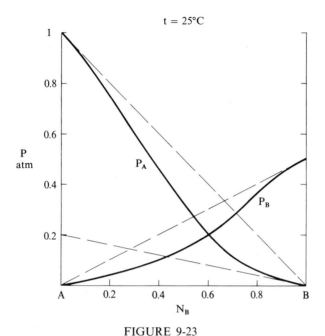

FIGURE 9-23

0.037. (e) By material balance, for example, the lever principle, N_B is given by

$$\frac{0.037}{0.2} = \frac{0.8 - N_B}{0.8 - 0.6}$$

or

$$N_B = 0.8 - 0.037 = \mathbf{0.763}$$

26. (a) We have to assume that by "boiling point" is meant "normal boiling point," that is, $P = 1$ atm. Then $760 = 1{,}190N_E + 574N_P = 574 + 616N_E$, so $N_E = \mathbf{0.30}$. (b) Similarly for the distillate, $1{,}066 = 574 + 616N_E$, $N_E = \mathbf{0.80}$. (c) By material balance, the moles of distillate $= (0.5 - 0.3)/(0.8 - 0.3) = 0.40$ (using the lever principle). The moles of ethanol in the distillate are then $0.40 \times 0.80 = \mathbf{0.32}$.

27. The ideal vapor pressure for component A would be $0.5 \times 0.75 = 0.375$, so the activity coefficient of A is $f_A = 0.25/0.375 = \mathbf{0.67}$. We then have

$$\log f_A = -\alpha N_B^2 = -\alpha(0.5)^2 = \log f_B$$

In this particular case, since $N_A = N_B = 0.5$, it follows that $f_A = f_B$, so

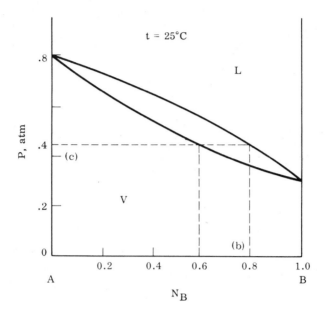

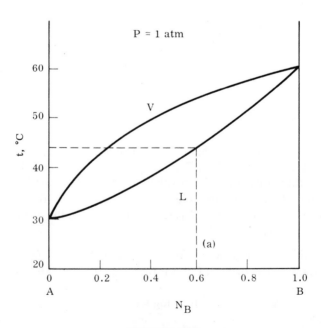

FIGURE 9-24

$f_B = 0.67$. The ideal vapor pressure for B would be $0.5 \times 1.5 = 0.75$, so the actual value will be $0.75 \times 0.67 = \textbf{0.50}$.

28. The average molar volume is the total volume divided by the number of moles, and it turns out (see your text) that the partial molar volumes are very easily obtained from the plot of V against mole fraction. One simply locates the tangent at the particular composition, and the two partial molar volumes are then given by the two intercepts. On doing this, one obtains the intercepts $\overline{V}_2 = \textbf{108}$ and $V_1 = \textbf{73 cc/mole}$. For the process

3 ethyl iodide + ethyl acetate = solution

the volumes are

$3 \times 70 + 100 = 4 \times V$ of a solution of $N_1 = 0.75$ or 4×81

ΔV is then $324 - 310$ or 14 cc.

29. $\overline{V} = (\partial V / \partial n)_{H_2O} = 35 + n$; n is so defined that, when $n = 0$, $m = 0$, and when $n = 1$, $m = 1$, so the two $\overline{V}$ values are **35** and **36 cc/mole**, respectively. For the process

$$55.5 \; H_2O + CH_3OH = \text{solution} \; (m = 1)$$

the volumes are   1,000   40 $(1,000 + 35 + 0.5)$

so $\Delta V = 1,035.5 - 1,040 = \textbf{-4.5 cc}$.

30.   (a) The tangent at $N_{HAc} = 0.2$ gives as intercepts $\overline{V}_{H_2O} = 19$ and $\overline{V}_{HAc} = 65$ (in cc/mole). (b) V may be read directly off the graph as **28 cc**. (c) The volume of the separate pure components would be $0.2 \times 55 + 0.8 \times 18$ or 25.6 cc. ΔV on mixing is then $28 - 25.6$ or about **2.4 cc**.

10

COLLIGATIVE
PROPERTIES

COMMENTS

Vapor-pressure lowering, boiling-point elevation, freezing-point depression, and the osmotic-pressure effects form a natural group. These phenomena have in common the situation consisting of an equilibrium between two phases in which two components (in the simplest case) are present in one phase, but only one of them is present in the second phase.

In the case of vapor-pressure lowering and boiling-point elevation, one of the components is nonvolatile. In that of freezing-point depression, the solid phase consists of one component only. Osmotic pressure develops if there is present a membrane permeable only to one of the components.

With binary systems in general, it is entirely arbitrary as to which component is designated as solvent. Thus a 50 mole % solution of water in alcohol could equally well be described as a 50 mole % solution of alcohol in water. Ordinarily one would be inclined to name as solvent that component present in the larger amount. Thus we would speak of a 40 mole % solution of water in alcohol.

For the purpose of discussing colligative-property effects, however, it is convenient to adapt a special definition of solvent as that component which is present in both phases. The solute component is then that which is present in only one of the phases. Thus, in the case of the 40 mole % solution of water in alcohol, the freezing-point-depression effect relates to the equilibrium between the solution and ice (since on cooling it would be ice, not solid alcohol, that would freeze out first). The solvent is therefore *water*, not alcohol. On the other hand, if we were dealing with the osmotic pressure of this solution, and had a membrane permeable only to alcohol, so that alcohol would be the species present in both phases, then, for the purpose of dealing with osmotic effect, *alcohol* would be the solvent.

It is important to keep in mind that the colligative-property equations are all derived by considering the equilibrium of a species between two phases, and that that species or component will be designated as solvent. The subscript one, as in N_1, then will always refer to solvent as defined in the above manner.

Another aspect of this matter occurs in relating solubility to freezing-point depression. If we saturate a liquid A with a solid B, we then speak of the equilibrium concentration of B in the liquid as the solubility of B. Implicitly, A is considered to be the solvent. However, the same equilibrium can also be thought of as having been reached by taking a solution of this particular composition, cooling it to the temperature in question, and observing that solid B just begins to freeze out. This is now a freezing-point-depression experiment; B is the component present in both phases, and from the colligative-property point of view, B is therefore the solvent. You will find several problems requiring you to relate solubility to freezing-point depression, or vice versa.

It will generally be assumed that the solutions are ideal. These are, after all, quiz problems, and it would usually have been asking too much in the time allowed to expect the student to handle nonideality corrections. It is not very difficult, however, to obtain the activity coefficient of the solvent from vapor-pressure data, and you should be able to do this. The principal potential source of confusion lies in the concept of standard state, and it might help to review briefly here just what the operational definition of activity coefficient is.

We can start with the chemical potential or partial molar free energy of a component of an ideal gas mixture, that is, $\mu_i(g) = \mu_i^0(g) + RT \ln P_i$. If now this species is in equilibrium with a solution, its chemical potential must be the same in the two phases, so $\mu_i(l) = \mu_i(g)$. If the solution were ideal, we could use Raoult's law to write $P_i = N_i P_i^0$ so that the expression for the chemical potential becomes

$$\mu_i(l) = \mu_i^0(g) + RT \ln N_i P_i^0 = \mu_i^0(l) + RT \ln N_i \tag{10-1}$$

It is convenient to retain this form for nonideal solutions, so we invent the activity a_i as the effective mole fraction of the component; thus $P_i = a_i P_i^0$. As a further matter of convenience, we retain mole fraction as the composition variable, and simply say that $a_i = f_i N_i$, where f_i is a correction factor and is known as the activity coefficient. Thus

$$\mu_i(l) \equiv \mu_i^0(l) + RT \ln f_i N_i \tag{10-2}$$

The standard state is that state for which $\mu_i = \mu_i^0$. In the above case, it would be the state for which $a_i = 1$; since Raoult's law is a limiting law for all solutions, it follows that as $N_i \to 1, f_i \to 1$, and $a_i \to 1$. The standard state is then the pure liquid species.

Now it is perfectly possible to invoke Henry's law to say that P_i should be equal to $N_i k_i$. This substitution then gives

$$\mu_i(l) = \mu_i^0(g) + RT \ln N_i k_i = \mu_i^{0\prime}(l) + RT \ln N_i \qquad (10\text{-}3)$$

The two μ_i^0's are not the same. Thus

$$\mu_i^0(l) = \mu_i^0(g) + RT \ln P_i^0 \quad \text{and} \quad \mu_i^{0\prime}(l) = \mu_i^0(g) + RT \ln k_i$$

Again, solutions in general will not obey Henry's law exactly, so we adapt the procedure of convenience of writing $P_i = a_i k_i$, and, further, of defining an activity coefficient y_i such that $P_i = y_i N_i k_i$. We now have

$$\mu_i(l) \equiv \mu_i^{0\prime} + RT \ln y_i N_i \qquad (10\text{-}4)$$

Bear in mind that y_i is *not* equal to f_i; it is the corrective factor to make the vapor pressure of the ith species conform to Henry's law; f_i is the factor to make the vapor pressure conform to Raoult's law.

The Henry's law standard state is again that state for which $a_i = 1$, but our state of ideal behavior is that for which $y_i = 1$. Now, Henry's law is approached at infinite dilution, that is, $y_i \to 1$ as $N_i \to 0$. The Henry's law standard state is then a hypothetical one in which the ith species has the chemical environment of its infinitely dilute solution, but a concentration such that its activity is unity. This state is therefore sometimes called the hypothetical mole-fraction-unity standard state. Were the molality concentration unit used rather than mole fraction, we would obtain a third equation

$$\mu_i(l) \equiv \mu_i^{0\prime\prime}(l) + RT \ln \gamma_i m_i \qquad (10\text{-}5)$$

where the standard state would now be called the hypothetical unit-molality standard state.

These "hypothetical" standard states may seem rather odd to you. Bear in mind that they are rigorously defined so that $\mu_i^{0\prime}(l)$ and $\mu_i^{0\prime\prime}(l)$ have definite values, even though the states are not physically obtainable.

EQUATIONS AND CONCEPTS

Vapor-Pressure Lowering

$$P_1 = N_1 P_1^0 \qquad N_2 = 1 - N_1 = \frac{\Delta P}{P_1^0} \qquad (10\text{-}6)$$

Boiling-Point Elevation

$$\ln N_1 = \frac{\Delta H_v}{R}\left(\frac{1}{T} - \frac{1}{T_b}\right) \tag{10-7}$$

where T_b is the boiling point of the solvent.
Dilute solution form:

$$\Delta T_b = K_b m \tag{10-8}$$

where

$$K_b = \frac{RT_b^2}{\Delta H_v}\frac{M_1}{1000}$$

Freezing-Point Depression

$$-\ln N_1 = \frac{\Delta H_f}{R}\left(\frac{1}{T} - \frac{1}{T_f}\right) \tag{10-9}$$

where T_f is the freezing point of the solvent.
Dilute solvent form:

$$\Delta T_f = K_f m \tag{10-10}$$

where

$$K_f = \frac{RT_f^2}{\Delta H_f}\frac{M_1}{1000}$$

Osmotic Pressure

$$RT\ln P_1^0 = RT\ln P_1 + \int \overline{V}_1 \, d\pi \tag{10-11}$$

Ideal solution form:

$$-RT\ln N_1 = \int \overline{V}_1 \, d\pi \cong \overline{V}_1 \pi \tag{10-12}$$

Dilute solution form:

$$\pi V = n_2 RT \tag{10-13}$$

Van't Hoff i factor: $\pi V = in_2 RT$. In the case of electrolytes, i is interpreted as giving the number of ions into which one formula weight dissociates.

Activity Coefficients

Mole-fraction-unity standard state: $P_i = f_i N_i P_i^0$.
Hypothetical mole-fraction-unity standard state: $P_i = y_i N_i k_i$, where k_i is the Henry's law constant for the ith species.

PROBLEMS

1. (15 min) The vapor pressure of water under 10 atm pressure is increased from 30 to 31 mm Hg at 25°C. How much NaCl would have to be added to 55.5 moles (1 liter) of water at 25°C so that the vapor pressure of the solution under 10 atm pressure would be 30 mm Hg? Calculate the osmotic pressure of this solution. (Assume that NaCl, although fully dissociated, gives otherwise ideal solutions.)

2. (24 min) (final examination question) Using the vapor pressure data shown in Fig. 10-1, (a) calculate the Henry's law constant for A, that is, k_A in the equation $P_A = k_A N_A$, where this equation is valid, (b) calculate f_A for $N_A = 0.6$, that is, the activity coefficient of A for this composition, assuming that the standard state of A is the pure liquid, and (c) calculate y_A for $N_A = 0.6$, where y_A is the activity coefficient taking the standard state to be hypothetical mole fraction unity.

3. (12 min) A 1.25 weight % solution of a substance of unknown molecular weight, in benzene as solvent, has a vapor pressure of 752.4 mm Hg at 80°C and a boiling point of 80.25°C. The normal boiling point of benzene is 80.00°C. Assuming the solute to be nonvolatile, calculate its molecular weight and the heat of vaporization (per gram) for benzene. The molecular weight of benzene is 78.

4. (3 min) Calculate ΔG per mole of NaCl for the process: NaCl (0.01 m, 25°C) = NaCl (0.001 m, 25°C) (neglect activity coefficients).

5. (24 min) (final examination question) The vapor pressures of solid and of liquid benzene are plotted on the P versus T diagram of Fig. 10-2. A 10% by weight solution of substance A in benzene is cooled until solid benzene just begins to freeze out. The vapor pressure of benzene above the mixture is then measured and found to be 20 mm Hg; A is nonvolatile. Assuming the liquid solution to be ideal, estimate the freezing point of the solution and molecular weight of A. The molecular weight of benzene is 78.

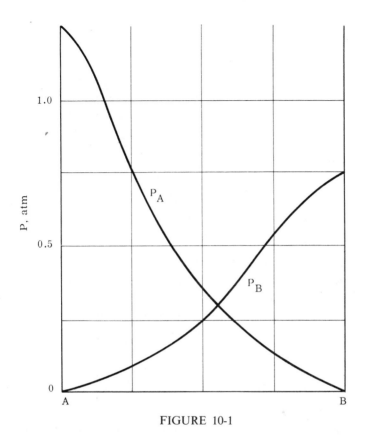

FIGURE 10-1

6. (24 min) (final examination question) The total vapor pressure of a 4 mole % solution of NH_3 in water at 20°C is 50.00 mm Hg; the vapor pressure of pure water is 17.00 mm Hg at this temperature. Apply Henry's and Raoult's laws to calculate the two partial pressures and the total vapor pressure for a 5 mole % solution.

7. (12 min) A 2% solution (by weight) of a substance of unknown molecular weight in toluene as solvent has a vapor pressure of 752.4 mm Hg at 110°C, and a normal boiling point of 110.25°C. The normal boiling point of toluene is 110.00°C; the solute is nonvolatile. Calculate the molecular weight of the solute and the heat of vaporization (per gram) of toluene ($C_6H_5CH_3$).

8. (12 min) A solution of a sugar (molecular weight unknown) and one of sodium chloride (both aqueous) are placed side by side in a closed container and left until equilibrium is reached. During the equilibration, water distills from one solution to the other until the two have the same vapor pressure. The two solutions are then analyzed and are found to contain 5% sugar

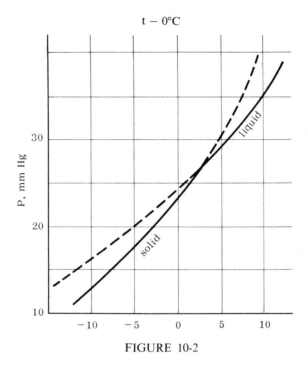

FIGURE 10-2

and 1% sodium chloride, by weight. Calculate the molecular weight of the sugar, assuming ideal solutions. The atomic weight of sodium is 23, and that of chlorine is 35.5.

9. (21 min) The molecular weight of a solid nonvolatile organic substance is to be determined by means of the following experiment:

Two open beakers are placed in a closed container. Beaker A initially contains 0.1 mole of naphthalene in 100 g of benzene, and beaker B initially contains 10 g of the unknown dissolved in 100 g of benzene. The beakers are allowed to stand side by side in the container until equilibrium is reached. Beaker A is then removed and weighed; it is found to have lost 8 g.

(a) Calculate the molecular weight of the unknown substance, assuming ideal solution behavior, and (b) state all the assumptions and approximations necessary to obtain the numerical answer in (a) with 1% accuracy.

---

10. (15 min) Given the following data: solubility of sucrose in water at 25°C is 6.2 m, molecular weight of sucrose is 342, melting point is 200°C, heat of fusion of water is 1,400 cal/mole, molecular weight and melting point of water, calculate the heat of fusion of sucrose assuming ideal solutions.

11. (18 min) Calculate the solubility of benzoic acid in ethanol at 80°C, and the vapor pressure of this saturated solution. You may assume the solution to be ideal, with benzoic acid nonvolatile and not associated or dissociated to any extent.

Compound	Mol. wt	mp (°C)	Normal bp (°C)	ΔH, cal/mole Fusion	Vaporiz.
Benzoic acid	122	122	249	4,000	12,000
Ethanol	46	−114	80	2,000	9,500

12. (13.5 min) A solution of sucrose in water freezes at −0.200°C. Calculate the vapor pressure of this solution at 25°C (accurate to 0.001 mm Hg). The vapor pressure of pure water at 25°C is 23.506 mm Hg and the molal freezing-point constant for water is 1.86°C/m.

13. (19.5 min) The normal boiling point of a saturated solution of benzanilide in ethanol is 82.00°C. The melting point of benzanilide is 161°C, the melting and normal boiling points of ethanol are −117°C and 80.00°C. K_b for ethanol is 1.22. Molecular weights are 46 for ethanol and 197 for benzanilide.

(a) Calculate the composition of this saturated solution of benzanilide.

(b) Calculate the molar heat of fusion of benzanilide. You may select the appropriate colligative-property equations so that your answers will be accurate only to about 1%, and you may assume ideal solution behavior.

14. (12 min) On cooling a 50 mole % solution of ethyl bromide and benzene, which substance should freeze out first? Show how you reach your conclusions. Assume the solution to be ideal.

	Mol. wt	K_f	T_f (°C)
Ethyl bromide	109	12.12	7
Benzene	78	5.0	7

(The melting points have been given as the same in order to facilitate your work.)

15. (24 min) (final examination question) The freezing point of a 0.02-mole-fraction solution of acetic acid in benzene is 4.4°C. Acetic acid exists partly as a dimer

$$2CH_3COOH = H_3C-C\begin{matrix} O-H-O \\ \diagup \qquad \diagdown \\ \diagdown \qquad \diagup \\ O-H-O \end{matrix}C-CH_3$$

Calculate the equilibrium constant for the dimerization. The melting point of pure benzene is 5.4°C, its heat of fusion is 2,400 cal/mole, and its molecular weight is 78. Assume the monomer and the dimer form ideal solutions.

16. (21 min) Some anthracene (molecular weight = 178) known to be contaminated with naphthalene (molecular weight = 128) is to be used in a research problem. In order to estimate the naphthalene content, the student takes 1.6 g, heats it until it is molten, then allows it to cool, and observes that the temperature of first appearance of solid is 175°C, 40°C below the melting point of pure anthracene. He now looks for a value of ΔH_f for anthracene (which he should have done first), and cannot find any.

Persisting with the freezing-point approach, the student then dissolves the 1.6 g in 100 g benzene (molecular weight = 78) and observes a freezing-point depression of 0.50°C. The freezing point of pure benzene is 5.4°C; its heat of fusion is known to be 2,240 cal/mole.
(a) Calculate the mole fraction of the naphthalene in the anthracene.
(b) Calculate the heat of fusion of anthracene.

17. (30 min) A fairly concentrated solution of naphthalene in *p*-xylene freezes at 7.4°C (first appearance of solid) and boils at 147°C. Pure *p*-xylene melts at 16°C and boils at 138°C (1 atm pressure) and its heat of *sublimation* is 12.80 kcal/mole. Calculate (a) the heat of fusion and the heat of evaporation of *p*-xylene, (b) the mole fraction of naphthalene in the solution, and (c) list all the assumptions and approximations embodied in the equations you use. These should not be such as to lead to more than 1% error if the solution is ideal.

18. (24 min) (final examination question) A solution comprising 0.1 mole of naphthalene and 0.9 mole of benzene is cooled until some solid benzene freezes out. The solution is then decanted off from the solid, and warmed to 80°C, at which temperature its vapor pressure is found to be 670 mm Hg. The freezing and normal boiling points of benzene are 5.5°C and 80°C, respectively, and its heat of fusion is 2,550 cal.

Calculate the temperature to which the solution was cooled originally and the number of moles of benzene that must have frozen out. Assume ideal solutions.

19. (8 min) (final examination question) Two organic substances, A and B, have the same solubility in benzene at 25°C in moles per 1,000 g benzene, but the melting point of pure A is 300°C while that of pure B is 350°C. Calculate the ratio of the heat of fusion of A to that of B. Assume their benzene solutions are ideal. State any other simplifying assumptions that you make.

20. (15 min) Substance A is a liquid at room temperature, with a normal boiling point of 60°C. Substance B is a solid at room temperature, is non-volatile, and melts at 80°C. A saturated solution of B dissolved in A has a vapor pressure of 0.80 atm at 60°C. The solution is ideal. Calculate (a) the composition of this solution, and (b) the heat of fusion of B.

21. (20 min) Substances A and B form ideal solutions, but are completely immiscible as solids; it happens that pure A and B have the same melting point of 25°C. A solution containing 60 mole % of A is cooled to 10°C, and, on filtering, it is found that 10 mole % of the system was in the form of solid A.

(a) Calculate the composition of the solution which was in equilibrium with the solid A at 10°C. (b) Calculate the heat of fusion of A. (c) Explain whether a maximum or minimum value can be given for the heat of fusion of B. If so, give this maximum or minimum value, specifying which it is. Your answer should be consistent with what you did in part (b). (If *both* a maximum and a minimum value exist, give both.)

---

22. (15 min) Figure 10-3 shows a schematic representation of an apparatus for determining the osmotic pressure of a solution. At equilibrium enough solvent has passed through the membrane to cause the solution to rise to a height *h*. At this point, the hydrostatic pressure on the solution equals its osmotic pressure. The vapor pressure of pure solvent is P^0, and that of the

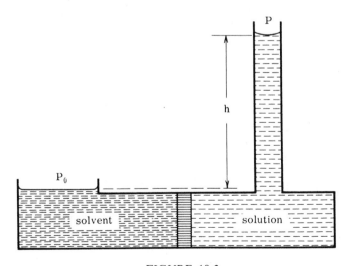

FIGURE 10-3

solvent above the solution is P. Considering the solution to be dilute, derive the equation:

$$RT \ln P/P^0 = -Mgh \qquad (M \text{ is the molecular weight of the solvent})$$

It is no coincidence that the above equation is just the barometric formula for the decrease in the pressure of a gas over a height h. Discuss why this is so.

23.   (12 min) A solution of NaCl of concentration m has an osmotic pressure of 2.0 atm at 25°C. Calculate ΔG_{298} for the process H_2O (solution, 25°C) = H_2O (pure, 25°C).

24.   (24 min) (final examination question) As shown in Fig. 10-4, pure solvent is separated from an ideal solution by a semipermeable membrane and, owing to the osmotic effect, the solution has risen to an equilibrium height h. The vapor pressure of the solution is P_1, and that of the pure solvent is P_1^0 (the solute being nonvolatile). Because of the barometric effect, P_1^0 decreases to $P_1^{0'}$ at height h. Derive the barometric formula by making use of the osmotic-pressure equation and the assumption that $P_1^{0'} = P_1$. Reasonable simplifying assumptions can be made to expedite the derivation.

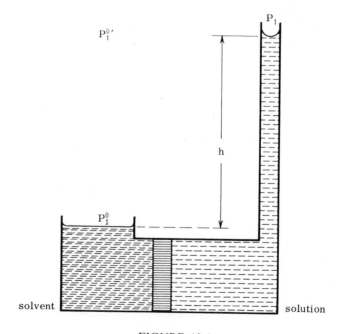

FIGURE 10-4

25. (10.5 min) The osmotic pressure of blood is 7 atm at 30°C. What is the molarity of the isotonic saline solution if the i factor for sodium chloride is taken to be 1.9?

26. (10 min) A membrane only to water separates a 0.01-M solution of sucrose from a 0.001-M one. On which solution must pressure be applied to bring the system into equilibrium? Calculate this osmotic pressure. Assume ideal solutions, 25°C.

ANSWERS

1. The osmotic pressure of the required solution is, of course, precisely 10 atm. The effective mole fraction of the NaCl must be 1/31 or 0.032, and the actual mole fraction, allowing for two ions per formula weight, will then be 0.016. The moles to be added to 55.5 of water are then given by 0.016 = $x/(55.5 + x)$ or $x \cong 0.016 \times 55.5$ or **0.88 mole.**

2. (a) k_A is most easily determined as the intercept at $N_A = 1$ of the straight line to which the P_A curve approaches as $N_A \to 0$. As indicated in Fig. 10-5,

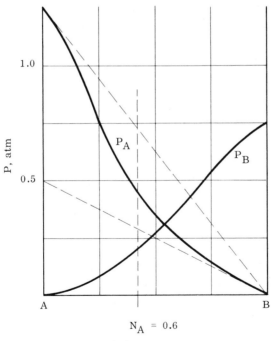

$$N_A = 0.6$$

FIGURE 10-5

this intercept is **0.5 atm.** (b) f_A is given by the equation $P_A = f_A N_A P_A^0$ or $f_A = P_A/P_{A,\,ideal}$. At $N_A = 0.6$, we find $f_A \cong 0.45/0.74 = $ **0.61.** (c) Similarly y_A is given by $P_A = y_A N_A k_A$ or $y_A = P_A/P_{A,\,ideal}$ where $P_{A,\,ideal}$ is now read off the Henry's law line. Thus $y_A \cong 0.45/0.30 = $ **1.50.**

3. The vapor-pressure lowering at 80°C is $760 - 752.4$ mm Hg or 7.6 mm Hg. The mole fraction of solute is then $N_2 = 7.6/760 = 0.01$. Taking as a basis a solution containing 100 g of benzene, $0.01 = n_2/(n_2 + 1.29)$, since 100 g corresponds to 1.29 mole. The $n \cong 0.0129$, and this must correspond to the 1.27 g of solute present, so its molecular weight is **approximately 100.** The solution is dilute enough that the simple boiling-point-elevation equation can now be used: $\Delta T = (RT_b^2/1000 h_f)m$ where m is the molality, here equal to 0.01×12.9 or 0.129, and ΔT is 0.25. Then

$$h_f = 1.98 \times 353^2 \times 0.129/1000 \times 0.25 = \textbf{129 cal/g}$$

4. Since NaCl is fully dissociated, the process is better written

$$Na^+(0.01\ m) + Cl^-(0.01\ m) = Na^+(0.001\ m) + Cl^-(0.001\ m)$$

G: $G_{Na^+}^0 + RT \ln 0.01$ $G_{Na^+}^0 + RT \ln 0.001$

$\quad\quad G_{Cl^-}^0 + RT \ln 0.01$ $G_{Cl^-}^0 + RT \ln 0.001$

or $\Delta G = 2RT \ln 0.001/0.01 = 2 \times 1.98 \times 298 \times (-1) \times 2.3 = $ **−2,710 cal.**

5. Since solid is in equilibrium with the solution, the solid vapor pressure must also be 20 mm Hg. From Fig. 10-2 the temperature is then **−3°C.** At this temperature, the pure (supercooled) liquid would have a vapor pressure of about 22 mm Hg. The mole fraction of A is then $N_A = \Delta P/P^0 = 2/22 = 0.091$.

100 g of solution contain 10 g of A and 90 g or 1.15 mole of benzene. The mole of A present is then

$$0.091 = n_A/(n_A + 1.15)$$

from which $n_A = 0.105$. The molecular weight of A is then $10/0.105$ or **95 g/mole.**

6. By Raoult's law, the partial pressure of the water must be 0.96×17 or **16.3 mm Hg.** The partial pressure of the NH_3 must then be **33.7 mm Hg,** from which the Henry's law constant for ammonia is $k = P/N = 33.7/4 = 8.42$.

For the 5 mole % solution, the partial pressure of the water will be $0.95 \times 17 = 16.1$, and that for the ammonia, $8.42 \times 5 = 42.1$; so the total pressure will then be **58.2 mm Hg.**

7. The vapor-pressure lowering at 110°C is $760 - 752.4$ or 7.6 mm Hg, so that $N_2 = 7.6/760 = 0.01$. Taking 100 g of solution as a basis, there are 2 g of solute and 98 g or $98/92 = 1.065$ mole of toluene. Then $0.01 = n_2/(n_2 + 1.065)$ or $n_2 = 0.0106$ and the molecular weight is $2/0.0106 = $ **189.**

The molality of the solution is $m = 0.01 \times 10.9 = 0.109$, and the boiling-point elevation is given as $0.25°C$; so from the dilute solution form of the boiling-point-elevation equation,

$$\Delta T = RT_b^2 m/1000 l_v \qquad l_v = 1.98 \times 383^2 \times 0.109/1000 \times 0.25$$
$$= 127 \text{ cal/g}$$

8. To have the same vapor pressure, the two solutions must have the same mole fraction of water and of solute. Taking 100 g as a basis, the sodium chloride solution has $1/58.5$ or 0.0171 mole of solute and $99/18$ or 5.50 mole of water. Since sodium chloride is fully dissociated, the effective mole fraction in solute is

$$N_2 = 0.0342/(0.0342 + 5.50) = 0.00619$$

This must then be equal to the mole fraction of the sugar, that is, $0.00619 = n_s/(n_s + 95/18)$, from which $n_s = 0.0324$. The molecular weight is then $5/0.0324 = $ **154.**

9. (a) At equilibrium the mole fraction of benzene must be the same in the two solutions and hence so must the mole fraction of solute. In the case of beaker A this last is $N_2 = 0.1/(0.1 + 92/78) = 0.0782$. Then for beaker B, $0.0782 = n_x/(n_x + 108/78)$ (since the 8 g lost from beaker A must have condensed into the solution in beaker B), or $n_x = 0.117$. The molecular weight of the unknown is then $10/0.117$ or **85.**

(b) We assume ideal solution, nonvolatile solutes, and, more trivially, that the vapor space is small enough that the entire 8 g lost from one beaker must have condensed in the other. Of course, the two beakers are assumed to be at the same temperature.

10. This problem is easy if you recognize that the situation can be treated as a freezing-point depression with sucrose as the solvent. The equation is then

$$-\ln N_s = \frac{\Delta H_f}{R}\left(\frac{1}{T} - \frac{1}{T_f}\right)$$

where ΔH_f and T_f are the heat of fusion and melting point of *sucrose*. N_s, for a 6.2-m solution, is $6.2/(6.2 + 55.5) = 0.10$, and $-\ln N_s$ is therefore 2.3. Then $\Delta H_f = 2.3 \times 1.98 \times 298 \times 473/175 = $ **3,650 cal.**

11. The solubility can be obtained easily if the situation is viewed as the freezing-point depression of benzoic acid by ethanol. Thus, in the equation

$$-\ln N_1 = \frac{\Delta H_f}{R}\left(\frac{1}{T} - \frac{1}{T_f}\right)$$

we use $\Delta H_f = 4{,}000$ and $T_f = 395$. Then

$$-\ln N_1 = \frac{4{,}000}{1.98}(1/353 - 1/395) = 0.61 \quad \text{and} \quad N_1 = 0.55$$

The saturated solution is then **0.55** mole fraction in benzoic acid.

The vapor pressure of the ethanol will then be $0.45 \times P^0$ or 0.45×1 atm, since at 80° we are at the boiling point of pure ethanol.

12. The molality of the solution must be $m = 0.2/1.86 = 0.108$. The mole fraction of the sucrose is then $0.108/(0.108 + 55.5) = 0.00192$, and this in turn is equal to $\Delta P/P^0$, so $\Delta P = 0.00192 \times 23.5 = 0.0452$ mm Hg. P is then $23.506 - 0.045 = \textbf{23.461 mm Hg.}$

13. (a) The boiling-point elevation is $82.00 - 80.00$ or $2.00°C$; hence $m = 2.00/1.22 = 1.64$, where m is the molality of the benzanilide.

(b) Since the solution is saturated with respect to benzanilide, the situation may now be viewed as a freezing-point-depression one, in which benzanilide is the solvent. The appropriate equation is then

$$-\ln N_b = \frac{\Delta H_f}{R}(1/T - 1/T_f)$$

where T and T_f are $355°K$ and $434°K$. N_b is given by $1.64/(1.64 + 1000/46) = 0.0705$. The heat of fusion is then $\Delta H_f = -2.3(\log 0.0705) \times 1.98 \times 355 \times 434/79 = \textbf{10.2 kcal.}$

14. One cannot draw a conclusion just by comparing the K_f values, since the actual solution is not dilute. Treating each component as potential solvent (that is, species that will freeze out first), we have

$$-\ln 0.5 = \frac{\Delta H_{f,eb}}{R}(1/T_{eb} - 1/300)$$

$$= \frac{\Delta H_{f,b}}{R}(1/T_b - 1/300)$$

Then

$$\frac{(1/T_{eb} - 1/300)}{(1/T_b - 1/300)} = \frac{\Delta H_{f,b}}{\Delta H_{f,eb}}$$

Since

$$K_f = \frac{RT_f^2 M}{1,000\Delta H_f}$$

the ratio of ΔH_f values becomes

$$\frac{\Delta H_{f,b}}{\Delta H_{f,eb}} = \frac{M_b}{M_{eb}} \frac{K_{f,eb}}{K_{f,b}} = \frac{78 \times 12.12}{109 \times 5} = 1.74$$

Therefore $(1/T_b - 1/300)$ is less than $(1/T_{eb} - 1/300)$, and T_b is then closer to 300 than is T_{eb}; that is, benzene will freeze out first. (For the purposes of answering the quiz question, it was not necessary to do more than set up the above equations and, by inspection, determine whether the ratio of temperature terms was greater or less than unity.)

15. The freezing-point-depression constant for benzene is

$$K_f = RT_f^2 M / 1,000 \Delta H = 1.98 \times 278^2 \times 78/1,000 \times 2,400 = 4.98$$

The molality of the solution was then $(5.4 - 4.4)/4.98 = 0.2\,m$, whereas if no dimerization were occurring, it should have been $m^0 = 0.02 \times 1,000/78 = 0.257$. If m_d is the molality of the dimer present, and m_m that of the monomer, then

$$m_m + m_d = 0.2 \quad \text{and} \quad m_m + 2m_d = 0.257$$

or $m_m = 0.143$ and $m_d = 0.057$, and the equilibrium constant for dimerization is $K = 0.057/(0.143)^2 = \textbf{2.80}$.

16. (a) The freezing-point-depression constant for benzene is

$$K_f = RT_f^2 M / 1,000 \Delta H = 1.98 \times 278^2 \times 78/1,000 \times 2,240 = 5.33$$

The benzene solution was then $0.5/5.33$ or $0.0937\,M$, and the solution contained $0.0937 \times 100/1,000$ or 0.00937 mole. The apparent molecular weight was then $1.6/0.00937$ or 170. We can now write

$$170 = N_a \times 178 + N_n \times 128 = 178 - 50N_n \quad \text{or} \quad N_n = \textbf{0.16}$$

(b) The dilute-solution equation cannot be used to treat the freezing-point depression of the anthracene, and we must write

$$-\ln N_a = \frac{\Delta H_f}{R}(1/T - 1/T_f)$$

or $\Delta H_f = 2.3(\log 0.84) \times 1.98 \times 448 \times 488/40$ or $\Delta H_f = \mathbf{1{,}870\ cal/mole}$.

17. (a) The solution, being concentrated, must be treated in terms of the nondilute forms of the freezing-point-depression and boiling-point-elevation equations. Thus

$$-\ln N_x = \frac{\Delta H_f}{R}(1/280.5 - 1/289)$$

$$= \frac{\Delta H_v}{R}(1/411 - 1/420)$$

On elimination of the log term,

$$\Delta H_f(8.6/280.5 \times 289) = \Delta H_v(9/411 \times 420)$$

or $\Delta H_v = 2.03\Delta H_f$. Since $\Delta H_s = 12.80 = \Delta H_v + \Delta H_f$, we can now solve simultaneously to get $\Delta H_v = \mathbf{8.6\ kcal/mole}$ and $\Delta H_f = \mathbf{4.2\ kcal/mole}$.
(b) $-\ln N_x = (4{,}200/1.98)(8.6/280.5 \times 289) = 0.226$ or $N_x = 0.80$ and $N_{naphth} = \mathbf{0.20}$.
(c) The various assumptions and approximations are as follow.
1. Those involved in the Clausius–Clapeyron equation, that is, first and second laws of thermodynamics, ideal gas behavior for the vapor, volume of vapor much larger than volume (per mole) of liquid, heat of vaporization independent of temperature, heat of sublimation independent of temperature, molar volume of solid much smaller than molar volume of vapor.
2. For the boiling-point-elevation equation: ideal solution, solute nonvolatile.
3. For the freezing-point-depression equation: ideal solution, solute insoluble in the solid solvent.

18. Since 80°C is its normal boiling point, the vapor pressure of pure benzene is 760 mm Hg as compared with 670 mm Hg for the solution. The vapor-pressure lowering is thus 90 mm Hg, and the mole fraction of benzene is $N_b = 670/760 = 0.882$. Then, according to the freezing-point-depression equation,

$$-\ln N_b = \frac{\Delta H_f}{R}(1/T - 1/278.6)$$

or $1/T = -2.3(\log 0.882) \times 1.98/2,550 + 1/278.6 = 9.62 \times 10^{-5} + 1/278.6$.
Since T is evidently close to 278.6, a way to retain precision in the calculation is to write $\Delta T/278.6^2 \cong 9.62 \times 10^{-5}$ or $\Delta T = 7.48$; a second round of approximation is then $\Delta T/271.1 \times 278.6 = 9.62 \times 10^{-5}$ or $\Delta T = 7.25$ and $t = 5.5 - 7.25 = -1.75°C$.
The mole fraction of the naphthalene in the freezing solution is 0.118. Then $0.118 = 0.1/(0.1 + n_b)$ or $n_b = 0.748$. Thus $0.9 - 0.748$ or about **0.15 mole** of benzene must have frozen out.

19. A and B act as the solvent in the two cases, and the freezing point depressions are large enough that the full Eq. (10-9) must be used. Since N_1 is the same for the two saturated solutions,

$$\Delta H_{f_A}(1/298 - 1/573) = \Delta H_{f_B}(1/298 - 1/623)$$

or

$$\frac{\Delta H_{f_A}}{\Delta H_{f_B}} = \frac{325}{298 \times 623} \frac{298 \times 575}{275} = 1.08$$

20. Since $P_A^0 = 1$ atm and $P_A = 0.8$ atm, $N_A = 0.8$ and $N_B = 0.2$. The solution is saturated with respect to B, which means that B is present as solvent in a freezing-point-depression situation; that is,

$$-\ln 0.2 = \frac{\Delta H_{f_B}}{R}(1/333 - 1/353)$$

or $\Delta H_{f_B} = 2.3 \times 0.70 \times 1.98 \times 353 \times 333/20 = 18,800$ cal/mole.

21. (a) Taking 100 moles as a basis, after cooling to 10°C, the residual solution contained 50 moles of A and still 40 moles of B, N_A was then **0.55**.
(b) The full Eq. (10-9) must be used

$$-\ln N_A = \frac{\Delta H_{f_A}}{R}(1/283 - 1/298)$$

Or $\Delta H_{f_A} = 2.3 \times 0.255 \times 1.98 \times 298 \times 283/15 = 6,530$ cal/mole.
(c) Suppose that B were just about to freeze out also; the condition almost met is that

$$-\ln N_B = \frac{\Delta H_{f_B}}{R}(1/283 - 1/298)$$

or $\Delta H_{f_B} = 2.3 \times 0.350 \times 1.98 \times 298 \times 283/15 = 9,000$ cal/mole.

Since B does not come out, the required N_B must be larger than 0.45, or $-\ln N_B$, smaller. The true ΔH_{f_B} must then be **less than 9,000 cal.**

22. The basic osmotic-pressure equation is $RT \ln P^0 = RT \ln P + \int V \, d\pi$ or $RT \ln P^0/P = \bar{V}\pi$, assuming the liquid to be incompressible. Here, π can be equated to the hydrostatic head $\rho g h$ or, alternatively, $\pi = Mgh/\bar{V}$. Thus substitution for π leads to the desired equation, $RT \ln P^0/P = Mgh$. If the system were placed in a closed box, one might expect a circulation of vapor, with solvent evaporating at P^0 and condensing on the solution whose vapor pressure is only P. This, if it did occur, would constitute a perpetual motion machine and thus a violation of the first law of thermodynamics. For it *not* to occur, the vapor pressure P^0 over the pure solvent must diminish to exactly the value P at a height h, which is just what the barometric formula would predict.

This analysis, incidentally, provides a way of proving that the molar volume V in the osmotic-pressure equation is that of a single molecule of solvent, and not of dimers, trimers, and so on, even though in the liquid state one may have largely such forms.

23. ΔG for the process can be formulated as follows:

	H_2O (solution)	H_2O (pure)
G:	$G^0 + RT \ln P$	$G^0 + RT \ln P^0$

so $\Delta G = RT \ln P^0/P$. The osmotic-pressure equation is $RT \ln P^0 = RT \ln P + \int \bar{V} \, d\pi$ or $RT \ln P^0/P \cong \bar{V}\pi = 18 \times 2 = $ **36 cc-atm.**

24. The general form of the osmotic-pressure equation is

$$RT \ln P_1^0 = RT \ln P_1 + \int \bar{V}_1 \, d\pi \quad \text{or} \quad RT \ln P_1^0/P_1 = V_1 \pi$$

if the simplifying assumption is made that the molar rather than the partial molar volume of the solvent may be used, and that the solvent is incompressible. Since π is equal to $\rho g h$, where ρ is the density of the solution, and since the solution is dilute ρ is essentially the density of the solvent, we can write for $V_1 \pi$, $\rho_1 g h V_1$. But $\rho_1 V_1 = M_1$, so we have $RT \ln P_1^0/P_1 = M_1 g h$. If $P_1 = P_1^{0'}$, the barometric formula has been derived.

Note that P_1 really should be equal to $P_1^{0'}$; if there were any difference in the two vapor pressures, it should be possible to devise a perpetual motion machine that would constitute a violation of the first law of thermodynamics.

25. The van't Hoff i factor is defined as $\pi = icRT$, where c is the molarity in formula weights per liter. Here, $c = 7/1.9 \times 0.082 \times 303 = $ **0.148 M.**

26. Both solutions can be treated as dilute. It is convenient to calculate for each the osmotic pressure relative to pure water. Thus, for the $0.01\text{-}M$ solution, π is $0.01RT$ or $0.01 \times 0.082 \times 298 = 0.25$ atm, while for the $0.001\text{-}M$ solution, it is $0.001RT$, or 0.025 atm. The more concentrated solution thus has the higher osmotic pressure; a differential pressure of 0.225 atm would have to be applied to it to bring equilibrium between the two solutions.

11

HETEROGENEOUS AND HOMOGENEOUS GAS EQUILIBRIUM

COMMENTS

There are three general types of problems in this chapter. First, you will find a number that involves equilibria between gaseous substances and requires that you calculate K_p from the data, or, given K_p, determine the extent of reaction at equilibrium. In most cases stoichiometry is very important—that is, you must utilize the equation for the over-all process to determine the mole ratios in which reactants disappear and products form. Often it will be convenient to set up stoichiometric relationships between the various mole numbers, and insert the results in the expression for K_n. K_p then follows, if the total pressure is known (and N, the total moles, can be obtained from the sum of the various mole numbers). If the amount of initial substance is not specified, it is often less confusing to assume some convenient amount as a basis; your choice should cancel out in the course of the calculation.

The second group of problems consists of those dealing with equilibria between solids and gases. Here, it is important to remember that, although the solids do not appear in the K_p expression, they must be present in the physical system for K_p to hold. Remember, too, that in a reaction of the type A(s) + B(g) = products, the amount of B needed to convert all of A to products is the sum of the equilibrium amount required by K_p, plus the amount needed by stoichiometry to convert A to products. Finally, if a solid dissociates into gaseous products, these will be formed in stoichiometric ratio; hence they, and K_p, can be expressed in terms of the total pressure if no other gases were present initially.

Last, you will find a number of simple thermodynamic calculations involving conversion of K_p to ΔG^0, the relation between ΔG^0, ΔH^0, and ΔS^0, and the variation of ΔG^0 and of K_p with temperature. Look for occasional shortcuts.

EQUATIONS AND CONCEPTS

Equilibrium-Constant Expressions

Given the general reaction involving only ideal gases at a given temperature,

$$aA + bB + \cdots = mM + nN + \cdots \tag{11-1}$$

$$K_p = \frac{P_M^m P_N^n \cdots}{P_A^a P_B^b \cdots} = \frac{n_M^m n_N^n \cdots}{n_A^a n_B^b \cdots} \left(\frac{P}{N}\right)^{\Delta n} = K_n \left(\frac{P}{N}\right)^{\Delta n} \tag{11-2}$$

Here, N denotes the total moles present, P the total pressure, n the moles of a particular species, and Δn the moles of products minus the moles of reactants.

If solids are involved in the reaction, K_p will contain only the terms that refer to gaseous species, and Δn will be the difference between moles of gaseous products and moles of gaseous reactants. The solid species must, of course, be present in the actual system if K_p is to hold.

Thermodynamic Relationships

$$\Delta G^0 = \Delta H^0 - T \Delta S^0 \tag{11-3}$$

$$\Delta G^0 = -RT \ln K_p \tag{11-4}$$

van't Hoff equation:

$$d \ln \frac{K_p}{dT} = \frac{\Delta H^0}{RT^2} \tag{11-5}$$

$$\ln \frac{K_2}{K_1} = -\frac{\Delta H^0}{R} \left(\frac{1}{T_2} - \frac{1}{T_1}\right) \tag{11-6}$$

PROBLEMS

1. (18 min) Suppose that the equilibrium $SO_3 = SO_2 + \frac{1}{2}O_2$ is such that when 8 g of SO_3 are introduced into a flask at 600°C, the equilibrium pressure and density are 1.8 atm and 1.6 g/liter, respectively. (a) Calculate K_p. (b) Calculate the moles of helium that would have to be introduced into the

flask containing the above equilibrium mixture in order to double the pressure.

2. (30 min) K_p is 10^5 atm^{-1} at 700°K for the reaction

$$2NO + O_2 = 2NO_2$$

and ΔS^0 is -40.7 cal/deg and ΔC_p^0 is zero.

(a) Some pure NO_2 is introduced into a 30-liter flask at 700°K until the equilibrium pressure is 0.2 atm. Calculate α, the degree of dissociation of the NO_2.

(b) If α were 0.3 under the above conditions, what would be the average molecular weight of the gas mixture? Show your calculation.

(c) Calculate ΔG^0 for the reaction at 25°C.

(d) Calculate the temperature coefficient of K at 700°K, that is, the percent change in K per degree temperature change. (You may neglect the $NO_2 - N_2O_4$ equilibrium, that is, assume no N_2O_4 is present.)

3. (24 min) K_p is 8×10^{-9} atm at 100°C for the equilibrium:

$$COCl_2(g) = CO(g) + Cl_2(g)$$

$\Delta S^0_{373°K}$ is 30 cal/deg.

(a) Calculate the degree of dissociation of phosgene at 100°C and 2 atm total pressure.

(b) Calculate $\Delta H^0_{373°K}$ for the reaction.

(c) Assuming ΔC_p^0 to be zero, at what temperature would the degree of dissociation of phosgene be 0.1%, again at 2 atm total pressure? Show your calculations.

4. (24 min) (final examination question) Gaseous COF_2 is passed over a catalyst at 1,000°C and comes to equilibrium according to the equation

$$2COF_2 = CO_2 + CF_4$$

The pressure is maintained at 10 atm. A sample of the equilibrium mixture is quickly cooled (which stops any shift in concentrations) and analysis shows that, out of 500 cc (STP) of the mixture, there are 300 cc (STP) of combined COF_2 and CO_2. (This is done by passing the mixture through barium hydroxide solution, which absorbs COF_2 and CO_2 but not CF_4.)

(a) Calculate K_p for the equilibrium.

(b) If K_p increases by 1% per degree around 1,000°C, calculate ΔH^0, ΔS^0, and ΔG^0 at this temperature.

5. (25 min) The following questions pertain to the equilibrium: $2 NOCl = 2 NO + Cl_2$. (a) A certain amount of NOCl is introduced into a flask at 200°C. At equilibrium the total pressure is 1 atm, and the partial pressure of NOCl is 0.64 atm. Calculate K_p. (b) K_p increases by 1.5% per degree around 200°C. Calculate ΔH^0 for the reaction. Assuming that K_p at 200°C is 0.1 atm, calculate ΔS^0. (c) Assuming that K_p is 0.1 atm at 200°C, calculate the pressure at which the degree of dissociation of NOCl will be 0.2.

6. (20 min) (final examination question) One and one-tenth grams of NOBr is placed in an evacuated 1-liter flask at -55°C. The flask is then warmed to 0°C, at which temperature the contents are gaseous and exert a pressure of 0.3 atm. On further warming to 25°C, the equilibrium total pressure rises to 0.35 atm. At both 0°C and 25°C the equilibrium

$$2NOBr = 2NO + Br_2$$

is present. Calculate K_p for the reaction at 0°C, and ΔH^0. The atomic weight of Br is 80.

7. (20 min) (final examination question) In a study of the equilibrium, $H_2 + I_2 = 2HI$, a certain number of moles x of HI are formed when 1 mole of H_2 and 3 moles of I_2 are introduced into a flask of volume V at temperature T. On introducing 2 additional moles of H_2, the amount of HI formed is found to be $2x$. Calculate K_p.

8. (20 min) (final examination question) For the reaction below K_p is 0.05 atm and ΔG^0 is 5.35 kcal at 900°K:

$$C_2H_6(g) = C_2H_4(g) + H_2(g)$$

If an initial mixture comprising 20 moles of C_2H_6 and 80 moles of N_2 (inert) is passed over a dehydrogenation catalyst at 900°K, what is the equilibrium percent composition of the effluent gas mixture? The total pressure is kept at 0.5 atm. Given that ΔS^0 is 32.3 cal/deg at 900°K, calculate ΔG^0 at 300°K (assume $\Delta C_p^0 = 0$).

9. (18 min) (final examination question) A 2-liter flask maintained at 700°K contains 0.1 mole of CO and a catalyst for the reaction

$$CO(g) + 2H_2(g) = CH_3OH(g)$$

Hydrogen is introduced until the equilibrium total pressure is 7 atm, at which point 0.06 mole of methanol is formed.
   (a) Calculate K_p.

(b) What would the final pressure be were the same amounts of CO and H_2 used but with no catalyst present so that no reaction occurs?

10.   (18 min) K_p has the value 10^{-5} for the equilibrium

$$CO_2 + H_2 = CO + H_2O(g)$$

at 25°C, and ΔS^0 is -10 cal/deg. (ΔH^0 and ΔS^0 do not change much with temperature.) One mole of CO, 2 moles of H_2, and 3 moles of CO_2 are introduced into a 5-liter flask at 25°C. Calculate (a) ΔG^0 at 25°C, (b) the equilibrium pressure, (c) the moles of each species present at equilibrium, and (d) K_p at 100°C.

11.   (30 min) ΔG^0 is 300 cal at 40°C for the reaction:

$$2NO_2 = N_2O_4$$

(a) Calculate K_p for this reaction at 40°C.
(b) The density of an equilibrium mixture of NO_2 and N_2O_4 gases is found to be 5.85 g/liter at 40°C and a certain pressure. Using the value in (a) for K_p, calculate the degree of dissociation of the N_2O_4, the average molecular weight of the mixture, and the total pressure of the mixture.

12.   (17 min) (final examination question) A container whose volume is V liters contains an equilibrium mixture that consists of 2 moles of PCl_5, 2 moles of PCl_3, and 2 moles of Cl_2 (all as gases). The pressure is 3 atm, and the temperature is $T°K$.
A certain amount of Cl_2 is now introduced, keeping the pressure and temperature constant, until the equilibrium volume is $2V$ liters. Calculate the moles of Cl_2 that were added, and the value of K_p for the equilibrium.

13.   (18 min) An equilibrium mixture for the reaction

$$CO + 2H_2 = CH_3OH$$

at 700°K consists of 2 atm of CH_3OH, 1 atm of CO, and 0.1 atm of H_2. The above mixture is allowed to expand to twice its original volume, still at 700°K. Calculate the new equilibrium pressures.

14.   (12 min) (final examination question) If a small amount of SO_2 is introduced into oxygen gas at 1 atm pressure, the equilibrium ratio of SO_3/SO_2 is 10^4 at 1,000°C. If this gaseous mixture is now allowed to expand to a volume four times the original one, still at 1,000°C, calculate what the new value of the SO_3/SO_2 ratio should be at equilibrium.

15.   (15 min) The value of K_p is 1×10^{-3} at 25°C for the reaction:

$2 NO_2 + Cl_2 = 2 NOCl$. A flask contains 0.02 atm of NO_2 at 25°C. Calculate the moles of Cl_2 that must be added if, at equilibrium, 1% of the NO_2 is to be converted to NOCl. The volume of the flask is such that 0.2 mole of gas produce 1 atm pressure at 25°C. (Ignore the possible association of NO_2 to N_2O_4.)

16. (18 min) The degree of dissociation, α, of PCl_5 into PCl_3 and Cl_2 is 0.1 at 250°C and 2 atm pressure. Also, α increases by 2% per degree (e.g., to 0.102 at 251°C) with pressure constant at 2 atm. Calculate ΔH^0 and ΔS^0 for the reaction at 250°C.

---

17. (10.5 min) K_p is 9 atm^2 for the reaction:

$$LiCl \cdot 3NH_3(s) = LiCl \cdot NH_3(s) + 2NH_3(g)$$

at 40°C. How many moles of ammonia must be added at this temperature to a 5-liter flask containing 0.1 mole of $LiCl \cdot NH_3$ in order to completely convert the solid to $LiCl \cdot 3NH_3$? Show your calculations.

18. (12 min) Ferrous sulfate undergoes a thermal decomposition as follows:

$$2FeSO_4(s) = Fe_2O_3(s) + SO_2(g) + SO_3(g)$$

At 929°K the total gas pressure is 0.9 atm with both solids present.
(a) Calculate K_p for this temperature.
(b) Calculate the equilibrium total pressure that will be obtained if excess ferrous sulfate is placed in a flask at 929°K, which contains an initial SO_2 pressure of 0.6 atm.

---

19. (24 min) (final examination question) One-tenth mole of H_2 and 0.2 mole of CO_2 are introduced into an evacuated flask at 450°C, and the reaction

(a) $\qquad H_2 + CO_2 = H_2O + CO$

occurs to give an equilibrium pressure of 0.5 atm. Analysis of the mixture shows that it contains 10 mole % water.
A mixture of CoO(s) and Co(s) is then introduced so that the additional equilibria (b) and (c) are established:

(b) $\qquad CoO(s) + H_2 = Co(s) + H_2O$

(c) $\qquad CoO(s) + CO = Co(s) + CO_2$

Analysis of the new equilibrium mixture shows it to contain 30 mole % water. Calculate the three equilibrium constants. If K_a increases by 1% per degree for temperatures around 450°C, calculate ΔH_a^0.

20. (15 min) K_p is 0.05 atm^2 at 20°C for the reaction:

$$NH_4HS(s) = NH_3(g) + H_2O(g)$$

0.06 mole of solid NH_4HS is introduced into a 2.4-liter flask at 20°C.

(a) Calculate the percent of the solid that will have decomposed into NH_3 and H_2S at equilibrium.

(b) Calculate the number of moles of ammonia that would have to be added to the flask to reduce the decomposition of the solid to 1%.

(c) Having reached the state of equilibrium described in (b), explain whether the addition of more $NH_4HS(s)$ would increase, decrease, or leave unchanged the ammonia pressure.

21. (30 min) Four and four-tenths grams of CO_2 are introduced into a 1-liter flask containing excess solid carbon, at 1,000°C, so that the equilibrium

$$CO_2 + C(s) = 2CO$$

is reached. The gas density at equilibrium corresponds to an average molecular weight of 36.

(a) Calculate the equilibrium pressure and the value of K_p.

(b) If, now, an additional amount of He (inert) is introduced until the total pressure is doubled, the equilibrium amount of CO will be (increased, decreased, unchanged, insufficient information to tell). If, instead, the volume of the flask were doubled, with He introduced to maintain the same total pressure, the equilibrium amount of CO would (increase, decrease, be unchanged, insufficient data to tell).

(c) If in (a) there were actually 1.2 g of C(s) present, how many moles of CO_2 would have to be introduced so that at equilibrium only a trace of carbon remained?

(d) If the K_p for the equilibrium doubles with a 10°C increase in temperature, what is ΔH^0 for the reaction?

22. (12 min) Calculate K_p for the reaction

$$S(s) + 2CO(g) = SO_2(g) + 2C(s)$$

At the temperature in question, 2 atm of CO are introduced into a vessel containing excess solid sulfur, and a final equilibrium pressure of 1.03 atm is observed.

23. (24 min) (final examination question) K_p has the value 10^{-6} atm^3 and 10^{-4} atm^3 at 25°C and 50°C, respectively, for the reaction

$$CuSO_4 \cdot 3H_2O = CuSO_4 + 3H_2O(g)$$

(a) What is the minimum number of moles of water vapor that must be introduced into a 2-liter flask at 25°C in order to completely convert 0.01 mole of $CuSO_4$ to the trihydrate. Show your calculations.
(b) Calculate ΔH^0 for the reaction.

24. (15 min) Ammonium chloride vaporizes according to the process:

$$NH_4Cl(s) = NH_3 + HCl$$

The vapor pressure of $NH_4Cl(g)$ is negligible. It is found that at 520°K the equilibrium dissociation pressure (P_{tot}) is 0.050 atm. In a second experiment, 0.02 mole of $NH_4Cl(s)$ and 0.02 mole of NH_3 is introduced into a 42.7-liter flask maintained at 520°K. Calculate the amount of each substance present at equilibrium and the partial pressures of the two that are gaseous.

25. (12 min) The standard free energy of formation of AgCl is -26 kcal/mole, and ΔH^0 is 30 kcal for the reaction $AgCl = Ag + \frac{1}{2}Cl_2$, both values for 25°C. (a) Calculate $\Delta G^0_{298°K}$ for the above reaction. (b) Calculate K_p and the dissociation pressure of chlorine over AgCl. (c) Obtain the value of $d(T \ln K_p)/dT$.

26. (12 min) (final examination question) K_p for the equilibrium $H_2 + I_2 = 2 HI$ is 20 at 40°C. 25 g (0.2 mole) of I_2 solid are placed in a 25.7-liter flask at 40°C; the vapor pressure of solid iodine being 0.1 atm at this temperature. How many moles of hydrogen gas should be introduced into the flask in order to convert all of the solid iodine present to HI? (*Note:* The volume 25.7 liter is chosen to facilitate calculation; at 40°C this is the molar volume of a gas at 1 atm.)

ANSWERS

1. (a) The 8 g corresponds to 0.1 mole. Let $x =$ moles SO_2, then $0.1 - x =$ moles SO_3, $x/2 =$ moles O_2, and the total moles is $0.1 + x/2$. The equilibrium constant is then

$$K_p = \frac{(x)(x/2)^{1/2}}{(0.1 - x)} \left(\frac{P}{0.1 + x/2} \right)^{1/2}$$

Also,

$$M_{av} = \rho RT/P = 1.6 \times 0.082 \times 873/1.8 = 63.8$$

but

$$63.8 = \frac{0.1 - x}{0.1 + x/2} 80 + \frac{x}{0.1 + x/2} 64 + \frac{x/2}{0.1 + x/2} 32$$

from which $x = 0.051$. K_p is then

$$K_p = \frac{(0.51)(0.0255)^{1/2}}{0.49} \left(\frac{1.8}{0.125}\right)^{1/2} = \mathbf{0.61}$$

(b) Adding inert gas at constant volume does not shift the equilibrium, so the moles of helium needed is just $0.1 + x/2$ or **0.125**, that is, the amount to double the moles present.

2.   (a) It is convenient to write K_p in the form:

$$K_p = \frac{n_{NO_2}^2}{n_{NO}^2 \times n_{O_2}} \frac{N}{P}$$

Let $n_{NO_2}^0$ be the number of moles of NO_2 before dissociation; then

$$n_{NO_2} = (1 - \alpha)n_{NO_2}^0$$
$$n_{NO} = \alpha n_{NO_2}^0$$
$$n_{O_2} = \frac{\alpha}{2} n_{NO_2}^0$$
$$N = (1 + \alpha/2)n_{NO_2}^0$$
$$10^5 = \frac{(1 - \alpha)^2(n_{NO_2}^0)^2}{\alpha^2(n_{NO_2}^0)^2 \times (\alpha/2)n_{NO_2}^0} \frac{(1 + \alpha/2)n_{NO_2}^0}{0.2}$$

Since K_p is so large, it should be safe to neglect α in comparison with unity, and we can then write

$$10^5 = \frac{1}{\alpha^2(\alpha/2) \times 0.2}$$

or $\alpha^3 = 10^{-4}$ and $\alpha = 0.046$.

As a second approximation we write $(1 - \alpha)$ as 0.954 and $(1 + \alpha/2)$ as 1.02. This leads to $\alpha^3 = 0.93 \times 10^{-4}$ and $\alpha = \mathbf{0.045}$.

(b) On substituting $\alpha = 0.3$ into the above equations, $n_{NO_2} = 0.7n_{NO_2}^0$, and so on, and the mole fractions become $N_{NO_2} = 0.7/1.15 = 0.61$, $N_{NO} = 0.3/1.15 = 0.26$ and $N_{O_2} = 0.13$. The average molecular weight of the mixture is then $M_{av} = 0.61 \times 46 + 0.26 \times 30 + 0.13 \times 32 = $ **40**.

(c) ΔG^0 at 700°K is $-1.98 \times 700 \times 2.3$ log $10^5 = -15.9$ kcal; hence $\Delta H^0 = -15.9 + 700 \times (-40.7)/1,000 = -44.4$ kcal. Then at 25°C: $\Delta G^0 = -44.4 - 298 \times (-40.7)/1,000 = $ **−32.4 kcal**.

(d) It is convenient here to use the van't Hoff equation in differential form:

$$d \ln K_p/dT = \Delta H^0/RT^2 = -44,400/1.98 \times 700^2 = -0.046$$

Since $d \ln K_p$ is the same as dK_p/K_p, the above is the desired temperature coefficient, or, in % per degree: **−4.6%**.

3. (a) $K_p = \dfrac{n_{CO} \times n_{Cl_2}}{n_{COCl_2}} \dfrac{P}{N}$. Assuming 1 mole of phosgene initially, $n_{COCl_2} = 1 - \alpha$, $n_{CO} = n_{Cl_2} = \alpha$, and $N = 1 + \alpha$. Then $8 \times 10^{-9} = [\alpha^2/(1 - \alpha)] \times [2/(1 + \alpha)]$ and, since α will be small, $8 \times 10^{-9} \cong 2\alpha^2$ or $\alpha = $ **6.3 $\times$ 10^{-5}**.

(b) $\Delta G^0_{373°K} = -1.98 \times 373 \times 2.3$ log $8 \times 10^{-9} = 13.7$ kcal, $\Delta H^0 = 13.7 + 373 \times 30/1,000 = $ **24.8 kcal**.

(c) If α is to be 0.001, then $K_p = 2 \times 10^{-6}$ and, from the van't Hoff equation, $\log(8 \times 10^{-9}/2 \times 10^{-6}) = -(24,800/2.3 \times 1.98)(1/373 - 1/T)$ or $(1/373 - 1/T) = 4.43 \times 10^{-4}$ or $T = $ **448°K**.

4. (a) There must have been $500 - 300$ or 200 cc (STP) of CF_4, and hence also 200 cc (STP) of CO_2 since this is formed in equimolar amounts. There were then 100 cc (STP) of COF_2. The mole fractions are then 0.4, 0.4, and 0.2 for these substances, respectively, and the partial pressures are then 4, 4, and 2 atm. K_p is thus $(4)(4)/(2)^2 = $ **4 atm**.

(b) We use the equation: $d \ln K_p/dT = \Delta H^0/RT^2$, where $d \ln K_p/dT$ is simply the fractional change in K_p per degree and is equal to 0.01. Then $\Delta H^0 = 0.01 \times 1.98 \times 1,273^2 = $ **32.1 kcal**. Also, $\Delta G^0 = -RT \ln K_p = -1.98 \times 1,273 \times 2.3$ log $4 = $ **−3.47 kcal**. Therefore $\Delta S^0 = (\Delta H^0 - \Delta G^0)/T = $ **27.9 cal/deg**.

5. (a) The pressure of the combined products must be $1 - 0.64$ or 0.36 atm; since $P_{NO} = 2P_{Cl_2}$, it follows that $P_{NO} = 0.24$ and $P_{Cl_2} = 0.12$. K_p is then $(0.24)^2(0.12)/(0.36)^2 = $ **0.053 atm**.

(b) The 1.5% increase per degree means that $d \ln K/dT = 0.015$, and therefore

$$0.015 = \frac{\Delta H^0}{RT^2}$$

or $\Delta H^0 = 1.98 \times 473^2 \times 0.015 = \textbf{6,650 cal.}$ $\Delta G^0 = -RT \ln K_p = -1.98 \times 473 \times 2.3 \times (-1) = 2,150$ cal, so that $\Delta S^0 = (6,650 - 2,150)/473 = \textbf{9.5 e.u.}$

(c) Let

$$n_{NOCl} = (1 - \alpha)n_0$$
$$n_{NO} = \alpha n_0$$
$$n_{Cl_2} = \alpha n_0/2$$

where n_0 is the initial moles as NOCl. The total moles at equilibrium is given by $N = n_0 + \alpha n_0/2$, and the expression for K_p becomes

$$K_p = \frac{(\alpha n_0)^2(\alpha n_0/2)P}{[(1 - \alpha)n_0]^2(n_0 + \alpha n_0/2)} = \frac{\alpha^3/2P}{(1 - \alpha)^2(1 + \alpha/2)} = 0.1$$

If α is to be 0.2, then $P = (0.1)(0.8)^2(1 + 0.1)/(0.2^3/2) = \textbf{17.6 atm.}$

6. The 1.1 g of NOBr corresponds to 0.01 mole, and if α denotes the degree of dissociation, then at equilibrium

$$n_{NOBr} \ (1 - \alpha)0.01$$
$$n_{NO} = 0.01\alpha$$
$$n_{Br_2} = 0.005\alpha$$
$$N = 0.01(1 + \alpha/2)$$

But

$$N = PV/RT = 0.3 \times 1/0.082 \times 273 = 0.0134$$

hence

$$1.34 = 1 + \alpha/2 \quad \text{or} \quad \alpha = 0.68$$

$$K_p = \frac{n_{NO}^2 n_{Br_2}}{n_{NOBr}^2} \frac{P}{N} = (0.68)^2(0.34)(0.3)/(0.32)^2(1.34) = \textbf{0.345 atm.}$$

Repeating the calculation for 25°C, $N = 0.0143$, $\alpha = 0.86$, and $K_p = 3.97$. Then

$$\log(3.97/0.345) = -\frac{\Delta H^0}{1.98 \times 2.3}(1/298 - 1/273)$$

and $\Delta H^0 = 1.061 \times 1.98 \times 2.3 \times 298 \times 273/25 = \textbf{15.7 kcal.}$

7. By material balance,

$$n_{HI} = x \qquad\qquad n_{HI} = 2x$$
$$n_{H_2} = 1 - x/2 \quad\text{and}\quad n_{H_2} = 3 - x$$
$$n_{I_2} = 3 - x/2 \qquad\quad n_{I_2} = 3 - x$$

Since there are an equal number of moles of gaseous species on both sides of the equation, $K_p = K_n$; hence

$$\frac{x^2}{(1 - x/2)(3 - x/2)} = \frac{(2x)^2}{(3 - x)^2}$$

On solving, $x = 3/2$ and $K_p = 3^2/(3/2)^2 = \mathbf{4.}$

8. Let x denote the moles of C_2H_4 present at equilibrium, then

$$n_{C_2H_4} = x \qquad n_{C_2H_6} = 20 - x \qquad N = 100 + x$$
$$n_{H_2} = x \qquad\quad n_{N_2} = 80 \qquad\qquad P = 0.5\,\text{atm}$$

Then

$$K_p = \frac{(x)(x)}{20 - x}\frac{0.5}{100 + x} = 0.05$$

and on solving the quadratic, $x = 10.3$ moles. The various mole fractions are then $N_{H_2} = N_{C_2H_4} = 10.3/110.3 = 0.093$ **(9.3%)**, $N_{C_2H_6} = 9.7/110.3 = 0.088$ **(8.8%)**, and $N_{N_2} = 0.72$ **(72%)**.

$\Delta H^0 = \Delta G^0 + T\Delta S^0 = 5.35 + 900 \times 32.3/1{,}000 = 34.4$ kcal; hence at $300°K$, $\Delta G^0 = 34.4 - 300 \times 32.3/1{,}000 = \mathbf{24.7\ kcal.}$

9. (a) The final total moles is $7 \times 2/0.082 \times 700 = 0.244$, of which 0.06 is of methanol, so 0.184 is the moles of CO and H_2. The moles of CO, however, must be $0.1 - 0.06 = 0.04$, so the moles of H_2 present is 0.144. Then $K_p = (0.06)(0.244)^2/(0.04) \times (0.144)^2(7)^2 = \mathbf{0.088\ atm^{-2}.}$

(b) Had no reaction occurred, the moles of CO would remain 0.1, and the moles of H_2 would be 0.144 plus twice the moles of methanol in (a), or $0.144 + 0.12 = 0.264$. The total moles are then 0.364 and $P = \mathbf{10.4\ atm.}$

10. (a) $\Delta G^0 = -RT \ln K_p = -1.98 \times 298 \times 2.3 \log 10^{-5} = \mathbf{6.78\ kcal.}$

(b) Since no change in the number of moles occurs on reaction, there will still be 6 moles of gas, and $P = 0.082 \times 298 \times 6/5 = \mathbf{29.3\ atm.}$

(c) Let $x = $ moles of H_2O, then since $K_p = K_n$ in this case, $10^{-5} = (1 + x)x/(3 - x)(2 - x)$. As a first approximation, then, $x = 6 \times 10^{-5}$ (no further approximations are needed). Then the moles of CO_2, H_2, CO, and H_2O are 3, 2, 1, and 6×10^{-5}, respectively.

(d) $\Delta H^0 = \Delta G^0 + T\,\Delta S^0 = 6.78 + 298 \times (-10)/1{,}000 = 3.80\,\text{kcal}$; so, at $100°C$, $\Delta G^0 = 3.80 - 373 \times (-10)/1{,}000 = 7.53\,\text{kcal}$. From this, $K_p = 3.8 \times 10^{-5}$.

11. (a) $\Delta G^0 = -RT\ln K_p$ or $\log K_p = -300/1.98 \times 313 \times 2.3 = -0.21$ or $K_p = \mathbf{0.62\,atm^{-1}}$.

(b) From the ideal gas law, $PM = \rho RT = 5.85 \times 0.082 \times 313 = 150$. Let n^0 denote the number of moles if the N_2O_4 were completely undissociated and α, the degree of dissociation. Then $n_{N_2O_4} = (1 - \alpha)n^0$, $n_{NO_2} = 2\alpha n^0$, and $N = (1 + \alpha)n^0$. The average molecular weight M is

$$M = 46N_{NO_2} + 92N_{N_2O_4} \qquad (N \text{ denotes mole fraction})$$

or

$$M = 46\frac{2\alpha}{1 + \alpha} + 92\frac{1 - \alpha}{1 + \alpha} = 92/(1 + \alpha)$$

Since $P = 150/M$, we find $P = (1 + \alpha)(150/92)$. Finally, the equilibrium constant is

$$K_p = \frac{n_{N_2O_4}}{n_{NO_2}^2}\frac{N}{P} = (1 - \alpha)(1 + \alpha)/4\alpha^2 P$$

or

$$0.62 = 92(1 - \alpha)/150 \times 4\alpha^2$$

Then $(1 - \alpha)/\alpha^2 = 4.04$ or $\alpha = \mathbf{0.39}$. M is then $92/1.39 = \mathbf{66}$, and $P = 150/66 = \mathbf{2.28\,atm.}$

12. Since

$$K_p = \frac{P_{PCl_3}P_{Cl_2}}{P_{PCl_5}}$$

for the reaction

$$PCl_5 = PCl_3 + Cl_2$$

we can write immediately that

$$K_p = \left(\frac{2 \times 2}{2}\right)\left(\frac{3}{6}\right) = 1\,\text{atm}$$

On introduction of more Cl_2, the qualitative effect will be to shift the equilibrium to the left, so let x denote the moles of PCl_5 formed as a result of the shift. Then

$$n_{PCl_5} = 2 + x \qquad n_{Cl_2} = 2 + n'_{Cl_2} - x$$
$$n_{PCl_3} = 2 - x \qquad N = 6 + n'_{Cl_2} - x$$

where n'_{Cl_2} denotes the moles of Cl_2 added. Since the new equilibrium volume is $2V$ liters, with P and T constant, it follows that the addition of Cl_2 doubled the number of moles present, so $N = 12$ and therefore $n'_{Cl_2} - x = 6$. We can now write

$$K_p = 1 = \frac{(2 - x)(8)}{(2 + x)} \frac{3}{12}$$

from which $x = \frac{2}{3}$. The moles of Cl_2 added were then $n'_{Cl_2} = 6 + 2/3 = \mathbf{20/3}$.

13.

$$K_p = \frac{P_{CH_3OH}}{P_{CO}P_{H_2}^2} = \frac{2}{1 \times 0.01} = 200 \text{ atm}^{-2}$$

A little reflection at this point shortens the problem considerably. The qualitative effect of expanding the mixture will be to shift the equilibrium to the left, but the shift cannot affect the CO and CH_3OH pressure very much since there is not much H_2 present. As a first approximation, then, consider that the expansion halves the CO and CH_3OH pressures, and find the H_2 pressure required by K_p:

$$200 = \frac{1}{0.5P_{H_2}^2} \quad \text{or} \quad P_{H_2} = 0.1$$

as before. Had no shift in equilibrium occurred, P_{H_2} would have been 0.05, so evidently as a second approximation we should take the CO pressure as 0.525 and the CH_3OH pressure as 0.975. On doing this, and recalculating P_{H_2}, we get **0.096 atm.**

14. K_p is given by

$$K_p = \frac{P_{SO_3}}{P_{SO_2}P_{O_2}^{1/2}}$$

Under the conditions given, oxygen is in great excess, and therefore at some constant pressure P^0 so that $K_p = 10^4/P^{0\,1/2}$. On expanding the mixture, the oxygen pressure drops to $P^0/4$, and since K_p does not change, the new value of SO_3/SO_2 must be 2×10^4.

15. The equilibrium pressure of $NOCl$ is to be 0.01×0.02, so the pressure of NO_2 remaining is 0.99×0.02. We thus have

$$K_p = 1 \times 10^{-3} = \frac{P_{NOCl}^2}{P_{NO_2}^2 P_{Cl_2}} = \frac{(0.01 \times 0.02)^2}{(0.99 \times 0.02)^2 P_{Cl_2}}$$

or $P_{Cl_2} = 0.102$ atm. This pressure corresponds to $0.102/0.2$ or 0.051 mole; in addition $0.01 \times 0.02/2$ or 10^{-4} moles have reacted, so the total moles needed would be **0.0511.**

16. Let n_0 be the initial moles of PCl_5; then

$$n_{PCl_5} = (1 - \alpha)n_0$$
$$n_{PCl_3} = n_{Cl_2} = \alpha n_0$$
$$n_{tot} = (1 + \alpha)n_0$$

Then

$$K_p = \frac{(\alpha n_0)^2 P}{(1 - \alpha)n_0(1 + \alpha)n_0} = \frac{\alpha^2 P}{1 - \alpha^2}$$

or $K_p = (0.1)^2 \times 2 \times P/(1 - 0.1^2) = 0.0202$ atm. ΔG^0 is then $-1.98 \times 523 \times 2.3 \log(0.0202) = 4{,}120$ cal.

Since Δ is small, $K_p \cong \alpha^2 P$, so $d \ln K_p/dT = 2 \, d \ln \alpha/dT = 0.04$. Then $0.04 = \Delta H^0/RT^2$ or $\Delta H^0 = 0.04 \times 1.98 \times 523^2 = $ **21,700 cal.** Also, $\Delta S^0 = (21{,}700 - 4{,}120)/523 = $ **33.7 e.u.**

17. Since $K_p = P_{NH_3}^2$, the equilibrium ammonia pressure must be 3 atm, and the amount of ammonia gas in the flask would then be $3 \times 5/0.082 \times 313 = 0.58$. In addition, 0.2 mole of ammonia is needed to effect the conversion to $LiCl \cdot 3NH_3$, so the total ammonia required is **0.78 mole.**

18. (a) $K_p = P_{SO_2} P_{SO_3}$. Since each gas is formed in equal amounts, their partial pressures are each 0.45 atm. K_p is then 0.45^2 or **0.203 atm^2.**

(b) Let P_{SO_3} be the equilibrium SO_3 pressure; then $P_{SO_2} = P_{SO_3} + 0.6$, so $0.203 = P_{SO_3}(P_{SO_3} + 0.6)$ or $P_{SO_3} = 0.24$, $P_{SO_2} = 0.84$, and $P_{tot} = $ **1.08 atm.**

19. Since the number of moles of gaseous species are the same on both sides of equation (a), $K_p = K_n = n_{H_2O} n_{CO}/n_{H_2} n_{CO_2}$, where n denotes number of moles. Then if $x = n_{H_2O}$,

$$n_{H_2O} = x \qquad n_{CO} = x \qquad n_{H_2} = 0.1 - x \qquad n_{CO_2} = 0.2 - x$$

The mole fraction of water is thus $x/0.3 = 0.1$, where $x = 0.03$. K_a is then $(0.03)^2/(0.07)(0.17) = \textbf{0.0757.}$

Neither reaction (b) nor (c) changes the total moles of gas present, which therefore remains at 0.3. According to the new equilibrium composition, 0.09 mole of water are present, or an increase of 0.06. This must have come about through reaction (b), so that 0.06 mole of H_2 were used up, and $0.07 - 0.06$ or 0.01 are left. K_b is then $0.09/0.01 = \textbf{9.}$ Since reaction (b) minus reaction (a) gives reaction (c), $K_c = K_b/K_a = \textbf{119.}$

From the van't Hoff equation, $d \ln K_p/dT = \Delta H^0/RT^2$, where $d \ln K_p/dT$ is the fractional change in K_p per degree and is given as 0.01. Then $\Delta H_a^0 = 0.01 \times 1.98 \times (723)^2 = \textbf{10.3 kcal.}$

20.   (a) $K_p = P_{NH_3}P_{H_2S} = \frac{1}{4}P^2$, since $P_{NH_3} = P/2$. Therefore $P^2 = 0.2$ and $P = 0.447$ atm. The moles of gas present are then $n = 0.447 \times 2.4/0.082 \times 293 = 0.0447$ mole. The moles of NH_3 or of H_2S are then 0.0223, and the moles of solid remaining are $0.06 - 0.0223 = 0.0377$ mole, or **37.2%** decomposed.

(b) If the decomposition is to be kept to 1%, the moles of H_2S present must be 0.0006, and $P_{H_2S} = 0.006$. From K_p, the pressure of NH_3 is then 0.05/0.006 or 8.33 atm. The moles of ammonia will then be **0.833.** (Notice that 2.4 liters is just 0.1 molar volume if P is 1 atm, so moles of gas $= 0.1 \times P$.)

(c) No change. Addition of more solid does not affect its thermodynamic activity.

21.   (a) From the definition of average molecular weight,

$$36 = 44N_{CO_2} + 28N_{CO} \qquad \text{or} \qquad N_{CO_2} = N_{CO} = \frac{1}{2}$$

The initial 4.4 g correspond to 0.1 mole of CO_2, and if x denotes the moles of CO formed, $0.1 - x/2 =$ moles of CO_2 remaining and $0.1 + x/2$ the total moles. Hence $\frac{1}{2} = x/(0.1 + x/2)$ or $x = 0.0667$ and the total moles are 0.133. The total pressure is then $P = 0.133 \times 0.083 \times 1,273/1$ or $P = \textbf{13.9 atm,}$ and $P_{CO_2} = P_{CO} = 6.95$ atm and $K_p = (6.95)^2/6.95 = \textbf{6.95 atm.}$

(b) Introducing inert gas at constant volume will not change the equilibrium partial pressures and hence will not change the position of equilibrium. Doing so at constant total pressure, however, dilutes the mixture, and the equilibrium will shift in the direction of forming more CO.

(c) The moles of CO formed must be 0.2 since the 0.1 mole of carbon is to be essentially used up. P_{CO} is then $0.2 \times 0.082 \times 1,273/1 = 20.9$, and, from the equilibrium constant, P_{CO_2} must be $(20.9)^2/6.95 = 62.9$ atm, so there

must be $62.9 \times 1/0.082 \times 1,273$ or 0.602 mole of CO_2 present. The total moles of CO_2 required will then be $0.1 + 0.602$ or **0.702.**
(d) From the van't Hoff equation,

$$\log 2 = \frac{\Delta H^0}{1.98 \times 2.3}(1/1,283 - 1/1,273)$$

or

$$\Delta H^0 = 0.3 \times 1.98 \times 2.3 \times 1,283 \times 1,273/10 = \textbf{22.3 kcal.}$$

22. Let α be the fraction of CO reacted; then $P_{SO_2} = 2(\alpha/2)$ and $P_{CO} = 2(1 - \alpha)$. The total pressure is $2(1 - \alpha/2) = 1.03$, whence $\alpha = 0.97$. Then $K_p = 0.97/(0.06)^2 = \textbf{270 atm}^{-1}.$

23. (a) Since $K_p = P_{H_2O}^3$, the equilibrium H_2O pressure at 25°C is 10^{-2} atm. The flask must then contain $0.01 \times 2/0.082 \times 298 = 8.2 \times 10^{-4}$ mole. The total amount of water needed is then $3 \times 0.01 + 8.2 \times 10^{-4} = \textbf{3.08} \times \textbf{10}^{-2}$ **moles.**
(b) From the van't Hoff equation,

$$\log 10^{-4}/10^{-6} = -\frac{\Delta H^0}{1.98 \times 2.3}(1/323 - 1/298)$$

or

$$\Delta H^0 = 2 \times 1.98 \times 2.3 \times 323 \times 298/25 = \textbf{35.1 kcal.}$$

24. Since $K_p = P_{NH_3} \cdot P_{HCl}$ and, in the first case, $P_{NH_3} = P_{HCl}$, then $K_p = (P/2)^2 = 0.025^2 = 6.25 \times 10^{-4}$ atm^2. The 0.02 mole of NH_3 corresponds to a pressure of $0.02 \times 0.082 \times 520/42.7$ or 0.02 atm. Then $K_p = P_{HCl}(P_{HCl} + 0.02) = 6.25 \times 10^{-4}$ or $P_{HCl} = \textbf{0.0169,}$ and $P_{NH_3} = \textbf{0.0369.}$ The moles of HCl are then

$$0.0169 \times 42.7/0.082 \times 520 = \textbf{0.0169}$$

and similarly the moles of NH_3 are **0.0369.** (Notice that the volume was chosen so that the pressure and mole number are equal.) The moles of NH_4Cl remaining are then $0.020 - 0.0169 = \textbf{0.0031.}$

25. (a) $\Delta G^0_{298°K}$ is just the negative of the standard free energy of formation, or **26 kcal.** (b) From Eq. (11-4)

$$\log K_p = \frac{-26,000}{2.3 \times 1.98 \times 298} = -19.1 \quad \text{or} \quad K_p = \textbf{8} \times \textbf{10}^{-20} \textbf{ atm}^{1/2}$$

The dissociation pressure of Cl_2 is just the square of K_p (since $K_p = P_{Cl_2}^{1/2}$), or **6.4 × 10^{-39} atm.** (c) The desired derivative is $(1/R) \, d(\Delta G^0)/dT$ or $-\Delta S^0/R$. Then

$$\Delta S^0 = \frac{\Delta H^0 - \Delta G^0}{T} = \frac{30{,}000 - 26{,}000}{298} = 13.5 \text{ cal/deg}$$

The answer is then $-13.5/R$ or **−6.8.**

26. As the last trace of solid I_2 disappears, the I_2 pressure will still be 0.1 atm, and 0.4 mole of HI, and hence 0.4 atm of HI will be present. The pressure of H_2 required for equilibrium is then

$$20 = \frac{(0.4)^2}{P_{H_2} \times 0.1} \quad \text{or} \quad P_{H_2} = 0.08 \text{ atm}$$

Then 0.08 mole of H_2 must be present as gas to maintain the equilibrium, plus 0.2 mole for the reaction producing 0.4 mole of HI, or **0.28 mole** total.

12

HETEROGENEOUS EQUILIBRIUM: PHASE DIAGRAMS

COMMENTS

The problems that follow are restricted to one-, two-, and three-component systems. In the first case, data are provided that allow the construction of a P versus T diagram. Such diagrams will consist of single-phase regions, lines of two-phase equilibria, and points of crossing of three such lines, or triple points at which three phases are in equilibrium. Note that the three lines that cross at the triple point for phases A, B, and C must comprise the lines for three two-phase equilibria A–B, A–C, and B–C. Remember, too, that the vapor-pressure curve for a solid will always be steeper than that for the liquid phase of the same substance, and use the Clapeyron equation to determine the slope of solid–solid and solid–liquid two-phase lines.

The two-component systems involved here are mostly freezing-point ones, and you will be dealing with T versus composition diagrams for the most part. There will now be one-phase regions and two-phase regions (alternating in any horizontal traverse of the diagram, if immiscible pure solids are taken to be narrow one-phase regions) and lines of three-phase equilibrium. In the case of cooling curves, there is a change of slope on passing from a one- into a two-phase region, and a halt at a line of three-phase equilibrium. Such lines are boundaries between the one-phase region below (above) and the two two-phase regions above (below) and are then eutectic (peritectic) in type.

There are a few diagrams dealing with salt hydrate systems, and you are occasionally asked to consider the vapor region in your labeling.

The three-component systems are mostly of the solubility type, that is, two salts plus water. Use triangular graph paper for these isobaric, isothermal diagrams. There will now be one-, two-, and three-phase regions, the latter always triangular in shape. Tie lines in the two-phase regions need not be parallel, of course.

By way of reassurance, the answers are not really as long as they may appear to be. Much of the reasoning has been given in considerably more detail than would be required on the actual examination.

EQUATIONS AND CONCEPTS

Phase Rule

$F + P = C + 2$, where $F =$ degrees of freedom, $P =$ phases present, and $C =$ components.

Material Balance Calculations

If a system of composition P_1^0 is partitioned into two phases of composition P_1 and P_2, the relative amounts of phase 1 and phase 2 will be in the proportion $(P_2 - P_1^0)/(P_1 - P_1^0)$ or the fraction of the system present as phase 1 will be $(P_2 - P_1^0)/(P_1 - P_2)$. P denotes any additive property, usually mole or weight fraction, in the case of phase diagrams. If P denotes mole (weight) fraction, then the ratio of amounts of phases will be in terms of moles (weights). It is often convenient to estimate this ratio by measuring on the graph the two portions of a tie line to which the above composition differences correspond. Because of the analogy that can be made, the above procedure is sometimes known as the "lever principle."

A similar procedure can be used in the case of triangular plots of three-component systems. As illustrated in Fig. 12-1, a system whose composition

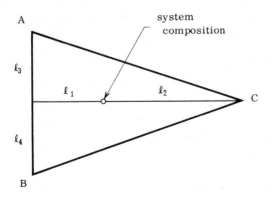

FIGURE 12-1

lies within a three-phase region will consist of the three phases A, B, and C in proportions that can be expressed in terms of certain ratios of lengths. Thus

$$\frac{C}{A + B} = \frac{l_1}{l_2} \qquad \frac{A}{B} = \frac{l_4}{l_3}$$

PROBLEMS

1. (13 min) Solid iodine is in equilibrium with liquid and vapor at 114°C and 90 mm Hg pressure. Let this ordinary solid form be called S_1, and suppose that a second crystalline modification, S_2, is found. At 114°C, S_2 has a vapor pressure of 110 mm Hg and its heat of sublimation is less than that of S_1. The density of S_2 is less than that of S_1; liquid is denser than S_1.

(a) Show how many triple points the complete phase diagram should in principle have. (b) Sketch the likely appearance of the phase diagram. Show metastable lines of two-phase equilibrium as dotted lines; label each line and each triple point. (Show at least three of the triple points.)

2. (12 min) White phosphorus melts at 44°C and 0.2 mm Hg pressure; red phosphorus melts at 490°C and 43 atm pressure. White phosphorus is more dense than the liquid, and the red form is less dense than the liquid. The vapor pressure of the white form is everywhere greater than that of the red. Sketch the *P* versus *T* diagram, label the areas, and explain which triple point(s) are stable and which metastable.

3. (15 min) Substance *X* exists in two crystalline modifications, here called A and B. At temperatures below 90°C, the vapor pressure of the A form is less than that of the B form; above 90°C, the reverse is true. The A form is 5% denser than liquid *X*, and the B form is 10% less dense than the A form. There is an A–L–V triple point at 110°C. Sketch the *P–T* diagram for *X*, filling in missing details in a plausible fashion. Label the phase regions, lines of two-phase equilibrium, and triple points. Discuss briefly whether the triple points you show represent stable, metastable, or entirely unstable equilibria.

4. (18 min) Suppose that a new form of ice is discovered, called ice IX, whose triple point (i.e., point of equilibrium with liquid water and water vapor) is 0.3 atm and 80°C. Ice IX is more dense than liquid water; its heat of sublimation is less than that of ordinary ice.

Draw the phase diagram for water substance, including ice I and ice IX. This should be done carefully enough that the various lines have the proper general appearance as to slope, curvature, and interrelationship consistent with general thermodynamic equations and with the above data. Use dotted lines for lines of metastable two-phase equilibria; show all triple points (insofar as possible).

---

5. (15 min) Table 12-1 gives the break and halt temperatures for the cooling curves of melts of metals A and B. Construct a phase diagram (use Fig. 12-2) consistent with these curves and label the phase regions. Give the probable formula of any compounds.

6. (15 min) Thermal analysis of the two-component system A–B (metals melting at 1,200°C and 600°C, respectively) shows that two solid phases of composition 40% B and 60% B, respectively, are in equilibrium at 800°C with liquid of composition 80% B. Construct the simplest melting-point diagram consistent with this information and label all phase regions. Sketch the cooling curves for compositions 20% B, 50% B, and 70% B and fully label what happens at each break and halt.

7. (24 min) (final examination question) Thermal analysis of the well-known system A–B gives the following information:

10 mole % B:	break at 900°C, second break at 650°C
30 mole % B:	break at 650, halt at 450
50 mole % B:	break at 550, halt at 450
60 mole % B:	break at 650, halt at 600, second halt at 450
80 mole % B:	break at 750, halt at 600
90 mole % B:	break at 780, halt at 600

A and B melt at 1,000°C and 850°C, respectively. Sketch the simplest phase

TABLE 12-1

Mole % A	First break (°C)	First halt (°C)	Second halt (°C)
100		1,100	
90	1,060	700	
80	1,000	700	
70	940	700	400
60	850	700	400
50	750	700	400
40	670	400	
30	550	400	
20		400	
10	450	400	
0		500	

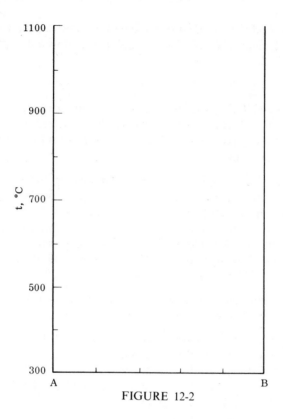

FIGURE 12-2

TABLE 12-2

Pt mole %	Break (°C)	Halt (°C)
0		960
10	1,025	
20	1,075	1,050
30	1,100	1,050
40	1,200	1,050
50	1,250	1,050
60	1,300	1,050
70	1,350	1,050
80	1,400	1,050
90	1,600	
100		1,770

diagram consistent with these data. Label all phase regions and give the formulas of any compounds.

8. (12 min) The cooling curve data of Table 12-2 are for the system silver–platinum; the second column gives the temperature at which a break or change in rate of cooling occurs, and the last column, the temperature of a halt.

Construct the simplest phase diagram (use Fig. 12-3) consistent with the above data, label each phase region, and give the phase reaction occurring during the 1,050°C halt.

9. (24 min) (final examination question) Na and K melt at 98°C and 65°C, respectively; they form one solid compound, NaK, which decomposes at

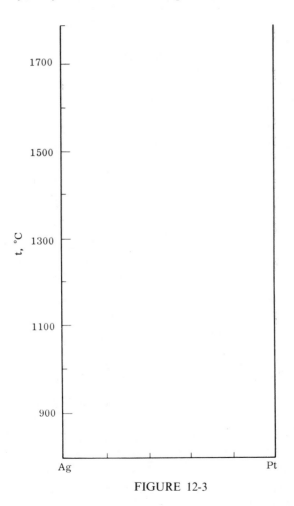

FIGURE 12-3

10°C to give a solid and a melt containing 60% K (mole %). There is a eutectic at −5°C. Sketch the simplest phase diagram consistent with the above data, label the phase regions, and draw cooling curves for melts containing 40% K, 55% K, and 90% K. Indicate the phases appearing or disappearing at each break or halt.

10. (18 min) The diagram shown in Fig. 12-4 is a melting-point diagram for components A and B. (a) Given that the melting point of B is 150°C, complete the diagram in a reasonable and simple manner. The line abc is known to be a line of three-phase equilibrium, of compositions as indicated by the location of the letters. (b) Label all phase regions and give the probable formulas for any compounds. (c) Sketch the cooling curves for melts of composition $N_B = 0.50$ and $N_B = 0.10$. State what phase(s) appear or disappear at each break or halt.

11. (12 min) (final examination question) The following information is obtained from cooling curve data on the system Sn–Mg:

Composition of melt (mole % Mg)	Temperature of break (if any) (°C)	Temperature of halt (if any) (°C)
0		250
10		200
40	600	200
67		800
80	610	580
90	610	580
100		650

Sketch the simplest melting-point diagram consistent with these data; label the phase regions and give the compositions of any compounds formed.

12. (18 min) Thermal analysis of the system A–B showed the presence of a line of three-phase equilibrium at 1,000°C, and one at 600°C. In no case, however, was there more than one halt for any one cooling curve. A and B melt at 800°C and 1,300°C, respectively, and the compound A_2B is known, with a melting point of 700°C. Sketch the simplest phase diagram that conforms with the above data, and label all the phase regions.

13. (24 min) (final examination question) Au and Sb melt at 1,060°C and 630°C, respectively, and form one compound, $AuSb_2$, which melts incongruently at 800°C. Sketch the simplest phase diagram consistent with this information and label all phase regions. Give the cooling curve for a melt containing 50% Au, and label each break and halt.

14. (30 min) Metals A and B form the compounds AB_3 and A_2B_3. Solids A, B, AB_3, and A_2B_3 essentially are immiscible in each other, but are com-

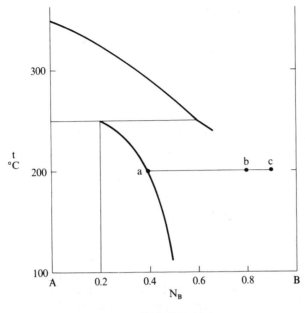

FIGURE 12-4

pletely miscible as liquids. A and B melt at 600°C and 1,100°C, respectively. Compound A_2B_3 melts congruently at 900°C and gives a simple eutectic with A at 450°C. Compound AB_3 decomposes at 800°C to give the other compound and a melt. There is a eutectic at 650°C.

(a) Draw the simplest phase diagram consistent with this information, and label all phase regions.

(b) Sketch cooling curves for melts of composition 90% A and 30% A, and label as to phases appearing or disappearing at each break and halt.

15. (24 min) (final examination question) Metals A and B melt at 1,200°C and 1,600°C, respectively. A thermal analysis shows the presence of the following three-phase equilibria

at 1,400°C: melt containing 10% B and two solid solutions containing 20% B and 30% B, respectively.

at 1,250°C: solid solution containing 65% B, melt containing 75% B, and solution containing 95% B.

There is one compound, A_2B_3, which melts at 1,700°C (congruently).

Construct the simplest phase diagram that will correspond to the above data and label the phase regions. Draw semiquantitative cooling curves for melts of composition 25% B and 90% B, and give the phases appearing or disappearing at each break or halt.

TABLE 12-3

mole % A	Break (°C)	Halt (°C)	Halt (°C)
100		1,000	
90	950	800	
80	900	800	
70	900	800	
60	1,000	800	
50		1,100	
40	1,000	700	
30	750	700	500
20	550	500	
10	575	500	
0		600	

16. (18 min) Table 12-3 gives the break and halt temperatures for the cooling curves of melts of metals A and B. Construct a phase diagram consistent with these curves and label each phase region. Give the probable formulas of any compounds.

17. (15 min) The cooling curves of Fig. 12-5 are obtained for the system CaF_2–$CaCl_2$ (compositions are in mole % of $CaCl_2$ in the melt). Construct the most reasonable semiquantitative freezing-point diagram, label all phase regions, and give the probable formulas of any compounds that are formed.

18. (12 min) The melting-point diagram for A and B has the appearance shown. Use "A" and "B" to denote pure solid A or B (and C_1, C_2, etc., for pure solid compounds), and α, β, and so on to denote solid solutions

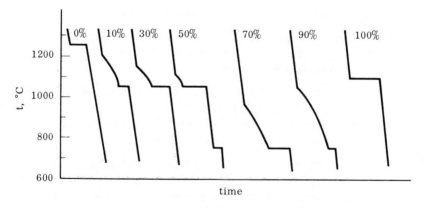

FIGURE 12-5

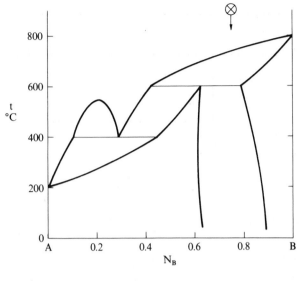

FIGURE 12-6

rich in A, B, and so on. Use L to denote a liquid solution (or L_1, L_2 if appropriate). (a) Label the diagram shown in Fig. 12-6 as to the phases present in each region. (b) Sketch the cooling curve for a system of the composition indicated in the figure; state what phase(s) appear or disappear at each halt or break.

---

19. (15 min) Given the solubility diagram of Fig. 12-7 for the salt MX and its hydrates, label each phase region and describe the sequence of phase

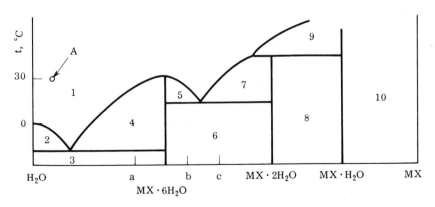

FIGURE 12-7

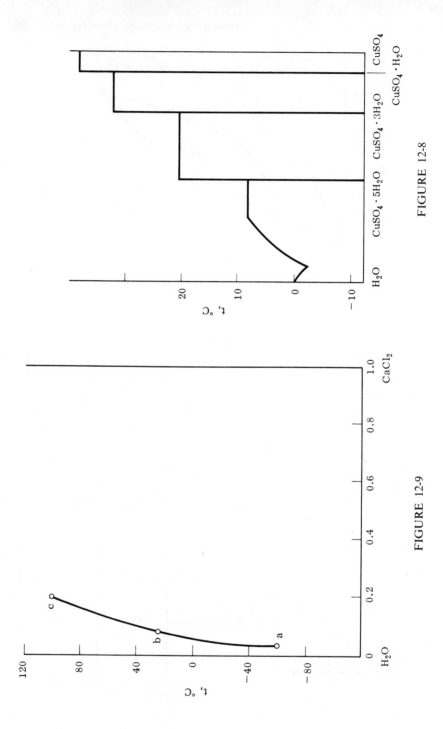

FIGURE 12-8

FIGURE 12-9

222

changes that would occur on isothermal evaporation at 30°C of a solution of composition A (indicated on the diagram) to dryness.

20. (24 min) (final examination question) The diagram of Fig. 12-8 is for the system H_2O–$CuSO_4$. Fill in the additional lines needed to define the various phase areas, and label each area as to the phase(s) present.

Describe the sequence of phase changes if a dilute solution of copper sulfate is dehydrated at 5°C, ending up with anhydrous copper sulfate.

21. (20 min) (final examination question) Given the solubility data for $CaCl_2$, where the curve ab of Fig. 12-9 gives the solubility of $CaCl_2 \cdot 6H_2O$, and the curve bc, that for $CaCl_2 \cdot H_2O$.

(a) Sketch the completed phase diagram between $-75°C$ and $100°C$ and label all phase regions.

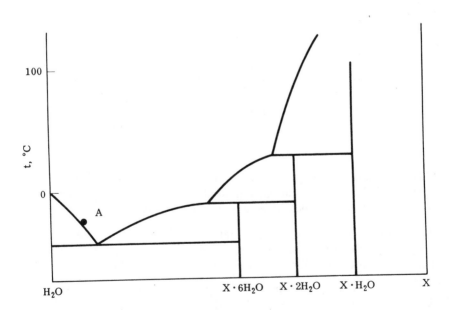

FIGURE 12-10

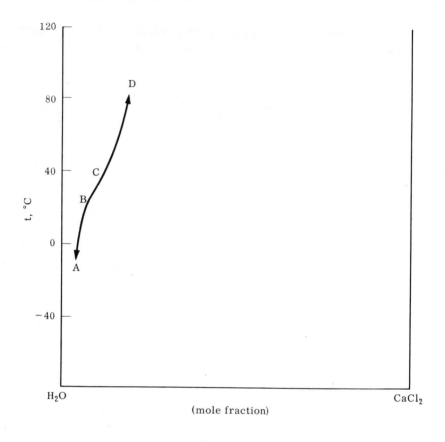

FIGURE 12-11

(b) One mole of $CaCl_2 \cdot H_2O$ and 9 moles of water are mixed and heated to 100°C. The resulting solution is allowed to cool to -75°C. Describe the sequence of phases that would appear and/or disappear during this cooling.

22. (15 min) Given the solubility diagram of Fig. 12-10 for the salt X and its hydrates, label each phase region and describe what sequence of phase changes would occur upon isothermal evaporation of solution A.

23. (15 min) A section of the solubility versus temperature curve for the system $CaCl_2$–water is shown in Fig. 12-11. Along AB, BC, and CD, the solutions are saturated with respect to $CaCl_2 \cdot 6H_2O$, $CaCl_2 \cdot 4H_2O$, and $CaCl_2 \cdot 2H_2O$, respectively.

(a) Sketch the complete temperature-composition diagram for 1 atm, including vapor-phase regions, filling in missing details in a plausible fashion.

(b) Describe in detail what would happen on heating at 1 atm a system initially consisting of 1 mole of ice and 1 mole of $CaCl_2 \cdot 6H_2O$ at $-50°C$. A closed container, such as a piston and cylinder arrangement, is used.

---

24. (15 min) The isobaric solubility diagram for the system $CHCl_3$–CH_3COOH–H_2O is shown in Fig. 12-12 with some representative tie lines. What phase(s) and of what composition(s) will be present if 1 mole of $CHCl_3$ is added to a system consisting of 0.6 mole of water and 0.4 mole of acetic acid? If more than one phase is present, give the relative amounts of each. Explain clearly your procedure. (The composition scale is in mole fraction.)

25. (24 min) (final examination question) In the triangular diagram of Fig. 12-13 shown for the system H_2O–Na_2SO_4–$NaCl$ at 25°C, the line ab gives

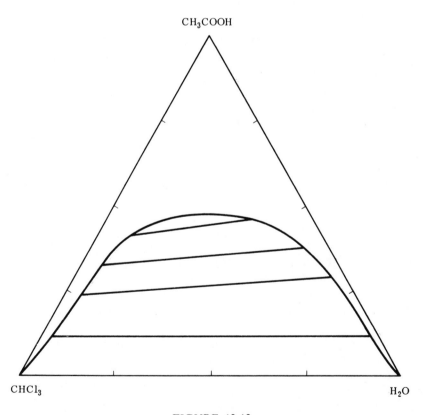

CH₃COOH

CHCl₃   H₂O

FIGURE 12-12

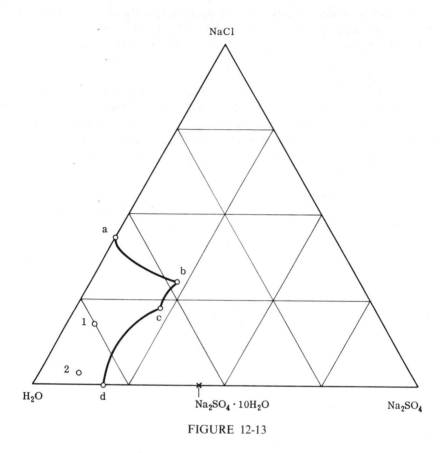

Nacl

a

b

1

c

2

H₂O   d

Na₂SO₄ · 10H₂O

Na₂SO₄

FIGURE 12-13

the compositions of solutions saturated with respect to NaCl, the line bc, those saturated with respect to Na_2SO_4, and the line cd, those saturated with respect to $Na_2SO_4 \cdot 10H_2O$. (The diagram is on a weight % basis.)

(a) Complete the diagram and label all phase regions.

(b) Describe carefully the sequence of events on complete dehydration at 25°C of solutions of compositions 1 and 2 as shown on the diagram.

26. (12 min) The isothermal, isobaric solubility diagram for the system $H_2O–KI–I_2$ is shown in Fig. 12-14. Compositions are in mole%; one compound, $KI \cdot I_2 \cdot H_2O$, is formed. Complete the diagram (if necessary) and label the various phase regions.

A solution containing 75 mole % H_2O, 20% KI, and 5% I_2 is evaporated at constant T and P. What phase or phases would be present and in what

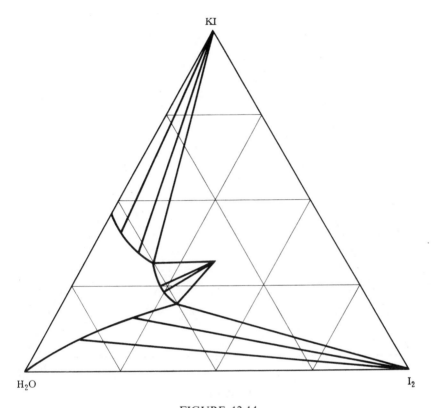

FIGURE 12-14

approximate amounts when the evaporation has proceeded until only 50 mole % of H_2O is present? Make your calculations clear.

27. (24 min) The following data are given for the system Na_2SO_4–$MgSO_4$–H_2O at 40°C and 1 atm.

Solution phase	*In equilibrium with solid phase*
1. 35% $NaSO_4$ 0% $MgSO_4$	Na_2SO_4
2. 0% Na_2SO_4 30% $MgSO_4$	$MgSO_4$
3. 25% Na_2SO_4 15% $MgSO_4$	$Na_2SO_4 + Na_2SO_4 \cdot MgSO_4 \cdot 4H_2O$
4. 5% Na_2SO_4 28% $MgSO_4$	$MgSO_4 + Na_2SO_4 \cdot MgSO_4 \cdot 4H_2O$

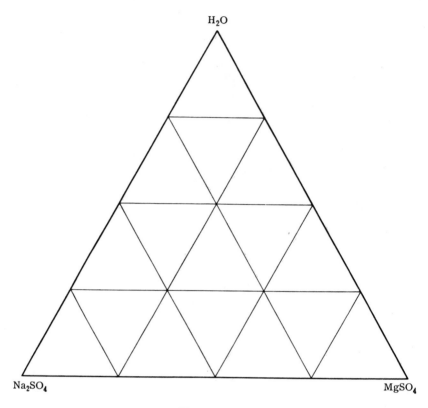

H_2O

Na_2SO_4 $MgSO_4$

FIGURE 12-15

The above are compositions of solutions saturated with respect to the solid phases given. Compositions are in weight percent ($Na_2SO_4 \cdot MgSO_4 \cdot 4H_2O$ is 42.4% Na_2SO_4 and 36% $MgSO_4$).

(a) Sketch the phase diagram in Fig. 12-15 and label all phase regions.

(b) If a solution containing 10% Na_2SO_4 and 10% $MgSO_4$ is evaporated, what solid will appear first, and what is the composition of the last solution to exist before the system becomes entirely solid?

(c) If equal weights of Na_2SO_4 and of solution (b) above are mixed, how many phases will be present at equilibrium and what will be their compositions? Make clear your procedures in obtaining the above answers.

28. (30 min) The isobaric, isothermal diagram for the system H_2O–Li_2SO_4–$(NH_4)_2SO_4$ is given in Fig. 12-16.

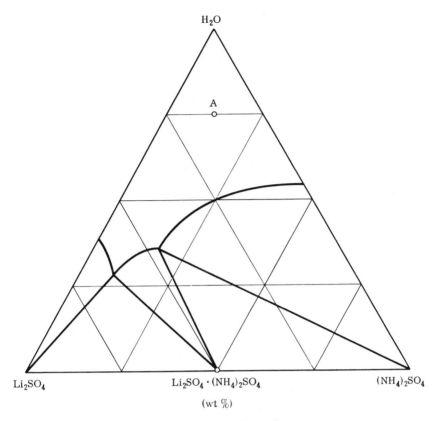

FIGURE 12-16

(a) Label all phase regions.

(b) Describe in detail the sequence of events on evaporation of a solution of composition A.

(c) Calculate the maximum number of grams of Li_2SO_4 that can be added to 100 g of solution A, if ammonium sulfate is still to be the first solid to separate out on evaporation.

(d) Show that, on evaporation of a dilute solution of the double salt $Li_2SO_4 \cdot (NH_4)_2SO_4$, one can never obtain a system consisting of the solid double salt in equilibrium with saturated solution.

29. (24 min) (final examination question) The data of Table 12-4 are for the isobaric isothermal system $H_2O-Na_2SO_4-NaCl$ (at 25°C and 1 atm). Sketch a reasonable phase diagram consistent with the above data and label all phase regions (use Fig. 12-17). Describe the sequence of events on evaporating a solution containing 5% NaCl, 5% Na_2SO_4, and 90% H_2O.

TABLE 12-4

Saturated solution	In equilibrium with solid phases
	(weight % compositions)
1. 80% H_2O, 0% Na_2SO_4	NaCl
2. 85% H_2O, 0% NaCl	$Na_2SO_4 \cdot 10H_2O$
3. 75% H_2O, 10% Na_2SO_4	$NaCl + Na_2SO_4$
4. 75% H_2O, 15% Na_2SO_4	$Na_2SO_4 + Na_2SO_4 \cdot 10H_2O$
	(formula weights)
$Na_2SO_4 = 142$; $Na_2SO_4 \cdot 10H_2O = 322$; $NaCl = 58$	

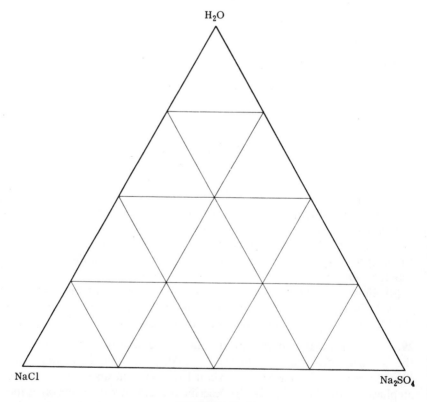

FIGURE 12-17

30. (12 min) In the triangular diagram of Fig. 12-18, the curve ab gives the compositions of solutions saturated with respect to NH_4Cl, and the curve bc gives the compositions of solutions saturated with respect to $(NH_4)_2SO_4$. Complete the diagram by drawing in the boundaries of the various phase regions, and label these regions. Compositions are in weight percent. Give the compositions and amounts of the phase or phases present in a system consisting of 20% water, 60% NH_4Cl, and 20% $(NH_4)_2SO_4$. Show your procedure.

31. (15 min) A portion of the isothermal, isobaric phase diagram for the system $H_2O–AgNO_3–KNO_3$ is shown (for 25°C, 1 atm) in Fig. 12-19. The lines ab, bc, and cd give the compositions of solutions saturated with respect to $AgNO_3$, the double salt $AgNO_3·KNO_3$, and KNO_3, respectively. The diagram is on a weight percent basis.
(a) Complete the diagram, and label the one-, two-, and three-phase

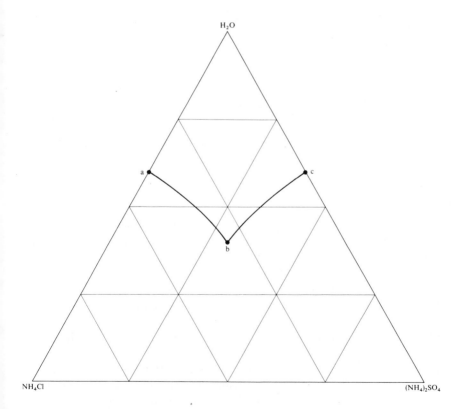

FIGURE 12-18

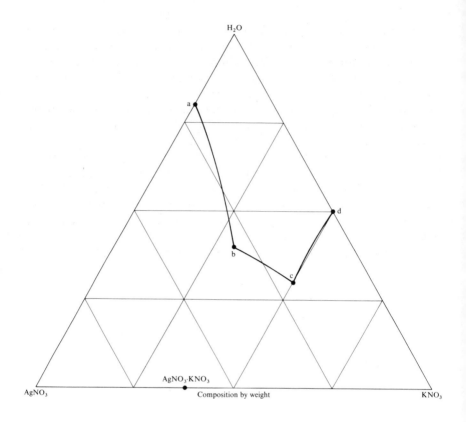

FIGURE 12-19

regions as to what phases are present. (b) A solution consisting of 80% water by weight, 5% KNO_3, and 15% $AgNO_3$ is evaporated at 25°C. What solid first appears and what is the composition of the solution at this point? What solid next appears, and what is the composition of the solution at this point? (c) 50 g of water and 50 g of $AgNO_3 \cdot KNO_3$ are mixed at 25°C. At equilibrium, what phases are present and in what amounts?

32. (12 min) The diagram shown in Fig. 12-20 is for composition in weight percent. Line ab gives compositions of solutions saturated with respect to solid A, and line bc, that of solutions saturated with respect to solid B.

(a) Complete the diagram in a reasonable and simple manner and label all phase regions. (b) A solution of composition 12.5% B and 25% A is evaporated to half of its original weight. Show what phases are present, of

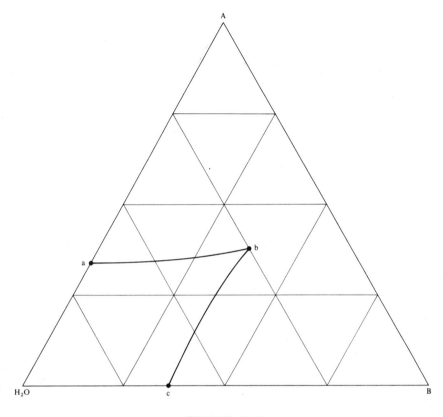

FIGURE 12-20

what composition, and in what proportions. (Make any graphical operations very clear.)

ANSWERS

1. (a) Four phases are observed, S_1, S_2, L, and V; there are **four** possible triple points, that is, four ways of taking three out of four phases. (b) See Fig. 12-21. Since S_2 has a higher vapor pressure than S_1, it is unstable with respect to it at 114°C; since its heat of sublimation is less, the slope of the S_2V line is generally less than that of the S_1V line and the two must meet above 114°C. As indicated in the figure, S_2 appears to be metastable under all conditions.

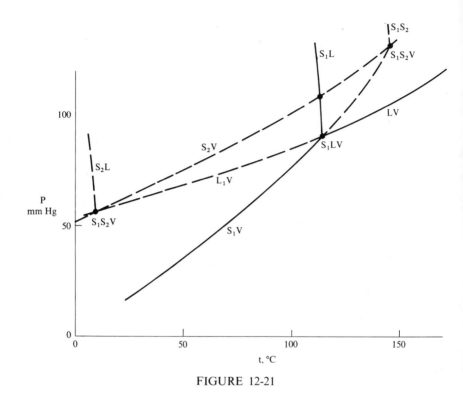

FIGURE 12-21

2. (a) Locates in Fig. 12-22 a metastable triple point, W (white phosphorus), L, and V; it is metastable since the vapor pressure of W is greater than that of R (red phosphorus).

(b) Locates a stable triple point, R–L–V.

(c) Locates the W–R–L triple point, which should be stable, but the fourth triple point, W–R–V, is totally unstable, as it probably lies above the melting point of the liquid, and one assumes that solids cannot be superheated.

3. Evidently, the vapor-pressure curve for A rises in Fig. 12-23 more steeply with temperature than does that for B, and this in turn is steeper than the vapor-pressure curve for the liquid. The AV and LV curves cross at the ALV triple point; the AV and BV curves at the ABV triple point at 90°C; and the BV and LV curves at the LBV triple point, which must lie above 110°C. A is then the low-temperature stable form.

From the relative densities, the AB line must slope to the right, as must the AL line, but less so, and the BL line must slope to the left. These lines must then meet at the ABL triple point. The triangular region enclosed by solid lines is then the region of stability of B.

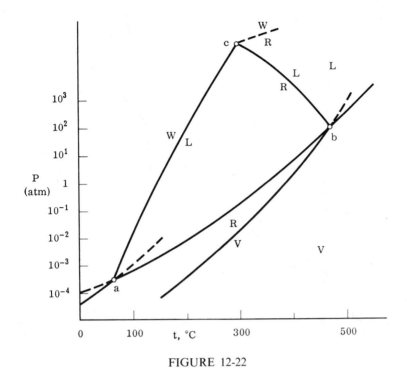

FIGURE 12-22

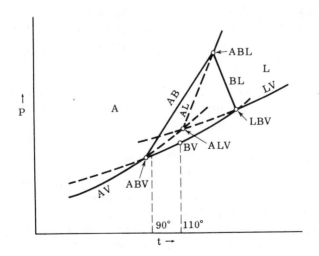

FIGURE 12-23

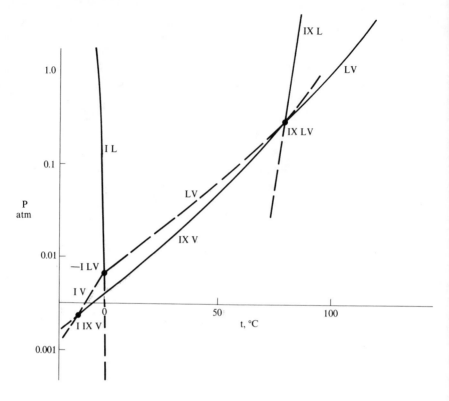

FIGURE 12-24

The stable triple points are those formed by the intersection of three solid (i.e., stable equilibrium) lines, that is, ABV, LBV, and ABL. The ALV triple point is metastable as phase A is unstable toward eventual transformation to B.

4. The possible diagram is shown in Fig. 12-24. The fourth triple point (I–IX–L) cannot conveniently be shown; it would lie somewhere near the bottom of the figure. Note that ice IX is more stable than ice I until very low temperatures. (The existence of an "ice IX" was the basis of a famous science fantasy story, *The Cat's Cradle* by K. Vonnegut, Jr.).

5. Each halt corresponds to a line of three-phase equilibrium and each break to a boundary between a one- and a two-phase region. With this guide, the simplest diagram is that shown in Fig. 12-25.

An unstable compound is indicated that must contain less than 80% and more than 70% A, since it is between these limits that the lower three-phase line appears. A good guess is 75% A, corresponding to the formula A_3B.

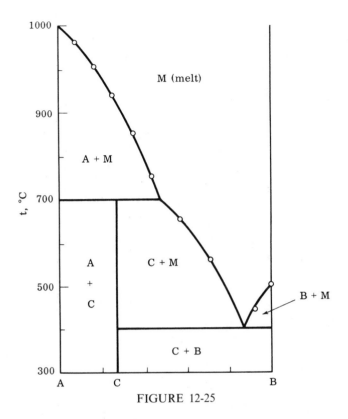

FIGURE 12-25

6. The diagrams of Fig. 12-26 are largely self-explanatory. The coexistence of three phases, two solid, at a temperature between the two melting points is a clear indication of a peritectic-type diagram.

7. The required diagram (Fig. 12-27) largely draws itself once the break and halt data are located. The simplest explanation for the two halts at 60% B is that an unstable compound forms. Its composition must lie between 60% and 80% B, and if it is 75% B, the formula would be AB_3. Alternatively, it might be AB_2 (67% B).

8. A halt temperature lying between the two melting points suggests a peritectic-type diagram (see Fig. 12-28). The limits of the α and β solid solutions are merely suggestive, as the data do not allow a precise locating of them. The phase reaction is

phase + melt $\rightarrow \alpha$ phase

For compositions to the left of point (a), β phase disappears before melt does, on cooling, whereas to the right of point (a), melt disappears before β phase.

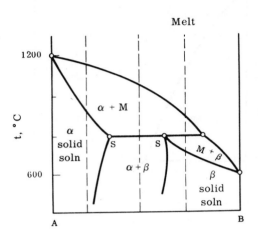

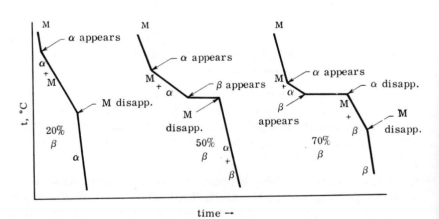

FIGURE 12-26

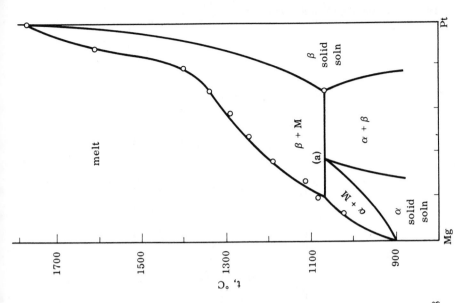

FIGURE 12-28

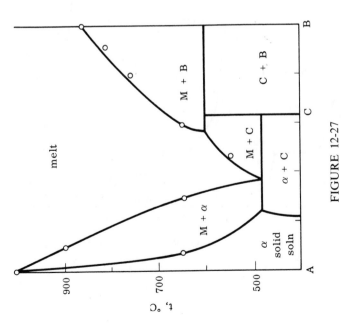

FIGURE 12-27

239

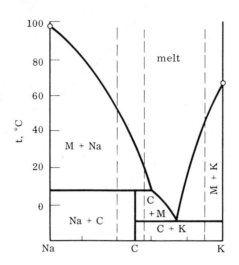

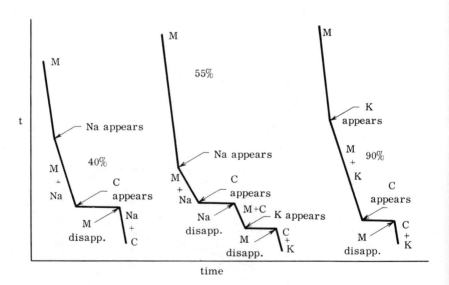

FIGURE 12-29

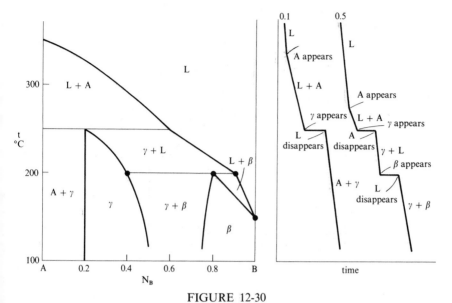

FIGURE 12-30

9. Since the compound decomposes into a melt richer in K than it is, the residual solid must be richer in Na. The simplest assumption is that the solid is pure Na. (See Fig. 12-29.)

10. The completed diagram is shown in Fig. 12-30. An unstable compound of $N_B = 0.2$ is indicated, corresponding to the formula A_4B. The cooling curves are also shown in the figure.

11. The answer is shown in Fig. 12-31. The presence of a congruent melting point at 67% indicates a compound of this composition, corresponding to formula Mg_2Sn.

12. The 600°C three-phase line is probably a simple eutectic line, as it lies below the melting point both of A and of A_2B. (See Fig. 12-32.)

   The other three-phase line, however, lies between the melting points of A_2B and B. An unstable compound is ruled out, since this would require that there be a cooling curve with halts at both 600°C and 1,000°C. The next simplest solution is that shown, that is, a peritectic, and limited solid solutions formed between A_2B and B.

13. Since the incongruently melting compound decomposes above the mp of Sb but below that of Au, the simplest supposition is that solid Au is one of the decomposition products, and a melt, of > 67% Sb, the other. The additional eutectic provides a simple way to connect up the rest of Fig. 12-33.

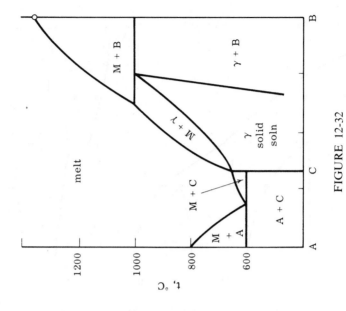

FIGURE 12-32

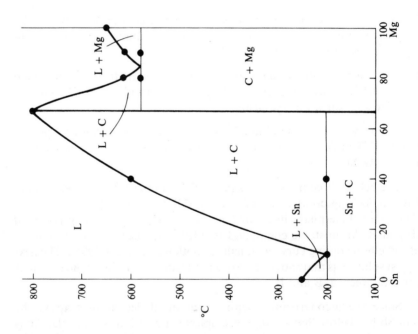

FIGURE 12-31

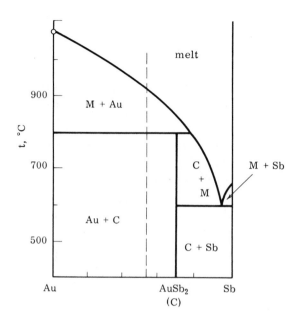

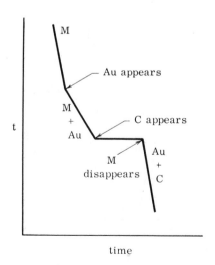

FIGURE 12-33

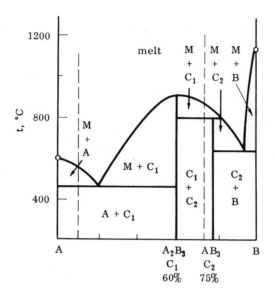

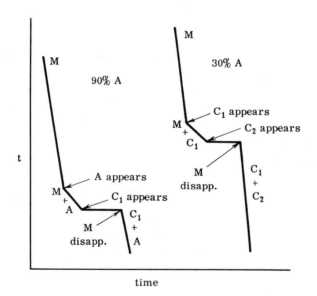

FIGURE 12-34

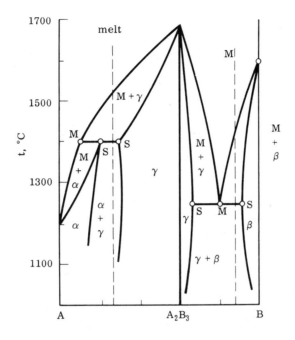

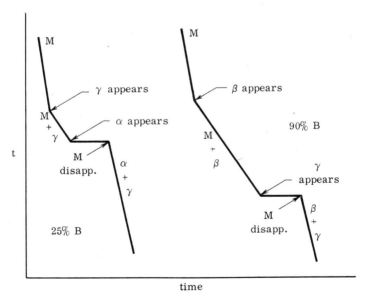

FIGURE 12-35

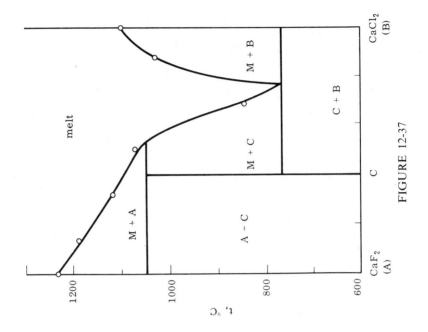

FIGURE 12-37

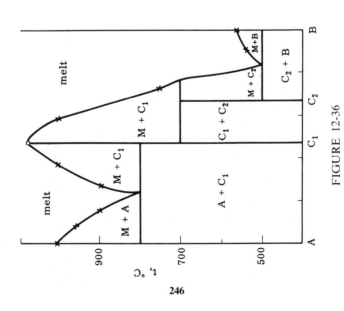

FIGURE 12-36

246

14. Figure 12-34 largely draws itself once the information is located on it. The composition of the first eutectic melt could lie either to the right or to the left of the 90% composition point. If it lay to the left, the order of appearance of A and C_1 would be the reverse of that shown in the cooling curve.

15. Note that a congruently melting compound essentially divides the phase diagram (Fig. 12-35) into two parts whose behavior can be quite disparate. The $A-A_2B_3$ system is most simply taken to be a peritectic type, since the three-phase line lies between the two melting points, whereas the A_2B_3-B system is obviously a simple eutectic.

16. (See Fig. 12-36.) There is evidently a congruently melting compound at 50% A, shown as a halt, but no break is observed; this corresponds to the formula AB. The double halt for the 30% melt suggests an unstable compound. Its composition must lie between 40% A and 30% A; it is, say 33% A; then the formula would be AB_2.

17. The curves of Fig. 12-5 at 0% and 100% obviously give the two melting points of the pure components. The two halts shown in the 50% curve fit the case of unstable compound formation, if the compound has a composition between 30% and 50% $CaCl_2$. The figure 40% is chosen in Fig. 12-37 corresponding to $(CaF_2)_3(CaCl_2)_2$.

18. The labeled diagram is shown in Fig. 12-38, as is the cooling curve.

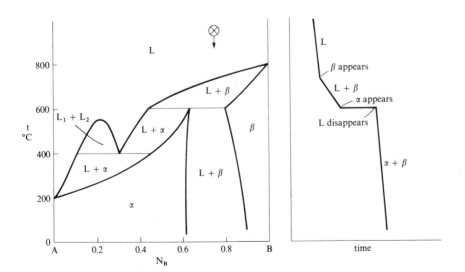

FIGURE 12-38

19.   The phase regions are as follow:

1. solution	6. $MX \cdot 6H_2O + MX \cdot 2H_2O$
2. solution + ice	7. solution + $MX \cdot 2H_2O$
3. ice + $MX \cdot 6H_2O$	8. $MX \cdot 2H_2O + MX \cdot H_2O$
4. solution + $MX \cdot 6H_2O$	9. solution + $MX \cdot H_2O$
5. solution + $MX \cdot 6H_2O$	10. $MX \cdot H_2O + MX$

To follow the evaporation, draw a horizontal line through point A.
(1) The solution increases in concentration to a value indicated by point a on the bottom of the graph. (2) At a, $MX \cdot 6H_2O$ begins to precipitate out and eventually the system consists of pure $MX \cdot 6H_2O$. (3) Further removal of water produces solution of composition b and eventually all the $MX \cdot 6H_2O$ disappears. (4) The solution concentrates to composition c, at which point solid $MX \cdot 2H_2O$ begins to come out, and eventually only pure solid $MX \cdot 2H_2O$ is present. (5) Further removal of water produces some $MX \cdot H_2O$; the proportion of this increases until the system consists entirely of $MX \cdot H_2O$. (6) Still further drying produces some MX, and eventually all water is removed and the system consists of pure MX.

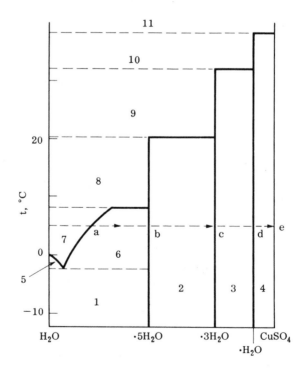

FIGURE 12-39

20. The added lines are shown in Fig. 12-39 as dotted, and the various phase regions are as follow:

1. ice + $CuSO_4·5H_2O$
2. $CuSO_4·5H_2O$ + $CuSO_4·3H_2O$
3. $CuSO_4·3H_2O$ + $CuSO_4·H_2O$
4. $CuSO_4·H_2O$ + $CuSO_4$
5. ice + solution
6. solution + $CuSO_4·5H_2O$

7. solution
8. H_2O vapor + $CuSO_4·5H_2O$
9. H_2O vapor + $CuSO_4·3H_2O$
10. H_2O vapor + $CuSO_4·H_2O$
11. H_2O vapor + $CuSO_4$

The horizontal dotted line shows the motion of the system composition on dehydration. At a, $CuSO_4·5H_2O$ precipitates out, and when system of composition b is reached, pure solid $CuSO_4·5H_2O$ is present. Further dehydration produces increasing amounts of $CuSO_4·3H_2O$, and at c only this is present. Next, $CuSO_4·H_2O$ begins to form, and at d pure monohydrate is present. Finally anhydrous $CuSO_4$ is formed, and when e is reached pure $CuSO_4$ is present.

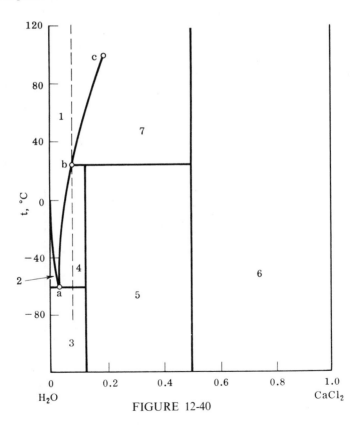

FIGURE 12-40

21. The phase regions (see Fig. 12-40) are as follow:

1. solution
2. solution + ice
3. ice + $CaCl_2 \cdot 6H_2O$ (14.3% $CaCl_2$)
4. solution + $CaCl_2 \cdot 6H_2O$
5. $CaCl_2 \cdot 6H_2O$ + $CaCl_2 \cdot H_2O$ (50% $CaCl_2$)
6. $CaCl_2 \cdot H_2O$ + $CaCl_2$
7. solution + $CaCl_2 \cdot H_2O$

The indicated solution consists of 1 mole of $CaCl_2$ and 10 moles of water and so is 9% $CaCl_2$, as shown by the dotted line. On cooling, $CaCl_2 \cdot 6H_2O$ will crystallize out (at about 20°C), and at about -60°C, ice will also freeze out. The temperature will then remain at -60°C until all the solution is converted to ice and $CaCl_2 \cdot 6H_2O$.

22. The phase regions are labeled in Fig. 12-41. Evaporation of solution A proceeds as follows. When composition 1 is reached, $X \cdot 6H_2O$ begins to crystallize out, and when the system reaches 2, pure hexahydrate is present. Further dehydration produces increasing amounts of $X \cdot 2H_2O$ and, when the system reaches 3, pure dihydrate is present. Further dehydration produces $X \cdot H_2O$, then, at 4, pure monohydrate, then X, and finally pure X.

23. In the completed diagram (Fig. 12-42), the dashed lines are plausibly located expected features, and the solid lines are those phase boundaries definitely required.

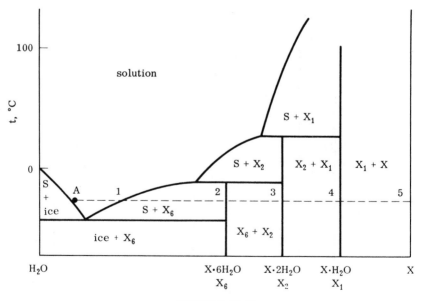

FIGURE 12-41

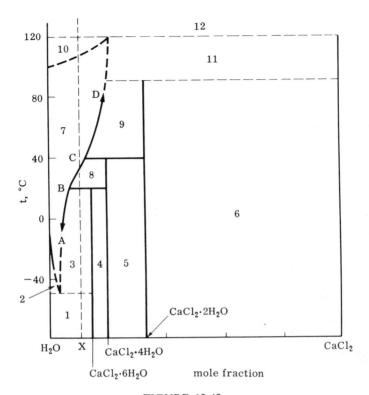

FIGURE 12-42

1. ice + $CaCl_2 \cdot 6H_2O$	7. solution
2. ice + solution	8. solution + $CaCl_2 \cdot 4H_2O$
3. solution + $CaCl_2 \cdot 6H_2O$	9. solution + $CaCl_2 \cdot 2H_2O$
4. $CaCl_2 \cdot 6H_2O$ + $CaCl_2 \cdot 4H_2O$	10. water vapor + solution
5. $CaCl_2 \cdot 4H_2O$ + $CaCl_2 \cdot 2H_2O$	11. solution + $CaCl_2$
6. $CaCl_2 \cdot 2H_2O$ + $CaCl_2$	12. water vapor + $CaCl_2$

The system to be heated is 12.5% $CaCl_2$, as indicated by point X. On heating, the system then moves along the dotted vertical line. At about $-45°C$, the ice melts and one has solution plus $CaCl_2 \cdot 6H_2O$; at about 30°C, $CaCl_2 \cdot 4H_2O$ begins to form, and above 30°C there is solution + $CaCl_2 \cdot 4H_2O$. At about 35°C, the $CaCl_2 \cdot 4H_2O$ is completely dissolved. There is no further change until about 105°C, when water vapor begins to form, and at 120°C, the solution precipitates $CaCl_2$. Above 120°C there exists only water vapor and $CaCl_2$.

24. The system composition is evidently 50% $CHCl_3$, 30% H_2O, and 20% acetic acid; this point is shown in Fig. 12-43, along with an intermediate

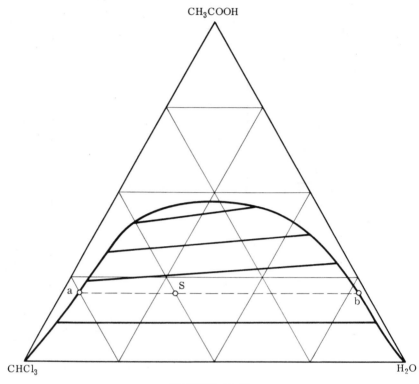

CH₃COOH

CHCl₃   H₂O

FIGURE 12-43

tie line. The ends of the tie line are at: (a) 75% $CHCl_3$, 5% H_2O, and 20% CH_3COOH; and (b) 4% $CHCl_3$, 76% H_2O, and 20% CH_3COOH. The relative amounts of the two phases are most easily obtained by the lever principle; that is, the ratio of moles of phase (a) to (b) is that of the distance sb to the distance sa, or about 1.75.

25. The phase regions are A: solution; B: solution of composition along ab plus solid NaCl; C: solution of composition along cd plus solid $Na_2SO_4·10H_2O$; D: solution of composition c plus solid $Na_2SO_4·10H_2O$ plus solid Na_2SO_4; E: solution of composition along bc plus solid Na_2SO_4; F: solution of composition b plus solid NaCl plus solid Na_2SO_4.

On dehydration of solution (1), the system moves along the dotted line of Fig. 12-44 away from the water corner. At e solid NaCl appears, and the solution composition moves toward b. When the system composition reaches f, solution is at b and solid Na_2SO_4 also begins to precipitate out. With further dehydration, the solution dries up leaving mixture g of solid NaCl and Na_2SO_4.

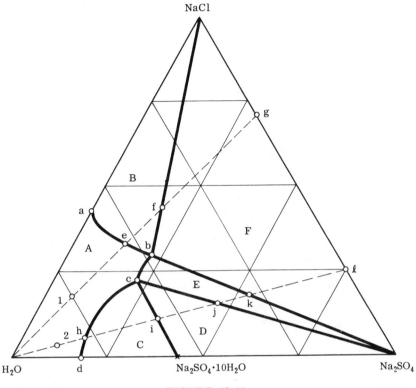

FIGURE 12-44

In the case of solution (2), $Na_2SO_4\cdot10H_2O$ begins to precipitate out when composition h is reached. The solution moves toward c and, when the system reaches i, solution of composition c is in equilibrium with $Na_2SO_4\cdot10H_2O$ and Na_2SO_4. Further dehydration reduces the amount of $Na_2SO_4\cdot10H_2O$ until at j only solution c and Na_2SO_4 are present. The solution composition then moves toward b, and, on reaching it, the further sequence is the same as for solution (1).

26. The labeling is given in Fig. 12-45. Note the additional lines drawn in to define the various three-phase triangular regions.

On evaporation, the system composition follows the dotted line drawn from the starting composition through the water corner. At 50% H_2O, the system will be at point X shown on the diagram, and the tie line through X goes from the KI corner to point (1) on the solution line. This point corresponds to about 55% H_2O, 35% KI, and 10% I_2, and the relative amounts of solution and of KI can be judged from the relative lengths of the two

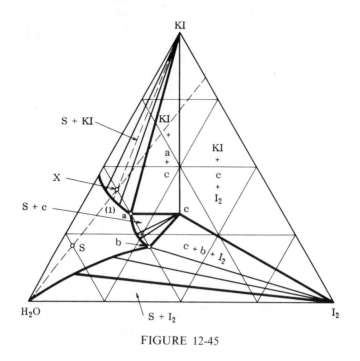

FIGURE 12-45

sections of the tie line, that is, the distance KI–X versus the distance X–solution, to be about 10 moles of solution to 1 of KI.

27. (a) The phase regions are as shown in Fig. 12-46. Note that solutions 3 and 4 mark the limits of the solutions' compositions in equilibrium with the compound.

(b) The system follows the dotted line drawn from the water corner through the solution composition, when evaporation occurs. At point *a*, the system enters the S + C two-phase region, so compound begins to crystallize out. When the system composition reaches *b* the solution has moved over to point 4, and both $MgSO_4$ and compound crystallize out; when the system reaches *c*, the last trace of solution (4) disappears.

(c) The point marked X locates the system composition (50% Na_2SO_4 and 15% $MgSO_4$); it lies in the A + 3 + C three-phase region, so these are the phases present. To estimate the relative amounts, the dotted line is drawn from 3 through X, intersecting the AC line at *d*. Point X lies about midway, so there are about equal weights of solution 3 and the mixed solids. Point *d* lies about midway between A and C, so there are about equal weights of the two solids. The system thus consists of about 50% solution 3, and 25% each of Na_2SO_4 and compound.

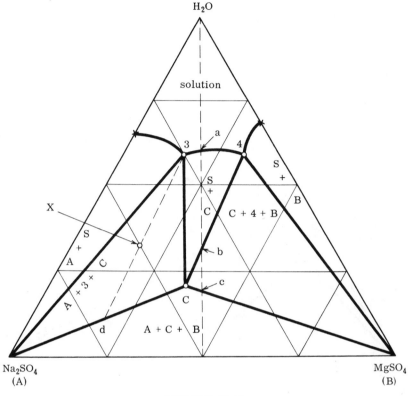

FIGURE 12-46

28. (a) The phase regions are labeled on Fig. 12-47.

(b) On evaporation of A, the system composition follows the dotted line drawn from the water corner through A. When its composition reaches 1, ammonium sulfate begins to crystallize out, and the solution composition moves toward c. When the system composition has reached 2, the solution is at c and the compound begins to crystallize out. Both solution c and $(NH_4)_2SO_4$ diminish in amount until finally the system dries up as pure $Li_2SO_4 \cdot (NH_4)_2SO_4$. (This is a special case, because solution A contains equal weights of the two salts.)

(c) For $(NH_4)_2SO_4$ to be the first salt to crystallize out, the evaporation line must intersect the cd solubility curve. The maximum possible Li_2SO_4 content that will still allow this to happen is that for a solution whose evaporation line just hits point c, as shown. The composition at c corresponds approximately to 15 g $(NH_4)_2SO_4$ and 47 g Li_2SO_4 or 12.5 g to 39. Solution A would have 12.5 g Li_2SO_4, 12.5 g $(NH_4)_2SO_4$ (and 75 g H_2O), so to shift

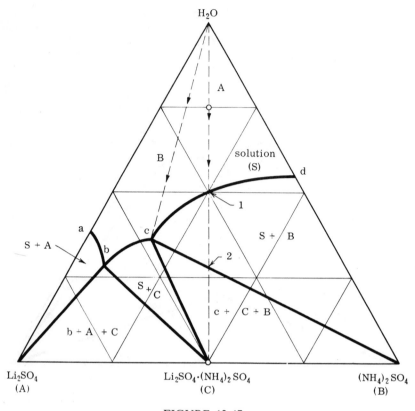

FIGURE 12-47

its proportion of salts to that of line B about 26 g of Li_2SO_4 should be added to the 100 g of A.

(d) It is evident from part (b) that solution A in fact corresponds to a solution of the compound, and it has already been found that the first solid to crystallize out is $(NH_4)_2SO_4$ and that the compound cannot exist with solution unless some ammonium sulfate is also present.

29. The diagram, with the phase regions labeled, is shown as Fig. 12-48. On evaporation of solution of the given composition (marked X on the diagram), the system composition moves along the dotted line drawn from the water corner through X. Na_2SO_4 starts crystallizing out when the system is at the point where the dotted line crosses the 2–3 solubility curve. The solution then moves in composition toward 2 and reaches point 2 when the system composition reaches point 4. NaCl now begins to crystallize out, and both salts continue to deposit until the system is dry.

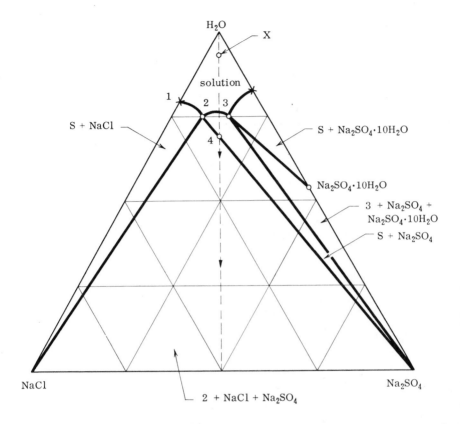

FIGURE 12-48

30. The labeled diagram is shown in Fig. 12-49. The given system composition is shown by point (1); this lies in the three-phase region, so solution
of composition (b), NH_4Cl, and $(NH_4)_2SO_4$ are the phases present. The tie
line from the $(NH_4)_2SO_4$ corner through point (1) indicates that about 10%
is present as this solid, and the rest as a mixture of (b) and NH_4Cl. A second
application of the lever principle, to point (2), indicates that of this 90%,
about 50% is present as (b) and the rest as NH_4Cl. The final percentages are
then 10% $(NH_4)_2SO_4$, 45% solution (b), and 45% NH_4Cl.

31. (a) The completed diagram is shown in Fig. 12-50. (b) The system
composition point is shown in the figure; evaporation causes the over-all
system composition to travel along the dashed line. At point (1), $AgNO_3$
appears, and the solution is about 72% water, 7% KNO_3, and 21% $AgNO_3$.
At point (2), $AgNO_3 \cdot KNO_3$ begins to form, and the solution is now at
composition (b). (c) The mixture must lie halfway along the line between the

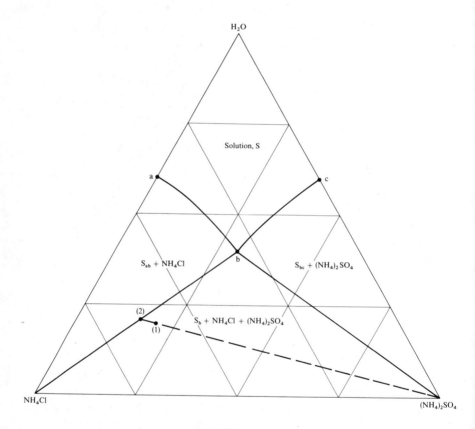

FIGURE 12-49

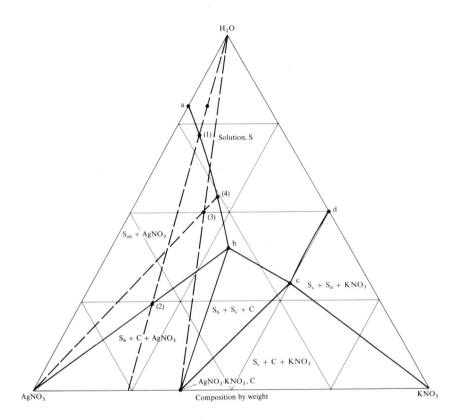

FIGURE 12-50

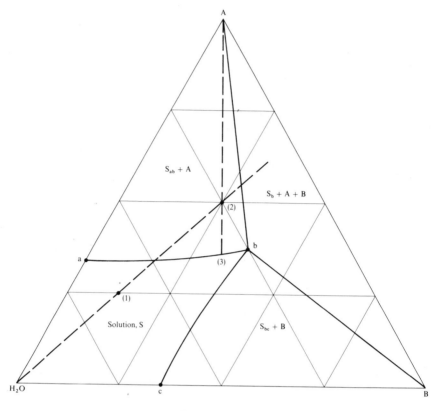

FIGURE 12-51

compound composition and the water corner, as shown by the dotted line, and point (3). The phases present are given by the tie line from the $AgNO_3$ corner through point (3), and consist of $AgNO_3$ and solution of composition (4). From the lever principle, there is 90–95% of solution, and the rest solid $AgNO_3$.

32. The complete diagram is shown in Fig. 12-51. The solution composition is shown by point (1); evaporation corresponds to the system composition moving away from the water corner, along the line between that corner and point (1). 100 g of original solution had 12.5 g B, 25 g A, and 62.5 g water; evaporation to half the weight, or 50 g, would mean that while 12.5 g of B and 25 g of A were still present, there was now only 12.5 g of water. The final system composition is then 25% B, 50% A, and 25% water, corresponding to point (2). This lies in a two-phase region, with solid A, and solution of composition (3) present; the proportions, from the lever principle, are about 70% (3) and 30% A.

13

ELECTROCHEMISTRY: CONDUCTANCE AND TRANSFERENCE

COMMENTS

The problems in this section should not be particularly difficult, since the subject is fairly straightforward and is somewhat limited in scope. About half of the problems deal with the conductivity of electrolyte solutions. Resistance, cell constant, specific conductivity, and equivalent conductivity are put through their paces. Remember that, in the case of solutions containing more than one electrolyte or where solvent conductance must be corrected for, such various contributions amount to a set of resistances in parallel. The reciprocal of the over-all resistance is therefore equal to the sum of the reciprocals of the resistances of the various contributing species. In other words, specific conductivities are additive.

We assume throughout that equivalent conductivities are independent of ionic strength. This is not so, of course, but the effect, although important, is not a major one for dilute solutions. To introduce it properly, however, would require a more thoroughgoing treatment of interionic attraction theory than seems appropriate here.

A few problems deal with the determination of a concentration by means of conductance measurements. Remember that, in the case of a weak electrolyte, only that portion which is dissociated contributes to the conductivity.

The molecular viewpoint of electrolytic conduction is that ions move in an electric field, quickly attaining a limiting speed that is a product of the force acting and the ionic mobility coefficient. Several problems do involve calculation of such ion-migration velocities.

The second general topic of this chapter, that of transference, is approached here mainly in terms of Hittorf transference experiments. Learn to do carefully the bookkeeping on what happens in the anode and cathode compartments. A variety of electrode reactions are made use of, partly to prepare you for Chapter 15 on the emf of cells.

In all the problems, the net change in an electrode compartment is determined by comparing the amount of electrolyte present with that originally associated with the given amount of solvent. The implicit assumption is then that solvent does not migrate during the experiment. This is not strictly true, since ions carry solvent of hydration with them, but the effect again is not a major one and is ignored here.

EQUATIONS AND CONCEPTS

Definitions

$$R = rl/A \tag{13-1}$$

where R = resistance (ohms, Ω), r = specific resistance (ohms-cm), l = path length, and A = cross section of the conductor.

$L = 1/r = k/R$, where L = specific conductivity (ohms^{-1}cm^{-1}) and $k = l/A$ (cm^{-1}) and is called the cell constant.

$$\Lambda = 1{,}000L/C \tag{13-2}$$

where Λ = equivalent conductivity (ohms^{-1}cm^{-1} $\times$ M^{-1}), and C = equivalents per liter.

$\Lambda = \lambda^+ + \lambda^-$, in the case of a single electrolyte, where λ^+ and λ^- are the individual ion equivalent conductivities. As a corollary, $\Lambda_{AB} = \Lambda_{AC} + \Lambda_{DB} - \Lambda_{DC}$. (Strictly speaking, this is true only if Λ^0's are used, i.e., the values of Λ at infinite dilution.)

Conductance of Mixtures

For a mixture L values are additive:

$$L = \Sigma L_i = L_{\text{solv}} + \Sigma C_i \lambda_i / 1{,}000 \tag{13-3}$$

Ionic Velocities

The velocity of an ion is proportional to V, the electric gradient in volts/cm: $v_i = \mu_i V$, where μ denotes the electrochemical mobility. μ is related to λ; through Ohm's law and the definition of equivalent conductivity; $\lambda_i = \mu_i \mathscr{F}$, where $\mathscr{F}$ is Faraday's number.

Transference Numbers

The transference number of an ion may be regarded as the fraction of the total current carried by that ion. If only the two ions of a simple electrolyte are involved, then $t^+ = \lambda^+/\Lambda$ and $t^- = \lambda^-/\Lambda$ (hence $t^+ + t^- = 1$). In general, however, $t_i = L_i/L$.

Hittorf Cells

In the case of Hittorf cell experiments, the anode is defined here as that electrode at which oxidation occurs. Then per faraday (96,500 coulombs), t^+ cations leave the anode region and t^- anions enter it; similarly, t^+ cations enter the cathode region and t^- anions leave it.

Weak Electrolytes

The degree of dissociation of a weak electrolyte can be obtained from conductance measurements by using them in a straightforward way to calculate the concentration of free ions and then comparing this to the total or stoichiometric concentration of the substance. Alternatively, one may define $\Lambda_{app} = 1,000L/C_s$ where C_s is the stoichiometric concentration. Then the degree of dissociation is given by Λ_{app}/Λ, where Λ is the sum of the ion equivalent conductivities and is therefore the expected value were the substance a strong electrolyte.

PROBLEMS

1. (10.5 min) The specific conductivity L is $0.0382\ \Omega^{-1}\,cm^{-1}$ for a solution 0.1 M in KCl and 0.2 M in NCl (a strong electrolyte). Calculate λ for N^+ (values for K^+ and Cl^- are 74 and 76, respectively).

2. (16 min) The solubility of $[Co(NH_3)_4Cl_2]ClO_4$ is measured by determining the conductance of a saturated solution. The cell constant is 0.20, and λ for $Co(NH_3)_4Cl_2^+$ is 50, and that for ClO_4^- is 70. The measured resistance was $33.5\ \Omega$. Calculate the solubility.

It is later realized that the above complex was dissolved in 0.01 M HCl, and not in pure water, so that part of the conductance was due to the acid present. With this added information, using data from your text as needed, calculate the correct solubility.

3. (24 min) A certain conductance cell has a resistance of $22\ \Omega$ when filled with solution A, $7.3\ \Omega$ when filled with solution B, and $16\ \Omega$ when filled with solution C (Table 13-1). Assuming equivalent conductivities to be independent of concentration, calculate

(a) the cell constant k,

(b) L and Λ for 0.1 M sodium acetate,

(c) L for solution B, and

(d) $\Lambda_{H^+,\ acetate^-}$ (i.e., $\lambda_{H^+} + \lambda_{acetate^-}$)

TABLE 13-1

Solution	Nature	Resistance
A	0.1 M sodium acetate	22
B	equal volumes of 0.1 M sodium acetate and 0.2 M HCl	7.3
C	0.1 M NaCl	16
Also Λ for NaCl is 126 and λ for Na$^+$ is 50		

4. (19.5 min) A student in the physical chemistry laboratory was assigned a conductimetric titration as an experiment. He was given 100 cc of HCl solution of unknown normality, and standardized 1.00-N NaOH solution, and was to obtain a plot of specific conductivity versus cc of NaOH added so as to determine the endpoint by locating the intersection of the two branches.

Using an immersion conductivity cell (whose k is given as 0.2), the student measured the resistance of the HCl solution, computed its specific conductivity (he failed to write it down, however, and forgot the value), and set up for the titration. He was somewhat distracted, however, by a neighboring student, a rather pretty girl, and overshot the endpoint on his first addition of NaOH. He did note that the volume added was 1.5 cc, and that the specific conductivity at this point was 0.00246.

He was resigned to repeating the experiment the next period, when the instructor came up, and, on seeing what happened, told the student that he did have enough data to compute the normality of the HCl, since he knew that the equivalent conductivities of H$^+$, Na$^+$, Cl$^-$, and OH$^-$ ions are 350, 50, 75, and 200, respectively. To make sure that he understood the material, the instructor asked the student to calculate:

(a) the normality of the HCl solution,

(b) the resistance he would have measured had he hit the endpoint exactly, and

(c) the transference number of Na$^+$ ion in the final solution (with the added NaOH), assuming the normality of the HCl to have been 0.005. Please help the student in his calculation by obtaining the answers to the above questions.

5. (15 min) A particular conductivity cell had a resistance of 468 Ω when filled with 0.001 M HCl, a resistance of 1,580 Ω when filled with 0.001 M

NaCl, and one of 1,650 Ω when filled with 0.001 M NaNO$_3$. The equivalent conductivity of NaNO$_3$ is 121. Neglecting changes in Λ values with concentration, calculate

(a) the specific conductivity of 0.001 M NaNO$_3$,

(b) the cell constant,

(c) the resistance of the cell when filled with 0.001 M HNO$_3$ and the equivalent conductivity of HNO$_3$.

6. (15 min) A conductivity cell has a resistance of 250 Ω when filled with 0.02 M KCl at 25°C, and one of 10^5 Ω when filled with 6 × 10^{-5}-M ammonium hydroxide solution. The specific conductivity of 0.02 M KCl is 0.00277 Ω cm^{-1}, and the equivalent conductivities for NH$_4^+$ and OH$^-$ are 73.4 and 198, respectively.

Calculate the cell constant and the degree of dissociation of the ammonium hydroxide in the 6 × 10^{-5}-M solution.

7. (24 min) The Λ values for HCl, NaCl, and NaAc (sodium acetate) are 420, 126, and 91, respectively. Calculate the Λ value for H$^+$ and Ac$^-$.

The resistance of a conductivity cell is 520 Ω when filled with 0.1-M acetic acid, and drops to 122 Ω when enough solid NaCl is added to make the solution 0.01 M in NaCl as well. Calculate the cell constant and the hydrogen ion concentration of the solution.

8. (15 min) It is found that at 25°C, the resistance of a certain conductivity cell is 220,000 Ω when it is filled with water, 100 Ω when filled with 0.02 M KCl, and 102,000 Ω when filled with water saturated with AgCl. The equivalent conductivity of AgCl is calculated to be 126.8 at 25°C, whereas that for KCl is known to be 138.3. Assume that the solutions are prepared with water of the same resistance as given above, and neglect the variation of equivalent conductivity with concentration. Calculate

(a) the cell constant,

(b) the specific conductivity of the saturated solution of AgCl, and

(c) the solubility of AgCl at 25°C.

---

9. (10.5 min) A Hittorf transference cell is filled with 0.1m PSO$_4$ (a strong electrolyte), and the electrodes are of the metal P. On passing a certain quantity of electricity through the cell, it is found that the solution in the anode compartment has gained twice as much weight as the anode has lost. Write the anode reaction and the gains and losses resulting from transference for the anode solution. If the atomic weight of P is 44 (SO$_4^{2-}$ is 96), calculate the transference number for P^{2+} in this solution.

10. (15 min) In a Hittorf transference experiment, the electrolyte was 0.1 M KClO$_4$ and the electrodes were platinum. After passage of current for a certain time, the gas evolved at the anode amounted to 22.4 cc (STP) as dry gas. The anode compartment was drained and rinsed with original solution, and the combined drain and rinse solutions totaled 50 cc and was 0.0732 M in $KClO_4$. Write the electrode reaction and the gain and loss by transference for the anode compartment. Calculate the transference number of Cl_4^- in this $KClO_4$ solution.

11. (13.5 min) The transference number of $Co(NH_3)_4Cl_2^+$ in a 0.02-M solution of $[Co(NH_3)_4Cl_2]$ Cl was determined by means of a Hittorf-type cell. The anode was silver and the anode solution, silver nitrate; the middle compartment contained 0.02m complex as did the cathode compartment; the cathode was a silver–silver chloride electrode. At the end of the experiment, the silver anode was found to have changed weight by 2 g, and the cathode compartment solution (150 cc volume) was 0.025m. Calculate the transference number. Write out also the electrode reactions and the gains and losses resulting from transference for each compartment.

12. (15 min) In an experiment to determine the equivalent conductance of the Reineckate ion R^-, two Hittorf cells were set up in series, as shown in Fig. 13-1. After passage of a certain amount of electricity, the two anode compartments were drained and analyzed. That from cell I was 0.1015 in R^- ion and contained 30 g water (or about 31 g of solution), whereas that from cell II was 0.0980m in KBr and contained 25 g water (or about 30.5 g of solution). [R^- denotes the ion $Cr(NH_3)_2(NCS)_4^-$.]

Write the anode reactions and the gains and losses by transference for each of the two anode sections. Calculate the number of faradays passed through the cells and the transference number of R^- ion.

13. (12 min) On electrolysis of 0.1m NaBr solution in a Hittorf cell, using a silver–silver bromide anode and a platinum cathode, it was found that it took 20 cc of 0.15 N HCl to neutralize the cathode solution, and that the anode solution weighed 85 g and contained 0.74 g or 7.3 × 10^{-3} mole of NaBr.

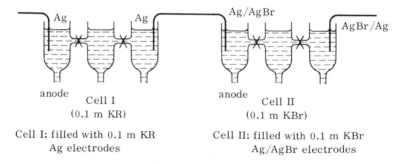

anode   Cell I
(0.1 m KR)

anode   Cell II
(0.1 m KBr)

Cell I: filled with 0.1 m KR
Ag electrodes

Cell II: filled with 0.1 m KBr
Ag/AgBr electrodes

FIGURE 13-1

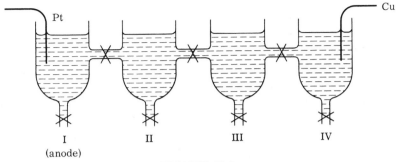

I   II   III   IV
(anode)

FIGURE 13-2

(a) Write the anode reaction and the gain or loss owing to transference for the anode compartment, per faraday.

(b) Calculate the transference number for bromide ion in $0.1m$ NaBr.

14. (12 min) The transference cell shown in Fig. 13-2 contained $0.1\text{-}M$ $CuSO_4$ solution initially. A certain quantity of electricity was passed through the cell and 4.48 cc (STP) of gas (dry) was collected at the anode. Analysis of solutions from compartments I, II, and III gave the following results:

Compartment	I	II	III
Volume, cc	25	30	30
Final Cu concentration, M	0.096	0.098	0.1

Calculate the transference number for Cu^{2+} ion. (At. wt Cu is 63, S is 32.) Give also the anode reaction.

15. (15 min) The transference number for Li^+ ion $LiNO_3$ is to be determined by the Hittorf method. The cell used consists of three compartments: anode, middle, and cathode. These compartments are connected by stopcocks ending in ground-glass joints so that they may be closed off from each other and then detached. The anode is silver, the cathode is platinum, and the electrolyte is $0.02m$ $LiNO_3$.

The following measurements are made: (1) weight of anode section filled with solution and including the silver electrode: 130 g initially and 130.865 g after the experiment; (2) weight of silver electrode separately: 12.16 g initially and 10.00 g after the run (at. wt Ag is 107.9, Li is 6.94).

(a) Write the electrode reactions and the gain or loss by transference for the anode and for the cathode sections, per faraday.

(b) Calculate the transference number for Li^+.

16. (24 min) (final examination question) The transference number for Na^+ in dilute NaCl solution is 0.4 and $\lambda_{Cl^-} = 75$. In addition, the following resistances have been measured for a conductivity cell filled successively with the indicated solutions: 0.1-M KCl solution: 7,000 Ω; a solution 0.1 M in KCl and 0.2 M in NaCl: 2,600 Ω.

Calculate the value of Λ_{KCl}.

17. (15 min) The transference numbers for 0.02m sodium sulfate are determined by means of a Hittorf-type transference cell, with platinum electrodes. After passage of a certain amount of electricity through the cell, the anode solution is drained out and the anode compartment rinsed with some of the original solution. Solution plus rinse weighed 150 g and was found to contain 0.002 equiv of hydrogen ion and 0.0036 mole of sulfate.

(a) Write the electrode reaction and the gain and loss by transference for the anode and for the cathode compartment.

(b) Calculate the transference number for sulfate ion in the sodium sulfate solution.

18. (14 min) A Hittorf cell contains 0.1-M $CuSO_4$ solution and has Cu electrodes. The anode compartment is so designed that the entire compartment, with solution and electrode, may be detached and weighed. (See Fig. 13-3.)

After passage of a certain amount of electricity, it is found that the anode solution gained 0.006 mole of $CuSO_4$, but that the total weight of the anode compartment (cell, solution, and electrode) had not changed in weight. Calculate the number of faradays passed through the cell and the transference number of SO_4^{2-} ion.

19. (10 min) A transference cell such as in Fig. 13-3 is filled with 0.2m $AgNO_3$ solution, and has a copper anode and a silver cathode. After passage of some current, the anode compartment is found to contain 0.03 mole of copper nitrate and the cathode compartment lost 0.04 mole of silver nitrate. Write the electrode reaction and the gain and loss by transference for each compartment and calculate the transference number for nitrate ion.

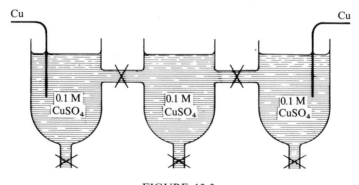

FIGURE 13-3

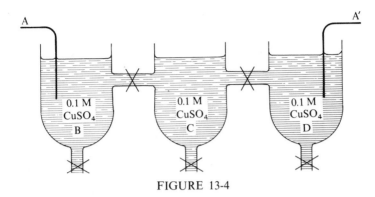

FIGURE 13-4

20. (10 min) (final examination question) The Hittorf cell in Fig. 13-4 has copper electrodes A and A', and three compartments filled with 0.1-M CuSO$_4$ solution. t_+ is 0.4 for Cu^{2+} in CuSO$_4$; in operation of the cell, the left electrode is the anode.

After passage of a certain amount of electricity, the cell is disassembled so that each electrode and each compartment (with its solution) can be weighed separately. For the following combinations state whether a gain, a loss, or no change in weight should be observed:

(a) A + A' + B + C + D (b) B + C (c) B + D (d) B + A'

---

21. (15 min) Solution A consists of 50 cc of 0.02 M NaOH and solution B, of 50 cc of 0.0466 M HCl. λ values are 350, 50, 200, and 75 for H$^+$, Na$^+$, OH$^-$, and Cl$^-$, respectively. The two solutions are then mixed; calculate the transference number for Na$^+$ in the mixture and the specific conductivity of the mixture.

22. (9 min) Calculate Λ^0_{XCl} and $\Lambda^0_{1/2\ YHSO_4}$. The ions X and Y are univalent, and all electrolytes involved are fully dissociated.

Electrolyte	YCl	YOH	$\frac{1}{2}$Y$_2$SO$_4$	HCl	$\frac{1}{2}$H$_2$SO$_4$	XOH
Λ^0	126	254	136	426	430	278

(Use only the data given.)

23. (15 min) The following data are obtained in an experiment to determine the dissociation constant of the weak acid HA

Solution	*Specific conductance*
0.01 M HA	3.8×10^{-5}
100 cc of 0.01 M HA + 1 cc of 1.0 M NaOH	80×10^{-5}
	(neglect the dilution)

Calculate, neglecting ionic-strength effects on Λ values and using data from your text as needed, Λ_{NaA}, λ_{A^-}, $\Lambda_{H^+A^-}$, and the degree of dissociation of 0.01 M HA.

24. (24 min) (final examination question) One hundred cc of 0.01 N H_2SO_4 is titrated with 1 N NaOH, and the conductance is followed by means of a dipping conductivity cell whose cell constant is 0.2. Given that the ionic conductivities are 350, 200, 80, and 50 for H^+, OH^-, $\frac{1}{2}SO_4^{2-}$, and Na^+ ions, respectively, and neglecting dilution effects and taking H_2SO_4 to be completely dissociated,

(a) calculate the specific conductivity of the initial H_2SO_4 solution,

(b) calculate the resistance that was measured for this solution,

(c) calculate the specific conductivity at the endpoint and when 10% excess NaOH has been added, and

(d) explain for which of the solutions in (a) and (c) the transference number of H^+ would be the greatest.

25. (18 min) Given that λ_{Li^+} is 40, λ_{Na^+} is 50, and $\lambda_{NO_3^-}$ is 70, and neglecting changes in λ values with concentration, calculate, for a solution which is 0.1 M in $LiNO_3$ and 0.2 M in $NaNO_3$ (both salts in the same solution),

(a) the specific conductivity,

(b) the transference number of Li^+ ion, and

(c) the distance Li^+ ions would move in 100 sec if some of this solution were placed in a tube of 5 cm^2 cross section and a current of 0.1 A were flowing.

26. (15 min) Given that $\Lambda_{NaCl} = 126$, $\Lambda_{KNO_3} = 145$, $\Lambda_{NaNO_3} = 121$, and $t^+_{KCl} = 0.50$ and assuming that these do not change with concentration,

(a) calculate the λ values for each of the above ions and

(b) calculate Λ_{HCl} if the resistance of 0.1 M HCl is 1/35 of that for 0.01 M NaCl, when measured in the same conductivity cell.

27. (18 min) The specific conductivity of a 0.1-M solution of NaOH is 0.0221. When an equal volume of 0.1 M HCl is added, the specific conductivity falls to 0.0056; on addition of a further volume of the HCl solution equal to that of the first portion added, the specific conductivity rises to 0.0170. Calculate

(a) Λ for NaOH,   (c) Λ for HCl,

(b) Λ for NaCl,   (d) Λ for H^+ + OH^-.

28. (24 min) (final examination question) The specific conductivity of 0.001-M Na_2SO_4 solution is 2.6×10^{-4} and rises to 7×10^{-4} if the solution is saturated with $CaSO_4$. The equivalent conductivities are 50 and 60 for Na^+ and $\frac{1}{2}Ca^{2+}$ ions, respectively. Calculate the solubility product for $CaSO_4$.

29. (12 min) Given the following equivalent conductivities at $25°C$: HCl, 426; NaCl, 126; NaC (sodium crotonate), 83; calculate the dissociation constant of crotonic acid. Additional information is that the specific conductivity of a 0.001-M solution of the acid is $3.83 \times 10^{-5} \, \Omega^{-1} \, cm^{-1}$. Neglect variation of equivalent conductivities with concentration.

30. (8 min) The salt $[Co(NH_3)_5Cl]Cl_2$ is a strong electrolyte, dissociating into the ion $[Co(NH_3)_5Cl]^{2+}$ and two Cl^- ions. Calculate the equivalent conductivity of the complex ion if the specific conductivity of a 0.001-M solution of the salt is 2.72×10^{-4}. The equivalent conductivity of chloride ion is 76. Using appropriate data from the text, calculate what the equivalent conductivity of the mixed salt $[Co(NH_3)_5Cl](Cl)(NO_3)$ should be.

---

31. (10.5 min) Calculate the velocity, in cm/sec, of Li^+ ions in $0.1 \, M$ LiCl, when a current of 1 A is passed through a column of the solution. The cross-sectional area of the column is $2 \, cm^2$, and the equivalent conductivities of Li^+ and Cl^- ions are 40 and 75, respectively.

32. (9 min) A cylindrical tube 100 cm long and $2 \, cm^2$ in cross section is filled with 0.1-M MCl (a strong electrolyte). A potential difference across the length of the cell is now applied such that a current of 0.01 A flows. If λ for M^+ is 60 (and Λ for MCl is 136) calculate the velocity of the ion M^+ in cm/sec.

33. (10 min) Calculate the velocity, in cm/sec, of Na^+ ions in $0.1 \, M$ $NaNO_3$, when a current of 1 A is passed through a column of the solution. Using the same apparatus and same applied potential, the velocity of K^+ ions in $0.2 \, M$ KCl is 3×10^{-2} cm/sec.

34. (10 min) (final examination question) A current of 0.1 A is carried by a tube of $10 \, cm^2$ cross section, filled with 0.01-N strong electrolyte AB (giving A^+ and B^- ions). Calculate the potential gradient in volts/cm if Λ_{AB} is 120.

35. (14 min) A solution $0.01 \, M$ in $LiNO_3$ and $0.02 \, M$ in KCl is placed in a conductivity cell whose electrodes are 3 cm apart. A potential difference of 6 V is maintained between the electrodes. Calculate the current carried by Li^+ ions. Use λ values as needed from the text. The measured conductivity in the experiment is $200 \, \Omega^{-1}$.

36. (12 min) A cylindrical tube of $10 \, cm^2$ cross section is filled with 0.01-M $LiNO_3$ solution. The electrodes are l cm apart and if a potential difference of 6 V is applied, a current of 0.003 A is found to flow. Calculate the velocity with which Li^+ ions move. Equivalent conductivities are 39 and 71 for Li^+ and NO_3^- ions, respectively.

ANSWERS

1. L is the sum of the various ionic conductances, so

$$L = \sum_i \frac{C_i \lambda_i}{1,000}$$

or $1,000 \times 0.0382 = 0.1 \times 74 + 0.3 \times 76 + 0.2\lambda_{N^+}$, from which

$\lambda_{N^+} = \mathbf{40}$

2. $L = k/R = 0.2/33.5 = 5.95 \times 10^{-3}$, and

$C = 1,000L/\lambda = 595/120 = \mathbf{0.0495N}$

Taking 425 as Λ for HCl, then $L_{HCl} = 0.01 \times 425/1,000 = 0.00425$. The net L due to the complex was then $0.00595 - 0.00425 = 0.00175$, and the concentration

$C = 1,000 \times 0.00175/120 = \mathbf{0.0145M}$

3. (a) L for 0.1 M NaCl is $126 \times 0.1/1,000$ or 1.26×10^{-2}; hence the cell constant must be

$k = 0.0126 \times 16 = \mathbf{0.201}$

(b) L for sodium acetate is then $0.201/22$ or 0.00915, and $\Lambda_{NaAc} = 100 \times 0.00915/0.1 = \mathbf{91.5.}$

(c) L for solution B is $0.201/7.3 = \mathbf{0.0277.}$

(d) After mixing, solution B evidently consists of 0.05 M acetic acid, 0.05 M sodium chloride, and 0.05 M HCl (remember there is a dilution effect). Neglecting the dissociation of acetic acid, L for this solution is the sum of the individual ion contributions:

$1,000L = \Sigma \lambda_i C_i = 0.05 \times 126 + 0.05\Lambda_{HCl} = 6.3 + 0.05\Lambda_{HCl}$

(d) From part (c), $1,000L$ is 27.7, then $\Lambda_{HCl} = 428$. Since Λ_{NaCl} is 126 and λ_{Na^+} is 50, λ_{Cl^-} is 76 and therefore λ_{H^+} must be 352. Finally, $\lambda_{acetate^-}$ is $91.5 - 50$ or 41.5, so $\Lambda_{H^+, acetate^-} = 352 + 41.5 = \mathbf{393.}$

4. We can neglect the dilution effect as being within slide rule error, so the formality (formula weights per liter) of added NaOH was 0.015. If C denotes the original HCl concentration, the final solution contained C concentration of NaCl and $(0.015 - C)$ of excess NaOH. Then $1,000L = C\Lambda_{NaCl} + (0.015 - C)\Lambda_{NaOH}$, or $2.46 = C \times 125 + (0.015 - C)250$, where

$$C = 1.29/125 = \mathbf{0.0103}M$$

(b) Had he hit the endpoint exactly, the solution would have consisted of $0.0103\ M$ NaCl, for which $L = 125 \times 0.0103/1,000 = 0.00129$. The resistance that would have been observed would then be

$$R = k/L = 0.2/0.00129 = \mathbf{155\ \Omega}$$

(c) Had the HCl normality been 0.005, the final solution would have consisted of $0.005\ N$ NaCl and $0.01\ N$ NaOH, for which L would be $1,000L = 0.005 \times 125 + 0.01 \times 250 = 3.13$. For the Na^+ ion alone, L is $1,000L = 0.015 \times 50 = 0.75$, so t_{Na^+} would be $0.75/3.13 = \mathbf{0.24}$.

5. (a) $L_{NaNO_3} = \Lambda_{NaNO_3}M/1,000 = 121 \times 0.001/1,000 = \mathbf{1.21 \times 10^{-4}}$.
(b) $k = L \times R = 1.21 \times 10^{-4} \times 1,650 = \mathbf{0.2}$.
(c) $\Lambda = 1,000L/C = 1,000k/C \times R$, so $\Lambda_{HCl} = 1,000 \times 0.2/0.001 \times 468 = \mathbf{428}$, $\Lambda_{NaCl} = 1,000 \times 0.2/0.001 \times 1,580 = \mathbf{127}$. Then $\Lambda_{HNO_3} = \Lambda_{HCl} + \Lambda_{NaNO_3} - \Lambda_{NaCl} = 428 + 121 - 127 = \mathbf{422}$. $L_{HNO_3} = 422 \times 0.001/1,000 = 4.22 \times 10^{-4}$, and $R_{HNO_3} = 473\Omega$.

6. The cell constant is equal to $L \times R$, so $k = 0.00277 \times 250 = 0.692$. The apparent Λ for the NH_4OH solution is $1,000L/C$ or $1,000k/CR$,

$$\Lambda_{app} = 1,000 \times 0.692/6 \times 10^{-5} \times 10^5 = 115$$

The degree of dissociation is them $\Lambda_{app}/\Lambda_{NH_4^+,OH^-} = 115/(73.4 + 198) = \mathbf{0.423}$.

7. $\Lambda_{H^+,Ac^-} = \Lambda_{NaAc} - \Lambda_{NaCl} = 420 + 91 - 126 = \mathbf{385}$.
The specific conductivity of the first solution L_1 is $L_1 = k/R_1 = k/520$, and the one for the second solution is $L_2 = k/122$. The difference gives the specific conductivity resulting from the 0.01 M NaCl, which should be equal to $L = \Lambda_{NaCl}C/1,000 = 126 \times 0.01/1,000 = 0.00126$. Then

$$0.00126 = k(1/122 - 1/520)$$

or

$$k = 1.26/6.28 = \mathbf{0.20}$$

The ion concentration in the acetic acid solution is then

$$C = 1{,}000 L_1/\Lambda_{H^+,Ac^-}$$
$$= 1{,}000 \times 0.2/520 \times 385 = \mathbf{0.0010\ M}$$

8. (a) L for the KCl solution is $\Lambda_{KCl}C/1{,}000 = 138.3 \times 0.02/1{,}000 = 2.77 \times 10^{-3}$; hence the cell constant is

$$k = L \times R = 2.77 \times 10^{-3} \times 100 = \mathbf{0.277}$$

(The conductivity of the water can be neglected here.)
  (b) L for the saturated solution is then

$$L = k/R = 0.277/102{,}000 = \mathbf{2.71 \times 10^{-6}}$$

(c) The water itself contributes a specific conductivity of

$$L_w = 0.277/220{,}000 = 1.26 \times 10^{-6}$$

The net specific conductivity owing to the AgCl is then $(2.71 - 1.26) \times 10^{-6} = 1.45 \times 10^{-6}$. The concentration is

$$C = 1{,}000 L/\Lambda = 1{,}000 \times 1.45 \times 10^{-6}/126.8 = \mathbf{1.15 \times 10^{-5}}$$

9. The anode reaction is

$$P = P^{2+} + 2e^-$$

and the anode compartment gains t^- equiv of SO_4^{2-} and loses t^+ equiv P^{2+} per faraday. The net is a gain of t^- equiv of PSO_4 per $\mathscr{F}$. Per $2\mathscr{F}$, the ratio of weight gained in the anode solution to weight lost by the anode is then $t^- \times 140/44$. This is equal to 2, from which $t^- = \mathbf{0.63}$.

10. The anode reaction is $H_2O = \frac{1}{2}O_2 + 2H^+ + 2e^-$, and the compartment gains t^- equiv of ClO_4^- and loses t^+ equiv of K^+ per faraday. An alternative statement is that the anode compartment gains 1 equiv of $HClO_4$ and loses t^+ equiv of $KClO_4$ per faraday.

The anode solution plus rinse contained 50×0.0732 or 3.66 meq of $KClO_4$ instead of the 5 meq it should have; hence 1.34 meq were lost. The 22.4 cc (STP) of oxygen corresponds to 1 mM or $4\ m\mathscr{F}$. t^- is then $1.34/4 = \mathbf{0.33}$.

11. The various changes are as follows:
 anode: $Ag = Ag^+ + e^-$
 solution: gains $t^- Cl^-$ (presumably precipitates as AgCl)
    loses $t^+ Ag^+$ (presumably precipitates as AgCl in
       the middle compartment)
 cathode: $AgCl + e^- = Ag + Cl^-$
 solution: gains t^+ complex ion
    loses t^- chloride ion
 net: gains t^+ complex salt

The cathode solution originally contained 150×0.02 or 3 meq of salt, and ended up with 0.025×150 or 3.75 meq, a gain of 0.75 meq. Since the number of faradays was $2/107$ or 0.0187, the value of t^+ must be $0.75/18.7 = \textbf{0.040}$.

12. Cell I: anode reaction: $Ag = Ag^+ + e^-$
    anode solution: gains $t^- R^-$
       loses $t^+ K^+$
 Cell II: anode reaction: $Ag + Br^- = AgBr + e^-$
    anode solution: gains $t^- Br^-$
       loses $t^+ K^+$
 Net: loss of $t^+ KBr$

The actual loss in cell II was $25(0.1 - 0.098) = 0.05$ meq, since t^+ for $0.1m$ KBr is about 0.48 (from your text), the number of faradays passed through the cell must have been $0.05/0.48 = 0.104$.
 The gain in R^- ion in cell I was $30(0.1015 - 0.1) = 0.045$, so $t^- = 0.045/0.104 = \textbf{0.43}$.

13. (a) The anode reaction is $Ag + Br^- = AgBr + e^-$, and there is a gain of t^- equiv of Br^- and a loss of t^+ equiv of Na^+ by transference; the net change is then a loss of t^+ equiv of NaBr.
 (b) The cathode reaction is

$$H_2O + e^- = \tfrac{1}{2}H_2 + OH^-$$

20×0.15 meq of HCl were required to neutralize the OH^- formed; hence $0.003 \,\mathscr{F}$ were passed through the cell. The anode compartment contained $85 - 0.74$ or 84.26 g of water and hence originally contained $84.26 \times 0.1/1,000$ or 8.43×10^{-3} mole of NaBr. There was then a loss of $8.43 \times 10^{-3} - 7.3 \times 10^{-3}$ or 1.13×10^{-3} mole, and t^+ is $1.13/3 = 0.377$, and t^- is **0.623**.

14. The anode reaction is $\tfrac{1}{2} H_2O = \tfrac{1}{4}O_2 + H^+ + e^-$. Anode solution gains t^- equiv SO_4^{2-}, loses t^+ equiv Cu^{2+}.

The solution in compartment I lost $25(0.1 - 0.096)$ or 0.1 millimole or 0.2 meq of Cu^{2+}, whereas that in compartment II lost $30(0.1 - 0.098)$ or 0.06 mM or 0.12 meq (evidently caused by some mixing between the two compartments). The total loss was then 0.32 meq (correction for the weight of $CuSO_4$ does not exceed slide rule error). The total number of faradays corresponds to 4.48 cc (STP) of O_2, which is 0.2 mM or 0.8 meq of O_2. Then $t_{Cu^{2+}} = 0.32/0.8 = \mathbf{0.4}$.

15. Anode reaction: $Ag = Ag^+ + e^-$; anode solution: gains $t^- NO_3^-$, loses $t^+ Li^+$. Net change for the anode solution: gains 1 equiv $AgNO_3$, loses $t^+ LiNO_3$.

Cathode reaction: $H_2O + e^- = \frac{1}{2}H_2 + OH^-$; cathode solution: gains $t^+ Li^+$, loses $t^- NO_3^-$. Net change for the cathode solution: gains 1 equiv $LiOH$, loses $t^- LiNO_3$.

(b) The silver electrode lost 2.16 g, or $2.16/107.9 = 0.02\ \mathscr{F}$. The anode section gained 0.864 g, which must correspond to $t_{NO_3^-}^- - t_{Li^+}^+$ (the electrode reaction does not affect the weight of electrode plus solution), hence, per faraday $0.864/0.02 = t^- \times 62 - (1 - t^-)6.94$ or $43.2 = 69\,t^- - 6.94$ or $t^- = 0.73$. Finally, $t^+ = \mathbf{0.27}$.

16. Λ for NaCl is evidently $75/0.6$ (0.6 is t^-) or 125. For the first solution $1{,}000\,L_1 = \Lambda_{KCl} \times 0.1$, and, for the second, $1{,}000 L_2 = \Lambda_{KCl} \times 0.1 + 125 \times 0.2$. Since the resistances are inversely proportional to the L values,

$$L_1/L_2 = R_2/R_1 = 2{,}600/7{,}000 = 0.1\Lambda_{KCl}/(0.1\Lambda_{KCl} + 25)$$

On solving for Λ_{KCl}, we find $\Lambda_{KCl} = \mathbf{147}$.

17. Anode reaction: $\frac{1}{2}H_2O = H^+ + \frac{1}{4}O_2 + e^-$; anode solution: gains t^- equiv of SO_4^{2-}, loses $t^+ Na^+$; cathode reaction: $H_2O + e^- = \frac{1}{2}H_2 + OH^-$; and cathode solution: gains $t^+ Na^+$, loses t^- equiv of SO_4^{2-}. Since 0.002 equiv of H^+ were produced in the anode section, 0.002 faradays must have been passed through the cell. Within slide rule error, the 150 g of solution contained 150 g of water, and hence originally contained 0.15×0.02 or 0.003 mole of SO_4^{2-}. The compartment then gained 0.0006 mole or 0.0012 equiv of SO_4^{2-} or $0.0012/0.002 = 0.6$ equiv per faraday. Then $t_{SO_4^{2-}} = \mathbf{0.6}$.

18. The anode reaction, per faraday, is

$$\tfrac{1}{2}Cu = \tfrac{1}{2}Cu^{2+} + e^-$$

and the anode compartment loses t^+ equivalents of Cu^{2+} and gains t^- equivalents of SO_4^{2-}; the net change in the compartment is then a gain of t^- equivalents of $CuSO_4$. However, if the anode plus anode compartment is weighed, the only change in weight that occurs is due to the loss of t^+Cu^{2+} and the gain of $t^-SO_4^{2-}$. Since no change in weight occurs, evidently

$$t^+\left(\frac{63.5}{2}\right) = t^-\left(\frac{96}{2}\right)$$

from which $t^+ = 0.60$ and $t^- = \mathbf{0.40}$. The net gain of 0.006 mole or 0.012 equivalents of $CuSO_4$ then corresponds to 0.012/0.40 or **0.030 faraday.**

19.   Anode reaction: $\frac{1}{2}Cu = \frac{1}{2}Cu^{2+} + e^-$
    Anode solution: loses t^+Ag^+
                                            gains $t^-NO_3^-$
    Cathode reaction: $Ag^+ + e^- = Ag$
    Cathode solution: gains t^+Ag^+
                                            loses $t^-NO_3^-$
    net for cathode compartment: loss of t^- $AgNO_3$

The presence of 0.03 mole or 0.06 equivalent of copper in the anode compartment means that 0.06 faraday was involved. The transference number for nitrate ion is then 0.04/0.06 or **0.067.**

20.   Anode, A: $\frac{1}{2}Cu = \frac{1}{2}Cu^{2+} + e^-$, so loses weight
                                            corresponding to one equiv of copper
    Anode solution, B: loses t^+Cu^{2+}, gains $t^-SO_4^{2-}$
    net, including Cu^{2+} from
    the anode reaction: gain of t^- equiv of $CuSO_4$
    Middle solution, C: no net change
    Cathode solution, D: gains t^+ Cu^{2+}, loses $t^-SO_4^{2-}$
    net, including loss of Cu^{2+}
    to the cathode: loss of t^- equiv of $CuSO_4$.
    Cathode, A':   gains weight corresponding to one
                                            equivalent of copper

(a) No change (i.e., the cell as a whole does not change weight!). (b) Gains the weight of t^- equivalents of $CuSO_4$. (c) The net reactions just cancel; no change. (d) Gains the weight of t^- equivalents of $CuSO_4$ and one equivalent of Cu.

21.   The total specific conductivity L equals $L_i = \sum \lambda_i C_i/1{,}000$. After mixing, the resulting solution will be $0.01.M$ in NaCl and 0.0133 M in HCl, so $1{,}000L = 0.01 \times 50 + 0.0233 \times 75 + 0.0133 \times 350 = 6.90$, or

$$L = \mathbf{6.9 \times 10^{-3}}$$

The transference number for Na^+ will be $L_{Na}/L = 0.5/6.9 = \mathbf{0.0725.}$

22. Λ^0_{XCl} is equivalent to

$$\Lambda^0_{XOH} + \Lambda^0_{YCl} - \Lambda^0_{YOH} = 278 + 126 - 254 = \mathbf{150}$$

$\Lambda^0_{\frac{1}{2}YHSO_4}$ is equivalent to

$$\tfrac{1}{2}(\Lambda^0_{\frac{1}{2}Y_2SO_4} + \Lambda^0_{\frac{1}{2}H_2SO_4}) = \mathbf{283}$$

23. The second solution evidently is just neutralized and so corresponds to 0.01 M NaA, for which Λ is then $1{,}000 \times 80 \times 10^{-5}/0.01 = \mathbf{80}$. λ_{Na^+} is 50, so λ_{A^-} must then be 30, and $\Lambda_{H^+A^-} = 350 + 30 = 380$. The apparent Λ_{HA} is $1{,}000 \times 3.8 \times 10^{-5}/0.01 = 3.8$, so the degree of dissociation is $3.8/380 = \mathbf{0.01}$ or $\mathbf{1\%}$.

24. (a) $1{,}000L = \sum \lambda_i C_i = 0.01(350 + 80) = 4.30$ or $L = 4.3 \times 10^{-3}$.
(b) $R = k/L = 0.2/0.0043 = \mathbf{46.6\ \Omega}$.
(c) Dilution effects can be neglected as being within slide rule error, so at the endpoint the solution consists of $0.01\ N$ Na_2SO_4, and $1{,}000\ L = 0.01(130) = 1.3$ or $L = \mathbf{1.3 \times 10^{-3}}$. When 10% excess of NaOH is present, there is in addition $0.001\ N$ NaOH, so $1{,}000L = 1.3 + 0.001 \times 250 = 1.55$, and $L = \mathbf{1.55 \times 10^{-3}}$.
(d) t_{H^+} would be greatest for solution (a) since H^+ ion concentration is greater, relative to other ions, in this solution than in the other two.

25. (a) $1{,}000L = \sum \lambda_i C_i = 0.1 \times 40 + 0.3 \times 70 + 0.2 \times 50 = 35$ or $L = \mathbf{0.035}$.
(b) For Li^+ ion alone, $1{,}000L$ is 0.1×40 or 4, so $t_{Li^+} = 4/35 = \mathbf{0.114}$.
(c) The current of 0.1 A corresponds to $0.1/96{,}500$ or 1.03×10^{-6} equiv/sec of which 0.114 or 1.18×10^{-7} is of Li^+ ions. This last is equal to (velocity)(equiv/cc)(area), so the velocity v of the Li^+ ions is

$$v = 1.18 \times 10^{-7}/0.1 \times 10^{-3} \times 5 = 2.36 \times 10^{-4}\ \text{cm/sec}$$

In 100 sec, then, Li^+ ions would move **0.0236 cm.**

26. (a) Since $t^+_{KCl} = 0.5, \lambda_{K^+} = \lambda_{Cl^-} = \tfrac{1}{2}\Lambda_{KCl}$. Also $\Lambda_{KCl} = \Lambda_{KNO_3} + \Lambda_{NaCl} - \Lambda_{NaNO_3} = 145 + 126 - 121 = 150$, so λ_{K^+} and λ_{Cl^-} equal **75.** Then $\lambda_{Na^+} = 126 - 75 = \mathbf{51}$, and $\lambda_{NO_3^-} = 145 - 75 = \mathbf{70}$.
(b) The specific conductivity of the HCl solution is then 35 times that of the NaCl solution, and since Λ is proportional to L/C, it follows that

$$\Lambda_{HCl} = \Lambda_{NaCl}(L_{HCl}/L_{NaCl})(C_{NaCl}/C_{HCl})$$

$$= 126(35)(0.01/0.1) = \mathbf{441}$$

27. (a) $\Lambda_{NaOH} = 1,000 \times 0.0221/0.1 = $ **221.**

(b) On neutralization with the HCl solution, the resulting solution is 0.05 M in NaCl, so $\Lambda_{NaCl} = 1,000 \times 0.0056/0.05 = $ **112.**

(c) The third mixture consists of 0.033 M NaCl and 0.033 M HCl, so $1,000L = 17 = 0.033 \times 112 + 0.033 \times \Lambda_{HCl}$, from which $\Lambda_{HCl} = $ **403.** Then $\Lambda_{H^+,OH^-} = \Lambda_{HCl} + \Lambda_{NaOH} - \Lambda_{NaCl} = 403 + 221 - 112 = $ **512.**

28. The increase in L is due to the $CaSO_4$ present, so $L_{CaSO_4} = 4.4 \times 10^{-4}$ and $C = 1,000L/\Lambda_{\frac{1}{2}CaSO_4} = 0.44/140 = 0.00315$. The value of $\Lambda_{\frac{1}{2}CaSO_4}$ was obtained as follows (remember, $C = 0.002 \, N$): $\Lambda_{\frac{1}{2}Na_2SO_4} = 1,000L/C = 0.26/0.002 = 130$; hence, $\Lambda_{\frac{1}{2}SO_4^{2-}} = 130 - 50 = 80$. Then $\Lambda_{\frac{1}{2}CaSO_4} = 60 + 80 = 140$. The K_{sp} for $CaSO_4$ is then

$$(0.00315/2)(0.00315/2) + 0.002/2 = \mathbf{4.0 \times 10^{-6}}$$

29. We need first the Λ value for the ions H^+, C^-: $\Lambda_{H^+C^-} = \Lambda_{HCl} + \Lambda_{NaC} - \Lambda_{NaCl} = 426 + 83 - 126 = 383$. The actual ion concentration in the 0.001-M solution of the acid is then $C = 1,000L/\Lambda = 1,000 \times 3.83 \times 10^{-5}/383 = 10^{-4} \, M$. The concentration of undissociated acid is then $9 \times 10^{-4} \, M$, and

$$K_{diss} = (10^{-4})^2/9 \times 10^{-4} = \mathbf{1.11 \times 10^{-5}}$$

30. $\Lambda = 1,000 \times 2.72 \times 10^{-4}/0.002 = 136$. Then λ for the complex ion is $136 - 76$ or **60.** λ for NO_3^- is 71; and by combining Eqs. (13-1) and (13-2),

$$\Lambda_{salt} = \lambda_{complex} + \tfrac{1}{2}\lambda_{Cl^-} + \tfrac{1}{2}\lambda_{NO_3^-}$$
$$= 60 + 76/2 + 71/2 = \mathbf{133.5}$$

31. (See also Problem 32 for an alternative approach.) This total current is 1 A or $1/96,500 = 1.03 \times 10^{-5}$ equiv/sec. Of this, the fraction $40/115$ is carried by Li^+ ions, or 3.59×10^{-6} equiv/sec. The current carried in equiv/sec is also given by (velocity) $\times$ (equiv/cc)(area), so the velocity v of Li^+ ions is

$$v = 3.59 \times 10^{-6}/0.1 \times 10^{-3} \times 2 = 1.8 \times 10^{-2} \text{ cm/sec}$$

32. The total current past a cross section is 0.01 A or $0.01/96,500 = 1.04 \times 10^{-7}$ equiv/sec. Of this, the fraction $60/136$ is carried by M^+, or 4.58×10^{-8} equiv/sec. This, in turn, is equal to: (velocity)(equiv/cc)(area). The velocity v is then

$$v = 4.58 \times 10^{-8}/0.1 \times 10^{-3} \times 2 = \mathbf{2.29 \times 10^{-4} \text{ cm/sec}}$$

Alternatively, the electrochemical mobility μ is equal to $\lambda/\mathscr{F}$, where $v = \mu V$ (V is volts/cm). But $V = iR/l$, $R = rl/a$, and $r = 1/L = 1,000/C\Lambda$. Then $V = 0.01(1,000/C\Lambda)(1/l)(l/a) = 0.01 \times 1,000/0.1 \times 136 \times 2$. Then

$$v = (60/96,500)(0.01 \times 1,000/0.1 \times 136 \times 2)$$
$$= 2.29 \times 10^{-4} \text{ cm/sec}$$

33. The main problem here is to eliminate the unnecessary information. Since the applied potential is the same in both cases, the potential gradient is the same, and the velocities of Na^+ and K^+ ions will be in proportion to their mobilities (which are, from your text, 52×10^{-5} and 76×10^{-5}, respectively, in cm/sec per V/cm). Then

$$v_{Na^+} = (52/76)(3 \times 10^{-2}) = \mathbf{0.0205 \text{ cm/sec}}$$

34. First, $L = c\Lambda/1,000 = 120 \times 0.01/1,000 = 0.0012$. Then, by Ohm's law, $\Delta V = iR = ilr/A = (il/A)(1/L)$. The desired V/cm, or $\Delta V/l = (i/A)(1/L) = (0.1/10)(1/0.0012) = \mathbf{8.3 \text{ V/cm}}$.

35. First, $i = \Delta V/R = 6/200 = 0.03$ A. For the solution $1,000L = \Sigma C_i \lambda_i = 0.01(38.7 + 71.4) + 0.02(73.5 + 76.3) = 4.1$, while for Li^+ ions only, $1,000L$ is $0.01 \times 38.7 = 0.387$. The transference number of Li^+ in the solution is then $0.387/4.1 = 0.094$, and the current carried by Li^+ ions is therefore $0.03 \times 0.094 = \mathbf{0.00283}$.

36. First, we need V, in V/cm, obtainable as follows:

$$\Delta V = iR = irl/A = il/AL,$$

so

$$V = \Delta V/l = (i/A)(1/L) = (0.003/10)(1/L)$$

But $L = C\Lambda/1,000 = 0.01(39 + 71)/1,000 = 1.1 \times 10^{-3}$; so $V = (0.003/10) \times (1/1.1 \times 10^{-3}) = 0.273$ V/cm. The mobility of Li^+ ion is $\lambda/\mathscr{F} = 39/96,500 = 4.03 \times 10^{-4}$; so the velocity of Li^+ ions will be $4.03 \times 10^{-4} \times 0.273 = \mathbf{1.1 \times 10^{-4} \text{ cm/sec}}$.

14

IONIC EQUILIBRIUM

COMMENTS

The topic of ionic equilibrium is one that you had in high school, in freshman chemistry, and in quantitative analysis. Perhaps it should not even appear in physical chemistry, unless, of course, you find the questions in this chapter hard to work!

Don't be surprised if you have a little trouble. Even for the case of a monobasic acid and its salt, there are four equations to deal with: the acid-dissociation-constant equation, a material-balance equation, a charge-balance equation, and the dissociation-constant equation for the solvent. The situations presented here will all be fairly simple, but will not necessarily be ones that you are used to. You will have to determine for yourself what approximations to use, from an analysis of the situation, rather than relying on cut-and-dried recipes.

In addition, it turns out that an explicit algebraic solution of the equations that are involved in the case of a solution of two or more weak electrolytes or of polybasic acids can be very complicated. The so-called logarithmic diagrams become a most powerful tool with which to handle such situations. These are simply plots (usually logarithmic, hence the name) of the fraction of a weak electrolyte, which is present in a given form, versus pH. In combination with the charge-balance equation, a quick round of approximations usually produces the desired solution to about 1% accuracy. This is ordinarily quite sufficient, since the limiting error will generally lie in the making of activity-coefficient corrections.

The general procedure for using these diagrams is as follows: Typically, one knows the formula weights per liter of the various substances that were used to make up the solution and wishes to calculate the equilibrium value of $[H^+]$. One first guesses (in an educated way) what $[H^+]$ might be, reads the values for the various fractions from the graph, and tests these values for consistency with the charge-balance equation. On this first trial, the sum of positive charges will fail to exactly equal that of the negative charges and, from the direction of the misfit, a second, closer estimate of $[H^+]$ is then made. With experience, the desired self-consistent set of values can be obtained in three or four rounds of approximation.

It is perhaps worth mentioning that the same procedure applies to the handling of solutions of complex ions. For example, the complex $Mn(C_2O_4)_3^{3-}$ dissociates in three stages to Mn^{2+} and oxalate ions, and there are, accordingly, three dissociation constants. The situation is perfectly analogous to that of a tribasic acid, with $[C_2O_4^{2-}]$ taking the place of $[H^+]$ in the logarithmic diagram.

Several problems describe a mixture in terms of formalities, that is, formula weights of various substances that were mixed, per liter of solution. These will sometimes be chemically incompatible, for example, H_2SO_4 plus NaOH, and it will then help to find an equivalent statement of composition that expresses it in terms of species more likely actually to be the dominant forms present at equilibrium. Thus a solution which is $0.1 f$ in H_2SO_4 and $0.1 f$ in NaOH is better expressed as being $0.1 f$ in $NaHSO_4$.

The complication of activity coefficient corrections is omitted from the above type of problems in order to make them manageable as examination questions. You are expected, however, to know about the Debye–Hückel limiting law, and to apply it in the simpler situation of solubility product calculations. Do so, however, only when it obviously is expected of you.

You will find that quoted dissociation and solubility product constants may vary somewhat from problem to problem, although they will never be completely unreasonable. Such adjustments are deliberate and were made mainly to avoid arithmetic complications and to assist the rapid working of the problem.

EQUATIONS AND CONCEPTS

Solubility Product

For a salt A_xB_y:

$$K_{sp} = [A]^x[B]^y = [xS]^x[yS]^y \tag{14-1}$$

where S denotes the solubility (in the absence of a common ion) in formula weights of the salt per liter.

Formality

Symbol, f. Denotes formula weights per liter, and is a useful way of expressing the make-up of a solution, if one wishes explicitly to avoid any suggestion that the substances added remain as such after mixing.

Thermodynamic Solubility Product

$$K_{th} = a_A^x a_B^y = K_{sp}\gamma_A^x\gamma_B^y = K_{sp}(\gamma_\pm)^{x+y} \tag{14-2}$$

Debye–Hückel Limiting Law

$$\log \gamma_\pm = -0.51 z^+ z^- \sqrt{\mu} \tag{14-3}$$

at 25°C and water as solvent, where z^+ and z^- are the ion charges or valences, and μ denotes the ionic strength. $\mu = \frac{1}{2}\Sigma\, C_i z_i^2$.

Dissociation Constant of Weak Electrolytes

For a weak acid HA, $K = [H^+][A^-]/[HA]$, and

$$K_a = \frac{[H^+](f_{MA} + [H^+] - [OH^-])}{f_{HA} - [H^+] + [OH^-]} \tag{14-4}$$

where f_{MA} and f_{HA} are the formalities of the salt of the acid MA, and of the acid HA, respectively. As an example of commonly made approximations, if only acid is present, that is, if $f_{MA} = 0$, then if $K f_{HA} > 10^{-12}$,

$$K \cong [H^+]^2/(f_{HA} - [H^+]) \tag{14-5}$$

If also $f_{HA} > 10^4 K$, then

$$K \cong [H^+]^2/f_{HA} \tag{14-6}$$

Logarithmic Diagrams

These are plots of F_{HA} and of F_{A^-} versus pH, where F denotes the fraction of total stoichiometric acid (i.e., $f_{MA} + f_{HA}$) actually present as undissociated acid HA and as A^- ion, respectively.

$$F_{HA} = \frac{[H^+]}{[H^+] + K} \quad \text{and} \quad F_{A^-} = 1 - F_{HA} = \frac{K}{[H^+] + K} \tag{14-7}$$

These plots are used in connection with the charge-balance equation:

$$[M^+] + [H^+] = [A^-] + [OH^-] \tag{14-8}$$

or

$$[M^+] + [H^+] = F_{A^-}(f_{MA} + f_{HA}) + [OH^-] \tag{14-9}$$

where M^+ is any nonhydrolyzing cation present.

For a dibasic acid,

$$F_{H_2A} = \frac{1}{1 + K_1/[H^+] + K_1K_2/[H^+]^2} \qquad (14\text{-}10)$$

$$F_{HA^-} = \frac{1}{[H^+]/K_1 + 1 + K_2/[H^+]} \qquad (14\text{-}11)$$

$$F_{A^{2-}} = 1 - F_{H_2A} - F_{HA^-} = \frac{1}{[H^+]^2/K_1K_2 + [H^+]/K_2 + 1} \qquad (14\text{-}12)$$

Associated charge-balance equation is

$$[M^+] + [H^+] = [HA^-] + 2[A^{2-}] + [OH^-] \qquad (14\text{-}13)$$

PROBLEMS

1. (15 min) Solid NaOH is added to 1 liter of 0.05-f solution of H_2S until the pH rises to 12. Taking K_1 and K_2 to be 10^{-9} and 10^{-14}, respectively, calculate (a) the initial pH and (b) the moles of NaOH added, and the concentrations of all species present in the solution. Neglect activity-coefficient effects.

2. (12 min) A 0.1-m solution of the acid H_2A is first made 0.06 f in NaOH, at which point the pH of the solution is 3, and is then made 0.16 f in NaOH, at which point the pH is 8. Calculate K_2 and K_1 for this acid.

3. (6 min) The various substances below may be acting as acids or bases, or neither, in the particular reactions given. Label each appropriately:

1. $HB + H_2O = B^- + H_3O^+$ (B denotes benzoate; water is the solvent)
2. $HB + HCl = H_2B^+ + Cl^-$ (B denotes benzoate; liquid HCl is the solvent)
3. $NH_3 + H_2O = NH_4^+ + OH^-$ (water is the solvent)
4. $(CH_3)_3N + PCl_3 = (CH_3)_3NPCl_3$ (gas phase)

4. (12 min) Calculate the moles per liter of acetic acid that must be added to a 0.01-M solution of Na_2CO_3 to bring the pH to 6. To simplify calculation take K for HAc to be 10^{-5}, and K_1 and K_2 for H_2CO_3 to be 3×10^{-7} and 1×10^{-10}, respectively.

5. (24 min) (final examination question) A solution is 0.1f in H_2SO_4, 0.1f in $NaHSO_4$, 0.1f in NaAc (sodium acetate), and 0.1f in NaOH. Calculate the actual concentrations of all species present in this solution. Sulfuric acid is a strong acid in the first stage, and the ionization constant for the second stage is to be taken as $K_2 = 0.01$. That for acetic acid is 2×10^{-5}.

6. (15 min) A solution that is $0.09 f$ in HCl, $0.09 f$ in Cl_2HCOOH, and $0.1 f$ in H_3CCOOH has a pH of 1.0. The second acid (dichloroacetic acid) has a dissociation constant K. That of acetic acid is 10^{-5}. Calculate the value of K, and the concentrations of all species present in this mixture.

7. (15 min) The first and second dissociation constants for oxalic acid are 10^{-2} and 10^{-5}, respectively. Calculate the concentrations of all species present in a solution made up by mixing 0.01 mole of NaOH, 0.02 mole of Na_2Ox, 0.03 mole of NaHOx, and 0.04 mole of H_2Ox with enough water to form 1 liter of solution.

8. (15 min) A solution which is $0.09 f$ in H_2SO_4 and $0.1 f$ in HAc (acetic acid) has a pH of 1. The first stage of dissociation of sulfuric acid is complete, but HSO_4^- is a weak acid having a dissociation constant K. The dissociation constant for HAc is 10^{-5}. Calculate the value of K, and the concentration of all species present in this solution.

9. (20 min) (final examination question) The two dissociation constants of H_2S are 10^{-7} and 10^{-15}, respectively, for K_1 and K_2. A $0.1\text{-}M$ solution of H_2S is "neutralized" with NaOH (i.e., the pH is 7 after the addition of the NaOH). Assume the NaOH solution was sufficiently concentrated so that the dilution effect was negligible. Calculate
  (a) the concentrations of all species present and
  (b) the number of equivalents of NaOH per equivalent of H_2S originally present.

10. (12 min) Calculate the concentrations of all species present in a solution which is $0.01 f$ in HD and $0.02 f$ in NaD. D denotes dichloroacetate, and K_a for HD is 0.002. (2% relative accuracy is sufficient in your answers.)

11. (15 min) Calculate the concentrations of all species present in a solution that is $0.03 f$ in acetic acid HAc, and $0.01 f$ in sodium cyanide. Dissociation constants are 10^{-5}, 10^{-10}, and 10^{-14} for HAc, HCN, and water, respectively.

12. (10 min) A weak acid HA is found to be 10% dissociated if the solution is $0.001 f$ in acid. Calculate the expected degree of dissociation for a solution $0.001 f$ in acid and $0.1 M$ in NaCl, allowing for activity coefficient effects.

---

13. (18 min) 0.02 mole of $Pb(IO_3)_2$ is suspended in 1 liter of water, and Na_2Ox (Ox = oxalate) is added until the lead iodate is just completely converted to lead oxalate, that is, just dissolves. Neglecting activity coefficient effects and any hydrolysis of oxalate or lead ions, calculate the moles of

Na_2Ox added and the final concentration of lead ions. K_{sp} values for $Pb(IO_3)_2$ and PbOx are 10^{-13} and 10^{-11}, respectively.

14. (18 min) Calculate the solubility of $PbCl_2$ in $0.1m$ NaCl, if the solubility in water is $0.01m$ at 25°C.
(a) Neglecting activity-coefficient effects and (b) including them, but using the simple limiting law.

15. (6 min) List in order of increasing solubility of AgCl: AgCl in
(a) $0.1\ M$ $NaNO_3$ (d) $0.1\ M$ $Ca(NO_3)_2$
(b) $0.1\ M$ NaCl   (e) $0.1\ M$ NaBr
(c) water

16. (18 min) 27.8 g (0.1 mole) of $PbCl_2$ is added to 1 liter of water, and then sufficient silver nitrate is added until, at equilibrium, half the $PbCl_2$ is dissolved. The solubility of $PbCl_2$ in water is 0.03 mole/liter. The temperature is 25°C.
(a) Neglecting activity-coefficient effects, calculate the equilibrium concentrations of all ions and the equilibrium amounts of any solids, per liter. Neglect possible hydrolysis of Pb^{2+}.
(b) If the above calculations were repeated recognizing activity-coefficient effects (assumed to be given by the Debye–Hückel limiting law), explain clearly whether the various concentrations and amounts of solids obtained above would come out smaller, larger, or be unchanged. Rough calculations will greatly assist in making your explanations clear.

17. (12 min) Calculate the solubility of AgX in $0.1\ M$ $NaNO_3$ if its solubility in water is $10^{-7}m$, where X denotes some monovalent anion.

18. (10.5 min) The solubility products for $Pb(IO_3)_2$ and $PbSO_4$ are 10^{-13} and 10^{-8}, respectively. It is desired to convert 0.1 mole of $PbSO_4$ to the iodate. The $PbSO_4$ is added to 1 liter of water and $NaIO_3$ is added until the solid is (just) entirely converted. Calculate the number of moles of sodium iodate added, and the final concentration of lead ions in the solution.

19. (4.5 min) Calculate the solubility of AgAc ($K_{th} = 4 \times 10^{-3}$) in water, at 25°C, including limiting-law activity-coefficient effects.

20. (19 min) (final examination question) The solubility of AgA is 0.0001 mole/liter (where A denotes the anion of a weak organic acid) in water at pH 7. Hydrolysis of A^- is negligible at this pH.
(a) Calculate the solubility of AgA in $0.1\ M$ $NaNO_3$ at pH 7.
(b) Calculate the dissociation constant of HA if the solubility in $10^{-3}\ M$ nitric acid is 0.00013 mole/liter.

21. (30 min) The thermodynamic solubility product of BaOx (Ox = oxalate) is 1×10^{-8}, and the first and second dissociation constants for oxalic acid (H_2Ox) are 1×10^{-3} and 1×10^{-5}, respectively.

(a) Neglecting activity-coefficient effects, calculate the solubility of BaOx in water.

(b) Calculate the pH of the solution in (a), still neglecting activity-coefficient effects.

(c) Including activity-coefficient effects, calculate the solubility of BaOx in 0.1-m $NaNO_3$ solution.

(d) 0.1 mole of solid BaOx is shaken with 1 liter of water, and concentrated HNO_3 gradually added until all the solid has just disappeared. Neglecting activity-coefficient effects, calculate the formality of acid added at this point, and the concentrations of all species present.

22. (15 min) 0.01 mole of $AgNO_3$ are added to 1 liter of solution which is 0.1 M in Na_2CrO_4 and 0.005 M in $NaIO_3$. Given that the K_{sp} values for Ag_2CrO_4 and $AgIO_3$ are 10^{-8} and 10^{-13}, respectively, calculate the moles of any solids formed at equilibrium and the concentrations of Ag^+, IO_3^-, and CrO_4^{2-} ions.

23. (12 min) Excess solid BaF_2 is shaken with 0.1 M sodium oxalate (Na_2Ox) solution until equilibrium is reached. K_{sp} values are 10^{-6} and 10^{-7} for BaF_2 and BaOx, respectively. Possible hydrolysis of F^- and Ox^{2-} ions may be neglected. Calculate the equilibrium concentrations of Ba^{2+}, F^-, and Ox^{2-} ions.

24. (15 min) The thermodynamic solubility product for PbF_2 is 4×10^{-9}. Including activity-coefficient effects where they are not negligible, calculate the solubility of PbF_2 in

(a) water,

(b) 0.1 M NaF, and

(c) 0.1 M HCl (you may neglect activity-coefficient effects in this case, but recognize that HF is a weak acid and is only 1% dissociated at a pH of 1).

25. (21 min) (final examination question) Given that K_{sp} is 1×10^{-6} for NaV and K_a for HV is 1×10^{-5}, where NaV is sodium valerate and HV is valeric acid $CH_3(CH_2)_3COOH$, calculate the solubility of NaV in the following media:

(a) 0.1 M NaCl (neglecting activity-coefficient effects).

(b) 0.1 M KNO_3 (including activity-coefficient effects).

(c) 0.001 M HCl (neglecting activity-coefficient effects).

26. (18 min) A solution is prepared that is $0.02 f$ in $Pb(NO_3)_2$, $0.01 f$ in Na_2SO_4, and $0.005 f$ in $NaIO_3$. Given that the solubility products are 1×10^{-8} and 1×10^{-13} for $PbSO_4$ and $Pb(IO_3)_2$, respectively, calculate the moles of the solid(s) present, per liter, and the concentrations of all species present in the equilibrium solution.

27. (10.5 min) The solubility of BaF_2 is 0.01 mole/liter in 0.5 M HCl. The dissociation constant for HF is 0.02.

(a) Calculate the K_{sp} for BaF_2, assuming ideal solutions.

(b) Assuming the value of the K_{sp} to be 10^{-8}, calculate the thermodynamic-solubility constant. The mean activity coefficient for the dissolved barium fluoride may be taken to be 0.60.

28. (12 min) The solubility of Ag_2SO_4 is 0.03 M at 25°C. Calculate the K_{sp} and, with allowance for activity-coefficient effects, the true or thermodynamic constant, K_{th}.

---

29. (18 min) Given the logarithmic diagram (see Fig. 14-1) for a monobasic acid HA, calculate

(a) the pK of the acid,

(b) the percent neutralization of a solution initially 0.01 M in HA if sufficient solid NaOH is added to bring the pH to 10, and

(c) the pH of a solution 0.01 M in HA which has been exactly neutralized with solid NaOH. Make your procedures clear. F in Fig. 14-1 denotes the fraction of acid present in a particular form.

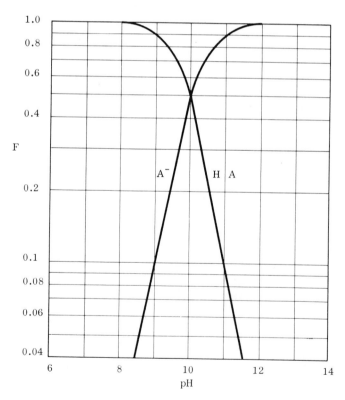

FIGURE 14-1

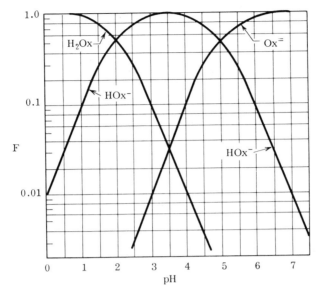

FIGURE 14-2

30. (18 min) Given the logarithmic diagram of Fig. 14-2 for oxalic acid H_2Ox,

(a) calculate K_1 and K_2 (these will not be quite the usual ones), and

(b) calculate the number of formula weights per liter of NaOH that must be added to 0.1 M H_2Ox to bring the pH to 5.5. Make your procedures clear. (F denotes the fraction of the acid present in a particular form.)

31. (24 min) (final examination question) The logarithmic diagram for fumaric acid $(C_4H_4O_4)$ is given in Fig. 14-3. Only curves for F_{H_2f} and $F_{f^{2-}}$ are given, however. Note that the F scale is linear rather than, as usual, logarithmic.

(a) At what pH will the fraction of fumaric acid present as Hf^- be at a maximum? What is this maximum value?

(b) Give the value for pK_1.

(c) Calculate the amount of solid NaOH that should be added to 1 liter of 0.1 M H_2f in order to bring the pH to 5. Make your work clear.

32. (24 min) (final examination question) Using the diagram shown in Fig. 14-4, which gives the variation of $F_{H_2CO_3}$ and $F_{HCO_3^-}$ with pH, calculate K_1 and K_2 for carbonic acid.

Assuming K_1 and K_2 to be 10^{-5} and 10^{-8}, respectively, calculate the equilibrium concentrations of all species present in a solution which is 0.2f in H_2CO_3 and 0.1f in Na_2CO_3.

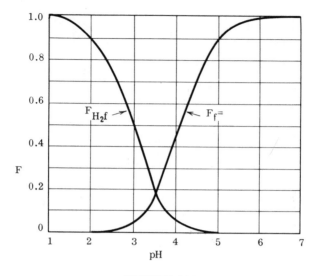

FIGURE 14-3

33. (18 min) The logarithmic diagram of Fig. 14-5 is for chloropropionic acid, HC.

(a) Draw in the plot for F_{C^-}.

(b) Calculate the pK.

(c) Calculate the pH of a 10^{-3}-M solution of the acid.

(d) Calculate the pH of 0.1 M HC, 10% neutralized by the addition of solid NaOH.

Make your procedures clear.

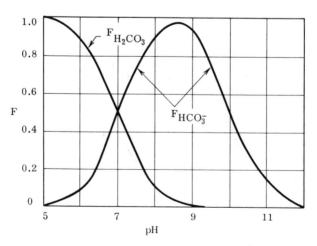

FIGURE 14-4

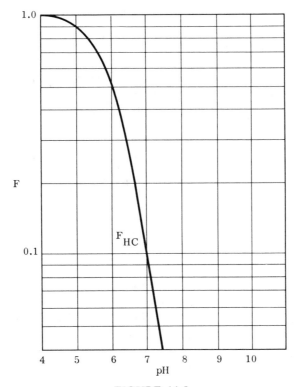

FIGURE 14-5

34. (15 min) A plot of F_{A^-} and F_{B^-} versus $[H^+]$ is given in Fig. 14-6, where A^- and B^- are the anions of the weak acids HA and HB.
(a) Calculate the dissociation constants.
(b) Calculate the pH of a solution made up by dissolving 0.01 mole of NaA and 0.02 mole of HB in 1 liter of water. (An accuracy of one part in 20 is sufficient.)

35. (12 min) (final examination question) Using Figs. 14-3 and 14-4, calculate the formality of NaHf that should be present in a 0.01-M solution of Na_2CO_3 if the pH is to be 7. (See Problem 31.)

36. (12 min) (final examination question) There is a class of weak dibasic acids for which $K_1 = 4K_2$, where K_1 and K_2 are the first and second acid dissociation constants. Show at what pH (expressed in terms of K_1 and/or K_2) the fraction of acid present in the HA^- form is at a maximum, and calculate the value of this fraction.

37. (15 min) (final examination question) For a certain hypothetical weak tribasic acid H_3A it is known that at pH 7 F_{H_3A} is 0.05 and $F_{A^{3-}}$ is 0.10.

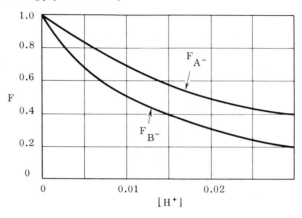

FIGURE 14-6

Also pK_1 is known to be 6. Calculate the values of pK_2 and pK_3 (or of K_2 and K_3).

38. (14 min) Under consideration are a weak acid HA (dissociation constant K_a), and a weak base BOH (dissociation constant K_b). It happens that the same curve, shown in Fig. 14-7, represents the variation of F_{HA} with pH as well as that of F_{B^+} with pH. (a) Give the values of K_a and K_b, and (b) calculate the pH of a solution $0.1 f$ in NaA and $0.2 f$ in BCl.

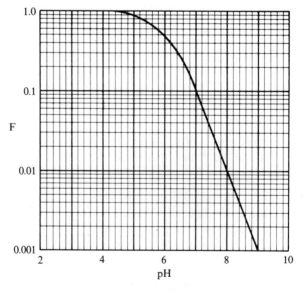

FIGURE 14-7

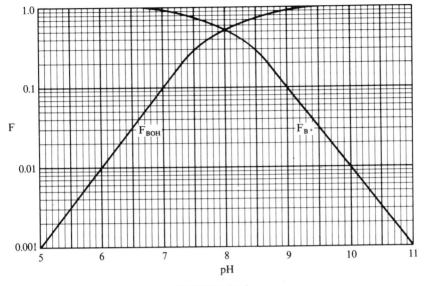

FIGURE 14-8

39. (12 min) The log diagram for a weak base, BOH, is shown in Fig. 14-8. Calculate or explain how you obtain (a) the pK of the base and (b) the percent neutralization of a 0.1-M solution of BOH when brought to a pH of 7 by the addition of HCl.

ANSWERS

1. (a) K_2 is so small compared with K_1 that only the first dissociation step need be considered. Then

$$10^{-9} = \frac{[H^+][HS^-]}{[H_2S]} \cong \frac{x^2}{0.05} \quad \text{or} \quad x = 7.1 \times 10^{-6}$$

The initial pH is then **5.15.**

(b) From K_1,

$$\frac{[HS^-]}{[H_2S]} = \frac{10^{-9}}{[H^+]} = 10^3$$

thus $[H_2S]$ is small at pH 12. From K_2,

$$\frac{[S^{2-}]}{[HS^-]} = \frac{10^{-14}}{[H^+]} = 0.01$$

Thus $[S^{2-}]$ is small also compared with $[HS^-]$. Then $[HS^-] = 0.05$, so $[H_2S] = 5 \times 10^{-5}$, $[S^{2-}] = 5 \times 10^{-4}$, and of course $[H^+] = 10^{-12}$ and $[OH^-] = 10^{-2}$. The charge-balance equation is

$$[H^+] + [Na^+] = [HS^-] + 2[S^{2-}] + [OH^-]$$

or $[Na^+] = 0.05 + 0.001 + 0.01 = 0.061$. Thus **0.061 mole** of NaOH was added.

2. The charge balance for the first condition,

$$[Na^+] + [H^+] = [HA^-] + 2[A^{2-}] + [OH^-]$$

becomes $0.06 + 10^{-3} = [HA^-]$, or $[HA^-] = 0.061$, neglecting $[A^{2-}]$, subject to later reconsideration. Then

$$[H_2A] = 0.039 \quad \text{and} \quad K_1 = \frac{(10^{-3})(0.061)}{0.039} = 1.57 \times 10^{-3}$$

After the second NaOH addition, the charge-balance equation becomes

$$0.16 + 10^{-8} = [HA^-] + 2[A^{2-}] + 10^{-6}$$

or $[A^{2-}] = 0.06$ and $[HA^-] = 0.04$. K_2 is then $10^{-8} \times 0.06/0.04 =$ **1.5×10^{-8}**.

K_2 is small enough that $[A^{2-}]$ could indeed be safely neglected in the first case.

3. 1. Acids: HB and H_3O^+. Bases: H_2O and B^-. (proton donors versus acceptors)

2. Acids: HCl and H_2B^+. Bases: HB and Cl^-. (proton donors versus acceptors)

3. Acids: H_2O and NH_4^+. Bases: NH_3 and OH^-. (proton donors versus acceptors)

4. Acid: PCl_3. Base: $(CH_3)_3N$. (electron-pair acceptor versus donor)

4. At pH 6, the ratio

$$\frac{[Ac^-]}{[HAc]} = \frac{10^{-5}}{10^{-6}} = 10$$

so $[Ac^-] = 0.91C$, where C denotes the added formality of HAc. Similarly,

$$\frac{[HCO_3^-]}{[H_2CO_3]} = \frac{3 \times 10^{-7}}{10^{-6}} = 0.3$$

and

$$\frac{[CO_3^{2-}]}{[HCO_3^-]} = \frac{10^{-10}}{10^{-6}} = 10^{-4}$$

Evidently $[CO_3^{2-}]$ is negligible, and therefore

$$[H_2CO_3] = 0.01/1.3 = 0.0077 \quad \text{and} \quad [HCO_3^-] = 0.0023$$

From the charge-balance equation,

$$[H^+] + [Na^+] = [Ac^-] + [HCO_3^-]$$

(neglecting $[CO_3^{2-}]$ and $[OH^-]$), we have $10^{-6} + 0.02 = 0.91C + 0.0023$, whence $C = \mathbf{0.0195}$.

5.   In the case of a problem of this type, it is desirable to decide how the various ingredients will react in a general way and then to express the composition in terms of new ingredients that will more nearly correspond to the final species present. Thus $0.1 f H_2SO_4$ and $0.1 f NaOH$ are better written as $0.1 f NaHSO_4$ (since they will react to yield this salt). The solution is now written as $0.2 f NaHSO_4$ and $0.1 f NaAc$. However, HSO_4^- is a stronger acid than is HAc, so it is better to represent $0.1 f NaHSO_4 + 0.1 f NaAc$ as $0.1 f HAc + 0.1 f Na_2SO_4$. The final phrasing of the composition is then $0.1 f HAc + 0.1 f NaHSO_4 + 0.1 f Na_2SO_4$.

Neglecting for the moment the dissociation of HAc, we have a buffer in sulfate–bisulfate, so the pH $= pK_2 = 2$. At this pH,

$$\frac{[Ac^-]}{[HAc]} = \frac{2 \times 10^{-5}}{10^{-2}} = 0.002 \quad \text{or} \quad [Ac^-] = 2 \times 10^{-4}$$

(and the neutralization of NaAc by $NaHSO_4$ was in fact essentially complete). The various concentrations are then

$$[Na^+] = 0.3 \qquad\qquad [HSO_4^-] = 0.1$$
$$[SO_4^{2-}] = 0.1 \qquad\qquad [HAc] = 0.1$$
$$[Ac^-] = 2 \times 10^{-4} \qquad [H^+] = 0.01$$
$$[OH^-] = 10^{-12}$$

6. At a pH of 1,

$$\frac{[\text{Ac}^-]}{[\text{HAc}]} = \frac{10^{-5}}{0.1} = 10^{-4}$$

so $[\text{HAc}] = \textbf{0.1 }\textit{M}$ and $[\text{Ac}^-] = \textbf{10}^{-5}\textbf{ }\textit{M}$.

Acetic acid thus contributes negligibly to the $[\text{H}^+]$. Since HCl is a strong acid, it contributes $0.09\ M\ [\text{H}^+]$, and $[\text{Cl}^-] = \textbf{0.09}$. Finally, by difference, the HD (dichloracetic acid) contributed $0.01\ M\ [\text{H}^+]$, so $[\text{D}^-]$ must be **0.01** and $[\text{HD}] = \textbf{0.08}$. K is then $(0.1)(0.01)/(0.08) = \textbf{0.0125}$.

7. First, we restate the composition in terms of constituents more likely to correspond to the major species actually present. 0.01 mole of NaOH and 0.01 mole of H_2Ox will form 0.01 mole of NaHOx, so the composition can be restated as 0.04 mole of NaHOx + 0.03 mole of H_2Ox + 0.02 mole of Na_2Ox. However, H_2Ox and Na_2Ox form 2NaHOx, so the final statement of composition becomes: 0.08 mole of NaHOx + 0.01 mole of H_2Ox.

The last statement of composition clearly shows the mixture to be a simple buffer and, to a good approximation,

$$K_1 = 10^{-2} = [\text{H}^+]\frac{0.08}{0.01}$$

or $[\text{H}^+] = \textbf{0.00125}$. Also $[\text{Na}^+] = \textbf{0.08 }\textit{M}$, $[\text{H}_2\text{Ox}] = \textbf{0.01 }\textit{M}$, and $[\text{HOx}^-] = \textbf{0.08 }\textit{M}$. Finally $[\text{Ox}^{2-}] = K_2[\text{HOx}^-]/[\text{H}^+] = 10^{-5} \times 0.08/0.00125 = \textbf{6.4} \times \textbf{10}^{-4}\textbf{ }\textit{M}$.

8. Acetic acid is weak enough that its dissociation does not contribute appreciably to $[\text{H}^+]$ at pH 1. The $0.09\ f\ \text{H}_2\text{SO}_4$ contributes $0.09\ M\ \text{H}^+$ from the first stage of dissociation, so the second stage must have supplied the missing $0.1\ M\ \text{H}^+$. Therefore $[\text{SO}_4^{2-}] = \textbf{0.01 }\textit{M}$, and $[\text{HSO}_4^-] = \textbf{0.08 }\textit{M}$, and $K = [\text{H}^+][\text{SO}_4^{2-}]/[\text{HSO}_4^-] = 0.1 \times 0.01/0.08 = \textbf{0.0125}$. $[\text{Ac}^-] = 10^{-5}[\text{HAc}]/[\text{H}^+] = 10^{-5} \times 0.1/0.1 = \textbf{10}^{-5}$, and, of course, $[\text{HAc}]$ is essentially $\textbf{0.1 }\textit{M}$. Finally, $[\text{OH}^-]$ is $\textbf{10}^{-13}$.

9. (a) pH 7 is the buffer point for the first stage, so

$$[\text{H}_2\text{S}] = [\text{HS}^-]$$

From K_2, $[\text{S}^{2-}]/[\text{HS}^-] = 10^{15}/10^{-7} = 10^{-8}$, so $[\text{S}^{2-}]$ is very small. Therefore, $[\text{H}_2\text{S}] = [\text{HS}^-] = \textbf{0.05}$ and $[\text{S}^{2-}] = \textbf{5} \times \textbf{10}^{-10}$.

(b) $0.05\ M$ NaOH is required to form the 0.05-M HS^-, or $0.05/0.2 = \textbf{0.25}$ mole/equiv H_2S, that is, one-fourth the amount for complete conversion to Na_2S.

10. Here is a case where the simple buffer calculation fails: according to it,

$$[H^+] = 2 \times 10^{-3}(0.01)/(0.02) = 0.001$$

yet 0.001 is an appreciable fraction of the added formality of HD. The appreciable dissociation of HD necessary to furnish the equilibrium $[H^+]$ diminishes the HD concentration and increases that of D^- relative to the initial formalities. Thus we have

$$K = \frac{[H^+](f_{NaD} + [H^+])}{f_{HD} - [H^+]}$$

Using the above calculation as a first approximation, the second becomes $0.002 = [H^+](0.02 + 0.001)/(0.01 - 0.001)$ or $[H^+] = 0.00086$.

Another round of approximation yields $[H^+] = 8.7 \times 10^{-4}$. Then $[HD] = \mathbf{0.00913}$, $[D^-] = \mathbf{0.0209}$ and, of course, $[Na^+] = \mathbf{0.02}$.

11. Since HCN is so very weak, the reaction

$$HAc + CN^- = Ac^- + HCN$$

should go almost to completion. We can therefore rephrase the composition as follows:

$$0.03 \, f \, HAc \rightarrow 0.02 \, f \, HAc$$
$$\underset{+}{} \qquad \searrow$$
$$0.01 \, f \, NaCN \rightarrow 0.01 \, f \, HCN + 0.01 \, f \, NaAc$$

The mixture is evidently an acetate buffer, so

$$[H^+] = 10^{-5}\frac{[HAc]}{[Ac^-]} = 10^{-5} \times \frac{0.02}{0.01} = \mathbf{2 \times 10^{-5}}$$

Then

$$[CN^-] = 10^{-10}\frac{[HCN]}{[H^+]} = \frac{10^{-10} \times 0.01}{2 \times 10^{-5}}$$
$$= \mathbf{5 \times 10^{-7}}$$

(and the assumption that most of the cyanide is present as HCN is confirmed; that is, the calculation is self-consistent).

Finally, $[OH^-] = \mathbf{5 \times 10^{-10}}$, and $[Na^+] = \mathbf{0.01}$.

12. We write

$$K = \frac{[H^+][A^-]}{[HA]} = \frac{\alpha^2 C}{(1-\alpha)}$$

From the given α and C values, $K = (0.1)^2 0.001/0.9 = 1.11 \times 10^{-5}$. The more correct equation is

$$K = \frac{[H^+][A^-]}{[HA]}(\gamma_{H^+})(\gamma_{A^-}) = \frac{\alpha^2 C}{(1-\alpha)}\gamma_\pm^2$$

$\gamma_\pm$ was essentially unity at the low ionic strength of the first calculation, so the above K is the thermodynamic one. In 0.1 M NaCl, however, application of Eq. (14-3) gives $\log \gamma_\pm = -0.051$ or $\gamma_\pm = 0.89$, and $\gamma_\pm^2 = 0.79$. Then

$$\frac{\alpha^2}{(1-\alpha)} = 1.11 \times 10^{-5}/0.001 \times 0.79 = 0.014$$

from which $\alpha = \mathbf{0.11}$.

13. Since the $Pb(IO_3)_2$ is just dissolved, $[IO_3^-] = 0.04$; the last trace of the solid established $[Pb^{2+}]$ as equal to

$$K_{sp}/[IO_3^-]^2 = 10^{-13}/(0.04)^2 = \mathbf{6.25 \times 10^{-11}}$$

Since the K_{sp} for PbOx also applies,

$$[Ox^{2-}] = 10^{-11}/6.25 \times 10^{-11} = \mathbf{0.16}.$$

The total sodium oxalate required is then 0.02 mole to form PbOx, plus 0.16 or **0.18**, and the sodium ion concentration is **0.36**. $K_{th} = [Pb^{2+}] \times [IO_3^-]^2 \gamma_{Pb^{2+}} \gamma_{IO_3^-}^2$ or $\log K_{th} = \log K_{sp} + 3 \log \gamma_\pm$. At 25°C, this last term is given by $\log \gamma_\pm = -0.51 z^+ z^- \sqrt{\mu}$ or, since μ will be essentially that due to the 0.01 m NaNO$_3$, $\mu = 0.01$, then $\log \gamma_\pm = 0.51 \times 2 \times 1 \times 0.1$ or $\log \gamma_\pm = -0.102$. Then $\log K_{sp} = -13 + 0.303$, and $K_{sp} = 2 \times 10^{-13}$. If S denotes the solubility, then $4S^3 = 2 \times 10^{-13}$ and $S = \mathbf{3.7 \times 10^{-5}}$.

14. (a) $K_{sp} = [S][2S]^2 = [0.01][0.02]^2 = 4 \times 10^{-6}$. Then, in 0.1 m NaCl, $4 \times 10^{-6} = [S][S + 0.1]^2$, or $S = \mathbf{4 \times 10^{-4}}$.

(b) The ionic strength of 0.01 m PbCl$_2$ is $\frac{1}{2}(0.01 \times 4 + 0.02) = 0.03$. Then $\log \gamma_\pm$ is $-0.51 \times 2 \times 1/\sqrt{0.03} = -0.176$ and $\log K_{th} = \log K_{sp} + 3(-0.176)$. Log K_{th} is then $-5.392 - 0.528 = -5.920$.

Turning to the 0.1-m NaCl solution, μ is now essentially 0.1, and the mean activity coefficient for $PbCl_2$ becomes

$$\log \gamma_\pm = -0.51 \times 2 \times 1\sqrt{0.1} = -0.322$$

Log K_{sp} is now log K_{th} − 3 log $\gamma_\pm$ or log K_{sp} = −5.920 + 0.966 = −4.954. Then K_{sp} = 1.11 × 10^{-5} = [S][S + 0.1]^2 or S = **1.11 × 10^{-3}**.

15.   Order of increasing solubility:
   (b) owing to common ion effect, solubility is very low
   (c) (solubility in H_2O)
   (a) more soluble than in water, owing to lowering of activity coefficient
   (d) effect larger than in (a), as the ionic strength is greater
   (e) the AgCl will dissolve to a high degree, since it is converted to AgBr

16.   (a) K_{sp} for $PbCl_2$ equals [S][$2S$]^2 where S = 0.03, so K_{sp} = 1.08 × 10^{-4}. Since half the $PbCl_2$ is to dissolve, [Pb^{2+}] = **0.05**, and therefore 1.08 × 10^{-4} = [Pb^{2+}][Cl^-]^2 = 0.05[Cl^-]^2; thus [Cl^-] = **4.65 × 10^{-2}**.

$$[Ag^+] = K_{sp}/[Cl^-] = 1.2 \times 10^{-10}/4.57 \times 10^{-2} = \mathbf{2.63 \times 10^{-9}}$$

(using the K_{sp} value from a textbook). This last is so small that the principal consumption of $AgNO_3$ is in forming the **0.1 mole** of AgCl from the Cl^- ion produced in the dissolving of the $PbCl_2$. (NO_3^-) is then **0.1 M,** and, of course, **0.05 mole** $PbCl_2$ remains.

   (b) The problem requires that half the $PbCl_2$ dissolve, so there must still be **0.05 mole** of $PbCl_2$ solid and **0.1 mole** of AgCl solid; hence [NO_3^-] must still be 0.1 M. The K_{sp} values will be increased, however, owing to the ionic strength effect, and that of $PbCl_2$ more than that of AgCl. Thus the solubility of $PbCl_2$ in water produces an ionic strength of 0.09; the limiting-law equation of

$$\log \gamma_\pm = -0.5 \times 2 \times 1\sqrt{0.09}$$

gives $\gamma_\pm$ = 0.5. K_{th} is then $K_{sp}(\gamma_\pm)^3$ or 0.125K_{sp}. The final solution is essentially 0.05 M $Pb(NO_3)_2$, so μ is 0.15; $\gamma_\pm$ for Pb^{2+}, $2Cl^-$ is then 0.4; and K_{sp} = $K_{th}/0.064$, or K_{sp} = (0.125/0.064)K_{sp} (for water), about a twofold increase.

   [Cl^-] is then about **40% larger than before.** In addition, K_{sp} for AgCl is changed. For a μ of 0.15, $\gamma_\pm$ for Ag^+Cl^- is 0.63, and the effective K_{sp} becomes 1.2 × 10^{-10}/(0.63)^2 or about 2.5 times larger. Even allowing for the 40% increase in [Cl^-], the new [Ag^+] will be **larger** than before.

17.   The solubility in water is low enough that activity–coefficient effects can be neglected; hence K_{sp} = S^2 = 10^{-14}. The ionic strength of 0.1 m

$NaNO_3$ is 0.1; hence $\log \gamma_\pm = -0.51(0.1)^{1/2} = -0.161$ or $\gamma_\pm = 0.69$ for Ag^+X^-. The K_{sp} for this ionic strength is then $10^{-14}/(0.69)^2 = 2.09 \times 10^{-14}$, and the solubility will then be **1.45×10^{-7}**.

18. If the $PbSO_4$ is to just dissolve, then $[SO_4^{2-}] = $ **0.1**, and, from the K_{sp}, $[Pb^{2+}] = 10^{-8}/0.1 = $ **10^{-7}**. Then, from the K_{sp} for $Pb(IO_3)_2$, $[IO_3^-]^2 = 10^{-13}/10^{-7}$ or $[IO_3^-] = $ **0.001**. The amount of $NaIO_3$ required is then 0.2 [to form 0.1 mole of $Pb(IO_3)_2$] + 0.001 = 0.201 M. Then $[Na^+] = $ **0.201.**

19. We have the relationship

$$K_{th} = K_{sp}\gamma_{Ag+}\gamma_{Ac-} = S^2\gamma_\pm^2$$

where S is the solubility. The resulting ionic strength μ is equal to S, so $\log \gamma_\pm = -0.51\sqrt{S}$. The final relationship is

$$\log(4 \times 10^{-3}) = 2\log S = 1.02\sqrt{S}$$

This can be solved by successive approximations. First

$$S_1 = (4 \times 10^{-3})^{1/2} = 0.0631$$

whence $\log \gamma_\pm = -0.51 \times 0.251$ and $\gamma_\pm = 0.75$.
As a second approximation,

$$K_{sp} = 4 \times 10^{-3}/(0.75)^2 = S_2^2 \quad \text{and} \quad S_2 = 0.084$$

$\text{Log } \gamma_\pm = -0.51 \times 0.289$ and $\gamma_\pm = 0.71$. S_3 then becomes **0.089.**

20. (a) K_{sp} is evidently $(0.0001)^2$ or 10^{-8}. The ionic strength of 0.1 m $NaNO_3$ is 0.1, so

$$\log \gamma_\pm = -0.51(0.1)^{1/2} = -0.161$$

and $\gamma_\pm$ for Ag^+A^- is 0.69. Then 10^{-8} (which must be close to the thermodynamic constant, in view of the low ionic strength of the saturated solution in water) equals $(S)^2(0.69)^2$, so $S = 10^{-4}/0.69 = $ **1.45×10^{-4}**.
(b) In 10^{-3} M HNO_3, the ionic strength of 10^{-3} leads to a $\gamma_\pm$ value of 0.96, so K_{sp} in this medium is 1.09×10^{-8}. Then $1.09 \times 10^{-8} = [Ag^+][A^-]$ or, since $[Ag^+] = S = 0.00013$, then $[A^-] = 8.4 \times 10^{-5}$. The acid-dissociation constant for HA is then $K = [H^+][A^-]/HA = (10^{-3})(8.4 \times 10^{-5})/(1.3 \times 10^{-4} - 8.4 \times 10^{-5}) = $ **1.83×10^{-4}**.

21. (a) The hydrolysis of Ox^{2-} is negligible at pH 7, so the solubility in water is simply **$10^{-4}M$.**

(b) There will be some hydrolysis, so the pH will change slightly. K_h is $K_w/K_a = 10^{-14}/10^{-5} = 10^{-9}$, and then $10^{-9} = [HOx^-][OH^-]/[Ox^{2-}]$ or $[OH^-]^2 = 10^{-8} \times 10^{-4} = 10^{-13}$. $[OH^-]$ is then 3.16×10^{-7} and $[H^+] = 3.16 \times 10^{-8}$ corresponding to a pH of **7.5**. (The correction to (a) is still negligible.)

(c) For $\mu = 0.1$, $\log \gamma_\pm = -0.51 \times 2 \times 2 \times \sqrt{0.1} = -0.645$, and $\gamma_\pm = 0.227$ for Ba^{2+}, Ox^{2-}. K_{sp} for this medium is then $10^{-8}/(0.227)^2$ and $S = 10^{-4}/0.227 = $ **4.42×10^{-4} M.**

(d) Since all the BaOx has dissolved, $[Ba^{2+}] = $ **0.1 M.** $[Ox^{2-}]$ at this point must be $10^{-8}/0.1 = $ **10^{-9} M.** The acidity is evidently high enough that most of the oxalate is present as H_2Ox and HOx^-, and, as a first guess, we assume that it is mostly as H_2Ox, so $[H_2Ox] = $ **0.1.** Then

$$K_1 K_2 = \frac{[H^+]^2[Ox^{2-}]}{[H_2Ox]}$$

so $[H^+] = (10^{-8} \times 0.1/10^{-9})^{1/2} = $ **1 M.** (On checking, it is indeed true that there is negligible HOx.) The total amount of HNO_3 needed is then 0.2 M for the H_2Ox and 1 M excess or **1.2 M** total. Then $[NO_3^-] = 1.2$ M and $[OH^-] = $ **10^{-14}.**

22. $AgIO_3$ is the less soluble of the two slightly soluble salts, so we assume that it precipitates first, until the IO_3^- concentration has dropped to a low value. We then have **0.005 mole** of $AgIO_3(s)$, and the rest of the silver ion precipitates as **0.0025 mole** of Ag_2CrO_4. This leaves **0.0975 M** CrO_4^{2-}, and therefore $[Ag^+] = (10^{-8}/0.0975)^{1/2} = $ **3.2×10^{-4}.** Finally $[IO_3^-] = 10^{-13}/3.2 \times 10^{-4} = $ **3.1×10^{-10} M** (and is indeed very low).

23. A rough calculation shows that BaOx is much less soluble than BaF_2, so a reasonable supposition is that all the oxalate is precipitated out. Then $[F^-] = 0.2$ and $[Ba^{2+}] = 10^{-6}/(0.2)^2 = 2.5 \times 10^{-5}$, so $[Ox^{2-}] = 10^{-7}/2.5 \times 10^{-5} = 0.004$.

As a second round of approximation, assume $[Ox^{2-}] = 0.002$, so $[F^-] = $ **0.196**, $[Ba^{2+}] = 10^{-6}/(0.196)^2 = $ **2.61×10^{-5}**, and $[Ox^{2-}] = 10^{-7}/2.61 \times 10^{-5} = $ **0.0038 M.** This is close enough.

24. (a) $4 \times 10^{-9} = [S][2S]^2$ or $S = $ **10^{-3}.** This is low enough that activity-coefficient effects are not very serious:

$$\log \gamma_\pm \cong -0.51 \times 2 \times 1(3 \times 10^{-3})^{1/2} = -0.056$$

and $\gamma_\pm = 0.88$ for Pb^{2+}, F^-. Using this approximation, the corrected K_{sp} becomes $4 \times 10^{-9}/(0.88)^3$ from which S is now **1.13×10^{-3}.**

(b) The ionic strength is now essentially 0.1, so $\log \gamma_{\pm} = -0.51 \times 2 \times 1\sqrt{0.1} = -0.322$ and $\gamma_{\pm} = 0.479$. K_{sp} is then $4 \times 10^{-9}/(0.479)^3$ and S is **3.64×10^{-6}**.

(c) Fluoride ion is now only 1% of the total fluoride; that is, $[F^-] = 0.02S$. Then $4 \times 10^{-9} = [S](0.02S)^2$ or $S = $ **0.0216**.

25. (a) $10^{-6} = [Na^+][V^-] = [S + 0.1][S]$ or $S = 10^{-5}$.

(b) At an ionic strength of essentially 0.1, $\log \gamma_{\pm} = -0.51 \times 1 \times 1\sqrt{0.1} = -0.16$ and $\gamma_{\pm} = 0.695$ for Na^+V^-. The K_{sp} for this medium is then $10^{-6}/(0.695)^2$ and $S = $ **1.44×10^{-3}**.

(c) In $0.001\ H^+$, $[V^-]/[HV] = 10^{-5}/0.001 = 0.01$, so only 1% of the total is present as V^-, or $[V^-] = 0.01\ S$. Then $10^{-6} = [S](0.01S)$ and $S = $ **$10^{-2}\ M$**.

26. Since there is excess Pb^{2+}, both SO_4^{2-} and IO_3^- should be largely removed from solution. The remaining $[Pb^{2+}]$ is then

$$0.02 - 0.01 - \frac{0.05}{2} = \mathbf{0.0075}$$

Then

$$[SO_4^{2-}] = \frac{10^{-8}}{0.0075} = \mathbf{1.33 \times 10^{-6}}$$

$$[IO_3^-]^2 = \frac{10^{-13}}{0.0075} = 1.33 \times 10^{-11}$$

or

$$[IO_3^-] = \mathbf{3.65 \times 10^{-6}}$$

There will be **0.01** mole of $PbSO_4$ and **0.0025** mole of $Pb(IO_3)_2$ per liter of solution and, of course, $[NO_3^-] = $ **0.04**.

27. (a) As a preliminary calculation, $[F^-]/[HF] = 0.02/0.5 = 0.04$, so about 96% of the fluoride will be converted to HF. Since 0.02 mole is present, this means that about 0.019 mole of H^+ ion is tied up, and, as a first approximation, $0.5 - 0.019$ or 0.48 remain. The ratio $[F^-]/[HF]$ is now $0.02/0.48$ or 0.0418 and $[F^-] = 0.02 \times 0.0418/1.041 = 8.0 \times 10^{-4}$. The K_{sp} is then

$$[Ba^{2+}][F^-]^2 = (0.01)(8 \times 10^{-4})^2 = \mathbf{6.4 \times 10^{-9}}$$

(b) $K_{th} = K_{sp}\gamma_{Ba^{2+}}\gamma_{F^-} = 10^{-8}\gamma_{\pm}^3 = \mathbf{2.16 \times 10^{-9}}$.

28. The thermodynamic solubility constant K_{th} is given by

$$K_{th} = a_{Ag}^2 \times a_{SO_4^{2-}} = (Ag^+)^2(SO_4^{2-})\gamma_{Ag}^2\gamma_{SO_4^{2-}}$$
$$= K_{sp}\gamma_\pm^3$$

K_{sp} is equal to $(0.06)^2(0.03) = \textbf{1.08} \times \textbf{10}^{-4}$, and $-\log \gamma_\pm = 0.51 \times 2 \times \mu^{1/2} = 0.51 \times 2[\frac{1}{2}(0.06 + 0.03 \times 4)]^{1/2} = 0.306$, so $\gamma_\pm = 0.495$ and $\gamma_\pm^3 = 0.121$. K_{th} is then $1.08 \times 10^{-4} \times 0.121 = \textbf{1.3} \times \textbf{10}^{-5}$.

29. (a) At the buffer point, $F_{HA} = F_{A^-}$ and $pH = pK$. Inspection of the diagram shows that this condition occurs at pH 10, so **$pK = 10$.**
  (b) The charge-balance equation is

$$[Na^+] + [H^+] = 0.01F_{A^-} + [OH^-]$$

At pH 10, $F_{A^-} = 0.5$, so $[Na^+] = 0.005$ and there is 50 percent neutralization. [This conclusion is fairly obvious from doing part (a).]
  (c) We now have a solution 0.01 M in NaA, and the charge-balance equation becomes $0.01 = 0.01F_{A^-} + [OH^-]$. Trial-and-error testing of various pH's yields pH 11 with $F_{A^-} = 0.9$ and $[OH^-] = 0.0001$.

30. (a) At pH 2, $F_{H_2Ox} = F_{HOx^-}$ so $pH = pK_1$ and **$pK_1 = 2$.** Similarly, from the crossing at pH 5, **$pK_2 = 5$.**
  (b) The charge-balance equation is

$$[Na^+] + [H^+] = [HOx^-] + 2[Ox^{2-}] + [OH^-]$$

$[H^+]$ and $[OH^-]$ will be negligible compared with the other terms, and the equation then becomes

$$[Na^+] = 0.1F_{HOx^-} + 0.2F_{Ox^{2-}}$$

and, reading off the graph at pH 5.5,

$$[Na^+] = 0.1 \times 0.3 + 0.2 \times 0.7 = \textbf{0.17}$$

which is the formality of NaOH required.

31. (a) This is a matter of summing F_{H_2f} and $F_{f^{2-}}$ at various pH's. A few trials establish the minimum at the crossing point, pH 3.5, at which point the value for this sum, and hence for $1 - F_{Hf^-}$, is 0.36, and F_{Hf^-} is then **0.64** at its maximum.
  (b) The pH will equal pK_1 if $F_{H_2f} = F_{Hf^-} = 1 - F_{H_2f} - F_{f^{2-}}$. A few trials establish this pH as very close to 3, so **$pK_1 = 3$.**

(c) The charge-balance equation is

$$[Na^+] + [H^+] = [Hf^-] + 2[f^{2-}] + OH^-]$$

Evidently, $[H^+]$ and $[OH^-]$ will be negligible compared with the other terms; since, at pH 5, $F_{Hf^-} = 0.1$ and $F_{f^{2-}} = 0.9$, we have $[Na^+] = 0.1 \times 0.1 + 0.2 \times 0.9$ or $[Na^+] = \mathbf{0.19}$, which is the formality of NaOH required.

32. The $F_{H_2CO_3}$ and $F_{HCO_3^-}$ curves cross at the value 0.5, so little CO_3^{2-} can be present at this pH, which is about 7. The $pK_1 = pH$ of crossing $= 7$. Similarly, $F_{HCO_3^-}$ drops to 0.5 at pH about 10, and since little H_2CO_3 is present, $F_{CO_3^{2-}}$ must also be 0.5, and this must be the pH equal to pK_2. $pK_2 = \mathbf{10.}$

H_2CO_3 and CO_3^{2-} will react to give $2HCO_3^-$, so the mixture is better given as $0.1\ f\ H_2CO_3 + 0.2\ f\ NaHCO_3$. This is in the buffer region, and to a good approximation then $[H^+] = K_1[H_2CO_3]/[HCO_3^-] = 10^{-5} \times 0.1/0.2 = \mathbf{5 \times 10^{-6}}$. Also $[Na^+] = \mathbf{0.2\ M}$, $[H_2CO_3] = \mathbf{0.1\ M}$, and $[HCO_3^-] = \mathbf{0.2\ M}$. Finally, $[CO_3^{2-}] = K_2[HCO_3^-]/[H^+] = 10^{-8} \times 0.2/5 \times 10^{-6} = \mathbf{4 \times 10^{-4}}$.

33. (a) $F_{C^-} = 1 - F_{HC}$, and the necessary points are easy to obtain.

(b) $pK = pH$ at the buffer point where $F_{HC} = F_{C^-} = 0.5$. This occurs at pH 6, so $\mathbf{pK = 6.}$

(c) Let $x = [H^+]$, then $10^{-6} = x^2/0.001 - x$. To a fair approximation, $x^2 = 10^{-9}$ and $x = \mathbf{3.16 \times 10^{-5}}$.

(d) The solution consists of 0.09 M HC and 0.01 M NaC, so $10^{-6} = [H^+](0.01)/(0.09)$ or $[H^+] = \mathbf{9 \times 10^{-6}}$.

34. (a) We know that $pK = pH$ at the buffer point, that is, the pH such that $F_{A^-} = F_{HA} = 0.5$ (and similarly for the second acid). $F_{A^-} = 0.5$ at $[H^+] = 0.02$, so $K_{HA} = \mathbf{0.02}$ and, similarly, $K_{HB} = \mathbf{0.01.}$

(b) The charge-balance equation is

$$[Na^+] + [H^+] = [A^-] + [B^-] \qquad (neglect\ [OH^-])$$

or

$$0.01 + [H^+] = 0.01F_{A^-} + 0.02F_{B^-}$$

This rearranges to

$$F_{A^-} + 2F_{B^-} = 1 + 100[H^+]$$

and a solution to this equation is found by successive approximations. Thus, a guess of $[H^+] = 0.01$, or $1 + 100[H^+] = 2$, gives $F_{A^-} + 2F_{B^-} \cong 0.7 + 2 \times 0.5 = 1.7$ (too small). A choice of $[H^+] = \mathbf{0.008}$ works fairly well. Thus $F_{A^-} + F_{B^-} \cong 0.75 + 2 \times 0.55 = 1.85$ as compared with $1 + 100[H^+] = 1.8$.

35. The charge-balance Eq. (14-13) becomes in the present case

$$Na^+ + H^+ = Hf^- + 2f^{2-} + HCO_3^- + 2CO_3^{2-} + OH^-$$

Since the pH is 7, $[H^+]$ and $[OH^-]$ can be neglected, and if C denotes the formality of added NaHf, the equation becomes

$$C + 0.02 = F_{Hf^-}C + 2F_{f^{2-}}C + 0.01F_{HCO_3^-} + 0.02F_{CO_3^{2-}}$$

From the figures cited, at pH 7, $F_{Hf^-} \cong 0$, $F_{CO_3^{2-}} \cong 0$, $F_{HCO_3^-} = 0.5$, and $F_{f^{2-}} \cong 1$, so the charge-balance equation reduces to

$$C + 0.02 = 2C + 0.005$$

or $C = \mathbf{0.015}$.

36. Substituting $K_1 = 4K_2$ into Eq. (14-11)

$$F_{HA^-} = \frac{1}{[H^+]^2/4K_2 + 1 + K_2/[H^+]}$$

$$= \frac{[H^+]}{[H^+]^2/4K_2 + [H^+] + K_2}$$

or differentiating and setting $dF_{HA^-}/[H^+] = 0$, we obtain

$$K_2 = \frac{[H^+]^2}{4K_2}$$

or $[H^+] = 2K_2$ when F_{HA^-} is at a maximum. On substituting this result back into the first equation, this maximum value of F_{HA^-} is easily found to be $\frac{1}{2}$.

37. Simple extension of Eqs. (14-10) and (14-12) gives

$$F_{H_3A} = \frac{1}{1 + K_1/[H^+] + K_1K_2/[H^+]^2 + K_1K_2K_3/[H^+]^3}$$

$$F_{A^{3-}} = \frac{1}{[H^+]^3/K_1K_2K_3 + [H^+]^2/K_2K_3 + [H^+]/K_3 + 1}$$

Examination of the above equations shows that

$$F_{H_3A}/[H^+]^3 = \frac{1}{[H^+]^3 + K_1[H^+]^2 + K_1K_2[H^+] + K_1K_2K_3}$$
$$= F_{A^{3-}}/K_1K_2K_3$$

or

$$K_2K_3 = \frac{F_{A^{3-}}[H^+]^3}{F_{H_3A}K_1} = \frac{0.10 \times 10^{-21}}{0.05 \times 10^{-6}} = 2 \times 10^{-15}$$

Then

$$F_{H_3A} = 0.05$$

$$= \frac{1}{1 + 10^{-6}/10^{-7} + 10^{-6}K_2/10^{-14} + 10^{-6} \times 2 \times 10^{-15}/10^{-21}}$$

or

$$1/0.05 = 20 + 1 + 10 + 10^8K_2 + 2 = 13 + 10^8K_2$$

and $K_2 = 7 \times 10^{-8}$. Then, $K_3 = 2 \times 10^{-15}/7 \times 10^{-8} = \mathbf{2.85 \times 10^{-8}}$.

38. (a) F_{HA} is 0.5 at pH 6, so $K_a = \mathbf{10^{-6}}$; F_{B^+} is likewise 0.5, and since the corresponding $[OH^-]$ is 10^{-8}, $K_b = \mathbf{10^{-8}}$.

(b) The charge-balance Eq. (14-13) is, for this case,

$$Na^+ + B^+ + H^+ = A^- + Cl^- + OH^-$$

First, it seems reasonable to assume (subject to later check) that $[H^+]$ and $[OH^-]$ will be negligible compared with the other terms. The equation then becomes

$$0.1 + 0.2F_{B^+} = 0.1F_{A^-} + 0.2$$

But $F_{A^-} = 1 - F_{HA} = 1 - F_{B^+}$, further simplification yields

$$0.3F_{B^+} = 0.2$$

or $F_{B^+} = \frac{2}{3}$. This will be true at **pH $\cong$ 5.5.**

39. (a) $F_{B^+} = 0.5$ at pH 8 or $[OH^-] = 10^{-6}$, so $pK_b = 6$.

(b) The charge-balance Eq. (14-13) becomes

$$B^+ + H^+ = Cl^- + OH^-$$

At pH 7, $[H^+]$ and $[OH^-]$ can be neglected, and $F_{B^+} = 0.9$; the equation then reduces to

$$0.1 \times 0.9 = Cl^-$$

or $[Cl^-] = 0.09$. The percent neutralization is then $0.09 \times 100/0.1 = \textbf{90\%}$.

15

ELECTROCHEMICAL CELLS

COMMENTS

You can expect to find the problems of this chapter relatively difficult, especially in terms of the times allowed. As you do them, it will become apparent that you are expected to have a fluent understanding of four or five basic types of situations.

First, and easiest, are the interrelations between $\mathscr{E}$, ΔG, ΔS, ΔH, and $d\mathscr{E}/dT$. Keep your signs straight, and remember that you can report energies in joules and do not have to convert them to calories.

A second group of problems [for example, number 19(a)] involves the combining of half-cell potentials to obtain an $\mathscr{E}^0$ for a redox reaction, and, from it, the value of the equilibrium constant. As part of the problem, you may have to obtain a desired half-cell emf by combining two others; remember the special procedure that is involved here.

There are a number of problems that require the use of the Nernst equation. Remember that the cell reaction and corresponding Nernst expression must be written in terms of the major species present; thus strong electrolytes must be written as the separate ions. Several problems require the relating of a simple metal–metal ion half-cell potential to one in which an insoluble salt of that ion appears, for example, Ag/Ag^+ versus $Ag/AgCl$. The answer to Problem 1 details two alternative approaches with which you should be familiar. A useful principle emerges—the standard potential of an $Ag/AgCl$ electrode must be the same as the potential of an Ag/Ag^+ electrode where the activity of Ag^+ ion is that which is in equilibrium with $AgCl$ in unit activity chloride ion solution.

Electrochemical cells provide a most important method of determining activity coefficients since the "Q" term (see below) of the Nernst equation contains the various activities. Often it will be convenient to write $\log Q = \log Q_C + \log Q_\gamma$, so as to separate concentration and activity-coefficient terms. However, unless explicitly required to do otherwise, you may neglect activity-coefficient effects.

Many of the above aspects will appear in connection with concentration cells (with or without transference). Here, $\mathscr{E}^0$ is zero, and $\mathscr{E}$ depends entirely on the $\log Q$ term of the Nernst equation.

EQUATIONS AND CONCEPTS

Cell Diagrams and Conventions

Consider a typical cell diagram

$$\overset{\overset{\displaystyle e^-}{\uparrow}}{\text{Hg}/\text{Hg}_2\text{Cl}_2(\text{s})/\text{NaCl}(0.1 \; m)} \underset{\underset{\displaystyle \text{anode} \quad +' \underset{\leftarrow}{\rightarrow} \; -}{}}{\diagup} \overset{\overset{\displaystyle e}{\downarrow}}{\diagup \text{NaCl}(0.001 \; m)/\text{Hg}_2\text{Cl}_2(\text{s})/\text{Hg}} \qquad \text{cathode} \qquad (15\text{-}1)$$

The sequence of phases is that through which current flows. Distinct phases are separated by a *full diagonal line*, whereas a junction, usually liquid, between miscible phases is shown by a *dashed diagonal*. Hg_2Cl_2 is given as $\text{Hg}_2\text{Cl}_2(\text{s})$ to make it clear that excess solid is present; the NaCl solutions are to be presumed to be saturated with respect to Hg_2Cl_2. In general, then, all phases not separated by a dashed line are assumed to be mutually saturated.

The convention as to current flow is that shown. A positive emf means that current flows spontaneously in the direction shown. Further, if the cell be regarded as made up of two half-cells, the convention means that the left-hand half-cell is functioning as the anode or electrode at which oxidation occurs, in reporting the sign of the emf.

Cell Reactions and the Nernst Equation

We obtain the cell reaction by writing the two half-cell processes (with the left-hand one written as an anode). If a liquid junction is present, t^+ and t^- equivalents of positive and negative ions, respectively, move from one concentration region to the other, per faraday (transference problems per se are taken up in Chapter 8). The directions of motion of positive and negative ions are those shown in the cell diagram above.

On adding up all processes, electrode plus transference, an equation of the type

$$aA + bB + \cdots = mM + nN + \cdots \qquad (15\text{-}2)$$

results, and the emf of the cell is then given by the Nernst equation:

$$\mathcal{E} = \mathcal{E}^0 - \frac{RT}{n\mathcal{F}} \ln Q \quad \text{where } Q = \frac{a_M^m a_N^n \cdots}{a_A^a a_B^b \cdots} \tag{15-3}$$

and a denotes activity and n the number of faradays. At 25°C, the above equation becomes

$$\mathcal{E} = \mathcal{E}^0 - \frac{0.059}{n} \log Q \tag{15-4}$$

It is essential that species that are in fact present as ions be so written, and it is wise to write after each what its concentration is.

Combining Half-Cells

Half-cell potentials are based on the convention that $\mathcal{E}^0$ for the half-cell $\frac{1}{2}H_2 = H^+ + e^-$ is zero, and it is perfectly equivalent to think of a half-cell as actually a complete cell whose other electrode is a standard hydrogen electrode. Half-cell potentials are reported here as oxidation potentials.

If two half-cell reactions can be combined to give a whole-cell reaction, that is, with cancellation of electrons, the $\mathcal{E}^0$ values are similarly combined to give the $\mathcal{E}^0$ of the cell.

If, on the other hand, two half-cell reactions are combined to give another half-cell reaction, the $\mathcal{E}^0$ values are similarly combined, but with each weighted according to the number of faradays of electricity involved. If in doubt, write the ΔG values for each process and combine them.

Thermodynamics of Cells

$\Delta G = -n\mathcal{F}\mathcal{E}$. $\mathcal{E}$ is simply the free energy in joules/coulomb for the cell process, that is, volts, with the convention that a positive sign indicates that the process as written is spontaneous.

Other relationships follow:

$$\Delta S = n\mathcal{F} \, d\mathcal{E}/dT \tag{15-5}$$

$$\Delta H = -n\mathcal{F}(\mathcal{E} - T \, d\mathcal{E}/dT) \quad \text{(Gibbs–Helmholtz equation)} \tag{15-6}$$

$$\mathcal{E}^0 = \frac{RT}{n\mathcal{F}} \ln K \quad \text{or at 25°C} \quad \mathcal{E}^0 = \frac{0.059}{n} \log K \tag{15-7}$$

where K is the equilibrium constant for the cell reaction.

Activity Coefficients

When dealing with electrolytes, it is convenient to consider that each ionic species not only has a concentration m_+ or m_-, but an activity a_+ or a_- and an activity coefficient γ_+ or γ_- as well. Thus $a_+ = m_+\gamma_+$ and $a_- = \gamma_-m_-$. However, solutions are always essentially electrically neutral, so that negative and positive ions must be produced or consumed in equivalent amounts. As a consequence, it is necessary to introduce a second set of quantities, $a_\pm$, $\gamma_\pm$, and $m_\pm$, called the mean activity, mean activity coefficient, and mean molality, respectively. We have the following relationships:

$$a_\pm = \gamma_\pm m_\pm \tag{15-8}$$
$$a_\pm = (a_+^{\nu_+} a_-^{\nu_-})^{1/\nu} \tag{15-9}$$
$$\gamma_\pm = (\gamma_+^{\nu_+} \gamma_-^{\nu_-})^{1/\nu} \tag{15-10}$$

and

$$m_\pm = (m_+^{\nu_+} m_-^{\nu_-})^{1/\nu} \tag{15-11}$$

Here ν_+ and ν_- denote the number of positive and negative ions, respectively, per formula, and $\nu = \nu_+ + \nu_-$.

Junction Potential

In the case of a cell with a liquid junction, for example,

$$\text{Pt/H}_2(1 \text{ atm})/\text{HCl}(m_1) \Big/ \text{HCl}(m_2)/\text{H}_2(2 \text{ atm})/\text{Pt} \tag{15-12}$$

it is sometimes convenient to write $\mathscr{E} = \mathscr{E}_e + \mathscr{E}_j$, where $\mathscr{E}_j$ is the junction potential. $\mathscr{E}_e$ denotes the potential that would be observed if there were no free-energy change resulting from transport of ions across the liquid junction, so that $\mathscr{E}_e$ can be thought of as the potential associated with the electrode reactions and $\mathscr{E}_j$ as that associated with the liquid junction.

PROBLEMS

1. (30 min) $\mathscr{E}_{298°K}^0$ is -0.627 V for the cell

$$\text{Ag/Ag}_2\text{SO}_4(\text{s})/\text{H}_2\text{SO}_4(m)/\text{H}_2(1 \text{ atm})/\text{Pt}$$

(a) Write the cell reaction (and each electrode reaction).
(b) Calculate $\mathscr{E}_{298°K}$ if $m = 0.1$ (neglect activity coefficients).

(c) Repeat the calculation in (b), but take the mean activity coefficient of 0.1 m H_2SO_4 to be 0.70. It is sufficient to set up an equation in which all quantities except the desired emf are replaced by appropriate numbers.

(d) Calculate the solubility product for Ag_2SO_4.

2. (30 min) The following information is given:

$$Cr + SO_4^{2-} = CrSO_4(s) + 2e^- \qquad \mathscr{E}^0_{298°K} = 0.4 \text{ V}$$
$$Cr = Cr^{3+} + 3e^- \qquad \mathscr{E}^0_{298°K} = 0.5 \text{ V}$$

K_{sp} for $CrSO_4$ is 10^{-6} at 25°C.

(a) Write the cell reaction for the cell

$$Cr/CrSO_4(s)/0.001 \ m \ H_2SO_4/H_2(1 \text{ atm})/Pt$$

(b) Calculate the emf for this cell at 25°C, neglecting activity coefficient corrections.

(c) Calculate $\mathscr{E}^0$ for this cell at 50°C if ΔH^0 is $-5,000$ J/faraday.

(d) Calculate the emf for this cell at 25°C, using activity coefficients from the Debye–Hückel limiting law.

3. (21 min) The following data are given:

$$Ag/AgBr(s)/HBr(0.1 \ m)/H_2(0.01 \text{ atm})/Pt$$
$$\mathscr{E}_{298°K} = -0.165 \text{ V} \qquad \Delta H^0 = -50,000 \text{ J}$$

K_{sp} for $AgBr$ is 1×10^{-12} and $\mathscr{E}^0_{298°K}$ for Ag/Ag^+ is -0.8 V. Using only the above data (plus general constants), (a) write the cell reaction, (b) calculate $\mathscr{E}^0_{298°K}$ for the cell [answer must be consistent with (a)], (c) calculate q, the heat absorbed per faraday when the cell operates reversibly, and (d) calculate $\gamma_\pm$ for 0.1 m HBr. [On part (d) it is sufficient to set up an equation in which all quantities besides $\gamma_\pm$ have been replaced by the appropriate numbers.]

4. (24 min) (final examination question) The potential of the cell

$$Cd/CdSO_4(s)/H_2SO_4(0.02 \ m)/H_2(1 \text{ atm})/Pt$$

is 0.38 V at 35°C. $\mathscr{E}^0$ at 35°C is 0.45 V for the half-cell Cd/Cd^{2+}. Calculate the solubility product for cadmium sulfate at 35°C. State the assumptions and approximations you need to use in obtaining a numerical answer from the above data.

5. (15 min) The K_{sp} for TlBr is 10^{-4} at 25°C, $\mathscr{E}^0$ for Tl/Tl$^+$ is 0.34 V, and $d\mathscr{E}^0/dT = -0.003$ V/deg.

(a) Calculate the emf at 25°C for the cell

Tl/TlBr/HBr(unit activity)/H$_2$(1 atm)/Pt

(b) Calculate ΔH for the cell reaction corresponding to

Tl/Tl$^+$(unit activity), H$^+$(unit activity)/H$_2$(1 atm)/Pt

6. (18 min) Given the cell

Cd/Cd(OH)$_2$(s)/NaOH(0.01 m)/H$_2$(1 atm)/Pt $\mathscr{E}_{298°K} = 0.000$ V

Cd/Cd^{2+} $\mathscr{E}^0_{298°K} = 0.4$ V

(a) Write the anode, cathode, and net cell reactions.
(b) Calculate the K_{sp} for Cd(OH)$_2$.
(c) Calculate ΔH for the cell reaction if $d\mathscr{E}/dT = 0.002$ V/deg.

7. (24 min) (final examination question) Given the cell

Au/AuI(s)/HI(m)/H$_2$(1 atm)/Pt   at   25°C

(a) Write the anode, cathode, and net cell reactions.
(b) $\mathscr{E}$ is -0.97 if m is 10^{-4} m and is -0.41 if m is 3.0 m. Calculate the activity coefficient for 3 m HI.
(c) If now informed that $\mathscr{E}^0$ for Au = Au$^+$ + e^- is -1.68, calculate, using this additional datum, the solubility product for AuI.

8. (24 min) (final examination question) Given the following data:

Tl/Tl$^+$ $\mathscr{E}^0_{298°K} = 0.34$ V

Cd/Cd^{2+} $\mathscr{E}^0_{298°K} = 0.40$ V

solubility product for TlCl = 1.6×10^{-3} at 25°C

Write the anode, cathode, and net cell reactions for the cell

Tl/TlCl(s)/CdCl$_2$(0.01 m)/Cd

Calculate $\mathscr{E}^0$ and $\mathscr{E}$ for this cell at 25°C.

9. (10.5 min) Given that $\mathscr{E}^0$ is 0.152 for Ag + I$^-$ = AgI + e^- at 25°C, and $\mathscr{E}^0$ for Ag = Ag$^+$ + e^- is -0.800 at 25°C, calculate K_{sp} for AgI.

10. (15 min) (final examination question) The ion M^+ of a certain metal M forms soluble nitrates but only a slightly soluble chloride. The cell

$$M/MNO_3(0.1\ m),\ HNO_3(0.2\ m)/H_2(1\ atm)/Pt$$

has an emf of -0.5 V at 25°C. On addition of sufficient solid KCl to make the solution of the cell 0.25 m in K^+ ion (with precipitation of MCl), the emf changes to -0.1 V at 25°C. Calculate the K_{sp} of MCl. (Neglect activity-coefficient effects.)

---

11. (30 min) (a) A solution is 0.01 m in $SnCl_2$ and 0.1 m in KCl. If $\gamma_\pm$ for $SnCl_2$ in *this solution* is 0.40, calculate $m_\pm$ and $a_\pm$ for $SnCl_2$.

(b) The above solution actually contains some $PbCl_2$ in addition to the $SnCl_2$ and KCl, and the lead content is determined by electrodepositing Pb, using a platinum cathode and a silver–silver chloride anode. Electrolysis is stopped when tin just starts to deposit along with the lead. Write the cell reaction that takes place during the deposition of lead and calculate the final Pb^{2+} concentration, neglecting activity-coefficient corrections.

(c) As a separate situation, and again neglecting activity-coefficient corrections, what should be the equilibrium Pb^{2+} and Sn^{2+} concentrations on adding excess solid lead to a solution initially 0.01 m in Sn^{2+}? Show your work.

(d) Would the Pb^{2+} concentration in (b) be greater, smaller, or the same, if the activity coefficient $\gamma_\pm$ of an MX_2-type salt is 0.4 for the solution in question? ($\mathscr{E}^0$ values at 25°C are 0.126 for Pb/Pb^{2+} and 0.140 for Sn/Sn^{2+}; assume 25°C throughout.)

12. (12 min) For the cell, $Cu/CuCl_2(m)/AgCl/Ag$, $\mathscr{E}_{298°K}$ is 0.191 and -0.074 for $m = 10^{-4}$ and $m = 0.2$, respectively. Write the electrode and net cell reactions and calculate the mean activity coefficient for 0.2 $m\ CuCl_2$.

13. (24 min) (final examination question) The emf at 25°C is -0.335 V for the cell

$$Pt/H_2(1\ atm)/HCl(0.1\ m)/AgCl/Ag \cdots Ag/AgCl/$$
$$HCl(10^{-4}\ m)/H_2(1\ atm)/Pt$$

Calculate (a) ΔG for the cell reaction at 25°C (write down the cell reaction), (b) the mean activity coefficient of 0.1 m HCl, and (c) ΔG for the cell reaction at 0°C, assuming ΔH to be zero.

14. (12 min) Given the cell

$$Ag/AgCl(s)/KCl(0.01\ m)\ /KCl(0.1\ m)/AgCl(s)/Ag$$

(where $/$ indicates a liquid junction), (a) write the cell reaction and (b) calculate the emf of the cell, assuming that t^- is 0.60 and that the mean activity coefficients are 0.90 and 0.80 for 0.01 and 0.1 m KCl, respectively.

15. (18 min) (final examination question) Given that $\mathscr{E}$ at 25°C is 0.5 V for the cell, $Ag/AgCl/MCl(m)/M(metal)$, when $m = 0.1$ (MCl is a strong electrolyte), calculate (a) $\mathscr{E}^0$ for this cell at 25°C, assuming that the Debye–Hückel law for activity coefficients applies, and (b) $\mathscr{E}$ for the cell (with $m = 0.1$) at 35°C if ΔH for the cell reaction is 2×10^5 J.

16. (15 min) (final examination question) The emf of the cell in Problem 10, -0.5 V, was computed neglecting activity-coefficient effects. Using the Debye–Hückel limiting law, calculate the mean activity coefficient for MNO_3 in the cell solution, and the "correct" emf of the cell, that is, corrected as necessary for activity-coefficient effects.

---

17. (9 min) Data are given in Table 15-1. Using only these data (plus general constants) calculate K for the reaction

$$2Fe^{2+} + Au^{3+} = 2Fe^{3+} + Au^+$$

TABLE 15-1

Half-cell	$\mathscr{E}^0_{298°K}$
$Au = Au^+ + e^-$	-1.68
$Au = Au^{3+} + 3e^-$	-1.50
$Fe^{2+} = Fe^{3+} + e^-$	-0.77

18. (24 min) (final examination question) Given the cell

$$Ag/Ag_2SO_4(0.02\ m)\ /salt\ bridge\ /CdSO_4(0.016\ m)/Cd$$

and the half-cell potentials at 25°C: Ag/Ag^+, -0.8 V; Cd/Cd^{2+}, 0.4 V.
(a) Write the anode, cathode, and net cell reactions.
(b) Calculate $\mathscr{E}$ and $\mathscr{E}^0$ for the cell, at 25°C.
(c) Calculate the equilibrium constant at 25°C for the reaction

$$2Ag^+ + Cd = 2Ag + Cd^{2+}$$

(d) If $\mathscr{E}^0$ does not change with temperature, explain whether $\mathscr{E}$ for the above cell would be greater, smaller, or unchanged at 35°C.

19. (30 min) (final examination question) (a) Excess solid silver is placed in a solution of ferric iron which is originally 0.1 M in Fe^{3+}, and contains no ferrous iron or silver ions. Calculate the equilibrium constant for the reaction $Ag + Fe^{3+} = Ag^+ + Fe^{2+}$ and the equilibrium constant of Ag^+.

(b) Calculate the voltage of the cell obtained by immersing a silver electrode in the equilibrium solution of (a) and connecting it to the electrode Fe^{2+}, Fe^{3+}/Pt. The cell is thus

$$\text{Ag} \Big/ \begin{matrix} [Ag^+, Fe^{2+}, Fe^{3+} \\ [\text{equil mixt of part (a)}] \end{matrix} \Big/ \begin{matrix} \text{salt} \\ \text{bridge} \end{matrix} \Big/ \begin{matrix} Fe^{2+}(a = 1) \\ Fe^{3+}(a = 1)/\text{Pt} \end{matrix}$$

Note: $\mathscr{E}^0$ values at 25°C are (Fe^{2+}, Fe^{3+}), -0.771 V, Ag/Ag^+, -0.799 V. Use 25°C in working the problem.

20. (16.5 min) Given that

Cell	$\mathscr{E}^0_{298°K}$V	$d\mathscr{E}^0/dT$
$Cu = Cu^+ = e^-$	-0.52	-0.002
$2NH_3 + Cu = Cu(NH_3)_2^+ + e^-$	0.11	-0.003
$Cu = Cu^{2+} + 2e^-$	-0.35	-0.0035

(a) Calculate K for $Cu + Cu^{2+} = 2Cu^+$, and the equilibrium concentration of Cu^+ ions when excess copper is added to 0.01-m Cu^{2+} solution.

(b) Calculate ΔG^0, ΔH^0, and ΔS^0 for the reaction $2NH_3 + Cu^+ = Cu(NH_3)_2^+$ at 25°C.

21. (15 min) (final examination question) Using only the data of Table 15-2 (see problem 37), calculate the equilibrium constant for the reaction

$$3Pb + 2Fe^{3+} = 3Pb^{2+} + 2Fe$$

22. (15 min) You are given the cell

$$Pt/H_2(1 \text{ atm})/HCl(0.01 \ m) \Big/ HCl(0.001 \ m)/H_2(0.01 \text{ atm})/Pt$$

(a) Omitting transference effects and considering just the electrode reactions, write these and calculate the corresponding emf.

(b) Write the changes that occur because of transference, and calculate the junction potential assuming t^+ to be 0.80. (The temperature is 25°C.)

23. (20 min) (final examination question) Calculate the emf at 25°C for the cell

$$Pt/H_2(1\ atm)/HCl(0.1\ m) \Big/ \begin{matrix} \text{membranes through} \\ \text{which only } H^+ \text{ ions} \\ \text{can pass} \end{matrix} \Big/ HCl(0.01\ m)/H_2(1\ atm)/Pt$$

(Write also the anode and cathode reactions and the gains or losses by transference.)

24. (12 min) Given the concentration cell

$$Hg/Hg_2Cl_2(s)/HCl(0.01\ m) \Big/ HCl(0.1\ m)/Hg_2Cl_2(s)/Hg$$

where the $\Big/$ sign denotes a membrane through which only H^+ ions can pass.

(a) Write the electrode reactions, the changes due to transference, and the net cell reaction.
(b) Calculate the emf of the cell at 25°C.

25. (16.5 min) $\mathscr{E}_{298°K}$ is 1.2 V for the cell

(A) $Na(Hg)/HCl(1\ M),\ NaCl(1\ M)/H_2(1\ atm)/Pt$

where Na(Hg) denotes a sodium amalgam of a certain specific composition. A second cell, cell (B), is then made up, using the same composition amalgam

(B) $Na(Hg)/NaOH(0.01\ M)/H_2(1\ atm)/Pt$

(a) Write the cell reactions for cells (A) and (B).
(b) Calculate the value of $\mathscr{E}$ at 25°C for cell (B).
(c) For cell (A), $d\mathscr{E}/dT$ is 0.002 V/deg. Calculate ΔH for the cell reaction. (Neglect activity-coefficient effects.)

26. (13.5 min) Calculate the emf at 25°C for the cell with transference

$$Ag/AgCl(s)/NaCl(0.1\ M) \Big/ NaCl(0.01\ M)/AgCl/Ag \qquad (t^+ = 0.4)$$

Write the electrode reactions and the changes resulting from transference, and then the net cell reactions.

27. (13.5 min) A student in the physical chemistry laboratory is told to make up a concentration cell using HCl and either Ag/AgCl *or* hydrogen electrodes. He is somewhat confused, and instead assembles the cell

$$Ag/AgCl/HCl(0.01\ m)\ /\ HCl(0.001\ m)/H_2(1\ atm)/Pt$$

However, given that $\mathscr{E}^0_{298\,^\circ K}$ is -0.22 V for the Ag/AgCl half-cell, he can still get the transference number for H^+ ion in HCl from his measured emf of -0.48 V. Help him by (a) writing the electrode reactions and the gain and loss by transference for the above cell and (b) calculating t^+. (Neglect activity coefficients and the variation of transference number with concentration.)

28. (30 min) Given the following cell

$$Hg/Hg_2Cl_2(s)/NaCl(0.1\ m)\ /\ NaCl(0.001\ m)/Hg_2Cl_2(s)/Hg$$

(where $/$ denotes a liquid junction).

(a) Write the anode and cathode reactions, and the net reaction for the cell.
(b) Write the equation for the emf of this cell at 25°C.
(c) Calculate the emf of this cell at 25°C, if t^+ is 0.40 for NaCl solutions, and the mean activity coefficient is 0.60 for 0.1 m NaCl (and unity for 0.001 m NaCl).
(d) Calculate what the emf of this cell would be if there were no liquid junction potential.

29. (14 min) (final examination question) The emf has been determined at 25°C for the two following cells:

(a) $Pt/H_2(1\,atm)/HCl(m_1)/AgCl/Ag \cdots Ag/AgCl/HCl(m_2)/H_2(1\ atm)/Pt$

(b) $Pt/H_2(1\ atm)/HCl(m_1)\ /\ HCl(m_2)/H_2(1\ atm)/Pt$

For each cell write the electrode reactions and the over-all cell reaction. Give the equation for each cell, whereby its emf can be calculated. Show how it would be possible to obtain the transference number of H^+ from the emf of the two cells. State briefly why the value of t^+ so calculated would be somewhat approximate.

30. (30 min) Given the cell

$$Cl_2(1\ atm)/NaCl(m_1)\ /\ NaCl(m_2)/Cl_2(1\ atm)$$

(for which the gas electrodes are reversible): (a) neglecting the liquid junction effects, write the net cell reaction per faraday, give the expression for the emf at 25°C, and calculate its value if $m_1 = 0.1$ and $m_2 = 0.01$.

(b) Including liquid-junction effects, write equations for all processes taking place per faraday, and the net cell reaction. Derive the expression for the emf of the cell, assuming that concentrations may be used instead of activities and that transference numbers are independent of concentration. If t^+ is 0.39, calculate the emf for the cell with transference $\mathscr{E}_t$, and the junction potential $\mathscr{E}_j$.

31. (15 min) (final examination question) Given the concentration cell

$$\text{Pt/H}_2(1 \text{ atm})/0.1 \ m \ \text{H}_2\text{SO}_4 \ / \ 0.2 \ m \ \text{H}_2\text{SO}_4/\text{H}_2(1 \text{ atm})/\text{Pt}$$

Write the electrode reactions, the changes due to transference, and the net cell reaction. Calculate the emf at 25°C, neglecting activity-coefficient effects.

32. (15 min) Consider the cell

$$\text{Ag/AgCl/NaCl(0.1 } m) \underset{A}{\overset{A}{\Big/}} \ \text{NaCl(0.2 } m) \underset{B}{\overset{B}{\Big/}} \ \text{NaCl(0.3 } m)/\text{AgCl/Ag}$$

where $A--A$ and $B--B$ denote semipermeable membranes. You have available either membranes that pass only cations or ones that pass only anions, and may select any combination for $A--A$ and $B--B$. With a proper combination, the net cell reaction will give a depletion of NaCl from the middle compartment so that a cell of this type can be used for desalination. (a) State the proper choice of membrane for $A--A$ and $B--B$. (b) Write the net cell reaction, and calculate the emf of the cell at 25°C (neglecting activity-coefficient effects).

---

33. (20 min) (final examination question) According to the Nernst equation, a piece of metal placed in water should show an infinite potential for the couple, metal/metal ion, if the water contained no metal ions at all. Actually, enough metal may be supposed to dissolve, by reducing water, for example,

$$\text{Pb} + \text{H}_2\text{O} = \text{Pb}^{2+} + \text{H}_2 + 2\text{OH}^-$$

until the metal ion concentration builds up to the point that $\mathscr{E}$ for the process becomes zero.

As a specific problem, calculate the equilibrium concentration of Pb^{2+} ion that should develop if lead were immersed in water. Assume the pH to be held at 6 and neglect any question of hydrolysis of plumbous ion. Use 25°C.

The hydrogen produced will remain in solution so long as its concentration does not exceed $10^{-3} \ M$, the solubility under 1 atm pressure. For concentrations less than or equal to $10^{-3} \ M$ the hydrogen activity will then be proportional to its concentration; that is, $a_{\text{H}_2} = k[\text{H}_2]$.

34. (19 min) (final examination question) Given the cell

$$\text{Pt/H}_2(1\text{ atm})\left/\begin{array}{c}\text{HNO}_3(m)\\\text{Hg}_2(\text{NO}_3)_2(m_1)\end{array}\right/\text{Hg}\left/\begin{array}{c}\text{HNO}_3(m)\\\text{Hg}_2(\text{NO}_3)_2(m_2)\end{array}\right/\text{H}_2(1\text{ atm})/\text{Pt}$$

(a) Write the cell reaction. What is ΔG^0?

(b) Calculate the emf at 25°C.

(c) Calculate the emf at 25°C, assuming that mercurous ion actually exists as Hg^+ rather than as Hg_2^{2+}. ($m = 0.1$; m_1 and m_2 are formula weights of $\text{Hg}_2(\text{NO}_3)_2$ per 1,000 g water and are 0.001 and 0.01, respectively.)

35. (18 min) Given the cell

$$\text{Pt/H}_2(1\text{ atm})/\text{NaOH}(m)/\text{HgO}(s)/\text{Hg} \qquad \mathscr{E}_{298°K} = 0.924 \text{ V}$$

(a) Write the electrode and net cell reactions.

(b) Calculate $\mathscr{E}^0_{298°K}$.

(c) Calculate $\mathscr{E}_{308°K}$ if ΔH for the cell reaction is -35 kcal.

(d) Explain briefly what would happen to the emf if small portions of nitric acid were added to partially neutralize the NaOH. (Data: Hg/Hg^{2+}, $\mathscr{E}^0_{298°K} = -0.854 \text{ V}$; 1 J = 0.239 cal.)

36. (19.5 min) Approximately what must be the concentration of Ag^+ in a solution that is also 0.1 m in Cu^{2+} in order that the two metals will plate out together in an electrodeposition? Copper and silver are assumed not to be soluble in each other, the temperature is 25°C, the activity coefficients of the Ag^+ and Cu^{2+} ions may be assumed to be the same, and equal to 0.80, and $\mathscr{E}^0$ values as given:

$$\text{Cu/Cu}^{2+} \qquad -0.345 \text{ V}$$
$$\text{Ag/Ag}^+ \qquad -0.800 \text{ V}$$

37. (18 min) Give cell diagrams and the corresponding cell reactions and Nernst equations such that, by means of the measured emf and the $\mathscr{E}^0$ values given in Table 15-2, one could determine (a) the dissociation constant of water, (b) the dissociation constant for the Ag(CN)_2^- complex into Ag^+ and 2CN^- ions. Do not use cells with transference or with salt bridges. Given that the two dissociation constants are 10^{-14} and 4×10^{-19}, respectively, insert reasonable concentrations, and calculate the expected emf for one of the two cells. Use only the half-cells listed.

38. (15 min) Given the cell $\text{Cu/0.02 } m$ Cu(II) in 0.5 m NH_3/calomel reference electrode ($\mathscr{E}$ is -0.22 V as a cathode). The $\mathscr{E}_{298°K}$ is -0.26 V for this cell, and $\mathscr{E}^0_{298°K}$ for Cu/Cu^{2+} is -0.34 V. Using only these data, calculate

TABLE 15-2*

Half-reaction	$\mathscr{E}^0_{298°K}$
$Fe^{2+} = Fe^{3+} + e^-$	-0.771
$Fe = Fe^{2+} + 2e^-$	0.440
$Ag + Cl^- = AgCl + e^-$	-0.222
$Hg + Cl^- = \frac{1}{2}Hg_2Cl_2 + e^-$	-0.280
(normal calomel half-cell)	
$Pb = Pb^{2+} + 2e^-$	0.126
$2Cl^- = Cl_2 + 2e^-$	-1.358
$2OH^- + 2Ag = Ag_2O + H_2O + 2e^-$	-0.344
$H_2 = 2H^+ + 2e^-$	0.000

* K_a for HCN is 1×10^{-10}.

(a) K for the equilibrium: $Cu^{2+} + 4NH_3 = Cu(NH_3)_4^{2+}$ (assume that no other copper ammonia complexes are present).

(b) $\mathscr{E}^0$ for the half-cell: $Cu/Cu(NH_3)_4^{2+}$

39. (17 min) Given the cell

$$Sb/Sb_2O_3 \left/ \begin{array}{c} KOH\,(10^{-3}) \\ KCl\quad(m) \end{array} \right/ H_2(1\ atm)/Pt$$

where $\mathscr{E}^0$ is 0.66 V at 25°C for $2Sb + 6OH^- = Sb_2O_3 + 3H_2O + 6e^-$.

(a) Write the anode and cathode and net cell reactions.

(b) Calculate the emf of the cell at 25°C, neglecting activity-coefficient effects.

(c) If the concentration of the KCl were increased from 0.01 m to 0.1 m, would the answer in (b) be increased, decreased, or not be changed?

40. (8 min) For the cell $Ag/Ag_2SO_4/H_2SO_4\,(0.1\ m)/H_2(1\ atm)/Pt$, $\mathscr{E}$ and $\mathscr{E}^0$ are -0.7 and -0.63 V at 25°C, respectively. Calculate the difference $(\mathscr{E} - \mathscr{E}^0)$ at 35°C (neglect any problems with nonideality).

ANSWERS

1. (a) The cell reaction is

Anode: $2Ag + SO_4^{2-}\,(m) = Ag_2SO_4 + 2e^-$

Cathode: $2H^+(2m) + 2e^- = H_2(1\ atm)$

Net reaction: $2Ag + 2H^+(2m) + SO_4^{2-}(m) = Ag_2SO_4 + H_2(1\ atm)$

(b)   The corresponding Nernst equation is then

$$\mathscr{E} = \mathscr{E}^0 + (0.059/2)\log([\text{H}^+]^2[\text{SO}_4^{2-}]) \qquad (\text{at } 25°\text{C})$$

so

$$\mathscr{E} = -0.627 + 0.0295\log[(0.2)^2(0.1)] = -0.627 - 0.0705$$

$$= -\mathbf{0.698}\ \mathbf{V}$$

(c) In the complete expression, each concentration is multiplied by its activity coefficient, or, separating out the activity coefficients in a separate term,

$$\mathscr{E} = -0.698 + 0.0295\log[(\gamma_{\text{H}^+})^2(\gamma_{\text{SO}_4^{2-}})]$$
$$\mathscr{E} = -0.698 + 0.0295\log(\gamma_\pm)^3$$
$$\mathscr{E} = -0.698 + 0.0295 \times 3\log 0.7$$
$$\mathscr{E} = -\mathbf{0.833}\ \mathbf{V}$$

(d) As a convenient approach, we rewrite the cell reaction as

$$2\text{Ag} + 2\text{H}^+(2m) = 2\text{Ag}^+(\text{in } m\ \text{H}_2\text{SO}_4) + \text{H}_2(1\ \text{atm})$$

Then

$$\mathscr{E} = \mathscr{E}^0_{(\text{Ag}=\text{Ag}^+ + e^-)} - 0.0295\log[(\text{Ag}^+ \text{ in } m\ \text{H}_2\text{SO}_4)^2/[\text{H}^+]^2]$$

or

$$\mathscr{E} = -0.799 - 0.0295\log[(K_{sp})/[\text{H}^+]^2[\text{SO}_4^{2-}]]$$

since $K_{sp} = [\text{Ag}^+]^2[\text{SO}_4^{2-}]$. We are told that when $[\text{H}^+]^2[\text{SO}_4^{2-}] = 1$, $\mathscr{E} = -0.627$, so $-0.627 = -0.799 - 0.0295\log K_{sp}$ or $\log K_{sp} = -5.88$ and $K_{sp} = \mathbf{1.3} \times \mathbf{10^{-6}}$.

An alternative method is the following. We have

$$2\text{Ag} + 2\text{H}^+ + \text{SO}_4^{2-} = \text{Ag}_2\text{SO}_4 + \text{H}_2 \quad \Delta G^0_1 = -2\mathscr{F}\mathscr{E}^0_1, \mathscr{E}^0_1 = -0.627$$
$$2\text{Ag} + 2\text{H}^+ = 2\text{Ag}^+ + \text{H}_2 \quad \Delta G^0_2 = -2\mathscr{F}\mathscr{E}^0_2, \mathscr{E}^0_2 = -0.799$$
$$\text{Ag}_2\text{SO}_4(\text{s}) = 2\text{Ag}^+ + \text{SO}_4^{2-} \quad \Delta G^0_3 = -RT\ln K_{sp}$$

Then $\Delta G^0_3 = \Delta G^0_2 - \Delta G^0_1$ or $(2.3RT/2\mathscr{F})\log K_{sp} = \mathscr{E}^0_2 - \mathscr{E}^0_1$ or $0.0295 \times$

$\log K_{sp} = -0.799 + 0.627$ and $\log K_{sp} = -5.88$.

$$K_{sp} = 1.3 \times 10^{-6}$$

Note that the two methods are completely equivalent mathematically; in subsequent problems one or the other will be used, but generally not both.

2. (a) The cell reaction is

$$Cr + 2H^+(0.002\ m) + SO_4^{2-}(0.001\ m) = CrSO_4(s) + H_2(1\ atm)$$

(b) The corresponding Nernst equation is

$$
\begin{aligned}
\mathscr{E} &= \mathscr{E}^0 - (0.059/2)\log\frac{1}{[H^+]^2[SO_4^{2-}]} \\
&= 0.4 + 0.0295\log[(0.002)^2(0.001)] \\
&= \mathbf{0.152\ V}
\end{aligned}
$$

(c) According to the Gibbs–Helmholtz equation, $\Delta H^0 = -\mathscr{F}(\mathscr{E}^0 - T\,d\,\mathscr{E}^0/dT)$, and, since $\mathscr{F}\mathscr{E}$ gives energy in joules per faraday, $5{,}000/96{,}500 = \mathscr{E}^0 - T\,d\,\mathscr{E}^0/dT$. As an approximation, take the average value of T as $37.5°C$ or $311°K$; then $d\mathscr{E}/dT = (0.4 - 0.052)/311 = 1.13 \times 10^{-3}$ and $\Delta\mathscr{E}^0 = 1.13 \times 10^{-3} \times 25 = 0.028$, so $\mathscr{E}^0$ at $50°C$ is approximately $0.4 + 0.028 = \mathbf{0.428\ V.}$

(d) If activity coefficients are included, the Nernst equation in part (b) becomes

$$
\begin{aligned}
\mathscr{E} &= 0.152 + 0.0295\log[(\gamma_{H^+})^2(\gamma_{SO_4^{2-}})] \\
&= 0.152 + 0.0295 \times 3\log\gamma_{\pm}
\end{aligned}
$$

The Debye–Hückel limiting law gives $\log\gamma_{\pm} = -0.51 \times 2 \times 1\sqrt{\mu}$, where μ in this case is $\frac{1}{2}(0.002 + 0.001 \times 4) = 0.003$. We then have

$$
\begin{aligned}
\mathscr{E} &= 0.152 - 0.0295 \times 3 \times 0.51 \times 2\sqrt{0.003} \\
&= 0.152 - 0.005 \\
&= \mathbf{0.147\ V.}
\end{aligned}
$$

3. (a) The cell reaction is

$$Ag + H^+(0.1\ m) + Br^-(0.1\ m) = AgBr + \tfrac{1}{2}H_2(0.01\ atm)$$

or, alternatively,

$$\text{Ag} + \text{H}^+(0.1\ m) = \text{Ag}^+(\text{in } 0.1\ m\ \text{HBr}) + \tfrac{1}{2}\text{H}_2(0.01\ \text{atm})$$

(b) We combine the equations:

$$\text{Ag} + \text{H}^+ = \text{Ag}^+ + \tfrac{1}{2}\text{H}_2 \quad \Delta G_1^0 = -\mathscr{F}\mathscr{E}_1^0,\ \mathscr{E}_1^0 = -0.8$$
$$(\text{subtract})\,\text{AgBr} = \text{Ag}^+ + \text{Br}^- \quad \Delta G_2^0 = -RT\ln K_{sp}$$
$$\text{Ag} + \text{H}^+ + \text{Br}^- = \text{AgBr} + \tfrac{1}{2}\text{H}_2 \quad \Delta G_3^0 = \mathscr{F}\mathscr{E}_3^0 = \Delta G_1^0 - \Delta G_2^0$$

Then

$$-\mathscr{F}\mathscr{E}_3^0 = -\mathscr{F}\mathscr{E}_1^0 + RT\ln K_{sp}$$

or

$$\mathscr{E}_3^0 = -0.8 + 0.059\log 10^{-12}$$
$$= -\mathbf{0.091\ V}$$

If the alternative method of writing the cell reaction is used, then $\mathscr{E}^0$ is simply that for Ag/Ag^+ or $-0.8\ \text{V}$!

(c) The reversible q is simply $T\Delta S$. Then $\Delta S = (\Delta H - \Delta G)/T$ or

$$\Delta S = [-50,000 + 96,500\,(-0.165)]/298$$
$$= -65,900/298$$
$$= -\quad \mathbf{221\ J/deg.}$$

(d) The Nernst expression is

$$-0.165 = -0.091 - 0.059\log\frac{(0.01)^{1/2}}{0.1\times 0.1} - 0.059\log\frac{1}{(\gamma_{\text{H}^+})(\gamma_{\text{Cl}^-})}$$

or

$$-0.165 = -0.150 + 0.059\times 2\log\gamma_{\pm}.$$

4. It is convenient to write the cell reaction as

$$\text{Cd} + 2\text{H}^+(0.04\ m) = \text{Cd}^{2+}\ (\text{in } 0.02\ m\ \text{H}_2\text{SO}_4) + \text{H}_2(1\ \text{atm})$$

The corresponding Nernst equation is then

$$\mathscr{E} = \mathscr{E}^0 - (0.059\times 308/298)(1/2)\log([\text{Cd}^{2+}]/[\text{H}^+]^2)$$

or

$$0.38 = 0.45 - 0.0305 \log(S/[H^+]^2)$$

where S is the solubility of $CdSO_4$ in this medium. Then

$$\log(S/[H^+]^2) = 0.07/0.0305 = 2.29$$

and

$$S = 195 \times (0.04)^2 = 0.312$$

The sulfate ion concentration is therefore $0.312 + 0.02 = 0.332$, and $K_{sp} = 0.312 \times 0.332 = \mathbf{0.103}$. We neglect activity coefficients and assume hydrogen is a perfect gas.

5. (a) The cell reaction is conveniently written in the form:

$$Tl + H^+ \text{ (unit act.)} = Tl^+ \text{ (in unit act. HBr)} + \tfrac{1}{2}H_2(1 \text{ atm})$$

The emf is then

$$\mathscr{E} = \mathscr{E}^0_{Tl/Tl^+} - 0.059 \log([Tl^+]/[H^+])$$
$$= 0.34 - 0.059 \log([K_{sp}/[H^+][Br^-]])$$
$$= 0.34 - 0.059 \log K_{sp} \quad \text{(since HBr is at unit activity)}$$
$$= 0.34 - 0.236 = \mathbf{0.58 \ V.}$$

(b) The desired ΔH is simply ΔH^0 for the Tl/Tl^+ couple, so

$$\Delta H^0 = -96{,}500[0.34 - 298(-0.003)] = \mathbf{-119{,}000 \, J.}$$

6. (a) Anode reaction: $\quad Cd + 2OH^- = Cd(OH)_2 + 2e^-$
or $\qquad\qquad\qquad\quad Cd = Cd^{2+} \text{ (in 0.01 } m \text{ NaOH)} + 2e^-$

Cathode reaction: $\quad 2H_2O + 2e^- = H_2 + 2OH^-$
or $\qquad\qquad\qquad\quad 2H^+ \text{ (in 0.01 } m \text{ NaOH)} + 2e^- = H_2$

Net reaction: $\qquad Cd + 2H_2O = Cd(OH)_2 + H_2$
or $\qquad\qquad\qquad\quad Cd + 2H^+ \text{ (in 0.01 } m \text{ NaOH)}$
$$= Cd^{2+} \text{ (in 0.01 } m \text{ NaOH)} + H_2$$

(b) The alternative method of writing the above reactions (equally acceptable) leads to the Nernst expression:

$$\mathscr{E}(= 0.000) = \mathscr{E}^0_{Cd/Cd^{2+}} - 0.0295 \log \frac{[Cd^{2+}]}{[H^+]^2}$$

or, since $K_{sp} = [Cd^{2+}][OH^-]^2$,

$$0 = 0.4 - 0.0295 \log[K_{sp}/K_w^2]$$

Then $K_{sp}/K_w^2 = 3.5 \times 10^{13}$ and $K_{sp} = \mathbf{3.5 \times 10^{-15}}$.

(c) $\Delta H = -96,500 \,(\mathscr{E} - T \, d\mathscr{E}/dT)$
$ = -96,500 \,(0 - 298 \times 0.002) = 57,500 \text{ J}$

To be consistent with the cell reaction, $\Delta H = \mathbf{115,000 \text{ J}}$.

7. (a) Anode reaction: $Au + I^-(m) = AuI + e^-$
Cathode reaction: $H^+(m) + e^- = \frac{1}{2}H_2$
Cell reaction: $Au + H^+(m) + I^-(m) = AuI + \frac{1}{2}H_2$

(b) The emf for the cell may be written in the form

$$\mathscr{E} = \mathscr{E}^0 - 0.059 \log \frac{1}{[H^+][I^-]} - 0.059 \log \frac{1}{(\gamma_{H^+})(\gamma_{I^+})}$$

or

$$-\mathscr{E}^0 = -\mathscr{E} + 0.118 \log m + 0.118 \log \gamma_\pm$$

It is reasonable to take $\gamma_\pm$ as nearly unity for the $10^{-4} \, m$ solution, and equating the right-hand sides of the two equations for the two conditions:

$$-(-0.97) + 0.118 \log 10^{-4} = -(-0.41) + 0.118 \log 3 + 0.118 \log \gamma_\pm$$

Then $0.118 \log \gamma_\pm = 0.031$ and $\gamma_\pm = \mathbf{1.83}$.

(c) It is convenient to rewrite the cell reaction as

$$Au + H^+(m) = Au(\text{in } m \text{ HI}) + \frac{1}{2}H_2$$

Then

$$\mathscr{E} = -0.97 = -1.68 - 0.059 \log \frac{[\text{Au}^+]}{10^{-4}} = -1.68 - 0.059 \log \frac{K_{sp}}{10^{-8}}$$

since $[\text{Au}^+] = K_{sp}/[\text{I}^-]$. Then $K_{sp}/10^{-8} = 10^{-12}$ and $K_{sp} = \mathbf{10^{-20}}$.

8. The reactions would normally be written as

> Anode reaction: $2\text{Tl} + 2\text{Cl}^-(0.02\ m) = 2\ \text{TlCl(s)}$
> Cathode reaction: $\text{Cd}^{2+}(0.01\ m) + 2e^- = \text{Cd}$
> (1) Net cell reaction: $2\text{Tl} + \text{Cd}^{2+}(0.01\ m) + 2\text{Cl}^-(0.02\ m)$
> $= 2\text{TlCl(s)} + \text{Cd}$

It is convenient for the present purpose (and was acceptable on the examination) to write the alternative formulation as

(2) $2\text{Tl} + \text{Cd}^{2+}(0.01\ m) = 2\text{Tl}^+(\text{in } 0.01\ m\ \text{CdCl}_2) + \text{Cd}$

The Nernst expression for this last formulation is then

(2) $\mathscr{E} = \mathscr{E}_2^0 - 0.0295 \log([\text{Tl}^+]^2/[\text{Cd}^{2+}])$

where

$$\mathscr{E}_2^0 = 0.34 - 0.40 = -\mathbf{0.06\ V}$$

$\mathscr{E}_1^0$ for the cell reaction as first written is easily obtained by noting that $[\text{Tl}^+] = K_{sp}/[\text{Cl}^-]$, so that

$$\mathscr{E} = -0.06 - 0.0295 \log(K_{sp}^2/\text{Cl}^{2+}][\text{Cl}^-]^2)$$

and $\mathscr{E}_1^0$ is the value of $\mathscr{E}$ when $a_{\text{CdCl}_2} = 1$. Then $\mathscr{E}_1^0 = -0.06 - 0.0295$ $\log(1.6 \times 10^{-3})^2 = \mathbf{0.105\ V}$. The value of $\mathscr{E}$ when $m = 0.01$ is then

$$\mathscr{E} = -0.06 - 0.0295 \log(K_{sp}^2/\text{Cd}^{2+}][\text{Cl}^-]^2)$$

$$= -\mathbf{0.054\ V}.$$

9. We write

> $\text{Ag} + \text{I}^- = \text{AgI} + e^-$ $\Delta G_1^0 = \quad 0.152\mathscr{F}$
> $\text{Ag} = \text{Ag}^+ + e^-$ $\Delta G_2^0 = -\mathscr{F}(-0.800)$

On subtracting,

$$\text{AgI} = \text{Ag}^+ + \text{I}^- \qquad \Delta G_3^0 = \Delta G_2^0 - \Delta G_1^0$$
$$= -RT \ln K_{sp}$$

Then $0.059 \log K_{sp} = -0.95$ and $K_{sp} = 1.2 \times 10^{-17}$.

10. The cell reaction is

$$\text{M} = \text{M}^+ (0.1 \, m) + e^-$$
$$\underline{\text{H}^+ (0.2 \, m) + e^- = \tfrac{1}{2}\text{H}_2 \, (1 \, \text{atm})}$$
$$\text{H}^+ (0.2 \, m) + \text{M} = \text{M}^+ (0.1 \, m) + \tfrac{1}{2}\text{H}_2 \, (1 \, \text{atm})$$

and the corresponding Nernst expression,

$$-0.5 = \mathscr{E}_{\text{M/M}^+}^0 - 0.059 \log \frac{0.1}{0.2} = \mathscr{E}_{\text{M/M}^+}^0 + 0.018$$

so

$$\mathscr{E}_{\text{M/M}^+}^0 = -0.5 - 0.018 = -0.518 \, \text{V}.$$

Addition of the KCl precipitates essentially all of the M^+, leaving 0.15 m excess Cl^- in solution. The cell reaction can still be written as above, but with M^+ now that concentration in equilibrium with MCl and 0.15 m Cl^-; that is, $(\text{M}^+) = K_{sp}/0.15$. The new Nernst expression is then

$$-0.1 = \mathscr{E}_{\text{M/M}^+}^0 - 0.059 \log \frac{K_{sp}/0.15}{0.2}$$

$$-0.1 = -0.518 + 0.059 \log (0.2 \times 0.15) - 0.059 \log K_{sp}$$

or $\log K_{sp} = (0.1 - 0.518 - 0.090)/0.059 = -8.62$ and $K_{sp} = 2.4 \times 10^{-9}$.

11. (a) From the definitions

$$m_\pm = [(m_+^{v^+})(m_-^{v^-})]^{1/v} = [(0.01)(0.12)^2]^{1/3}$$

(i.e., all the Cl^- contributes to m_-)

$$m_\pm = (1.44 \times 10^{-4})^{1/3}$$
$$= 0.0525.$$

Then

$$a_{\pm} = \gamma_{\pm} m_{\pm}$$
$$= 0.4 \times 0.0525$$
$$= 0.021$$

(b) The cell reaction is $Pb^{2+} + 2Ag + 2Cl^- = Pb + 2AgCl(s)$, and the electrolysis is stopped when the emf for the reaction becomes equal to that for the corresponding reduction of Sn^{2+}. We can then equate the two Nernst expressions:

$$-\mathscr{E}^0_{Pb/Pb^{2+}} - (0.059/2) \log \frac{1}{[Pb^{2+}][Cl^-]^2}$$

$$= -\mathscr{E}^0_{Sn/Sn^{2+}} - 0.0295 \log \frac{1}{[Sn^{2+}][Cl^-]^2}$$

Or, on rearranging and cancelling out $[Cl^-]$,

$$\log[Pb^{2+}]/[Sn^{2+}] = (0.126 - 0.140)/0.0295 = -0.475$$
$$[Pb^{2+}] = 0.335 \times 0.01 = \mathbf{3.35 \times 10^{-3} \, m}$$

(c) $\mathscr{E}^0$ for the reaction $Pb + Sn^{2+} = Pb^{2+} + Sn$ is $0.126 - 0.140 = -0.014$, whence $\log K = 0.014/0.0295 = -0.475$ and $K = 0.335$ [same arithmetic as in (b)]. Let x denote the equilibrium $[Pb^{2+}]$, so $0.335 = x/(0.01 - x)$ or $x = \mathbf{2.52 \times 10^{-3} \, m}$.

(d) The answer in (b) will be **the same** since activity coefficients will enter as the ratio of $\gamma_{\pm}$ for $PbCl_2$ to that for $SnCl_2$, which will be equal to unity.

12. Anode reaction: $\quad Cu = Cu^{2+}(m) + 2e^-$

Cathode reaction: $\quad 2AgCl + 2e^- = 2Ag + 2Cl^-(2m)$

Net cell reaction: $\quad Cu + 2AgCl = 2Ag + Cu^{2+}(m) + 2Cl^-(2m)$

From the Nernst equation

$$\mathscr{E} = \mathscr{E}^0 - 0.0295 \log[a_{Cu^{2+}} a_{Cl^-}^2]$$
$$= \mathscr{E}^0 - 0.0295 \times 3 \log \gamma_{\pm} - 0.0295 \log[(Cu^{2+})(Cl^-)^2]$$

It is reasonable to assume that $\gamma_{\pm}$ is unity for the case of $m = 10^{-4}$; hence $0.191 = \mathscr{E}^0 - 0.0295 \log[(10^{-4})(2 \times 10^{-4})^2] = \mathscr{E}^0 + 0.335$, whence $\mathscr{E}^0 =$

-0.144. Then, for the second case we have

$$-0.074 = -0.144 - 0.0295 \log[(0.2)(0.4)^2] - 0.0885 \log \gamma_\pm$$

from which

$$\log \gamma_\pm = -0.294$$

$$\gamma_\pm = \mathbf{0.51}$$

13.   On combining the two anode and two cathode reactions, we obtain

$$H^+ (10^{-4}\, m) + Cl^- (10^{-4}\, m) = H^+ (0.1\, m) + Cl^- (0.1\, m)$$

(a) $\Delta G = -96,500 \times (-0.335) = \mathbf{32,200\ J.}$

(b) $\mathscr{E} = -0.059 \log[(0.1)^2/(10^{-4})^2] - 0.059 \log \gamma_\pm^2$ (taking $\gamma_\pm$ to be unity for the $10^{-4}\, m$ solution). Then $\log \gamma_\pm^2 = -0.32$ and $\gamma_\pm = \mathbf{0.69.}$

(c) Since $\Delta G = \Delta H - T \Delta S$, if ΔH is zero, then $\Delta S = -\Delta G/T$, and, neglecting any change in ΔS with temperature, $\Delta G_1/T_1 = \Delta G_2/T_2$, so $\Delta G_{273°K} = (273/298)(32,200) = \mathbf{29,600\ J.}$

14.   Anode: $Ag + Cl^- (0.01\, m) = AgCl + e^-$
Cathode: $AgCl + e^- = Ag + Cl^- (0.1\, m)$
By transference: $t^- Cl^- (0.1\, m) = t^- Cl^- (0.01\, m)$
$t^+ K^+ (0.01\, m) = t^+ K^+ (0.1\, m)$
Net cell reaction: $t^+ K^+ (0.01\, m) + t^+ Cl^- (0.01\, m)$
$= t^+ K^+ (0.1\, m) + t^+ Cl^- (0.1\, m)$

Then

$$\mathscr{E} = -0.059 \log \left[\frac{(0.1)^2}{(0.01)^2} \frac{(\gamma_{\pm 0.1m})^2}{(\gamma_{\pm 0.01m})^2} \right]^{t^+}$$

$$= -0.059 t^+ \log[100 \times 0.8^2/0.9^2]$$

$$= \mathbf{-0.045\ V}$$

15.   (a) The cell reaction is

$$Ag + M^+(m) + Cl^-(m) = AgCl + M$$

and the corresponding Nernst equation is

$$\mathscr{E} = \mathscr{E}^0 - 0.059 \log \frac{1}{a_{M^+} a_{Cl^-}}$$

This may be arranged to

$$\mathscr{E} = \mathscr{E}^0 + 0.118 \log m + 0.118 \log \gamma_\pm$$

By Debye–Hückel theory, $\log \gamma_\pm = -0.51 (0.1)^{1/2} = -0.161$, and inserting this value and solving, $\mathscr{E}^0 = 0.50 + 0.118 + 0.021 = \mathbf{0.639\ V.}$

(b) First, $\Delta G = -96,500 \times 0.5 = -4.8 \times 10^4$ J; ΔS is then $(2 \times 10^5 + 4.8 \times 10^4)/303 = 819$ J/deg (an average temperature of 30°C has been used). Then $d\mathscr{E}/dT = 819/96,500 = 8.5 \times 10^{-3}$ V/deg, and $\mathscr{E}$ at 35°C is then $0.50 + 0.085 = \mathbf{0.585\ V.}$

16. Referring to the solution to Problem 10, the cell reaction is

$$H^+ (0.2\ m) + M = M^+ (0.1\ m) + \tfrac{1}{2} H_2 (1\ atm)$$

The complete Nernst expression is then

$$\mathscr{E} = \mathscr{E}^0_{M/M^+} - 0.059 \log \frac{a_{M^+}}{a_{H^+}}$$

$$\mathscr{E} = \mathscr{E}^0_{M/M^+} - 0.059 \log \frac{0.1}{0.2} - 0.059 \log \frac{\gamma_{M^+} \gamma_{NO_3^-}}{\gamma_{H^+} \gamma_{NO_3^-}}$$

The last term above is obtained by multiplying numerator and denominator of the argument of the logarithm by $\gamma_{NO_3^-}$, to show that the term involves the ratio $\gamma_\pm^2$ (or M^+, NO_3^-) to $\gamma_\pm^2$ (of H^+, NO_3^-). However, these electrolytes are in the **same** solution, and by the **limiting** law, the two $\gamma_\pm$ values are the **same**! Their ratio is thus unity, and the log term is therefore zero. The "corrected" emf is thus still -0.5 V. (Sorry—this is one of the very few booby traps in the book!)

17. First,

$$Au = Au^+ + e^- \qquad \mathscr{E}^0_1 = -1.68$$
$$Au = Au^{3+} + 3e^- \qquad \mathscr{E}^0_2 = -1.50$$
$$Au^+ = Au^{3+} + 2e^- \qquad \mathscr{E}^0_3 = (-\mathscr{E}^0_1 + 3\mathscr{E}^0_2)/2$$
$$= (1.68 - 4.50)/2$$
$$= -1.42$$

(remember that when combining half-reactions, each is weighted according to the number of equivalents involved when the $\mathscr{E}$'s are combined). Then for

$$2Fe^{2+} + Au^{3+} = 2Fe^{3+} + Au^{+} \qquad \mathscr{E}^0 = -0.77 - (-1.42)$$
$$= 0.65$$

Then $0.65 = (0.059/2) \log K$ or

$$\log K = 22$$
$$K = 10^{22}$$

18. (a) Anode reaction: $2Ag = 2Ag^{+} + 2e^{-}$
Cathode reaction: $Cd^{2+} + 2e^{-} = Cd$
Net cell reaction: $2Ag + Cd^{2+}(0.016\,m) = 2Ag^{+}(0.04\,m) + Cd$
(b) The Nernst expression is

$$\mathscr{E} = \mathscr{E}^0_{Ag/Ag^+} - \mathscr{E}^0_{Cd/Cd^{2+}} - 0.0295 \log \frac{[Ag^+]^2}{[Cd^{2+}]}$$

or

$$\mathscr{E} = 0.8 - 0.4 - 0.0295 \log [(0.04)^2/(0.016)]$$
$$= -1.2 + 0.0295$$
$$= -1.17\,V$$

$\mathscr{E}^0$, of course, is -1.2 V.

(c) The equilibrium constant is given by $0.0295 \log K = \mathscr{E}^0$ (in terms of the cell reaction written for 2 faradays), or $\log K = -40.8$ or $K = 1.6 \times 10^{-41}$.

(d) Although $\mathscr{E}^0$ does not change with temperature, the coefficient of the log term in the Nernst equation is proportional to T, and this term will increase in magnitude with temperature. On taking signs into account, the effect will be to make $\mathscr{E}$ less negative, that is, **greater** algebraically.

19.   (a) $\mathscr{E}^0$ for the reaction is $\mathscr{E}^0_{Ag/Ag^+} - \mathscr{E}^0_{Fe^{2+}/Fe^{3+}}$ or $-0.799 - (-0.771) = -0.028$ V. Then $-0.028 = 0.059 \log K$, whence $K = 0.335$. Then $0.335 = [Ag^+][Fe^{2+}]/[Fe^{3+}] = x^2/(0.1 - x)$ and $x = [Ag^+] = 0.080$ m.

(b) Since the solution in (a) is an equilibrium one, then Ag/Ag^+ and the Fe^{2+}/Fe^{3+} potentials must be the same; hence either could be used to give the anode potential of the cell. Using the former, the Nernst equation is

$$\mathscr{E} = \mathscr{E}^0_{\text{Ag/Ag}^+} - \mathscr{E}^0_{\text{Fe}^{2+}/\text{Fe}^{3+}} - 0.059 \log \left(\frac{[\text{Fe}^{2+}][\text{Ag}^+]}{[\text{Fe}^{3+}]} \right)$$

$$= -0.799 - (-0.771) - 0.059 \log \left[\frac{(1)(0.080)}{(1)} \right]$$

$$= -0.028 - 0.059(-1.095)$$

$$= \mathbf{0.037 \ V.}$$

20. (a) We want to combine the half reactions

$$\text{(1)} \qquad \text{Cu} = \text{Cu}^+ + e^-$$
$$\text{·(2)} \qquad \text{Cu} = \text{Cu}^{2+} + 2e^-$$

$$2(1) - (2) \quad \text{Cu} + \text{Cu}^{2+} = 2\text{Cu}^+$$

and $\mathscr{E}^0 = 2(\mathscr{E}^0_1 - \mathscr{E}^0_2) = \mathbf{-0.34 \ V.}$

Perhaps a better way of seeing how the factor of two comes in is as follows: first,

$\text{Cu} = \text{Cu}^+ + e^-$ $\qquad \mathscr{E}^0_1$	Since half-reactions are being added,
$\text{Cu}^+ = \text{Cu}^{2+} + e^-$ $\qquad \mathscr{E}^0_3$	each $\mathscr{E}^0$ is weighted according to the
$\overline{\text{Cu} = \text{Cu}^{2+} + 2e^- \qquad \mathscr{E}^0_2}$	number of electrons, so $2\mathscr{E}^0_2 = \mathscr{E}^0_1 + \mathscr{E}^0_3$ and $\mathscr{E}^0_3 = 2(-0.35) - (-0.52) = -0.18$

Then

$$\text{Cu} = \text{Cu}^+ + e^- \qquad \mathscr{E}^0_1$$
$$\text{(subtract)} \quad \text{Cu}^+ = \text{Cu}^{2+} + e^- \qquad \mathscr{E}^0_3$$

$$\overline{\text{Cu} + \text{Cu}^{2+} = 2\text{Cu}^+} \qquad \mathscr{E}^0 = \mathscr{E}^0_1 - \mathscr{E}^0_3 = -0.52 - (-0.18)$$
$$= \mathbf{-0.34 \ V}$$

The equilibrium constant is now given by $-0.34 = 0.059 \log K$ and $K = \mathbf{1.7 \times 10^{-6}}$. Then

$$1.7 \times 10^{-6} = \frac{[\text{Cu}^+]^2}{[\text{Cu}^{2+}]} = x^2/(0.01 - x/2)$$

from which $x = \mathbf{1.3 \times 10^{-4}} = [\text{Cu}^+]$.

(b) We now combine

$$2NH_3 + Cu = Cu(NH_3)_2^+ + e^-$$

(subtract) $$Cu = Cu^+ + e^-$$

$$2NH_3 + Cu^+ = Cu(NH_3)_2^+$$

$$\mathscr{E}^0 = 0.11 - (-0.52) = 0.63 \text{ V}$$

also

$$d\mathscr{E}^0/dT = -0.003 - (-0.002) = -0.001$$

Then

$$\Delta G^0 = -n\mathscr{F}\mathscr{E}^0 = -96,500 \times 0.63 = -\textbf{ 61,000 J}$$
$$\Delta S^0 = -d(\Delta G^0)/dT = \mathscr{F}(d\mathscr{E}^0/dT) = -96,500 \times 0.001$$
$$= -95.5 \text{ J/deg}$$
$$\Delta H^0 = \Delta G^0 + T\Delta S^0 = -61,000 - 298 \times 96.5 = -\textbf{89,700 J.}$$

21. First it is necessary to obtain $\mathscr{E}^0$ for $Fe = Fe^{3+} + 3e^-$:

$$Fe^{2+} = Fe^{3+} + e^- \qquad -0.771$$
$$Fe = Fe^{2+} + 2e^- \qquad 0.440$$

$$Fe = Fe^{3+} + 3e^- \qquad \mathscr{E}^0 = (-0.771 + 2 \times 0.440)/3 = 0.036$$

Then

$$Pb = Pb^{2+} + 2e^- \qquad 0.126$$
$$Fe^{3+} + 3e^- = Fe \qquad -\textbf{0.036}$$

$$3Pb + 2Fe^{3+} = 3Pb^{2+} + 2Fe \qquad \mathscr{E}^0 = 0.126 - 0.036 = 0.090$$

Finally, $\log K = n\mathscr{E}^0/0.059 = 6 \times 0.090/0.059 = 9.17$ and $K = 1.5 \times 10^9$. (Note that $n = 6$ for this reaction.)

22. (a) The sum of the two electrode reactions is

$$\tfrac{1}{2}H_2(1 \text{ atm}) = H^+(0.01 \text{ m}) + e^-$$
$$H^+(0.001 \text{ m}) + e^- = \tfrac{1}{2}H_2(0.01 \text{ atm})$$

$$\tfrac{1}{2}H_2(1 \text{ atm}) + H^+(0.001 \text{ m}) = \tfrac{1}{2}H_2(0.01 \text{ atm}) + H^+(0.01 \text{ m})$$

$$\mathscr{E} = -0.059 \log \frac{(0.01)^{1/2}(0.01)}{(0.001)} = 0$$

(b) The changes owing to transference are

$$t^+ H^+(0.01\ m) = t^+ H^+(0.001\ m)$$
$$t^- Cl^-(0.001\ m) = t^- Cl^-(0.01\ m)$$

The emf associated with these changes, which is the junction potential, is

$$\mathscr{E} = 0.059 \log \left[\frac{(0.001)^{0.8}}{(0.01)^{0.8}} \right] - 0.059 \log \left[\frac{(0.01)^{0.2}}{(0.001)^{0.2}} \right]$$
$$= -0.59(0.8 - 0.2) \log [(0.001/(0.01)]$$
$$= \mathbf{0.035\ V}$$

23. Anode reaction: $\frac{1}{2}H_2(1\ atm) = H^+(0.1\ m) + e^-$

Cathode reaction: $H^+(0.01\ m) + e^- = \frac{1}{2}H_2(1\ atm)$

Transference process: $H^+(0.1\ m) = H^+(0.01\ m)$

Net reaction: Nil!

The $\mathscr{E}$ for this cell is then **zero**.

24. (a) Anode reaction: $Hg + 2Cl^-(0.01\ m) = Hg_2Cl_2 + 2e^-$

Cathode reaction: $Hg_2Cl_2 + 2e^- = 2Cl^-(0.1\ m) + 2Hg$

Transference process: $2H^+(0.01\ m) = 2H^+(0.1\ m)$

Net cell reaction: $2H^+(0.01\ m) + 2Cl^-(0.01\ m)$
$$= 2H^+(0.1\ m) + 2Cl^-(0.1\ m)$$

(b) The Nernst expression is then

$$\mathscr{E} = 0 - 0.0295 \log \left(\frac{[H^+]^2[Cl^-]^2}{[H^+]^2[Cl^-]^2} \right)$$
$$= -0.0295 \log(0.1)^4/(0.01)^4$$
$$= \mathbf{-0.118\ V}$$

25. (a) Cell A: $Na(Hg) + H^+(1\ M) = Na^+(1\ M) + \frac{1}{2}H_2(1\ atm)$

Cell B: $Na(Hg) + H_2O = Na^+(0.01\ M) + OH^-(0.01\ M)$
$$+ \frac{1}{2}H_2(1\ atm)$$

or $Na(Hg) + H^+ \text{(in 0.01 } M \text{ NaOH)} = Na^+(0.01\ M)$
$$+ \frac{1}{2}H_2(1\ atm)$$

(b) The alternative way of writing the reaction for cell (B) has the advantage of being in the same form as that for the reaction for cell (A) so that the Nernst equation is the same for both cells.

$$\mathscr{E} = \mathscr{E}^0 - 0.059 \log \frac{[\mathrm{Na}^+]}{a_{\mathrm{Na(Hg)}}[\mathrm{H}^+]}$$

Then for cell (A), since Na^+ and H^+ are at unit concentration,

$$\mathscr{E}^0 + 0.059 \log a_{\mathrm{Na(Hg)}} = 1.2$$

Then, for cell (B) we can write

$$\mathscr{E} = \mathscr{E}^0 - 0.059 \log \left(\frac{[\mathrm{Na}^+]}{[\mathrm{H}^+]} \right) + 0.059 \log a_{\mathrm{Na(Hg)}}$$

$$= 1.2 - 0.059 \log \left[\frac{0.01}{10^{-12}} \right]$$

$$= \mathbf{0.61\ V}$$

(c) By the Gibbs–Helmholtz equation,

$$\Delta H = -96{,}500\,(\mathscr{E} - Td\mathscr{E}/dT)$$

$$= -96{,}500\,(1.2 - 298 \times 0.002)$$

$$= \mathbf{-58{,}500\ J}$$

26. Anode reaction: $\mathrm{Ag} + \mathrm{Cl}^+(0.1\ M) = \mathrm{AgCl} + e^-$

Cathode reaction: $\mathrm{AgCl} + e^- = \mathrm{Ag} + \mathrm{Cl}^-(0.01\ M)$

Transference change: $t^-\mathrm{Cl}^-(0.01\ M) = t^-\mathrm{Cl}^-(0.1\ M)$
$t^+\mathrm{Na}^+(0.1\ M) = t^+\mathrm{Na}^+(0.01\ M)$

Net reaction: $(1 - t^-)(\mathrm{or}\ t^+)\mathrm{Cl}^-(0.1\ M) + t^+\mathrm{Na}^+(0.1\ M)$
$= (1 - t^-)(\mathrm{or}\ t^+)\mathrm{Cl}^-(0.01\ M)$
$+ t^+\mathrm{Na}^+(0.01\ M)$

The $\mathscr{E}$ for the cell is then given by

$$\mathscr{E} = -0.059 \log \frac{(0.01)^{2t^+}}{(0.1)^{2t^+}}$$

$$= 0.118t^+$$

$$= \mathbf{0.0472\ V}$$

27. (a) Anode reaction: $Ag + Cl^-(0.01\ m) = AgCl + e^-$

Cathode reaction: $H^+(0.001\ m) + e^- = \frac{1}{2}H_2(1\ atm)$

Transference changes: $t^+ H^+(0.01\ m) = t^+ H^+(0.001\ m)$

$t^- Cl^-(0.001\ m) = t^- Cl^-(0.01\ m)$

(b) The net cell reaction is then

$Ag + t^+ H^+(0.01\ m) + t^- H^+(0.001\ m)$
$+ t^+ Cl^-(0.01\ m) + t^- Cl^-(0.001\ m) = AgCl + \frac{1}{2}H_2$

The emf is then

$$\mathscr{E} = \mathscr{E}^0_{Ag/AgCl} - 0.059 \log \left[\frac{1}{(0.01)^{t^+}(0.001)^{t^-}(0.01)^{t^+}(0.001)^{t^-}} \right]$$
$$= -0.22 \overset{\centerdot}{+} 2 \times 0.059[t^+ \log(0.01) + t^- \log(0.001)]$$

Then $-0.48 = -0.22 + 0.118[-2t^+ - 3(1 - t^+)]$ or $t^+ - 3 = -2.20$, and $\mathbf{t^+ = 0.80.}$

28. (a) Anode reaction: $Hg + Cl^-(0.1\ m) = \frac{1}{2}Hg_2Cl_2 + e^-$

Cathode reaction: $\frac{1}{2}Hg_2Cl_2 + e^- = Hg + Cl^-(0.001\ m)$

By transference: $t^- Cl^-(0.001\ m) = t^- Cl^-(0.1\ m)$

$t^+ Na^+(0.1\ m) = t^+ Na^+(0.001\ m)$

Net cell reaction per $\mathscr{F}$: $t^+ Na^+(0.1\ m) + t^+ Cl^-(0.1\ m)$
$= t^+ Na^+(0.001\ m) + t^+ Cl^-(0.001\ m)$

(b) and (c)

$$\mathscr{E} = -0.059 \log \left[\frac{(0.001)^2(\gamma_{\pm 0.001})^2}{(0.1)^2(\gamma_{\pm 0.1})^2} \right]^{t^+}$$
$$= -0.059 \times 0.4 \log[10^{-6}/0.01 \times 0.6^2]$$
$$= \mathbf{0.084\ V}$$

(d) If we neglect that part of the cell reaction resulting from transference (which is equivalent to neglecting the junction potential), then

$$\mathscr{E} = -0.059 \log[0.001/0.1] = \mathbf{0.118\ V}$$

(It is necessary to neglect activity coefficients, since we do not have single ion values.)

29. Cell (a): Anode reaction: $\frac{1}{2}H_2(1\ \text{atm}) = H^+(m_1) + e^-$

Cathode reaction: $AgCl + e^- = Cl^-(m_1) + Ag$

Anode reaction: $Ag + Cl^-(m_2) = AgCl + e^-$

Cathode reaction: $H^+(m_2) + e^- = \frac{1}{2}H_2(1\ \text{atm})$

Net: $H^+(m_2) + Cl^-(m_2)$

$\qquad = H^+(m_1) + Cl^-(m_1)$

$$\mathscr{E}_a = -0.059 \log[m_1^2/m_2^2] - 0.059 \log[\gamma^2_{\pm\,m_1}\gamma^2_{\pm\,m_1}\gamma^2_{\pm\,m_2}]$$

Cell (b): Anode reaction: $\frac{1}{2}H_2(1\ \text{atm}) = H^+(m_1) + e^-$

Cathode reaction: $H^+(m_2) + e^- = \frac{1}{2}H_2(1\ \text{atm})$

By transference: $t^-Cl^-(m_2) = t^-Cl^-(m_1)$

$\qquad\qquad\qquad t^+H^+(m_1) = t^+H^+(m_2)$

Net: $t^-H^+(m_2) + t^-Cl^-(m_2)$

$\qquad = t^-H^+(m_1) + t^-Cl^-(m_1)$

$$\mathscr{E}_b = -0.059 \log[m_1^2/m_2^2]^{t^-} - 0.059 \log[\gamma^2_{\pm\,m_1}/\gamma^2_{\pm\,m_2}]^{t^-}$$

or $\mathscr{E}_b/t^- = \mathscr{E}_a$, $t^- = \mathscr{E}_b/\mathscr{E}_a$, so the transference numbers can be obtained easily from the ratio of the two emf's. The result would be approximate in that t^- (and $t^+ = 1 - t^-$) would be an average between the values at m_1 and at m_2.

30. (a) Anode reaction: $Cl^-(m_1) = \frac{1}{2}Cl_2 + e^-$

Cathode reaction: $\frac{1}{2}Cl_2 + e^- = Cl^-(m_2)$

Net reaction: $Cl^-(m_1) = Cl^-(m_2)$

$$\mathscr{E} = -0.059 \log(m_2/m_1)$$
$$= -0.059 \log(0.01/0.1)$$
$$= \mathbf{0.059\ V}$$

(b) In addition to the above, changes from transference are

$t^-Cl^-(m_2) = t^-Cl^-(m_1)$

$t^+Na^+(m_1) = t^+Na^+(m_2)$

In combination with the net electrode reaction given in (a),

$t^+Na^+(m_1) + t^+Cl^-(m_1) = t^+Na^+(m_2) + t^+Cl^-(m_2)$

Then

$$\mathscr{E}_t = -0.059 \log[m_2^2/m_1^2]^{t^+}$$
$$= -0.118t^+ \log[0.01/0.1]$$
$$= \mathbf{0.046 \ V}$$

Also, $\mathscr{E}_t = \mathscr{E}_{\text{electrode}} + \mathscr{E}_j$, so $\mathscr{E}_j = -\mathbf{0.013 \ V}$.

31. Anode reaction : $\quad \frac{1}{2}H_2(1 \text{ atm}) = H^+(0.2 \ m) + e^-$

Cathode reaction : $\quad H^+(0.4 \ m) + e^- = \frac{1}{2}H_2(1 \text{ atm})$

By transference : $\quad t^+ H^+(0.2 \ m) = t^+ H^+(0.4 \ m)$

$\quad\quad\quad\quad\quad\quad\quad\quad t^- SO_4^{2-}(0.2 \ m) = t^- SO_4^{2-}(0.1 \ m)$

Net cell reaction : $\quad t^- H^+(0.4 \ m) + t^- SO_4^{2-}(0.2 \ m)$

$\quad\quad\quad\quad\quad\quad\quad\quad = t^- H^+(0.2 \ m) + t^- SO_4^{2-}(0.1 \ m)$

(all entries refer to equivalents). The emf is then

$$\mathscr{E} = -0.059 \, t^- \, \log \frac{(0.2)(0.1)^{1/2}}{(0.4)(0.2)^{1/2}}$$

or $\mathscr{E} = 0.059 \times 0.449t^- = 0.0265t^-$. t^- is not given, but if taken from your text (as infinite dilution values), it would be about $80/430$, so $\mathscr{E} = \mathbf{0.0050 \ V}$.

32. (a) It is desired that Na^+ ions leave, but not enter the middle compartment, so A--A must be permeable to anions only, and B--B permeable to cations. This combination also ensures that Cl^- ions can also leave but not enter the middle section. (It helps to keep Eq. (15-1) in mind, as it shows the direction of motion of cations and anions.)

(b) Anode reaction $\quad\quad Ag + Cl^-(0.1 \ m) = AgCl + e^-$

Cathode reaction : $\quad AgCl + e^- = Ag + Cl^-(0.3 \ m)$

Net : $\quad\quad\quad\quad\quad\quad Cl^-(0.1 \ m) = Cl^-(0.3 \ m)$

By transference : $\quad Cl^-(0.2 \ m) = Cl^-(0.1 \ m)$

$\quad\quad\quad\quad\quad\quad\quad Na^+(0.2 \ m) = Na^+(0.3 \ m)$

Over-all net : $\quad\quad Na^+(0.2 \ m) + Cl^-(0.2 \ m) = Na^+(0.3 \ m) +$

$\quad\quad\quad\quad\quad\quad\quad Cl^-(0.3 \ m)$

The emf is then

$$\mathscr{E} = -0.059 \log \frac{(0.3)^2}{(0.2)^2} = -\mathbf{0.021 \ V}$$

33. First, since we shall want to use the standard potential for Pb/Pb^{2+}, the equation should be rewritten as $Pb + 2H^+ = Pb^{2+} + H_2$. Assuming that the hydrogen concentration remains less than $10^{-3} M$, the Nernst expression is

$$\mathscr{E}(= 0) = \mathscr{E}^0_{Pb/Pb^{2+}} - 0.0295 \log \left(\frac{[Pb^{2+}]k[H_2]}{[H^+]^2} \right)$$

From the table in your text, $\mathscr{E}^0$ is 0.126 V. Also, since we want a_{H_2} to be unity when dissolved hydrogen is in equilibrium with gas at 1 atm, it follows that $k = 10^3$. In addition, $[H_2] = [Pb^{2+}]$ and $[H^+] = 10^{-6}$, so

$$0 = 0.126 - 0.0295 \log \frac{[Pb^{2+}]^2(10^3)}{10^{-12}}$$

Solving for $[Pb^{2+}]$, we obtain $1.87 \times 10^4 = 10^{15}[Pb^{2+}]^2$ or $[Pb^{2+}] =$ **4.3 $\times$ 10^{-6}**.

34. (a) First anode reaction: $H_2 = 2H^+(m) + 2e^-$

First cathode reaction: $Hg_2^{2+}(m_1) + 2e^- = 2Hg$

Second anode reaction: $2Hg = Hg_2^{2+}(m_2) + 2e^-$

Second cathode reaction: $2H^+(m) + 2e^- = H_2$

Net reaction: $Hg_2^{2+}(m_1) = Hg_2^{2+}(m_2)$

$\mathscr{E}^0$ and hence ΔG^0 is zero, for the over-all cell.

(b) The emf is then given by

$$\mathscr{E} = -0.0295 \log(m_2/m_1)$$
$$= -0.0295 \log 10$$
$$= \mathbf{-0.0295 \ V}$$

(c) If mercurous ion actually is present as Hg^+, the cell reaction per 2 faradays becomes

$$2Hg^+(2m_1) = 2Hg^+(2m_2)$$

The emf is then

$$\mathscr{E} = -0.0295 \log[(2m_2)^2/(2m_1)^2]$$
$$= \mathbf{-0.059 \ V}$$

35. (a) Anode reaction: $\frac{1}{2}H_2 + OH^-(m) = H_2O + e^-$

Cathode reaction: $e^- + \frac{1}{2}HgO + \frac{1}{2}H_2O = \frac{1}{2}Hg + OH^-(m)$

Net cell reaction: $\frac{1}{2}HgO + \frac{1}{2}H_2 = \frac{1}{2}Hg + \frac{1}{2}H_2O$

(b) Since all the species involved in the over-all reaction are pure sub-stances (neglecting the dilution of the water by dissolved NaOH), there are no concentration terms and $\mathscr{E}^0_{298} = \mathscr{E} = \textbf{0.924 V.}$
(c) If $\Delta H = -35,000$ cal or $-147,000$ J, then

$$T\Delta S = -147,000 - 96,500 \times 0.924$$
$$= -58,000$$

and

$$\Delta S = -58,000/298$$
$$= -195 \text{ J}$$

Then at $308°K$,

$$\Delta G = -147,000 - 308 \times 195$$
$$= -87,500 \text{ J}$$

and

$$\mathscr{E}_{305\,°K} = \textbf{0.91 V}$$

Alternatively, if ΔS is -195 J/deg, then $d\mathscr{E}/dT$ is -0.00202 V/deg and $\mathscr{E}$ at $308°K$ is $0.924 - 10 \times 0.00202 = \textbf{0.91 V}$. (Answers somewhat different will result if you happened to write the cell reaction for 2 faradays, since ΔG will now be $-2 \times 96,500 \times 0.924$, etc.)
(d) So long as HgO(s) remains, partial neutralization of the NaOH will not affect the over-all reaction (note that $[OH^-]$ cancels out in the over-all cell reaction).

36. We want the reduction (or oxidation) potentials to be the same; that is,

$$\mathscr{E} = \mathscr{E}^0_{Cu/Cu^{2+}} - 0.0295 \log a_{Cu^{2+}}$$
$$\mathscr{E}_2 = \mathscr{E}^0_{Ag/Ag^+} - 0.0296 \log a^2_{Ag^+}$$
$$\mathscr{E}_1 = \mathscr{E}_2$$

Then

$$-0.345 + 0.800 = 0.0295 \log[a_{Cu^{2+}}/a_{Ag^+}^2]$$
$$= 0.0295 \log([Cu^{2+}]/[Ag^+]^2)$$
$$+ 0.0295 \log(1/0.80)$$

or

$$\log([Cu^{2+}]/[Ag^+]^2) = 15.4$$
$$[Cu^{2+}]/[Ag^+]^2 = 2.5 \times 10^{15}$$
$$[Ag^+] = \mathbf{6.3 \times 10^{-9}}$$

(The result would be quite different were it recognized that silver would be soluble in the copper, since then the activity or, roughly, the mole fraction of silver in the copper would appear in the Nernst equation.)

37. (a) One should look for half-cells involving H^+ and OH^- ions, which narrows the choice to $Ag/Ag_2O/NaOH(m)/H_2(1 \text{ atm})/Pt$
  Cell reaction:

$$2Ag + 2H^+ (\text{in } m \text{ NaOH}) + 2OH^-(m) = Ag_2O + H_2(1 \text{ atm}) + H_2O$$

Nernst equation:

$$\mathscr{E} = -0.344 - 0.0295 \log \left[\frac{1}{[H^+]^2[OH^-]^2} \right]$$
$$= -0.344 + 0.0295 \log K_w^2$$

  (b) We want a half-cell involving one of the ions of the equilibrium, hence

  $Ag/AgNO_3(f_1), NaCN(f_2)$ buffered to pH $7/H_2(1 \text{ atm})/Pt$

[Other cathodes could be used; the particular solution is merely a convenient one since, if f_2 is much larger than f_1, the concentration of $Ag(CN)_2^-$ is essentially equal to f_1, where f denotes formula weight per liter; also $[CN^-] = 10^{-3} f_2$ at pH 7.]

Cell reaction: $Ag + H^+(\text{pH } 7) = Ag^+$ (in this solution) $+ \frac{1}{2}H_2(1 \text{ atm})$
Nernst equation: $\mathscr{E} = -0.799 - 0.059 \log([Ag^+]/[H^+])$

but $[Ag^+] = K[Ag(CN)_2^-]/[CN^-]^2$ and $K_a = [H^+][CN^-]$, so

$$\mathscr{E} = -0.799 - 0.059 \log \left[\frac{K[Ag(CN)_2^-]}{K_a[CN^-]} \right]$$
$$= -0.799 - 0.059 \log K - 0.059 \log[f_1/10^{-13} f_2] \qquad (f_2 \gg f_1)$$

In the case of (a), the NaOH concentration cancels out, and

$$\mathscr{E} = -0.344 - 0.059 \log 10^{-14} = \mathbf{0.481\ V}$$

In the case of (b), assume f_1 is 0.001 formal, and f_2 is 0.1 formal, then

$$\mathscr{E} = -0.799 - 0.059 \log 4 \times 10^{-19} - 0.059 \log[0.001/10^{-14}]$$
$$= -\mathbf{0.36\ V}$$

38. (a) The anode reaction is

$$Cu = Cu^{2+}\ (\text{in } 0.5\,m\ NH_3) + 2e^-$$

and the Nernst equation is then

$$\mathscr{E} = \mathscr{E}^0_{Cu/Cu^{2+}} - \mathscr{E}_{ref} - \frac{0.059}{2} \log Cu^{2+}$$

or

$$-0.26 = -0.34 - 0.22 - 0.0295 \log Cu^{2+}$$

from which $\log Cu^{2+} = -10$ or $(Cu^{2+}) = 10^{-10}$. The equilibrium constant is then

$$K = \frac{[Cu(NH_3)_4^{2+}]}{[Cu^{2+}][NH_3]^4} = \frac{0.02}{10^{-10} \times 0.5^4} = \mathbf{3.2 \times 10^9}$$

(b) The cell reaction may alternatively be written as

$$Cu + 4NH_3 = Cu(NH_3)_4^{2+} + 2e^-$$

and the Nernst expression,

$$\mathscr{E} = \mathscr{E}^0_{Cu/Cu(NH_3)_4^{2+}} - \mathscr{E}_{ref} - \frac{0.059}{2} \log \frac{[Cu(NH_3)_4^{2+}]}{[NH_3]^4}$$

$\mathscr{E}$ is the same, regardless of the choice of convention, so that

$$\mathscr{E}^0_{Cu/Cu(NH_3)_4^{2+}} = -0.26 + 0.22 + 0.0295 \log \frac{0.02}{0.5^4}$$

$$= -0.04 - 0.015 = -\mathbf{0.055}\ \mathbf{V}$$

(*Note*: The concentration of $Cu(NH_3)_4^{2+}$ is taken to be 0.02 on the grounds that essentially all of the Cu^{2+} is complexed; the K value of 10^{10} confirms the validity of this assumption.)

39. (a) Anode reaction: $2Sb + 6OH^- = Sb_2O_3 + 3H_2O + 6e^-$

Cathode reaction: $6H^+$ (in m OH^-) $+ 6e^- = 3H_2(1\ atm)$

Net: $2Sb + 6OH^-(m) + 6H^+$ (in m OH^-)

$$= Sb_2O_3 + 3H_2O + 3H_2$$

(b) The corresponding Nernst expression is

$$\mathscr{E} = \mathscr{E}^0_{Sb/Sb_2O_3} - \frac{0.059}{6} \log \frac{1}{(H^+)^6(OH^-)^6}$$

or

$$\mathscr{E} = 0.66 + 0.059 \log K_w = -\mathbf{0.17}\ \mathbf{V}$$

(c) Increasing the KCl concentration will (by the limiting Debye–Hückel law) reduce the activity coefficient $\gamma_{\pm}$ (of H^+, OH^-). Assuming K_w is essentially the thermodynamic constant, then in 0.1 m electrolyte, the concentration product $(H^+)(OH^-)$ should be greater than 1×10^{-14}. The answer to (b) would then be more positive than -0.17 V.

40. The cell reaction is

$$2Ag + 2H^+(0.2\ m) + SO_4^{2-}(0.1\ m) = Ag_2SO_4 + H_2(1\ atm)$$

and the corresponding Nernst equation is then

$$\mathscr{E} = \mathscr{E}^0 - \frac{RT}{n\mathscr{F}} \ln Q$$

where $Q = P_{H_2}/(H^+)^2(SO_4^{2-})$. Q itself does not change with temperature, so $\mathscr{E} - \mathscr{E}^0$ is proportional to T. Then

$$\frac{(\mathscr{E} - \mathscr{E}^0)_{298}}{(\mathscr{E} - \mathscr{E}^0)_{308}} = \frac{298}{308}$$

and $(\mathscr{E} - \mathscr{E}^0)_{308} = (-0.7 + 0.63) \times 308/298 = -\mathbf{0.072}\ \mathbf{V}.$

16

CHEMICAL KINETICS

COMMENTS

Chemical kinetics, even as limited by the scope of the problems that follow, is a fairly complex topic. There are several somewhat distinct disciplines within it, however, and it is helpful to consider these separately.

Typically, the rate of a known over-all reaction is being studied; we know, therefore, the stoichiometric relationships between reactants and products; thus, $aA + bB + \cdots = mM + nN + \cdots$. The rate, expressed perhaps as $d[A]/dt$, will be some function of the concentrations of A, B, M, and N, and of any catalyst present (activity-coefficient effects will be neglected here). This function is usually of the type $d[A]/dt = -k[A]^x[B]^y[M]^z$ and so on, as far as the present problems are concerned, and there are two general approaches for finding x, y, z, and so on. First, we may determine an initial rate, or the amount of reaction per unit time, where the experimental time interval is short enough that no appreciable change in concentrations has occurred. Several experiments are made in which the initial concentration of each species is varied in turn. Thus, if the rate quadruples on doubling A (keeping all other concentrations constant), then $x = 2$. Once each exponent has been determined, the specific rate constant k is obtained by substituting into the rate equation the data from any one of the experiments.

The alternative approach is to measure the amount of reaction over an extended period of time, in which case the data must be fitted to an integrated rate law. With respect to the present problems, it will not generally be necessary to consider more than the simple first- and second-order integrated forms. As you study the answers, you will encounter various shortcuts to testing whether the data fit one or the other form. Often one or more of the reactants will be in great excess so that their concentrations will not change. Thus if, in the case of the example above, the rate depends only on A and B, and B is in considerable excess, we can write

$$d(A)/dt = -k_{app}[A]^x \qquad (16\text{-}1)$$

where $k_{app} = k[B]^y$. Since B is constant, the system acts as if it were following the simpler rate law, but with a rate constant given as k_{app}. A second special situation is that which results when the initial concentrations of A and B are in the ratio required by the over-all stoichiometry. Thus if $b[A]^0 = a[B]^0$, then $b[A] = a[B]$ throughout the course of the reaction, and the rate law reduces to $d[A]/dt = -k(b/a)^y[A]^{x+y}$. There is a fair amount of emphasis on these devices since they, in fact, represent desirable experimental procedure.

The experimental determination of $d[A]/dt$, or of the extent of reaction as a function of time, requires some type of analysis for one or another species. Often, it is possible to measure some physical property of the system that is an additive one, that is, one that can be written as a sum of contributions from each species. The pressure of a mixture of ideal gases at constant volume and temperature and the optical density at a given wavelength for a solution containing solutes obeying Beer's law would be examples of an additive property. When a property $\mathscr{P}$ is additive, the degree of advancement of the reaction is given by $(\mathscr{P} - \mathscr{P}^\infty)/(\mathscr{P}^0 - \mathscr{P}^\infty)$. Thus for a reaction of the type $A \rightarrow$ products,

$$[A]/[A]^0 = (\mathscr{P} - \mathscr{P}^\infty)/(\mathscr{P}^0 - \mathscr{P}^\infty) \tag{16-2}$$

if the reaction goes to completion.

A second general aspect of chemical kinetics is that of interconnecting the experimental rate law and a reaction mechanism. With the possible exception of the case where solvent is a reactant, it is safe to assume that the actual chemical processes consist of one or more bimolecular reaction steps. There may be a rapid equilibrium (or preequilibrium) to form an intermediate that is then involved in the rate-determining step. With more complex mechanisms there may be several intermediates; they may be formed, react, and be regenerated in sequential steps so as to give a chain reaction.

There are several problems involving a two-step mechanism, with the first step reversible, for example,

$$A + B \underset{k_2}{\overset{k_1}{\rightleftharpoons}} C + I \tag{16-3}$$

$$I + A \overset{k_3}{\rightarrow} \text{products} \tag{16-4}$$

Here I denotes a reactive intermediate and hence one present only in very small concentrations. Then $d[I]/dt = k_1[A][B] - k_2[C][I] - k_3[I][A]$, and as an approximation $d[I]/dt$ is set equal to zero so that

$$[I] = \frac{k_1[A][B]}{k_2[C] + k_3[A]} \tag{16-5}$$

The rate of formation of products is $R = k_3[A][I]$, and insertion of the expression for $[I]$ gives the final form of the rate law. In other words, it is customary that a rate law should be expressed in terms of those species that are in fact present in the over-all equation, in so far as possible. Incidentally, the above type of approximation is called the *steady-state approximation*.

Notice that, in the above example, if $k_2[C]$ is much greater than $k_3[A]$, the rate law reduces to the case of a preequilibrium followed by a rate-determining step. Notice, too, that the concentration of one of the products, C, appears in the rate law; this is a sure indication that the mechanism does involve a reversible step preceding the rate-determining one.

The third major discipline in chemical kinetics is that of the theory of reaction rates. We are concerned here with treating the rate of the rate-determining (and usually bimolecular) step; the theory is then essentially that of bimolecular reactions. The two approaches that are dwelt on here are the classical Arrhenius and the transition-state theories. According to the first one, we view the reaction as a collision in which molecules are required to have between them a certain minimum energy E^*, called the activation energy. Thus

$$k = Ae^{-E^*/RT} \tag{16-7}$$

where A is the frequency factor and contains the terms (except for concentrations) that determine the collision frequency.

Transition-state theory holds that the reactants are in equilibrium with a special kind of intermediate: $A + B = (AB)^{\ddagger}$. The transition state $(AB)^{\ddagger}$ has the property of possessing the structure and the energy such that it can go over to products within the time of about one vibration or about 10^{-13} sec. We then write

$$k = 10^{13}e^{-\Delta G^{0\ddagger}/RT} = 10^{13}e^{\Delta S^{0\ddagger}/R}e^{-\Delta H^{0\ddagger}/RT} \tag{16-8}$$

For simplicity, we shall subsequently write $\Delta G^{\ddagger}$, $\Delta S^{\ddagger}$, and $\Delta H^{\ddagger}$; also, in comparing the two theories, the distinction between energy and enthalpy will be ignored.

Either of the above formulations can be applied to the experimental specific rate constant, but the resulting E^* or $\Delta H^{\ddagger}$ value will have the above theoretical significance *only* if the experimental rate constant is the specific rate constant for the rate-determining single step. When a two-or-more step mechanism is involved, as in the above example of a steady-state treatment, the experimental reaction rate constant is composite. For example, in the simple case of a pre-equilibrium followed by a rate-determining step, we might have $k_{app} = kK$, where K is the equilibrium constant for the pre-equilibrium. The apparent activation energy will now be the sum of a true activation energy and the ΔH^0 for the equilibrium.

EQUATIONS AND CONCEPTS

Mass Action Law

For a reaction $aA + bB \rightarrow$ products,

$$d[A]/dt = -k[A]^a[B]^b \qquad (16\text{-}9)$$

First-Order Rate Equation

$d[A]/dt = -k[A]$. $[A] = [A]^0 e^{-kt}$ if the reaction goes to completion. Half-life, $T_{1/2}$, is given by $T_{1/2} = 0.69/k$. In successive half-life times, $[A]/[A]^0 = \frac{1}{2}, \frac{1}{4}, \ldots, (\frac{1}{2})^n$ where n is the number of half-lives. The time for $[A]/[A]^0$ to equal to some other fraction F is $T_F = (-\ln F)/k$. Again, $[A]/[A]^0 = F, F^2, \ldots, F^n$, where n is the number of time intervals T_F.

Second-Order Rate Equation

$$d[A]/dt = -k[A]^2 \qquad 1/[A] - 1/[A]^0 = kt \qquad (16\text{-}10)$$

Half-life: $T_{1/2} = 1/[A]^0 k$. At the beginning of each succeeding half-life, $[A]^0$ would be the concentration at that point, so that the time for successive decreases in $[A]$ by a half doubles each time. Thus $T_{1/4} = 3T_{1/2}$.

Other Rate Equations

Integrated forms for rate equations of the type $d[A]/dt = -k[A]^a[B]^b$ where $[A] \neq [B]$, will not be required here. However, if the reaction stoichiometry is $aA + bB =$ products, and $b[A]^0 = a[B]^0$, then $b[A] = a[B]$ at all times, and the rate law reduces to $d[A]/dt = -k(b/a)^y[A]^{x+y}$. The general form is then $d[A]/dt = -k[A]^n$, which on integration gives $1/[A]^{n-1} - 1/[A^0]^{n-1} = (n-1)kt$.

Units

$[A]$ may be the concentration of species A or, in the case of gases, the partial pressure of A. The dimensions should be given for rate constants.

Mass-Action Equilibrium

On equating the forward and reverse rates for an equilibrium system, we obtain $K = k_f/k_b$, where k_f and k_b denote the reaction rate constants for the forward and reverse processes. Some complications can develop if this procedure is applied to the case of complex mechanisms in that one may obtain K^n, where n is a rational number rather than just K.

Arrhenius Equation

$$k = Ae^{-E^*/RT} \quad \text{or} \quad \log k_2/k_1 = \frac{E^*}{2.3R}(1/T_1 - 1/T_2) \tag{16-11}$$

where A is the frequency factor, E^* is the activation energy, and k_1 and k_2 are the rate constants at T_1 and T_2.

Transition-State Theory

$$k = 10^{13}e^{\Delta S^{\ddagger}/R}e^{-\Delta H^{\ddagger}/RT} \quad \text{or} \quad k = 10^{13}\,e^{-\Delta G^{\ddagger}/RT} \tag{16-12}$$

where k must now be in $\sec^{-1}$. Again,

$$\log\frac{k_2}{k_1} = \frac{\Delta H^{\ddagger}}{2.3R}\left(\frac{1}{T_1} - \frac{1}{T_2}\right) \tag{16-13}$$

Heterogeneous Catalysis

In some cases of heterogeneous catalysis it is useful to assume that reaction occurs on the surface and that the adsorbed species obey the Langmuir adsorption isotherm. For the ith species of a gaseous mixture,

$$\theta_1 = \frac{b_i P_i}{1 + \Sigma b_j P_j} \tag{16-14}$$

where θ denotes the fraction of surface occupied, and the b's are constants of the form $b_i = b_i^0\,e^{Q_i/RT}$, where Q is the heat of adsorption. We further assume that the mass-action rate law applies on the surface, so that, for a reaction $A + B \rightarrow$ products, $dP_A/dt = -k\theta_A\theta_B$. For the simplest case of a first-order surface decomposition, $A \rightarrow$ products,

$$\frac{dP_A}{dt} = -k\frac{bP_A}{1 + bP_A} \tag{16-15}$$

The high- and low-pressure limiting forms are then $dP_A/dt = -k$ and $dP_A/dt = -kbP_A$.

Apparent Activation Energy

If the experimental reaction rate constant is actually composite, that is, a k_{app}, and is equal to some product of temperature dependent quantities, application of the Arrhenius equation gives only an apparent activation energy. If $k_{app} = kK$, where K is an equilibrium constant, then $E^*_{app} = E^* + \Delta H^0$, where ΔH^0 is the enthalpy change for the equilibrium. Similarly, if in heterogeneous catalysis,

$$k_{app} = kb \tag{16-16}$$
$$= kb^0 e^{Q/RT}$$
$$E^*_{app} = E^* - Q \tag{16-17}$$

PROBLEMS

1. (30 min) The kinetics of the reaction: $2Fe(CN)_6^{3-} + 2I^- = 2Fe(CN)_6^{4-} + I_2$ was studied by determining the initial rate of iodine production for mixtures of various compositions, as given below, at 25°C. None of the solutions initially contained any iodine.

(a) The rate law can be expressed in the form

$$d(I_2)/dt = k[Fe(CN)_6^{3-}]^a[I^-]^b[Fe(CN)_6^{4-}]^c[I_2]^d$$

Show what can be deduced about the values of a, b, c, and d. Calculate the value of k (give its dimensions as well).

(b) The data in Table 16-1, in terms of absolute reaction rate theory, lead to a free energy of activation of 18,000 cal at 25°C. Data at 35°C lead to a value of 18,200 cal. Calculate $\Delta H^{\ddagger}$ and $\Delta S^{\ddagger}$ of activation.

TABLE 16-1   DATA AT 25°C

Run no.	Composition, M			Initial rate (moles I_2/liter/h)*
	$Fe(CN)_6^{3-}$	I^-	$Fe(CN)_6^{4-}$	
1	1×10^{-3}	1×10^{-3}	1×10^{-3}	1×10^{-3}
2	2×10^{-3}	1×10^{-3}	1×10^{-3}	4×10^{-3}
3	1×10^{-3}	2×10^{-3}	2×10^{-3}	1×10^{-3}
4	2×10^{-3}	2×10^{-3}	1×10^{-3}	8×10^{-3}

* Actually determined from the amount of I_2 produced in the first few seconds.

(c) Assuming that the values of a, b, c, and d are 2, 2, -1 and 0, respectively, suggest a mechanism and show briefly that it does lead to a rate law consistent with these values. This can be brief. It *is* essential that you make it clear that you understand how to get the desired rate law from your mechanism.

TABLE 16-2

	Experiment 1			
	$(P^0_{Cl_2} = 400$ mm Hg, $P^0_{CO} = 4$ mm Hg)			
Time, min	0	34.5	69.0	138	∞
P_{COCl_2}, mm Hg	0	2.0	3.0	3.75	4.0
	Experiment 2			
	$(P^0_{Cl_2} = 1,600$ mm Hg, $P^0_{CO} = 4$ mm Hg)			
Time, min	0	34.5	69.0	∞
P_{COCl_2}, mm Hg	0	3.0	3.75	4.0

2. (30 min) The data in Table 16-2 are obtained for the reaction CO(g) + Cl_2(g) = $COCl_2$(g) at 25°C (which goes to completion). In each case the indicated initial pressures of reactants are introduced into a flask at 25°C.

(a) The rate law for the reaction has the form $dP_{COCl_2}/dt = kP^a_{CO}P^b_{Cl_2}$. Determine the values of a and b.

(b) Calculate the numerical value for k. Give the units of k.

(c) Suppose that the rate law were actually $dP_{CO}/dt = -kP_{CO}$, with $k = 0.01$ min^{-1}, but that through a misprint it were written $dP_{CO}/dt = -kP^{1/2}_{CO}$ [same value of k numerically, but with units min^{-1}(mm Hg)$^{1/2}$]. Assuming P^0_{CO} to be 4 mm Hg, show whether the half-life calculated by the erroneous rate law would be longer, shorter, or the same as the correct half-life. (This may be done by qualitative reasoning.)

3. (15 min) (final examination question) At 25°C and at a constant pH of 5, the inversion of sucrose proceeds with a constant half-life of 500 min. At this same temperature, but at a pH of 4, the half-life is constant at 50 min. Explain what the exponents a and b must be in the rate law $d(\text{sucrose})/dt = -k \times (\text{sucrose})^a[H^+]^b$.

4. (22.5 min) Some PH_3(g) is introduced into a flask at 600°C containing inert gas. The PH_3 proceeds to decompose into P_4(g) and H_2(g) and the reaction goes essentially to completion. The total pressure is given below as a function of time.

Time, sec	0	60	120	∞
P, mm Hg	262.40	272.90	275.53	276.40

(a) Show the order of the reaction and calculate the reaction rate constant. Give its units.

(b) The back reaction obeys a rate law of the form

$$dP_{PH_3}/dt = k_2 P_{P_4}^a P_{H_2}^b$$

Show what the values of a and b must be.

(c) Assuming the initial pressure of PH_3 was 18 mm Hg and that the half-life was 20 sec, calculate the value of the reaction rate constant for the decomposition if it is first order in PH_3. Do likewise, assuming the decomposition to be second order. State the units of k_1 in both cases.

5. (12 min) The Pt-catalyzed decomposition of HI obeys the rate law $dP_{HI}/dt = k_1$, at high pressures, with $k_1 = 500$ mm Hg/sec at 100°C. At low pressures, the rate law becomes $dP_{HI}/dt = k_2 P_{HI}$, with $k_2 = 50 \sec^{-1}$ at 100°C. Calculate the HI pressure at which the value of dP_{HI}/dt should be 250 mm Hg/sec at 100°C.

6. (24 min) (final examination question) A gaseous substance decomposes according to the over-all equation $AB_3 = \frac{1}{2}A_2 + \frac{3}{2}B_2$. The variation of the partial pressure of AB_3 with time (starting with pure AB_3) is given below for 200°C.

Time, h	0	5.0	15.0	35.0
P_{AB_3}, mm Hg	660	330	165	82.5

The decomposition is irreversible. Show what the order of the reaction is and calculate the rate constant (give its units).

7. (30 min) The reaction $2NO + H_2 = N_2O + H_2O$ goes to completion and is known to follow the rate law

$$dP_{N_2O}/dt = k P_{NO}^2 P_{H_2}$$

The data in Table 16-3 have been obtained.

TABLE 16-3

Run	P_{NO}^0 (mm Hg)	$P_{H_2}^0$ (mm Hg)	Half-time (sec)	Temperature (°C)
1	600	10	19.2	820
2	600	20	.	820
3	10	600	830	820
4	20	600		820
5	600	10	10	840

(The half-times are the time for half completion of the reaction, i.e., for the species not in excess to fall to half of its initial pressure. The runs are at constant temperature and volume.)

(a) Insert the missing half-times, with explanation. (b) Calculate k at 820°C, in $(mm\ Hg)^{-2}\ sec^{-1}$. (c) Calculate the activation energy. It is sufficient to set up an equation in which E^* is the only nonnumerical quantity. (d) Suggest a mechanism giving the above rate law that involves only bimolecular steps or reaction sequences. (e) Calculate the half-life at 820°C if $P^0_{NO} = 20\ mm\ Hg$ and $P^0_{H_2} = 10\ mm\ Hg$. (*Note:* The rate law reduces to a special case in this instance.)

8. (24 min) (final examination question) The reaction $CH_3CONH_2 + HCl + H_2O = CH_3COOH + NH_4Cl$ may be followed by measuring the conductivity of the system. On mixing equal volumes of 2-N solutions of acetamide and HCl, at 63°C, the following conductivities are observed:

Time, min	0	13	34	52
$L,\ ohm^{-1}cm^{-1}$	0.409	0.374	0.333	0.310

The ionic equivalent conductances at 63°C are 515, 133, and 137 $cm^2equiv^{-1} ohm^{-1}$ for H^+, Cl^-, and NH_4^+, respectively. Neglecting nonideality effects, determine the order of the reaction and calculate the value of the reaction rate constant.

9. (24 min) (final examination question) The reaction $Co(NH_3)_5F^{2+} + H_2O = Co(NH_3)_5(H_2O)^{3+} + F^-$ is acid catalyzed and proceeds according to the rate law

$$rate = -d[Co(NH_3)_5F^{2+}]/dt$$
$$= k[Co(NH_3)_5F^{2+}]^a[H^+]^b$$

The times for half and for three-quarters of the complex to react are given in Table 16-4 for the indicated temperatures and initial concentrations.

(a) Show what the values of the exponents a and b must be. (b) Calculate the value of k. (c) Calculate the activation energy. You may set up an equation in which E^* is the only unknown.

TABLE 16-4

$[Co(NH_3)_5F^{2+}]$, M	$[H^+]$, M	t (°C)	$T_{1/2}$ (h)	$T_{3/4}$ (h)
0.1	0.01	25	1	2
0.2	0.02	25	0.5	1
0.1	0.01	35	0.5	1

10. (20 min) (final examination question) The rate of the acid catalyzed lactonization of hydroxyvaleric acid is given by the rate law

$$R = -d[\text{HVA}]/dt = k[\text{HVA}][\text{HCl}]$$

The net reaction is

$$\text{CH}_3\text{CHOHCH}_2\text{CH}_2\text{COOH} = \text{CH}_3\text{—CH—CH}_2\text{—CH}_2\text{—CO} + \text{H}_2\text{O}$$
$$\underset{\text{(HVA)}}{\lfloor\text{————O————}\rfloor}$$

and k has the value of $4.0 \, M^{-1} \, \text{min}^{-1}$ at 25°C.
(a) Neglecting any back reaction, calculate the time required for half the HVA to react, if the initial concentrations are $0.01 \, M$ for HVA and for HCl (at 25°C). (b) For the same initial concentrations as in (a), at what temperature should the half-life be halved if the activation energy is 20 kcal?

11. (30 min) The decomposition of gaseous paraldehyde into gaseous acetaldehyde, which may be represented by the equation, $P = 3A$, has been followed at 260°C by observing the change in total pressure with time (see Table 16-5).

TABLE 16-5

Time (h)	P_{tot}(mm Hg)	Fraction P decomposed
0	100	
1	173	
2	218	
3	248	
4	266	
∞	300	

(a) Calculate the fraction of paraldehyde decomposed at each time. (b) Show by means of appropriate numerical calculations whether the reaction is first or second order. (c) Calculate the reaction rate constant at 260°C. (d) Assuming that the reaction rate constant at 260°C is $1.0 \, \text{h}^{-1}$, and that the entropy of activation is zero, calculate the energy of activation and the value of the specific rate constant at 280°C.

12. (30 min) The reaction $2\text{Fe}^{2+} + 2\text{Hg}^{2+} = \text{Hg}_2^{2+} + 2\text{Fe}^{3+}$ has been followed spectrophotometrically. The measurement is that of the optical density D, versus time, of a solution containing initially only ferrous and mercuric ions. (D is an additive property.) Given in Table 16-6 the two runs

TABLE 16-6

	Run A *Initially:* $[Fe^{2+}] = 0.1\ M$ $[Hg^{2+}] = 0.1\ M$		Run B *Initially:* $[Fe^{2+}] = 0.1\ M$ $[Hg^{2+}] = 0.001\ M$	
Time, sec	D	$[Hg^{2+}]/0.1$	Time, sec	$[Hg^{2+}]/0.001*$
0	0.1		0	1
1×10^5	0.4		0.5×10^5	0.585
2×10^5	0.5		1×10^5	0.348
3×10^5	0.55		1.5×10^5	0.205
∞	0.7		2×10^5	0.122
			∞	0

* The actual concentration ratios are given for this run, instead of the
D values, to save your time.

at 80°C, (a) calculate the ratios $([Hg^{2+}]/0.1)$ for run A. (b) What is the order of
the reaction in run A, and what is it in run B? Show how you reached your
conclusions. (c) If the rate equation is written $R = k[Fe^{2+}]^p[Hg^{2+}]^q$ show
what the values of p and q must be.

13. (12 min) (final examination question) The rate of the acid catalyzed
hydrolysis of ethylacetate in hydrochloric acid solution obeys the following
rate law

$$R = -d(\text{ester})/dt = k(\text{ester})(\text{HCl})$$

where $k = 0.1\ M^{-1}\ h^{-1}$ at 25°C. Neglecting any back reaction, calculate
the time required for half of the ester to be hydrolyzed if the initial concentrations of ester and of HCl catalyst are $0.02\ M$ and $0.01\ M$, respectively.

14. (12 min) A first-order reaction, for example $(Co(NH_3)_5Br^{2+} + H_2O =$
$Co(NH_3)_5(H_2O)^{3+} + Br^-)$, is occurring in solution and its progress is
followed by periodic measurements of the optical density of the solution, and
additive property. The reaction goes to completion. Given that the optical
density D is 0.80 at 20 min elapsed time, 0.35 after 40 min and 0.20 at infinite
time, calculate the specific rate constant and the optical density at zero time.

15. (9 min) In the study of a first-order reaction, $A \rightarrow$ products, it is found
that A/A_0 is 0.125 after 1 h. The system initially consisted of 0.2 mole of
gaseous A at STP. Calculate the initial rate of reaction in moles of A reacting
per second.

16. (30 min) Smith and Daniels [*J. Am. Chem. Soc.*, **69** 1735 (1947)] studied
the kinetics of the reaction (which goes to completion) $N_2O_5 + NO =$

$3NO_2(25°C)$. When the initial pressures of N_2O_5 and NO were 1 mm Hg and 100 mm Hg, respectively, a plot of log $P_{N_2O_5}$ versus time gave a straight line with a slope corresponding to a half-life of 2.0 h. In a second experiment, with initial pressures of N_2O_5 and NO each 50 mm Hg, the following data were obtained:

P_{tot} (mm Hg)	100	115	125
Time (hours)	0	1	2

(a) Assuming the experimental rate law to be expressible in the form: rate $= kP_{N_2O_5}^x P_{NO}^y$, show from the above data what x and y must be, and calculate k.

(b) The authors propose the mechanism

$$N_2O_5 \underset{k_2}{\overset{k_1}{\rightleftharpoons}} NO_2 + NO_3$$

$$NO + NO_3 \overset{k_3}{\rightarrow} 2NO_2$$

Using the stationary-state assumption, derive the rate law for this mechanism, that is the expression for $dP_{N_2O_5}/dt$. Relate k above to one or more of the specific rate constants in the mechanism. Discuss what can be said about the relative magnitudes of k_2 and k_3.

(c) Calculate the half-life if the initial pressures of N_2O_5 and NO were 100 and 1 mm Hg, respectively, that is, the time for half of the NO to react.

17. (30 min) Higginson and coworkers (Manchester University, England) have proposed the following mechanism for one of the redox reactions that they studied (between iron and vanadium):

(1) $Fe(III) + V(IV) \underset{k_{-1}}{\overset{k_1}{\rightleftharpoons}} Fe(II) + V(V)$

(2) $V(V) + V(III) \overset{k_2}{\rightarrow} 2V(IV)$ (rate-determining)

(a) To what over-all reaction does this mechanism correspond?

(b) Derive the rate law according to the mechanism, assuming reaction (1) to be a rapid equilibrium. Show what the actual activation energy is for reaction (2) if the ΔH^0 for (1) is -5 kcal and the apparent activation energy for the over-all reaction is 12 kcal.

(c) Vanadium (V) acts as a trace intermediate in this case. Using the stationary-state hypothesis, derive an expression for the concentration of $V(v)$ in terms of the concentrations of the other species.

(d) For the reverse reaction, the slow step is $2V(\text{IV}) \xrightarrow{k_2'} V(\text{V}) + V(\text{III})$. If k_{-2} is found to be $0.01\ M^{-1}\ \text{sec}^{-1}$ at $25°C$, and from theoretical considerations, the entropy of activation is estimated to be $-5\ \text{cal/deg}$, calculate the energy of activation and the collision-frequency factor A in the Arrhenius theory $(k = Ae^{-E^*/RT})$. It is sufficient to set up equations in which numerical values have been inserted for all quantities except E^* and A.

18. (7.5 min) The following mechanism is proposed for the alkaline hydrolysis of $Co(NH_3)_5Cl^{2+}$:

$$Co(NH_3)_5Cl^{2+} + OH^- \overset{K}{=} Co(NH_3)_4(NH_2)(Cl)^+ + H_2O$$
$$\text{(rapid equilibrium)}$$

$$Co(NH_3)_4(NH_2)(Cl)^+ \xrightarrow{k_1} Co(NH_3)_4(NH_2)^{2+} + Cl^- \qquad \text{(slow)}$$

$$Co(NH_3)_4(NH_2)^{2+} + H_2O \xrightarrow{k_2} Co(NH_3)_5(OH)^{2+} \qquad \text{(fast)}$$

Derive the rate law for this mechanism.

19. (18 min) The acid-catalyzed esterification reaction $RCOOH + R'OH \xrightarrow{H^+} RCOOR' + H_2O$ is studied under conditions such that it goes nearly to completion, and the rate law is $d[\text{ester}]/dt = k_1[H^+][A][B]$, where A and B denote $RCOOH$ and $R'OH$, respectively. Furthermore, with the specific conditions of pH 2 and initial concentrations $[A]^0 = [B]^0 = 0.01\ M$, the half-life is found to be 1 h at $25°C$, and 2 h at $15°C$.

(a) Calculate the true and the apparent reaction rate constants at $25°C$. Specify their units. (b) Calculate the activation free energy at $25°C$, and the activation energy and entropy. (c) Discuss whether the activation energy found in (b) is probably a true one, in the collision-theory sense of being the minimum total combined kinetic energies that the two reactants must possess in order to react.

20. (30 min) The homogeneous decomposition of ozone, $2O_3 = 3O_2$, has a number of complications to its kinetics, one of which is featured here. The decomposition is catalyzed by various gases, one of which is CO_2, and the data of Table 16-7 give the variation with time of the total pressure of a mixture kept at $50°C$ and consisting initially of O_3 and CO_2.

(a) Show what the apparent order is for run 1, and calculate k_{app}.

(b) Assume the complete rate law to be of the form $d[O_3]/dt = -k[O_3]^a[CO_2]^b$, show what the value of the exponent b must be, and calculate k.

(c) The value of k quadruples with a $10°C$ rise in temperature. Show how to calculate E^* by setting up an equation in which this is the only non-numerical quantity.

TABLE 16-7

Run 1 $[CO_2] = 0.01$ mole/liter		Run 2 $[CO_2] = 0.005$ mole/liter	
Time (min)	P_{tot} (mm Hg)	*Time* (min)	P_{tot} (mm Hg)
0	400	0	300
30	450	30	330
60	475	60	350
∞	500	120	375
		∞	400

(d) A proposed mechanism for the decomposition in the absence of CO_2 is

$$2O_3 \underset{k_2}{\overset{k_1}{\rightleftharpoons}} O_3 + O_2 + O$$

$$O_3 + O \overset{k_3}{\rightarrow} 2O_2$$

Derive the rate law corresponding to this mechanism, assuming stationary state conditions.

21. (30 min) Nitrous oxide decomposes according to the reaction $2N_2O = 2H_2 + O_2$. The rate is quite small, however, unless halogens are present as catalyst. Specifically, Cl_2 catalyzes the above decomposition, and the rate under this circumstance is dependent on both the N_2O and the Cl_2 pressure; that is,

$$R = kP_{N_2O}^a P_{Cl_2}^b$$

The course of the decomposition can be followed by measuring the increase in pressure with time (at constant temperature), and in some recent work the procedure was to determine the initial rate of the reaction by obtaining the increase in pressure over a short enough time that the value of P_{N_2O} was still essentially at its initial value. Some of the results are as in Table 16-8.

(a) Determine from the above data the values of a and b in the rate law.

(b) Calculate the value for the reaction rate constant k at 800°K.

(c) Calculate the half-life for the reaction (time for half of the initial N_2O to be decomposed) under the conditions of the first line of the table of data.

(d) Calculate the activation energy for the reaction. Also, set up the equation whereby the frequency factor may be calculated. It is sufficient to obtain an equation containing the frequency factor as the *only* nonnumerical quantity.

TABLE 16-8

$T,°K$	Initial pressure (cm Hg) P_{N_2O}	Initial pressure (cm Hg) P_{Cl_2}	Initial rate (in cm Hg increase in pressure per min)
800	30	4	0.3
	15	4	0.15
	30	1	0.15
810	30	4	0.6

(3 minutes extra) Set up a chemically reasonable mechanism for the Cl_2 catalyzed reaction, which will lead to the observed rate law.

22. (30 min) The reaction $C_3H_7Br + S_2O_3^{2-} = C_3H_7S_2O_3^- + Br^-$ is bimolecular and the second-order specific rate constant is $1.64 \times 10^{-3} \, \text{sec}^{-1} \, M^{-1}$ at 37°C. The initial concentrations in a particular experiment are $A = 0.1 \, M$ and $[S_2O_3^{2-}] = 0.1 \, M$ (where A denotes the C_3H_7Br concentration).

(a) Calculate the initial rate dA/dt. (b) Derive the integrated rate law for this experiment, that is, find A as $f(t)$. (c) Calculate the time for the rate dA/dt to drop to one fourth of its initial value. (d) If $\Delta H^‡$ is 20 kcal, calculate the temperature coefficient, that is, the specific rate constant ratio for a 10° rise in temperature. (e) Calculate $\Delta S^‡$. To save time, a numerical value need not be obtained; write an equation in which $\Delta S^‡$ is the only nonnumerical quantity.

(10% extra credit) Derive the stationary state rate law for the reaction scheme

$$CH_3COCH_3 + OH^- \underset{k_2}{\overset{k_1}{\rightleftharpoons}} CH_3COCH_2^- + H_2O$$

$$CH_3COCH_2^- + Br_2 \overset{k_3}{\rightarrow} CH_3COCH_2Br + Br^-$$

$$(k_2 \gg k_3; k_1 \gg k_3)$$

23. (30 min) At the moment of mixing, a solution contained $1 \times 10^{-2} \, M$ $S_2O_8^{2-}$ and $2 \times 10^{-2} \, M \, Mo(CN)_8^{4-}$, which then reacted (at 20°C) as follows:

$$S_2O_8^{2-} + 2Mo(CN)_8^{4-} = 2SO_4^{2-} + 2Mo(CN)_8^{3-}$$

The concentrations of $Mo(CN)_8^{4-}$ (in moles/liter) after times t were

Conc.	0.02	0.015	0.010	0.005	0.0025
t, hours	0	30	90	270	630

(a) The rate law is of the form

$$R = -d[\text{Mo(CN)}_8^{4-}]/dt = k[\text{S}_2\text{O}_8^{2-}]^a[\text{Mo(CN)}_8^{4-}]^b$$

Show the value of $(a + b)$; that is, what is the over-all order of the reaction.
(b) Calculate the value of k consistent with your answer to (a).
(c) Assume that $a = 1$ and $b = 1$. If k is $2 \times 10^{-4} \, M^{-1} \, \text{h}^{-1}$ at 20°C and $2 \times 10^{-3} \, M^{-1} \, \text{h}^{-1}$ at 45°C, calculate the values of $\Delta H^{\ddagger}$ and $\Delta S^{\ddagger}$.
(d) Derive the rate equation that would be predicted by the mechanism

$$\text{S}_2\text{O}_8^{2-} = 2\text{SO}_4^{-} \qquad \text{(rapid equilibrium)}$$

$$\text{SO}_4^{-} + \text{Mo(CN)}_8^{4-} \xrightarrow{k_1} \text{SO}_4^{2-} + \text{Mo(CN)}_8^{3-} \qquad \text{(slow)}$$

24. (6 min) The forward rate for the reaction $2\text{NO} + \text{O}_2 = 2\text{NO}_2$ obeys the rate law: rate $= k(\text{NO}^2(\text{O}_2))$. The specific rate constant k is found to decrease with increasing temperature. Give a reasonable explanation for this unusual behavior.

25. (12 min) (final examination question) For the reaction $\text{H}_2 + \text{I}_2 \xrightarrow{k_f} 2 \text{HI}$, the forward reaction is second order with k_f having the value 1×10^{-6} $\text{sec}^{-1}\text{atm}^{-1}$ at 400°K. In a particular kinetic study, the reaction is followed by introducing an initial pressure of 0.5 atm of hydrogen into a flask that contains excess solid iodine. It may be assumed that the solid iodine rapidly establishes and remains in equilibrium with its own vapor pressure of 1.2 atm, and that there is no back reaction.

Calculate the time for half of the added hydrogen to react. Discuss whether the actual mechanism of the reaction could be (insofar as the above information is concerned)

$$\text{I}_2(\text{g}) \overset{K}{=} 2\text{I} \qquad \text{(rapid equilibrium)}$$

$$\text{H}_2 + 2\text{I} \xrightarrow{k} 2 \text{HI} \qquad \text{(slow step)}$$

26. (12 min) (final examination question) The decomposition of N_2O is catalyzed by solid In_2O_3, and the rate law is $d(\text{moles of } \text{N}_2\text{O})/dt = -k\theta_{\text{N}_2\text{O}}$, where $\theta_{\text{N}_2\text{O}}$ is the fraction of surface covered by adsorbed N_2O; the adsorption obeys the Langmuir isotherm. The heat of adsorption of N_2O on In_2O_3 is 10 kcal. *Explain* whether the reaction rate should be more or less temperature dependent (i.e., should vary more or less rapidly with temperature) if $P_{\text{N}_2\text{O}}$ is low (and constant) as opposed to being high (and constant).

---

27. (15 min) (final examination question) A certain reaction may proceed by a direct path or by a catalyzed one. The specific rate constants for the two mechanisms are k and k'. If $\Delta S^{\ddagger}$ is 10 cal/deg greater than $\Delta S^{\ddagger'}$, and $\Delta E^{\ddagger}$ is 5 kcal greater than $\Delta E^{\ddagger'}$, show which specific rate constant is larger at 25°C.

28. (21 min) The reaction $RI + HI = RH + I_2$ is second order (first order in each reactant) and occurs with an activation energy of 25 kcal. If the concentration of each reactant were $1 M$, the collision frequency would be 0.5×10^{11} moles of collisions/liter/sec. The specific rate constant should then be given by $k(M^{-1} \sec^{-1}) = 0.5 \times 10^{11} e^{-E*/RT}$.

(a) The actual value of k is 1,000 times less than that given by this formula, however. Calculate the actual value of k at 25°C and the actual value of $\Delta S^{\ddagger}$ for the reaction. It is sufficient to give the answers in the form of a product of numbers and of ten to some power, if more convenient to do so.

TABLE 16-9

Experiment (25°C)	C_{RI}^{0} (M)	C_{HI}^{0} (M)
1	1	1
2	1	0.01
3	0.1	10
4	0.01	0.01

(b) Arrange the experiments in Table 16-9 in the order of increasing half-life (i.e., time for half of that reactant not in excess to be gone).

29. (19 min) (final examination question) Suppose that substance A may react to give products by either of two quite different reaction paths, both first order:

(1) $\quad A \xrightarrow{k_1}$ products B and C

(2) $\quad A \xrightarrow{k_2}$ products D and E

Products B and C might be the desired ones, whereas D and E would then be unwanted by-products. Assume that the frequency factors for the two reactions are identical (and independent of temperature) but that the activation energy for (1) is greater than that for (2). (a) Make a semiquantitative plot of $\log k$ versus $1/T$ for k_1 and k_2 (both on the same plot). (b) Explain whether the rate of (1) can ever be faster (or slower) than that of (2).

30. (30 min) The reaction $cis\text{-}Cr(en)_2(OH)_2^{+} \underset{k_2}{\overset{k_1}{\rightleftharpoons}} trans\text{-}Cr(en)_2(OH)_2^{+}$ is reversible and first order in both directions. The values of k_1 and of K (the equilibrium constant) are $0.02 \min^{-1}$ and 0.16 at 25°C and 0.08 and 0.16 (unchanged) at 35°C. (a) In an experiment starting with the *cis* form only, calculate how long it would take for half of the equilibrium amount of *trans* isomer to be formed, at 25°C. (b) Calculate E^* and $\Delta S^{\ddagger}$ for the forward and for the reverse reaction.

31. (18 min) The following data were obtained on the rate of hydrolysis of methyl acetate (to methanol and acetic acid) in approximately $1\ N$ HCl at 25°C. Aliquots of equal volume were removed at intervals and titrated with a solution of NaOH. Show what the order of the reaction is under these conditions, and calculate the rate constant and the half-life.

Time (min)	0	5	15	25	∞
Volume NaOH Solution (cc)	24.00	27.00	31.40	34.35	40.00

32. (12 min) The reaction $CH_3CH_2NO_2 + OH^- = H_2O + CH_3CHNO_2^-$ obeys the rate law $d(A)/dt = -k(A)(OH^-)$, where A denotes $CH_3CH_2NO_2$. It takes 0.5 min for one percent of the A to react at 25°C, in the case of a solution $0.002\ M$ in A and $0.3\ M$ in NaOH. Calculate k. Calculate also how long it would take for half of A to react.

ANSWERS

1. (a) First, it is evident that the exponent d must be zero. Since no I_2 is present initially, there could not otherwise be a nonzero, finite initial rate. Exponent a must be 2 since by comparison of runs 1 and 2, doubling $[Fe(CN)_6^{3-}]$ quadruples the rate. Exponent b must be 1 since, from runs 2 and 4, doubling $[I^-]$ doubles the rate. Finally, exponent c must be -1 since doubling both $[I^-]$ and $[Fe(CN)_6^{4-}]$ (runs 1 and 3) leaves the rate unchanged so that the doubling that should occur from the increase in $[I^-]$ must be just compensated. Then

$$d[I_2]/dt = k[Fe(CN)_6^{3-}]^2[I^-][Fe(CN)_6^{4-}]^{-1}$$

On inserting the data of run 1, we find

$$1 \times 10^{-3} = k(1 \times 10^{-3})^2(1 \times 10^{-3})(1 \times 10^{-3})^{-1}$$

$$k = 10^3\ M^{-1}\ h^{-1}$$

(b) Since $\Delta G^{\ddagger} = \Delta H^{\ddagger} - T\Delta S^{\ddagger}$, the two sets of data give

$$\Delta H^{\ddagger} = 18,000 + 298\ \Delta S^{\ddagger} = 18,200 + 308\ \Delta S^{\ddagger}$$

whence

$$\Delta S^{\ddagger} = -20\ \textbf{cal/deg}$$

Then

$$\Delta H^{\ddagger} = 18,000 - 298 \times 20 = \mathbf{12,040\ cal}$$

(c) We are to assume the rate law

$$d[I_2]/dt = k[Fe(CN)_6^{3-}]^2[I^-]^2[Fe(CN)_6^{4-}]^{-1}$$

The negative exponent can only mean that $Fe(CN)_6^{4-}$ is involved in a pre-equilibrium, and the following is a possible scheme:

$$Fe(CN)_6^{3-} + 2I^- \overset{K}{=} Fe(CN)_6^{4-} + I_2^-$$
$$I_2^- + Fe(CN)_6^{3-} \overset{k_1}{\rightarrow} Fe(CN)_6^{4-} + I_2 \qquad \text{(slow step)}$$

Then

$$d[I_2]/dt = k_1[I_2^-][Fe(CN)_6^{3-}] = k_1 K[Fe(CN)_6^{3-}]^2[I^-]^2[Fe(CN)_6^{4-}]^{-1}$$

2. (a) Since Cl_2 is in large excess, the rate data show only the dependence on P_{CO}. In the first experiment, it is observed that half of the CO has reacted in the first time interval, three-quarters in the next equal interval, and so on. The reaction is then first order in P_{CO}. From experiment 1, the half-life is 34.5 min, whereas in experiment 2, three-quarters of the CO has reacted in this time, so the half-life must be 17.2 min. The rate has thus been doubled on quadrupling the value of P_{Cl_2}, so b must be $\frac{1}{2}$. Then

$$dP_{COCl_2}/dt = kP_{CO}P_{Cl_2}^{1/2}$$

From experiment 1, k_{app} is $0.69/34.5 = 0.02\ \text{min}^{-1}$.
  (b) Since $k_{app} = kP_{Cl_2}^{1/2}$, $k = 0.02/20 = \mathbf{1 \times 10^{-3}\ min^{-1}(mm\ Hg)^{-1/2}}$.
  (c) By the correct rate law, the initial rate is 0.04 mm Hg/min, whereas by the misprinted one, it is given as 0.02 mm Hg/min. This comparison suffices to show that the calculated half-life would be longer as determined from the erroneous rate law. (Actual values, which might be worked out as a separate exercise, are 69 min by the correct rate law and 116 min by the erroneous one.)

3. A constant half-life means that the reaction is first order; hence $a = 1$. Since a tenfold increase in $[H^+]$ decreases the half-life and hence increases the rate tenfold, b must also be one.

4. (a) Since pressure is an additive quantity, we can write $A/A^0 = (P - P^\infty)/(P^0 - P^\infty)$ where A denotes P_{PH_3}. Then at the two intermediate times, A/A^0 is $3.50/14 = 0.25$ and $0.87/14 = 0.0625$, or $\frac{1}{4}$ and $\frac{1}{16}$, respectively.

Evidently A/A^0 decreases by the same fraction ($\frac{1}{4}$) in each 60-sec interval; hence the reaction is first order. If 60 sec is the quarter-life, 30 sec would be the half-life, and $k_1 = 0.69/30 = \textbf{0.023 sec}^{-1}$.

(b) The over-all reaction is $4PH_3 = P_4 + 6H_2$. Then

$$K = \frac{P_{P_4}P_{H_2}^6}{P_{PH_3}^4} = \left(\frac{k_1}{k_2}\right)^n = \left[\frac{P_{P_4}^a P_{H_2}^b}{P_{PH_3}}\right]^n$$

The simplest (but not the only solution) is to identify denominators, so that n must be 4. Then $4a = 1$ and $a = \frac{1}{4}$, and $4b = 6$, so $b = \frac{3}{2}$. Remember that the quotient of the forward and reverse rate laws gives K^n, where n is a power not *a priori* predictable.

(c) If the reaction is first order with a half-life of 20 sec, $k_1 = 0.69/20 = \textbf{0.035 sec}^{-1}$. If second order, the half-life equals $1/k_1 P^0$ or $k_1 = 1/(20 \times 18) = \textbf{0.00277 sec}^{-1}\textbf{ mm Hg}^{-1}$.

5. We assume HI is adsorbed according to the Langmuir equation and that the surface reaction rate is proportional to the amount adsorbed, hence

$$\frac{dP_{HI}}{dt} = \frac{kP_{HI}}{1 + bP_{HI}}$$

At large P_{HI}, $dP_{HI}/dt = k/b = k_1 = 500$. At low P_{HI}, the rate equals $k = k_2 = 50$, so $b = 0.1$. Then for dP_{HI}/dt to be 250, we want

$$250 = \frac{50P_{HI}}{1 + 0.1P_{HI}}$$

whence $P_{HI} = \textbf{10 mm Hg}$.

6. The successive values of P/P^0 for AB_3 are 0.5, 0.25, and 0.125, while the successive intervals of time are 5, 10, and 20 h. Thus the half-life doubles each time, characteristic of a second-order reaction. Then, $dP/dt = -kP^2$ and $k = 1/(P^0 \times T_{1/2}) = 1/(660 \times 5) = \textbf{3.03} \times \textbf{10}^{-4}\textbf{ h}^{-1}\textbf{ mm Hg}^{-1}$.

7. (a) Since NO is in excess, in the case of runs 1 and 2, the rate will be first order; that is, rate $= k_{app}P_{H_2}$, and the half-life is then independent of $P_{H_2}^0$. Then for run 2, $T_{1/2} = \textbf{19.2 sec}$. In the case of runs 3 and 4, H_2 is in excess, so the rate will be second order; that is, rate $= k'_{app}P_{NO}^2$. We then have the relationship, $T_{1/2} = 1/(k_{app}P_{NO}^0)$. Since P_{NO}^0 is twice as large in run 4 as it is in run 3, the half-life will be halved, and $T_{1/2} = \textbf{415 sec}$.

(b) Using run 1 (or 2) as a basis, $k_{app} = 0.69/19.2 = 0.036 \text{ sec}^{-1}$. But, $k_{app} = kP_{NO}^2$, so $k = 0.036/(600)^2 = \textbf{1} \times \textbf{10}^{-7}\textbf{ mm Hg}^{-2}\textbf{ sec}^{-1}$.

(c) From a comparison of runs 1 (or 2) and 5, the ratio of half-lives is 10/19.2; hence, that of the k values is 19.2/10. Then

$$2.3 \log \left(\frac{19.2}{10}\right) = \frac{E^*}{R}\left(\frac{1}{1,093} - \frac{1}{1,113}\right) \quad (R = 1.98)$$

(d) A likely possibility as to mechanism is

$$2NO = N_2O_2 \quad \text{(rapid equilibrium)}$$
$$N_2O_2 + H_2 = N_2O + H_2O \quad \text{(slow step)}$$

(e) Since, from the stoichiometry of the reaction, NO is used up twice as fast as H_2, the 2:1 ratio of pressures applies throughout the reaction, and the rate law reduces to

$$\frac{dP_{N_2O}}{dt} = k[2P_{H_2}]^2[P_{H_2}] = 4kP_{H_2}^3 = -\frac{dP_{H_2}}{dt}$$

On integration,

$$\frac{1}{P_{H_2}^2} - \frac{1}{P_{H_2}^{0\,2}} = 8kt$$

and

$$8kT_{1/2} = \frac{3}{P_{H_2}^{0\,2}}$$

Then

$$T_{1/2} = \mathbf{3.8 \times 10^4 \, mm \, Hg^{-2} \, sec^{-1}}$$

8. Since the initial conductivity is essentially all due to the HCl, and the final is all due to NH_4Cl, the ratio of initial to final values should be (515 + 133) to (137 + 133) or 2.38. The final specific conductivity should then be 0.409/2.38 = 0.172. Conductivities are additive, so if A denotes the acetamide concentration, $A/A^0 = (L - L^\infty)/(L^0 - L^\infty)$. A/A^0 is thus 0.85, 0.68, and 0.58 at the three times.

If the reaction were first order, $\log(A/A^0)/t$ should be constant. A check shows that this ratio is not constant. If the reaction were second order (of the simple type $dA/dt = -kA^2$, expected here since the reactants are in equal concentrations and react in 1:1 proportion), then $(A^0/A - 1)/t$ should be a constant. This ratio is constant, hence the reaction *is* second order. The value of $(A^0/A - 1)/t$ is 0.0137, and this is equal to kA^0; since A^0 is 1 M (remember the dilution), $k = \mathbf{0.0137\,M^{-1}\,min^{-1}}$.

9. (a) Since the half-life is constant, that is, $T_{3/4} = 2T_{1/2}$, the reaction must be first order in complex, so $a = 1$. Since $T_{1/2}$ is halved on doubling $[H^+]$, $b = 1$.

(b) Since $[H^+]$ is constant, $k_{app} = k[H^+]$. $k_{app} = 0.69/1 = 0.69\,h^{-1}$, so $k = 0.69/0.01 = \mathbf{69\,h^{-1}\,M^{-1}}$.

(c) On comparing lines 1 and 3, it is seen that the rate doubles, so

$$\log 2 = \frac{E^*}{2.3 \times 1.98}\left(\frac{1}{298} - \frac{1}{308}\right)$$

or

$$E^* = 0.3 \times 2.3 \times 1.98 \times 298 \times \frac{308}{10} = \mathbf{12.5\,kcal}$$

10. (a) Since $[HCl]$ is constant, the reaction is first order, with $k_{app} = 4.0 \times 0.01 = 0.04\,min^{-1}$. Then $T_{1/2} = 0.69/0.04 = \mathbf{17.3\,min}$.

(b) If $T_{1/2}$ is halved, k_{app} and hence k is doubled, so

$$-\log 2 = \frac{20,000}{2.3 \times 1.98}\left(\frac{1}{T} - \frac{1}{298}\right)$$

or $\Delta T = 298 \times T \times 6.85 \times 10^{-5} \cong 298^2 \times 6.85 \times 10^{-5} = 6.0°$. As a second approximation, $\Delta T = 298 \times 304 \times 6.85 \times 10^{-5} = 6.1°$ or $T = \mathbf{304°K}$.

11. (a) The fraction of paraldehyde decomposed will be given by $(P - P^0)/(P^\infty - P^0)$. ($P$ here denoting pressure.) The values are thus 0.36, 0.59, 0.74, and 0.83; the fractions remaining are then 0.64, 0.41, 0.26, and 0.17.

(b) Observe that $0.41 = (0.64)^2$ and $0.17 = (0.41)^2$, that is, that the amount of paraldehyde is reduced by 0.64 each hour. This constant "fraction-life" is characteristic of a **first-order reaction.**

(c) If we take the data at 4 h, $2.3 \log 0.17 = -4k$ and $k = \mathbf{0.44\,h^{-1}}$.

(d) From $k = 10^{13}\,e^{\Delta S^{\ddagger}/R}\,e^{-\Delta H^{\ddagger}/RT}$, with $k = 1/3,600\,sec^{-1}$, $2.3 \log(1/3,600) = 2.3 \log 10^{13} - \Delta H^{\ddagger}/RT$ or $\Delta H^{\ddagger} = 38 \times 1.98 \times 533 = \mathbf{40.2\,kcal}$. Then $\log k'/k = (40,200/2.3 \times 1.98) \times (1/533 - 1/553) = 0.597$ or $k' = \mathbf{4.0\,h^{-1}}$.

12. (a) The desired ratio is simply the measure of the amount of reactants remaining and so is given by $(D - D^\infty)/(D^0 - D^\infty)$. The values are then 0.5, 0.33, and 0.25.

(b) It thus takes 1×10^5 sec for half of the Hg^{2+} to react, and 3×10^5 sec for half of that (or three-quarters total) to react. This doubling of half-times is indicative of a second-order reaction. Furthermore, since the initial concentrations are equal and Fe^{2+} and Hg^{2+} react in $1:1$ ratio, the two concentrations will remain equal, and in run A, the rate law reduces to $R = k[Fe^{2+}]^{(p+q)}$. Thus $(p+q) = 2$. Turning to run B, observe that $0.348 = (0.585)^2$ and that $0.122 = (0.348)^2$. Thus, the time for the fraction of Hg^{2+} remaining to diminish by 0.585 is a constant. Such a constant fraction-time indicates a first-order reaction. In this case Fe^{2+} is in great excess, so the experimental reaction order is given by the order in Hg^{2+}.

(c) Thus $q = 1$ and, hence, $p = 1$.

13. This one is easy. Since HCl is a catalyst, its concentration does not change, and the rate law reduces to

$$R = -d(\text{ester})/dt = k_{\text{app}}(\text{ester})$$

where k_{app} is 0.1×0.01 or $10^{-3} \, h^{-1}$. The half-life is then $0.69/10^{-3} = $ **690 h.**

14. Since D is an additive property, $A/A_0 = (D - D_\infty)/(D_0 - D_\infty)$, where A denotes the concentration of the reactant, $Co(NH_3)_5Br^{2+}$ in this case. Then at $t = 20 \, \text{min}$, $A/A_0 = (0.80 - 0.20)/(D_0 - D_\infty)$ and at $t = 40 \, \text{min}$, $A/A_0 = (0.35 - 0.20)/(D_0 - D_\infty)$. Since the reaction is first order, A/A_0 after 40 min must be the square of A/A_0 after 20 min; that is, there must be a constant "fraction" life. Then

$$\frac{(0.6)^2}{(D_0 - D_\infty)^2} = \frac{0.15}{(D_0 - D_\infty)}$$

from which $(D_0 - D_\infty) = 0.36/0.15 = 2.40$. D_0 is then $2.40 + 0.20$ or **2.60.**

15. 0.125 corresponds to $(\frac{1}{2})^3$ so 1 h corresponds to three half-lives and one half-life is therefore 20 min. The rate constant is then $0.69/20 = 3.45 \times 10^{-2} \, \text{min}^{-1}$. The initial rate R is given by

$$R = -\frac{dA}{dt} = kA$$

so $R = 0.0345 \times 0.2 = 0.0060$ mole of A per minute or **1.15×10^{-4} mole/ sec.**

16. (a) The linear log $P_{N_2O_5}$ versus time plot when NO was in excess means that the rate is first order in N_2O_5; that is, $x = 1$. For the second experiment, P^∞ is, from the reaction stoichiometry, $\frac{3}{2}P^0$ or 150 mm Hg. Then $P_{N_2O_5}/P^0$ was $(115 - 150)/(100 - 150)$ or 0.7 after 1 h and $(125 - 150)/(100 - 150)$ or 0.5 after 2 h. Thus 0.7 of what was present after 1 h was left after 2 h; that is, the 0.7 life is constant; hence the experimental order is again one. Since

N_2O_5 and NO were present in equal initial amounts, and react in one-to-one mole ratio, the order observed would be $x + y$. Then $x + y = 1$, and, since $x = 1$, $y = 0$. The rate law is then $dP_{N_2O_5}/dt = -kP_{N_2O_5}$. The half-life of 2 h (from either experiment) then gives $k = 0.693/2 = \mathbf{0.35\ h^{-1}}$.

(b) By the stationary state assumption the rate of change of trace species is taken to be zero. Thus $dP_{NO_3}/dt = 0 = k_1 P_{N_2O_5} - k_2 P_{NO_2} P_{NO_3} - k_3 P_{NO} P_{NO_3}$, from which

$$P_{NO_3} = \frac{k_1 P_{N_2O_5}}{k_2 P_{NO_2} + k_3 P_{NO}}$$

Then, since $dP_{NO}/dt = dP_{N_2O_5}/dt = -k_3 P_{NO} P_{NO_3}$, we obtain

$$\frac{dP_{N_2O_5}}{dt} = -\left(\frac{k_1 k_3}{k_2 P_{NO_2} + k_3 P_{NO}}\right) P_{N_2O_5} P_{NO}$$

The apparent reaction rate constant k is then equal to the expression in parentheses. Moreover, since the experimental rate did not depend on P_{NO}, $k_3 P_{NO}$ must dominate in the denominator above so the rate law reduces to $dP_{N_2O_5}/dt = -k_1 P_{N_2O_5}$ and $k = k_1$.

(c) If N_2O_5 is in excess, the initial rate will be $dP_{N_2O_5}/dt = dP_{NO}/dt = -0.35 \times 100 = -35$ mm Hg/h. It would then take $\frac{1}{35}$ h for all the NO and $\frac{1}{70}$ h for half of the NO to be consumed.

17. (a) Addition of (1) and (2) gives Fe(III) + V(III) = Fe(II) + V(IV).

(b) For the slow step, $d[V(IV)]/dt = k_2[V(V)][V(III)]$. [There is no factor of two as for each two V(IV) produced by (2), one has been removed by (1).] If $[V(V)]$ is eliminated by means of the equilibrium-constant expression for (1):

$$\frac{d[V(IV)]}{dt} = k_2 K_1 \frac{[Fe(III)][V(IV)][V(III)]}{[Fe(II)]}$$

Then $k_{app} + k_2 k_1$ and $E_{app}^* = E_2^* + \Delta H_1^0$ (neglecting the difference between E and H). Then $E_2^* = 12 + 5 = \mathbf{17\ kcal.}$

(c) We set $d[V(V)]/dt = 0$; that is,

$$0 = k_1[Fe(III)][V(IV)] - k_{-1}[Fe(II)][V(V)] - k_2[V(V)][V(III)]$$

or

$$[V(V)] = \frac{k_1[Fe(III)][V(IV)]}{k_{-1}[Fe(II)] + k_2[V(III)]}$$

(d) From absolute rate theory, $k = 10^{13} e^{\Delta S^{\ddagger}/R} e^{-\Delta E^{\ddagger}/RT}$, or $0.01 = 10^{13} e^{-5/1.98} e^{-\Delta E^{\ddagger}/(1.98)(298)}$, from which $\Delta E^{\ddagger}$ (taken to be the same as $\Delta H^{\ddagger}$) can be calculated. Then, from the Arrhenius equation, $0.01 = Ae^{-\Delta E^{\ddagger}/1.98)(298)}$, from which A can be calculated.

18. The slow step is rate controlling; hence

$$\text{rate} = k_1[\text{Co(NH}_3)_4(\text{NH}_2)\text{Cl}^+]$$

It is customary, however, to express a rate law in terms of the major species actually present. Thus, making use of the expression for K, we obtain

$$\text{rate} = k_1 K \frac{[\text{Co(NH}_3)_5\text{Cl}^{2+}][\text{OH}^-]}{[\text{H}_2\text{O}]}$$

(The concentration of solvent usually is omitted, i.e., incorporated into K.)

19. (a) Since $[A]^0 = [B]^0$ and they react in $1:1$ stoichiometry, $[A] = [B]$ at all times, and the rate law is second order: $d[\text{ester}]/dt = k_{\text{app}}[A]^2$. $k_{\text{app}} = k_1[\text{H}^+]$. $[\text{H}^+]$ is constant during the reaction. For a second-order reaction, $k = 1/(C^0 \times T_{1/2})$ so $k_{\text{app}} = 1/(1 \times 0.01) = \textbf{100 h}^{-1}\,\textbf{M}^{-1}$. Then $k_1 = 100/[\text{H}^-] = \textbf{10}^4\,\textbf{h}^{-1}\,\textbf{M}^{-2}$.

(b) From absolute rate theory $k = 10^{13} e^{\Delta S^{\ddagger}/R} e^{-\Delta E^{\ddagger}/RT} = 10^{13} e^{-\Delta G^{\ddagger}/RT}$. Then, $2.3 \log(10^4/3{,}600) = 2.3 \log 10^{13} - \Delta G^{\ddagger}/RT$, from which $\Delta G^{\ddagger} = \textbf{17 kcal}$. Since the half-life is doubled and hence the rate is halved on going to $15°\text{C}$, we have

$$2.3 \log 2 = \frac{\Delta E^{\ddagger}}{R}\left(\frac{1}{288} - \frac{1}{298}\right)$$

or

$$\Delta E^{\ddagger} = 2.3 \times 0.3 \times 1.98 \times 288 \times \frac{298}{10} = \textbf{11.7 kcal}$$

Then, since $\Delta G^{\ddagger} = \Delta E^{\ddagger} - T\,\Delta S^{\ddagger}$, we find $\Delta S^{\ddagger} = (11{,}700 - 17{,}000)/298$ or about **18 cal/deg.**

20. (a) $P_{\text{O}_3}/P_{\text{O}_3}^0$ is given by $(P - P^\infty)/(P^0 - P^\infty)$, and the values at 30 and 60 min are then 0.5 and 0.25 in the case of run 1. The half-life is thus constant so the reaction is **first order**. $k_{\text{app}} = 0.69/30 = \textbf{0.023 min}^{-1}$.

(b) Similarly, for run 2, $P_{\text{O}_3}/P_{\text{O}_3}^0$ is 0.7, 0.5, and 0.25 for 30, 60, and 120 min. The half-life is thus 60 min, so the rate has been halved on halving the CO_2 concentration. The exponent b is then **unity**. Since $k_{\text{app}} = k[\text{CO}_2]$, $k = 0.023/0.005 = \textbf{4.6 (mole/liter)}^{-1} \times \textbf{min}^{-1}$.

(c) From the Arrhenius equation,

$$2.3 \log 4 = \frac{E^*}{R}\left(\frac{1}{323} - \frac{1}{333}\right) \qquad (R = 1.98 \text{ cal/deg-mole})$$

(d) Atomic oxygen is clearly the trace species here, and by the stationary-state hypothesis:

$$\frac{d[O]}{dt} = 0 = k_1[O_3]^2 - k_2[O_3][O_2][O] - k_3[O_3][O]$$

Then

$$[O] = \frac{k_1[O_3]^2}{k_2[O_3][O_2] + k_3[O_3]} = \frac{k_1[O_3]}{k_2[O_2] + k_3}$$

The rate is given by

$$\frac{d[O_3]}{dt} = -2k_3[O_3][O]$$

(The factor of two enters since, every time the second reaction occurs, one O_3 must have been consumed in the first reaction.) On eliminating $[O]$

$$\frac{-d[O_3]}{dt} = \frac{2k_1k_3[O_3]^2}{k_2[O_2] + k_3}$$

21. (a) Comparing lines 1 and 2, we see that halving the N_2O pressure halves the rate, so $a = 1$; that is, the reaction is first order in N_2O. Comparing lines 1 and 3, we see that reducing P_{Cl_2} by a factor of four, halves the rate, so $b = \frac{1}{2}$. (The ratio of pressures to the a or b power must equal the ratio of the rates.)

(b) Using the data of line 1, we have $0.3 = k(30)(4)^{1/2}$ or $k = 0.005 \text{ min}^{-1}$ cm $Hg^{-1/2}$. Strictly speaking, $dP_{N_2O}/dt = 2dP_{tot}/dt$, since for each cm Hg of N_2O disappearing, 1 cm Hg of N_2 forms plus $\frac{1}{2}$ cm Hg of O_2, so the net pressure change is $\frac{1}{2}$ cm Hg. If R is defined as dP_{N_2O}/dt (as it should be), k should be 0.01 min^{-1} cm $Hg^{-1/2}$. Then, in (c), the half-life is **34.5 min.**

(c) Since P_{Cl_2} is a constant during the reaction, any given run is pseudo first order, with $k_{app} = kP_{Cl_2}^{1/2}$. In the case of line 1, $k_{app} = 0.01$, whence $T_{1/2} = 0.69/0.01 = $ **69 min.**

(d) On comparing the first and last lines of data, we find that the rate doubles with a 10° rise in temperature, so

$$\log 2 = \frac{E^*}{2.3 \times 1.98}\left(\frac{1}{800} - \frac{1}{810}\right)$$

or

$$E^* = 0.3 \times 2.3 \times 1.98 \times 800 \times 810/10$$
$$= \textbf{88.5 kcal}$$

According to the Arrhenius equation, $k = Ae^{-E^*/RT}$, so $A_{800°K} = (0.005/60)e^{88,500/1.98 \times 800}$. (The factor of 1/60 converts to seconds, the usual time unit. E^* and A are, of course, apparent values as the mechanism evidently is complex.)

A possible mechanism would be the following:

$Cl_2 = 2Cl$	(rapid equilibrium) (so $P_{Cl} = KP_{Cl_2}^{1/2}$)
$Cl + N_2O \rightarrow N_2 + ClO$	(slow step)
$ClO + N_2O = N_2 + ClO_2$	(fast)
$ClO_2 + Cl = Cl_2 + O_2$	(fast)

22.   (a) Since $dA/dt = -k[A][S_2O_3^{2-}]$, by substitution $dA/dt = -1.64 \times 10^{-3} \times 0.1 \times 0.1 = -\textbf{1.64} \times \textbf{10}^{-5} M \, \textbf{sec}^{-1}$.

(b) Initially $A = [S_2O_3^{2-}]$ and since they react in 1:1 ratio, this condition applies throughout the course of the reaction. The rate law thus reduces to $dA/dt = -kA^2$. Integration gives $1/A - 1/A^0 = kt$.

(c) If the rate drops to one-fourth, then $A = \frac{1}{2}A^0$ and $t = T_{1/2} = 1/kA^0 = \textbf{6.1} \times \textbf{10}^3 \textbf{ sec.}$

(d) $\log k'/k = (20,000)/(2.3 \times 1.98)(1/310 - 1/320) = 0.445$, whence $k'/k = \textbf{2.78.}$

(e) By Eq. (16-8),

$$1.64 \times 10^{-3} = 10^{13} \, e^{\Delta S^{0\ddagger}/R} \, e^{20,000/-1.98 \times 300}$$

(Extra credit) Let B denote $[CH_3COCH_2^-]$; then, since B is a trace species, $dB/dt = 0 = k_1A[OH^-] - k_2[B] - k_3B[Br_2]$. ($A = [CH_3COCH_3]$.) The rate is given by the slow step,

$$dC/dt = k_3B[Br_2] = k_3[Br_2]\frac{k_1A[OH^-]}{k_2 + k_3[Br_2]}$$

$$(C = [CH_3COCH_2Br])$$

Since $k_2 \gg k_3$, it should be safe to neglect $k_3[Br_2]$ in the denominator, so

$$\frac{dC}{dt} = (k_1 k_3/k_2)[A][Br_2][OH^-]$$

23. (a) Let $A = [S_2O_8^{2-}]$ and $B = [Mo(CN)_8^{4-}]$. Since $A^0 = \frac{1}{2}B^0$, and A and B react in this ratio, $A = \frac{1}{2}B$ at all times, and the rate law reduces to

$$\frac{dB}{dt} = -k(\tfrac{1}{2}B)^a(B)^b = -k'B^{(a+b)}$$

The reaction will thus follow a simple first-, second-, and so on, order type of rate law where the order equals $(a + b)$. Inspection of the data shows that it took 90 min for half of B to react and 180 min more for half of that to react; that is, the half-life doubles each time. The order is thus two, so $(a + b) = 2$.

(b) The integrated form for a second-order reaction is $1/B - 1/B^0 = k't$, or $T_{1/2} = 1/k'B^0$. Then $k' = 1/(0.02 \times 90) = \mathbf{0.55\,M^{-1}\,h^{-1}}$. Since we do not know the separate value of a, the actual k value cannot be obtained.

(c) We use the equation

$$\log \frac{2 \times 10^{-3}}{2 \times 10^{-4}} = \frac{\Delta H^{\ddagger}}{2.3 \times 1.98}\left(\frac{1}{293} - \frac{1}{318}\right)$$

from which $\Delta H^{\ddagger} = \mathbf{16.9\,kcal}$. Also $k = 10^{13}\ e^{\Delta S^{\ddagger}/R}\ e^{-\Delta H^{\ddagger}/RT}$. At 20°C, $k = 2 \times 10^{-4}/3{,}600 = 5.55 \times 10^{-8}\,M^{-1}\,sec^{-1}$, so $2.3 \log 5.55 \times 10^{-8} = 2.3 \times 13 + \Delta S^{\ddagger}/R - 16{,}900/1.98 \times 293$. Then $\Delta S^{\ddagger}/R = -17.5$ and $\Delta S^{\ddagger} = \mathbf{-34.6}$ **cal/deg.**

(d) The rate is $R = k_1[SO_4^-][Mo(CN)_8^{4-}]$, but $[SO_4^-]^2 = K[S_2O_8^{2-}]$, so the rate law becomes

$$R = k_1 K^{1/2}[S_2O_8^{2-}]^{1/2}[Mo(CN)_8^{4-}]$$

24. The rate law suggests the following mechanism:

$$2NO \overset{K}{=} N_2O_2 \qquad \text{(rapid equilibrium)}$$

$$N_2O_2 + O_2 \overset{k}{\rightarrow} 2NO_2 \quad \text{(slow step)}$$

The rate would then be $R = k[N_2O_2][O_2] = kK[NO]^2[O_2]$, and the apparent activation energy would be the sum of E^* for the slow step and ΔH^0 for the preequilibrium. ΔH^0 should be negative (since we have the association of NO to give N_2O_2 and a bond must be formed) and, if it is sufficiently negative, the apparent activation energy could be negative. The apparent specific rate constant would then decrease with increasing temperature.

This example should serve as a warning that apparent activation energies in general do not necessarily relate to a single step and that it is dangerous to draw conclusions from activation energy values unless a fairly good idea exists as to what the actual mechanism is for the reaction.

25. According to the rate law $dP_{H_2}/dt = -k_f P_{H_2} P_{I_2}$. Since P_{I_2} is constant at 1.2 atm, the rate law for this special case becomes $dP_{H_2}/dt = -k_{app} P_{H_2}$, that is first order, with $k_{app} = 1.2 \times 10^{-6}$. The half-life is then $0.69/k_{app}$ or **5.8×10^5 sec.**

In terms of the proposed mechanism, $dP_{H_2}/dt = -kP_{H_2}P_I^2$, but $P_I^2 = KP_{I_2}$, so we obtain $dP_{H_2}/dt = -kKP_{H_2}P_{I_2}$. This last conforms to the experimental rate law, and the proposed mechanism is thus a possible one.

26. According to the Langmuir isotherm

$$\theta_{H_2O} = \frac{bP_{H_2O}}{1 + bP_{H_2P}}$$

and the rate law is then

$$R = \frac{-d(\text{moles of } H_2O)}{dt} = k\frac{bP_{H_2O}}{1 + bP_{H_2O}}$$

At low P_{H_2O} the rate expression reduces to $R = kbP_{H_2O}$, or is first order. Since b contains the term $\exp(Q/RT)$, where Q, the heat of adsorption, is 10 kcal, the apparent activation energy will then be $E^*_{app} = E^*_{surf} - Q$. At high P_{H_2O}, the rate law reduces to $R = k$, and the observed activation energy is E^*_{surf}. The observed activation energy is thus **higher** if P_{H_2O} is **high.**

27. From absolute rate theory, $k = 10^{13} e^{\Delta S^{\ddagger}/R} e^{-\Delta E^{\ddagger}/RT}$. Then

$$\ln \frac{k}{k'} = \frac{\Delta S^{\ddagger} - \Delta S^{\ddagger'}}{R} - \frac{\Delta E - \Delta E^{\ddagger'}}{RT}$$

$$= \frac{10}{1.98} - \frac{5,000}{1.98 \times 298}$$

$$= -\frac{6.8}{1.98}$$

Thus k' must be larger than k.

28. (a) The actual value of k should be

$$k = 10^{-3} \times 0.5 \times 10^{11} e^{-25,000/RT}$$
$$= 10^{-3} \times 0.5 \times 10^{11} 10^{-25,000/2.3 \times 1.98 \times 298}$$

From the point of view of absolute rate theory, $k = 10^{13} e^{\Delta S^{\ddagger}/R} \times e^{-\Delta E^{\ddagger}/RT}$, so comparing equations gives: $10^{13} e^{\Delta S^{\ddagger}/R} = 10^{-3} \times 0.5 \times 10^{11}$ or $e^{\Delta S^{\ddagger}/R} = 5 \times 10^{-6}$ or $10^{\Delta S^{\ddagger}/2.3 \times 1.98} = 5 \times 10^{-6}$. (We take advantage of the instructions permitting answers to be left in such a form.)

(b) The experiments fall into two categories: those with equal initial concentrations and, therefore, simple second order, and those with one species in large excess, and, therefore, pseudo first order. In the first case, $T_{1/2} = 1/C^0 k$, whereas in the second, $T_{1/2} = 0.69/k_{app}$, where $k_{app} = kC_1^0$ if C_1^0 is the concentration of the species in excess. We then have:

Experiment	$T_{1/2}$
1	$1/k$
2	$k_{app} = k$; $T_{1/2} = 0.69/k$
3	$k_{app} = 10k$; $T_{1/2} = 0.69/10k$
4	$T_{1/2} = 1/0.01k = 100/k$

The order of experiments of increasing half-life is then **3, 2, 1, 4.**

29. (a) At infinite temperature $e^{-E^*/RT}$ becomes unity, so both plots in Fig. 16-1 must have the same intercept. Since E_1^* is greater than E_2^*, the slope

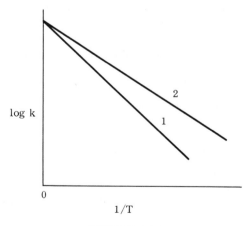

log k

0

1/T

FIGURE 16-1

for $\log k_1$ will be greater, as shown in the plot. (b) It then follows that the rate of 1 must be slower than that of 2 at any finite temperature.

30. (a) Since $K = k_1/k_2$, k_2 is $0.02/0.16 = 0.125\ \text{min}^{-1}$ at 25°C. In the case of a reversible first-order reaction, the integrated rate law is

$$\frac{A - A_\infty}{A_0 - A_\infty} = e^{-kt}$$

where $k = k_1 + k_2$. $(A_0 - A)$ is the amount reacted, and $(A_0 - A_\infty)$ is the equilibrium amount reacted, or the equilibrium amount of product, *trans* isomer in this case. We want the time for $(A_0 - A)/(A_0 - A_\infty) = \frac{1}{2}$. Since

$$\frac{A - A_\infty}{A_0 - A_\infty} = 1 - \frac{A_0 - A}{A_0 - A_\infty}$$

then at this time, $(A_0 - A_\infty)/(A_0 - A_\infty)$ is also $\frac{1}{2}$ (as might be guessed), and hence $\exp(-kt) = \frac{1}{2}$. The time required is then $0.69/k$, where $k = 0.02 + 0.125 = 0.145$, so $t = $ **4.8 min.**

(b) For the forward reaction

$$\log\left(\frac{0.08}{0.02}\right) = \frac{E^*}{2.3R}\left(\frac{1}{298} - \frac{1}{308}\right)$$

or $E_1^* = 2.3 \times 1.98 \times 0.602 \times 298 \times 308/10 = $ **25.2 kcal.**

Since K does not change with temperature, ΔE^0 (or actually ΔH^0) is zero, and $E_2^* = E_1^* = $ **25.2 kcal.** k_1 is $0.02/60$ or $3.33 \times 10^{-4}\ \text{sec}^{-1}$, so we can write

$$3.33 \times 10^{-4} = 10^{13}\ e^{\Delta S_1^\ddagger/R}\ e^{-\Delta H_1^\ddagger/RT}$$

or

$$2.3 \log(3.33 \times 10^{-4}) = 2.3 \times 13 + \frac{\Delta S_1^\ddagger}{R} - \frac{\Delta H_1^\ddagger}{298R}$$

and $\Delta S_1^\ddagger/R = -8.0 - 29.9 + 42.7 = 4.8$. $\Delta S_1^\ddagger$ is then $4.8 \times 1.98 = $ **9.5 e.u.** (Note that the distinction between E and H has not been bothered with.) Finally, for the reverse reaction, k_2 is $0.125/60$ or $2.09 \times 10^{-3}\ \text{sec}^{-1}$, $2.3 \log(2.09 \times 10^{-3}) = -6.2$, and repetition of the above calculation gives $\Delta S_2^\ddagger/R = -6.2 - 29.9 + 42.7 = 6.6$ and $\Delta S_2^\ddagger = $ **13 e.u.**

31. The amount of acid present is an additive property with respect to the degree of reaction, so A/A_0 is given by $(V - V_\infty)/(V_0 - V_\infty)$, where A denotes the amount of methyl acetate and V, the volume of NaOH solution used. The values of A/A_0 are then $13/16$, $8.6/16$, and $5.65/16$ for 5, 15, and 25 min, respectively, or 0.81, 0.54, and 0.35. A test shows that these last numbers correspond to 0.81, $(0.81)^3$, and $(0.81)^5$, so that the time for A/A_0 to decrease by 0.81 is a constant, 5 min, and the reaction is therefore first order. It then follows that $\ln(0.81) = -5t$, or $k = \mathbf{0.041\ min^{-1}}$. The half-life is $0.69/k$ or **16.9 min.**

32. Since OH^- is in great excess, the reaction will be first order, with $k_{app} = k(OH^-)$ or $0.3k$. The first-order rate law, $dA/dt = -k_{app}A$, can be written $dA/A = -k_{app}\,dt$ from which it follows that $0.01 = k_{app} \times (0.5)$ and $k_{app} = 0.02$. The second-order rate constant is then $0.02/0.3 = \mathbf{0.067}$ $\mathbf{min^{-1}M^{-1}}$. With respect to half-life, the reaction is still first order, to $T_{1/2} = 0.69/0.02 = \mathbf{35.4\ min.}$

17

COLLOID AND SURFACE CHEMISTRY

COMMENTS

This short chapter on colloid and surface chemistry should constitute a bit of a breather after the rather heavy going of the past two chapters. Unfortunately, in my opinion, there simply is not much time available in the normal one-year course of physical chemistry to do these subjects justice. The problems that follow deal only with a very few and very limited aspects. These are the Gibbs adsorption equation and such transport-type processes as diffusion, sedimentation, and viscosity.

The Gibbs equation is one of the four principal equations of surface chemistry (the other three are the Laplace equation, the Kelvin equation, and the Young and Dupré equation for contact angles). It connects surface tension, concentration (or partial pressure), and surface excess. Through its use, data on any two gives information on the third. Thus if surface tension is known as a function of concentration, the surface concentration of adsorbed material can be calculated. If the surface concentration is known as a function of bulk concentration or pressure, as in the case of adsorption, the change in surface tension can be calculated. Several problems deal with these types of interconversions.

The viscosity of solutions is considered here only in terms of the Einstein equation for the case of solute molecules treated as rigid spheres in a uniform medium. Diffusion and sedimentation can be thought of as manifestations of applying a force to a particle in a viscous medium. This force is the chemical potential gradient in the first instance, and that due to gravity, in the second. In each case the particle acquires a net velocity proportional to the force, $v = F/f$, where f is called the friction coefficient. This coefficient should be the same for both phenomena, and, if the solute particles behave as hard spheres, f is given by Stokes' law. You will find several problems requiring you to be familiar with various interrelations between these two phenomena.

Finally, a concept important to colloid chemistry is that of average molecular weight. Where solutes of differing molecular weight are present, as in the case of a solution of polymer molecules of differing degrees of polymerization, it turns out that more than one kind of average molecular weight must be considered. The two important averages are the ordinary or number-average molecular weight, and the weight-average molecular weight. One obtains the first from colligative-property measurements, for example, and the second from viscosity measurements. If both types of average are known, one obtains valuable information about the degree of spread of solute molecular weights.

EQUATIONS AND CONCEPTS

Gibbs Equation

$d\gamma = -\Gamma_2 d\mu_2 = -\Gamma_2 RT \, d \ln a_2$. Here γ denotes surface tension (dynes/cm or ergs/cm^2), Γ_2 denotes surface excess in moles/cm^2, R should be in ergs/deg-mole, and a_2 is the activity of the solute. The surface excess, incidentally, is defined as the moles of solute per cm^2 of surface region in excess of the number of moles that would be associated with a portion of bulk solution having the same number of moles of solute. For ideal solutions, alternative forms are

$$\Gamma_2 = -\frac{C}{RT}\frac{d\gamma}{dC} \quad \text{or} \quad \Gamma_2 = \frac{C}{RT}\frac{d\pi}{dC} \tag{17-1}$$

π is given by $\gamma^0 - \gamma$, γ^0 being the surface tension of the pure substrate, and is called the surface pressure. In the case of dilute solutions, π may be proportional to solute concentration, in which case the Gibbs equation yields

$$\pi = \frac{RT}{\Gamma} \quad \text{or} \quad \pi\sigma = RT \tag{17-2}$$

where σ denotes the surface area per mole.

If adsorption occurs from the gas phase, C is replaced by P. The film pressure may then be obtained by integration:

$$\pi = RT \int \Gamma_2 \, d \ln P \tag{17-3}$$

Viscosity

$$\eta_{sp} = \left(\frac{\eta}{\eta_0} - 1 \right)$$ (17-4)

where η_0 is the viscosity of the pure solvent. By the Einstein equation

$$\eta_{sp} = \frac{5}{2}\phi$$ (17-5)

where $\phi = c\bar{v}$ and ϕ is the volume fraction of solute, c is its concentration in g/cc, and $\bar{v}$ is its specific volume. The reduced viscosity is $\eta_{sp}/100c$ and its limiting value as c approaches zero is called the intrinsic viscosity $[\eta]$. This last is related to the weight-average molecular weight by the semi-empirical equation $[\eta] = KM_w^a$ where K and a are constants.

Diffusion

The net force F is given by

$$F = -\frac{kT \, d \ln C}{dx}$$ (17-6)

where dC/dx is the concentration gradient (assuming ideal solutions); this gives rise to a net diffusion velocity v; $v = F/f$, where f is the friction coefficient. The permeation or flow across unit area P is

$$P = vC = -\frac{kT}{f} dC/dx$$ (17-7)

and, since by Fick's law $P = -\mathscr{D} \, dC/dx$, we have

$$\mathscr{D} = \frac{kT}{f}$$ (17-8)

In the case of rigid spheres, f is given by Stokes' law as $f = 6\pi\eta r$, where η is the viscosity of the solvent and r is the radius of the particle. The combined equation $\mathscr{D} = kT/6\pi\eta r$ is known as the Stokes–Einstein equation. On a molecular basis, diffusion results from the drift from its original position of a particle as a consequence of Brownian motion. $\mathscr{D}$ is in fact related to the average net displacement x in time τ, by

$$\mathscr{D} = \frac{x^2}{2\tau}$$ (17-9)

Sedimentation

The force is now $F = \frac{4}{3}\pi r^3(\rho - \rho_0)$.

$$g = \frac{M}{A}(1 - \bar{v}\rho_0)g \tag{17-10}$$

where $\bar{v}$ is the specific volume of the particle and M, its molecular weight. The sedimentation coefficient S is defined as the velocity per unit acceleration

$$S = \frac{(dx/dt)}{g} = \frac{1}{f}\frac{M}{A}(1 - \bar{v}\rho_0) \tag{17-11}$$

In a centrifugal field

$$S = \frac{(dx/dt)}{\omega^2 x} \tag{17-12}$$

where ω is the angular velocity.

Average Molecular Weight

The number-average molecular weight M_n is defined as the total weight divided by the number of particles, that is

$$M_n = \frac{\Sigma M_i n_i}{\Sigma n_i} \tag{17-13}$$

The weight-average molecular weight M_w is given by

$$M_w = \frac{\Sigma M_i w_i}{\Sigma w_i} \tag{17-14}$$

where w_i is the weight of particles M_i molecular weight.

PROBLEMS

1. (15 min) The surface tension of solutions of RSO_3H in water is found to obey the equation $\gamma = \gamma^0 - bC^2$. R denotes a long-chain hydrocarbon group and RSO_3H is a strong acid; C is in moles/liter and the temperature is 25°C.

(a) Derive the corresponding equation of state of the adsorbed film, that is, obtain π as a function of σ.

(b) Give a brief explanation of how it might be that γ depends linearly on C^2 rather than just on C.

2. (12 min) By means of a mechanical scoop, a very thin layer of surface is skimmed off a dilute soap solution. By this means, 2 cc of solution (containing essentially 2 g of water) are skimmed off of 300 cm^2 of surface. The amount of soap is found to be 4.013×10^{-5} mole, as compared to 4.000×10^{-5} mole in the same amount of bulk solution. Making (and stating) reasonable assumptions, calculate the surface tension of this solution.

3. (24 min) (final examination question) A 2×10^{-4}-M aqueous solution of a surface-active agent is being studied. With a device known as a microtome, it is possible to skim off a thin surface layer of known area, and, by this means, it is determined that the surface excess of the surfactant is 3×10^{-10} mole/cm^2. Calculate the surface tension of the solution, making reasonable assumptions. The temperature is 25°C.

4. (9 min) The surface tension of ethanol–water mixtures follows the equation $\gamma = 72 - 0.5C + 0.2C^2$, where C is the ethanol concentration in moles/liter. The temperature is 25°C. Calculate the surface excess of ethanol in moles/cm^2 for a 0.5-M solution.

5. (14 min) (final examination question) The surface tension of a 1% by weight solution of a surfactant is 70 dyne/cm, and that of a 2% solution is 68 dyne/cm (water is 72 dyne/cm). Show that the adsorbed film obeys the two-dimensional ideal gas law, and calculate the molecular weight of the surfactant, if it is known that the 2% solution had 20×10^{-9} g of surface excess of surfactant per cm^2. Assume 25°C.

6. (10 min) Show that the equation $\gamma/\gamma^0 = 1 - B \ln(1 + C/a)$, where B and a are constants, corresponds to a Langmuir adsorption isotherm.

7. (9 min) A 0.002 m aqueous solution of a mild surfactant has a surface tension of 68 dyne/cm at 25°C, and from other measurements it is known that the surface film obeys the two-dimensional ideal gas law. Calculate or explain clearly at what concentration the surface tension should be 69 dyne/cm.

8. (15 min) The diffusion coefficient of a monodisperse sulfur sol is 10^{-7} cm^2 sec^{-1}, and its sedimentation coefficient is 10^{-7} sec, both at 25°C.

(a) Calculate how far a sulfur particle would diffuse in 10 sec (neglecting any gravitational settling) and how far it would settle, owing to gravity, in this same time (neglecting any diffusional motion).

(b) By means of a rough calculation (i.e., one good only to an order of magnitude) explain or show clearly whether the value given for the sedimentation coefficient is reasonably consistent with that for the diffusion coefficient, or whether it is much too large or much too small.

9. (18 min) A colloidal particle is 0.2μ in diameter and has a density of 1.15 g/cc. Using suitable assumptions or approximations, calculate how long it should take the particle to move 0.2 mm if (a) diffusion alone is involved and (b) it is sedimenting under the influence of gravity, and diffusion is neglected. The medium is water at 25°C.

10. (18 min) The sedimentation coefficient for a protein molecule of molecular weight 60,000 and density 1.3 is 4×10^{-13} sec. The density of the medium (essentially water) is 1 and the temperature is 25°C. (a) Calculate the apparent and the Stokes' law friction factors. (b) Discuss briefly possible reasons for any discrepancy between the two values.

11. (9 min) Calculate the molecular weight of a protein whose sedimentation coefficient at 25°C is 1×10^{-12} sec and whose diffusion coefficient at 25°C is 4×10^{-7} cm²/sec. The density of the substrate is 1.02 g/cc and that of the protein is 1.3 g/cc.

12. (30 min) An aqueous solution contains 2% protein by weight, and, from an electrophoresis study, it is found that there are only two protein species, one of molecular weight 100,000 and the other of molecular weight 60,000. The two are present in *equimolar* concentrations. It may be assumed that the proteins behave as rigid spheres of density 1.3 g/cc. The substrate (solvent) has a density of 1 g/cc and a viscosity of 0.01 poise. Assume 25°C.

(a) Calculate the number-average and the weight-average molecular weights.

(b) Estimate the viscosity of the solution.

(c) Estimate the ratio of the diffusion coefficient for the higher-molecular-weight protein to that for the lower-molecular-weight one.

(d) Estimate the ratio of the sedimentation coefficients.

(e) If the protein contained in 1 cc of the solution were spread as a gaseous monolayer covering 10,000 cm², what would its film pressure be? (Numerical answers obtained by specific equations are wanted. The term "estimate" merely refers to the fact that there may be some latitude as to what equations are the most appropriate.)

13. (12 min) Solution A consists of a 1% by weight solution of a polystyrene polymer in toluene, of molecular weight 20,000 g/mole. Solution B consists of a 1% by weight solution also, but of polymer of molecular weight 60,000. Equal volumes of A and B are mixed.

(a) Calculate $\overline{M}_n$ and $\overline{M}_w$ for the mixture. (b) Calculate the viscosity of the mixture. Assume $[\eta] = 10^{-4} M^{1/2}$ and that the viscosity of pure toluene is 0.006 poise.

14. (12 min) The adsorption of hydrogen on nickel is found to obey the equation $x/m = aP^{1/2}$ where $x =$ moles adsorbed and $m =$ g nickel. (Note that $x/m = \Sigma \Gamma$, where Σ is the cm^2/g of surface and $\Gamma = moles/cm^2$. Also, $\Gamma = 1/\sigma A$, where $\sigma = cm^2$ of surface per molecule.)

Derive the corresponding two-dimensional equation of state for the adsorbed gas; that is, obtain π as a function of σ or of Γ. (The modified form of the two-dimensional ideal-gas law that is obtained suggests that the adsorbed gas is dissociated on the surface of the metal.)

15. (18 min) The adsorption of benzene vapor on graphite at $25°C$ follows the Langmuir equation fairly well, and at a pressure of 0.3 mm Hg, the fraction of surface covered is 0.05. The molecular area of the benzene molecule, assuming that it lies flat on the surface, is $30\ A^{0\,2}$.

(a) Calculate the pressure at which the surface should be half covered.

(b) Calculate π, the reduction in the graphite surface free energy, when a clean graphite surface is exposed to benzene vapor maintained at 0.3 mm Hg. Neglect the denominator in the Langmuir equation, $1 + bP$, for this part of the problem.

(c) Calculate the specific surface area of the graphite sample, Σ, if 2 g adsorbed 1 mM of benzene at a pressure such that the surface was fully covered.

ANSWERS

1. (a) According to the Gibbs equation, $\Gamma = (C/RT)\,d\pi/dC$. Here, $\pi = bC^2$, $d\pi/dC = 2bC$, and $\Gamma = 2bC^2/RT = 2\pi/RT$. Since $\Gamma = 1/\sigma$, $\pi\sigma = \frac{1}{2}RT$.

(b) C^2 would be proportional to the activity of undissociated RSO_3H or alternatively phrased, the concentration of RSO_3H would be equal to $[H^+][RSO_3^-]/K$ where K is the acid-dissociation constant. One possibility, then, is that it is the undissociated acid species that actually is the main surface adsorbed constituent.

2. Evidently, the surface excess is 0.013×10^{-5} mole on the basis of the choice of convention that there be no surface excess of solvent. Γ_2 is then 4.33×10^{-10} mole/cm^2. From the Gibbs equation, $\Gamma_2 = (C/RT)\,d\gamma/dC$. If it is assumed that at these dilutions, surface tension is a linear function of concentration, that is, $\gamma = \gamma^0 - bC$, then $d\gamma/dC = -b$ or $\Gamma_2 = bC/RT = (\gamma^0 - \gamma)/RT$. Then $(\gamma^0 - \gamma) = 4.33 \times 10^{-10} \times 8.31 \times 10^7 \times 298 = 10.7$. At the assumed temperature of $25°C$, γ^0 is 72, so γ would be about **61 dyne/cm.**

3. The solution is dilute enough that a reasonable assumption is that the surface tension is linear in concentration, which means that the adsorbed film obeys the two-dimensional ideal gas law. Then $\pi\sigma = RT$ or $\pi = RT\Gamma = 8.31 \times 10^7 \times 298 \times 3 \times 10^{-10} = 7.45$. Then $\gamma = 72 - 7.5 = $ **64.5 dyne/cm.**

4. From the Gibbs equation $\Gamma = -(C/RT)(d\gamma/dC)$, and here $d\gamma/dC = -0.5 + 0.4C = -0.3$. The $\Gamma = 0.5 \times 0.3/RT = 0.15/8.31 \times 10^7 \times 298 = $ **6.05×10^{-12} mole/cm^2**.

5. Evidently surface tension is linear with concentration, $\gamma = \gamma^0 - bC$, and differentiation gives $d\gamma/dC = -bC = \gamma - \gamma^0 = -\pi$. Substitution in Eq. (17-1) then yields $RT\Gamma = \pi$ or $\pi\sigma = RT$. For the 2% solution, $\pi = 4$, so $\Gamma = \pi/RT = 4/8.3 \times 10^7 \times 298 = 1.62 \times 10^{-10}$ mole/cm. The molecular weight is the $20 \times 10^{-9}/1.62 \times 10^{-10} = $ **123 g/mole**.

6. On differentiation,

$$\frac{d\gamma}{dC} = -B\gamma^0 \frac{1/a}{1 + C/a}$$

It then follows from Eq. (17-1) that

$$\Gamma = B\gamma^0 \frac{C/a}{1 + C/a}$$

Since Γ is proportional to θ, the above equation is in the form of a Langmuir equation, that is, $\theta = bC/(1 + bC)$.

7. As explained in connection with Eq. (17-2), if the two-dimensional ideal gas law is obeyed, the surface tension is linear with concentration, that is, $\gamma = \gamma^0 - bC$, or $\pi = bC$. Evidently, $b = (72 - 68)/0.002 = 4/0.002 = 2,000$. Then if π is to be $(72 - 69)$ or 3, C must be 3/2,000 or **0.0015**.

8. (a) First, $\mathcal{D} = x^2/2\tau$, so $x^2 = 2 \times 10 \times 10^{-7}$ or $x = $ **1.4×10^{-3} cm**. The sedimentation coefficient is the velocity per unit acceleration, so $dx/dt = Sg = 980 \times 10^{-7}$. Then Δx in 10 sec is $10 \times 980 \times 10^{-7} = $ **10^{-3} cm**.

(b) Since $\mathcal{D} = (1/f)kT$, the friction coefficient f is equal to $kT\mathcal{D}$. Thus $f = 1.38 \times 10^{-16} \times 298/10^{-7} = 10^{-6}$. The sedimentation coefficient is $S = (1/f)(4/3)\pi r^3$ (neglecting the density correction). We can estimate r from Stokes' law: $f = 6\pi\eta r$, or $r = 10^{-6}/6 \times 3 \times 0.01 \cong 10^{-7}$ cm. Then $S = (1/10^{-6})(4)(10^{-7})^3 \cong 10^{-15}$. Thus the value of S given is much **too large**.

9. (a) Assuming the particle to be spherical, the Stokes' law friction coefficient is $f = 6\pi\eta r = 6 \times 3.14 \times 0.01 \times 10^{-5} = 1.89 \times 10^{-6}$. Then $\mathcal{D} = kT/f = 1.38 \times 10^{-16} \times 298/1.89 \times 10^{-6} = 2.17 \times 10^{-8}$ cm^2/sec. From the Einstein equation, $\mathcal{D} = x^2/2\tau$, $\tau = 0.02^2/2 \times 2.17 \times 10^{-8} = $ **9.3×10^3 sec**.

(b) The sedimentation coefficient is given by

$$S = \tfrac{4}{3}\pi r^3 \frac{\rho - \rho^0}{f} = \frac{4 \times 3.14 \times (10^{-5})^3(1.15 - 1)}{3 \times 1.89 \times 10^{-6}}$$
$$= 3.31 \times 10^{-10}$$

Then $dx/dt = Sg = 3.25 \times 10^{-7}$, and $t = 0.02/3.25 \times 10^{-7} = $ **6.2 $\times$ 10^4** sec.

10. (a) $S = (1/f)(M/A)(1 - \bar{v}\rho^0)$ where $1/v = 1/1.3$. Then

$$f = \frac{60,000(1 - 0.77)}{4 \times 10^{-13} \times 6.02 \times 10^{-23}}$$

or $f = $ **5.72 $\times$ 10^{-8} g/sec.** The Stokes' law value is given by $f = 6\pi\eta r$. The radius is found from the relationship $(4/3)\pi r^3 = M\bar{V}/A$ or $r^3 = 60,000 \times 0.77 \times 3/[4 \times 3.14 \times 6.02 \times 10^{23}] = 1.85 \times 10^{-20}$ cm^3, and $r = 2.63 \times 10^{-7}$ cm. Then $f = 6 \times 3.14 \times 0.01 \times 2.63 \times 10^{-7} = $ **4.92 $\times$ 10^{-8} g/sec.**

(b) This is fairly close agreement and the discrepancy could, for example, be due to the departure of the protein molecule from a spherical shape, since the apparent f is larger than the Stokes' law value.

11. We have the two relationships $S = (M/fA)(1 - \bar{v}\rho^0)$ and $\mathcal{D} = kT/f$. Then $S = (M\mathcal{D}/RT)(1 - \bar{v}\rho^0) = (M\mathcal{D}/RT)(1 - 1.02/1.3)$. Thus $M = 1 \times 10^{12} \times 8.31 \times 10 \times 298/4 \times 10^{-7} \times 0.215 = $ **2.8 $\times$ 10^5.**

12. (a) Let n denote the number of moles of each protein in 100 g of solution; then $2 = 100,000n + 60,000n = 160,000n$, so $n = 2/160,000$. The number-average molecular weight is $M_n = $ (total weight)/(total moles) $= 2/2n = 2 \times 160,000/2 \times 2 = $ **80,000.** The weight-average molecular weight is

$$M_w = \frac{\Sigma W_i M_i}{\text{total weight}} = \frac{100,000n \times 100,000 + 60,000n \times 60,000}{160,000n}$$

$$= \textbf{85,000}$$

(b) We can use the equation $\eta_{sp} = \frac{5}{2}\phi$ where ϕ is the volume fraction of the polymer, in this case $(2/1.3)/100$ or 0.0154. Then $\eta_{sp} = 0.0385$, and $\eta/\eta_0 = 1.0385$ so $\eta = 0.010385 = $ **0.0104.**

(c) From the Stokes–Einstein equation $\mathcal{D}$ is inversely proportional to r and hence to the cube root of the molecular weight. $\mathcal{D}_2/\mathcal{D}_1 = (60,000/100,000)^{1/3} = $ **0.84.**

(d) S is proportional to the volume and inversely proportional to the friction coefficient, so $S \propto r^3/r \propto r^2$. The ratio S_2/S_1 is then $(100,000/60,000)^{2/3} = $ **1.41.**

(e) The 0.02 g in 1 cc of solution amounts to $0.02/80,000$ or 2.5×10^{-7} mole. Then $\pi A = nRT$ and $\pi = 2.5 \times 10^{-7} \times 8.31 \times 10^7 \times 298/10^4 = $ **0.62 dyne/cm.** (A denotes area.)

13. (a) If we take as a basis $200\,g$ of mixture, there will be $1\,g$ of each polymer type, and $1/20,000$ and $1/60,000$ mole of each type. Then

$$\overline{M}_n = \frac{2}{1/20,000 + 1/60,000} = \mathbf{30,000}$$

$$\overline{M}_w = \frac{1 \times 20,000 + 1 \times 60,000}{2} = \mathbf{40,000}$$

(b) The intrinsic viscosity is $[\eta] = 10^{-4}(40,000)^{1/2} = 0.02$ (the weight-average molecular weight should be used). Then $\eta_{sp} = 1 \times [\eta] = 0.02$, $\eta/\eta_0 = 1.02$, and $\eta = \mathbf{0.00612\ poise}$.

14. We use the Gibbs equation in the form $d\pi = RT \int \Gamma\, d\ln P$. The adsorption isotherm may be written $\Gamma = (a/\Sigma)P^{1/2}$, and substituting for Γ in the Gibbs equation,

$$d\pi = \frac{RTa}{\Sigma} \int P^{1/2}\left(\frac{dP}{P}\right) \qquad \pi = \frac{2RTa}{\Sigma}P^{1/2}$$

But $aP^{1/2}/\Sigma = \Gamma = 1/\sigma A$, so $\pi = 2RT/A\sigma$ or $\pi\sigma = \mathbf{2kT}$.

15. (a) The Langmuir equation can be written in the form $bP = \theta/(1 - \theta)$, so $bP = 0.05/(1 - 0.05) = 0.53$, and $b = 0.53/0.3 = 1.75$. For $\theta = 0.5$, $bP = 1$; hence $P = \mathbf{1.75\ mm\ Hg}$.

(b) First, Γ and θ are related by the equation

$$\Gamma = \frac{\theta}{\sigma^0 A}$$

where A is Avogadro's number and σ^0 is the actual area per molecule. Since, further, θ can be taken as equal to bP, substitution into Eq. (17-3) gives

$$\pi = \frac{RTb}{\sigma^0 A}P\,d\ln P = \frac{RTb}{\sigma^0 A}P$$

Then $\pi = 8.3 \times 10^7 \times 298 \times 1.75 \times 0.3/30 \times 10^{-16} \times 6 \times 10^{23} = \mathbf{7.2}$ **dyne/cm**. (c) The area covered by $2\,g$ would be $0.001 \times 6 \times 10^{23} \times 30 \times 10^{-16} = 1.8 \times 10^6\ cm^2$. The specific surface area is then $\mathbf{9 \times 10^5\ cm^2/g}$.

18
CRYSTAL STRUCTURE

COMMENTS

The subject of x ray crystallography constitutes a rather extended discipline, which to some extent sits alongside rather than astride the mainstream of physical chemistry. Some rather elegant mathematical methods have been devised for processing the usually many hundreds of $\sin \theta$ values obtained from an x-ray diffraction pattern so as to obtain a structure consistent with them. Nowadays most of the mathematical chore work is turned over to computers and the distinguishing trait of the great crystallographer is now not so much ability to make the calculations as the ability to perceive which crystal structures are more worthwhile doing, and why.

Crystals can be classed by symmetry properties, and the rules for doing so, along with familiarity with various point groups and space groups, are very important. We prefer, however, to get into the matter of symmetry in a later chapter, where applications to individual molecules can be considered.

After duly acknowledging the great scope of the topic, we shall follow the usual procedure of dealing with a rather restricted set of concepts, but shall expect you to deal with them fluently. Since these are examination questions, nearly all crystals encountered will be cubic; you are expected to know the three basic types, plus those for NaCl and CsCl. The Bragg equation is much used, and you should be adept at calculating interplanar distances. You will find reference to, say, 100 type and 100 actual planes. The latter are planes spaced a unit cell length apart; that is, they are the actual planes indicated by the Miller indices. The former simply refers to planes parallel to 100 planes, as a means of identifying from which crystal face a Bragg reflection has been obtained.

There are several problems involving a calculation of a crystal density. Be sure to count properly the number of atoms in the unit cell. Remember that, if the primitive unit cell or actual repeating unit is given, the number of atoms in it is the number to be used. If, however, as will often be the case, a unit cell is shown that is really the smallest piece of the crystal showing the symmetry, each atom is counted only to the extent it belongs to the piece shown. Thus the actual repeating unit for a simple cubic crystal is a single atom; by suitable translations the whole crystal can be generated. The simplest piece of the crystal showing the symmetry is, however, a cube of side a, with an atom at each corner. There are thus eight atoms shown, but each is shared with eight cells, so counts only one-eighth and the total is again one whole atom.

Generally speaking, the data given for real systems are correct. Sometimes numbers have been rounded to facilitate calculations, and in the case of certain problems there may be hidden complications that do not hinder the working of the problem but do prevent you from cribbing an answer (e.g., a density) merely by looking it up.

As a final point, remember that the concept of atomic radius in crystallography rests on the arbitrary but reasonable notion that atoms that are nearest neighbors can be thought of as spheres that are touching. Actually such an equilibrium distance is simply the balance point between forces of attraction and of repulsion. In the case of ionic crystals, oppositely charged ions will be nearest neighbors, and the attractive force is largely coulombic; it is then balanced by the repulsion of the electron clouds as they begin to overlap.

EQUATIONS AND CONCEPTS

Types of Cubic Crystals

The three types are simple, body-centered, face-centered. In the case of ionic crystals, additional types to know are the NaCl-type (two interpenetrating face-centered lattices) and the CsCl-type (Cs^+ ions at the corners of a cube and a Cl^- ion at the center).

Miller Indices

A simple way of obtaining Miller indices is to count the number of planes of the particular type that are crossed in going one lattice distance along each of the three crystallographic axes.

Interplanar Distances

For cubic crystals, the ratio of interplanar distances, $d_{100}:d_{110}:d_{111}$ is $1:1/\sqrt{2}:1/\sqrt{3}$ for simple cubic, $1:2/\sqrt{2}:1/\sqrt{3}$ for body-centered, and $1:1/\sqrt{2}:2/\sqrt{3}$ for face-centered. Remember that the indices as used above refer to *type* planes not actual planes (see Comments).

Again for cubic crystals, the general formula for the interplanar distance is

$$\frac{1}{d_{hkl}^2} = \frac{1}{a^2}(h^2 + k^2 + l^2) \tag{18-1}$$

where (hkl) is the Miller index. d_{hkl} is now the distance between actual (hkl) planes.

When a powder pattern is obtained, the geometry of the film and apparatus allows each spot to be assigned a corresponding $\sin \theta$ value. These in turn are proportional to $1/d$, so the set of $(\sin \theta)^2$ values must correspond to a set of $(h^2 + k^2 + l^2)$ values. In the case of cubic crystals $(h^2 + k^2 + l^2)$ can have the values 1, 2, 3, 4, 5, 6, 8, 9, and so on (7 being missing), in the case of a simple cubic structure. For a body-centered cubic (bcc) one, the allowed numbers are 2, 4, 6, 8, 10, 12, 14, 16, and so on, which are in the ratio 1, 2, 3, 4, 5, 6, 7, 8, and so on, so that 7 appears in the series. Finally, for a face-centered cubic (fcc) crystal, the numbers are 3, 4, 8, 11, ...

A general way of determining the intensity of the reflection from a given set of planes can be obtained from a detailed analysis of the phase changes that occur. This is given by a function P for each atom of the unit cell.

$$P_j = 2\pi(hx + ky + lz) \tag{18-2}$$

where x, y, and z are the locations of the jth atom of the unit cell, expressed in terms of the unit cell length (of a cubic crystal). The amplitude of the reflection is then determined by the structure factor F for the particular (hkl) plane:

$$F(hkl) = \sum_j f_j e^{2\pi i P_j} \tag{18-3}$$

Here, f_j is the atomic scattering factor for the j atom. Equation (18-3) can be simplified, since $e^{ix} = \cos x + i \sin x$, and in the problems encountered here, it will turn out that $\sin x$ will always be zero.

Bragg Equation

$$n\lambda = 2s \sin \theta \tag{18-4}$$

Density of a Crystal

The density is given by

$$\rho = \frac{\Sigma n_i M_i}{A(a)^3} \tag{18-5}$$

where n denotes the number of atoms of the ith kind in the unit cell, M its molecular weight, A is Avogadro's number, and a is the side of the unit cell.

Double Repulsion

In the case of ionic crystals, there can be a critical value of the ratio of the radii of the two types of ions such that *like* charged ions are in "contact," that is, are at a distance equal to the sum of their radii. In such a case two types of repulsive forces have become strong: the coulomb repulsion and the repulsion of the electron shells.

PROBLEMS

1. (18 min) The element polonium (atomic weight 210) crystallizes in the cubic system. Bragg first-order reflections, using x rays of wavelength 1.54 Å, occur at sin θ values of 0.225, 0.316, 0.388 for reflections from 100-, 110-, and 111-type planes.
  (a) Show whether the unit cell is simple, face-centered, or body-centered.
  (b) Calculate the value of a, the side of the unit cell.
  (c) Calculate the density of polonium.
  (d) Calculate the number of atoms per cm^2 for 110 (actual) planes. (Multiplying this number by the bond energy gives an estimate of the surface tension.)

2. (15 min) (final examination question) Calculate the percent void space in crystalline polonium metal. This element crystallizes in the simple cubic system.

3. (15 min) (final examination question) The element gallium crystallizes in the simple tetragonal structure with $a = b = 4.51$ Å and $c = 7.51$ Å. Calculate the density of gallium, and the 110-interplanar distance.

4. (30 min) CaF$_2$ is cubic. The primitive unit cell contains four calcium atoms at $(0, 0, 0)$, $(0, \frac{1}{2}, \frac{1}{2})$, $(\frac{1}{2}, 0, \frac{1}{2})$, and $(\frac{1}{2}, \frac{1}{2}, 0)$, and eight fluorine atoms at $(\frac{1}{4}, \frac{1}{4}, \frac{1}{4})$, $(\frac{1}{4}, \frac{3}{4}, \frac{1}{4})$, $(\frac{3}{4}, \frac{1}{4}, \frac{1}{4})$, $(\frac{3}{4}, \frac{3}{4}, \frac{1}{4})$, $(\frac{1}{4}, \frac{1}{4}, \frac{3}{4})$, $(\frac{3}{4}, \frac{1}{4}, \frac{3}{4})$, $(\frac{1}{4}, \frac{3}{4}, \frac{3}{4})$, and $(\frac{3}{4}, \frac{3}{4}, \frac{3}{4})$.
  (a) Locate all the atoms associated with the full unit cell.
  (b) Calculate the density of CaF$_2$. The length of the edge of the unit cube is 5.45 Å. (Atomic weights Ca $= 40$, F $= 19$.)
  (c) Show in Fig. 18-1 and calculate the shortest Ca–F distance in the crystal.
  (d) Describe the coordination of each kind of atom.
  (e) Give the relative intensities of the first four orders of (100) as obtained in an x-ray diffraction experiment.
  (f) If the experiment in (e) above is carried out with x rays of wavelength 1.5 Å, at what values of sin θ are these four orders predicted to occur?

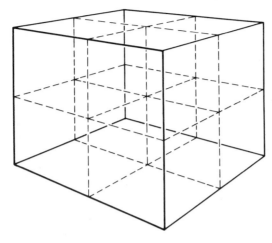

FIGURE 18-1

5. (30 min) Cs metal (atomic weight 133) crystallizes in a cubic structure. Using x rays of 0.8 Å wavelength, the sin θ values for the first-order reflections from 100-, 110-, and 111-type planes are 0.133, 0.094, and 0.230, respectively. (a) Explain which type of cubic unit is present. (b) Calculate the length a of the side of the unit cell. (c) Calculate the density of Cs metal. (d) Give the coordination number for Cs in this structure. (e) Calculate the radius of Cs atoms. (f) Pick a type of cubic system other than that stated in (a), and show whether the density of Cs metal would be greater or less if it crystallized in this second type, assuming that the radius of the Cs atom remained the same.

6. (5 min) (final examination question) KCl crystallizes in the NaCl-type structure and CsF in the CsCl-type structure. The molecular weight of CsF is twice that of KCl, and the a value for KCl is 1.5 times that for CsF. Calculate the ratio of the density of CsF to that of KCl.

7. (20 min) (final examination question) KCl crystallizes in the same type of lattice as does NaCl. The ionic radius of Na^+ is 0.5 of that of Cl^-, and is 0.7 of that of K^+. Calculate (a) the ratio of the side of the unit cell for KCl to that for NaCl, (b) the ratio of the density of NaCl to that of KCl (atomic weights: K, 39; Na, 23; Cl, 35.5).

(c) Sketch the projected appearance of a group of unit cells as they would appear on looking down the z axis, and draw the lines corresponding to the projection of 200 and 230 planes.

8. (3 min) (final examination question) Mn crystallizes in the same type of cubic unit cell as does Cu, but with an a value 5% larger. Calculate the density of Mn if the density of Cu is 8.92, and the atomic weights are 63.6 and 54.9 for Cu and Mn, respectively.

9.   Iron (atomic weight 56) has a fcc structure in the γ form, with $a = 3.68$ Å, and a bcc structure in the β form, with $a = 2.9$ Å. Calculate the ratio of the densities in the two crystalline modifications.

10.   (12 min) (final examination question) An element crystallizes in the bcc structure. Calculate the percent void space.

11.   (15 min) (final examination question) Element X crystallizes in a cubic system; the unit cell has atoms at positions (000), $(\frac{1}{2}, \frac{1}{2}, \frac{1}{2})$, (101), (110), (011). Of the various types of planes which contain all the atoms, the set with the largest interplanar distance is one with distance 1.5 Å. Calculate the density of X; its atomic weight is 25. (*Note:* The "unit cell" given is actually a portion $(\frac{1}{8})$ of the true unit cell; however, use the cell given as though it were the complete unit cell.)

---

12.   (30 min) A hypothetical salt A_xB_y crystallizes in the cubic system (see Fig. 18-2). The unit cell has atoms of A at each corner plus one at $(\frac{1}{2}, \frac{1}{2}, \frac{1}{2})$. It has atoms of B at the positions $(0, \frac{1}{2}, \frac{1}{2})$ and $(1, \frac{1}{2}, \frac{1}{2})$.
(a) Locate the atoms of A and B in the unit cell.
(b) Give the values of x and y in A_xB_y.
(c) If the side of the unit cell is 6 Å, and the atomic weights of A and B are 40 and 120, respectively, calculate the density of A_xB_y.
(d) Considering only the B atoms, calculate the interplanar distances for 100-, 110-, and 111-*type* planes.
(e) Recognizing both A and B atoms, assign intensities to the reflections from the following planes (these are now actual, not type planes).

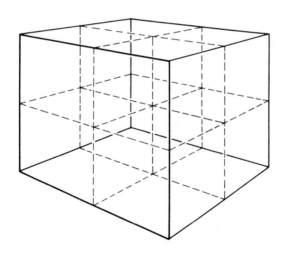

FIGURE 18-2

Plane	Intensity	Plane	Intensity
100		110	
200		220	

Give intensities as absent, weak, or strong.

(f) Give the coordination number for *A* and for *B*, and describe the geometric figure that corresponds to the arrangement of *A* atoms around a *B* atom.

13. (30 min) The unit cell (cubic) for an oxide of copper is shown in Fig. 18-3, where atoms of one kind are located in a body-centered lattice, and the others are at the centers of alternate small cubes.

(a) Explain whether the circles probably denote copper or oxygen atoms, and the formula for the oxide. (b) Calculate the density of the crystal if $a = 4.26$ Å. (c) Consider the first eight values for the sum $h^2 + k^2 + l^2$, and specify whether the reflections in each case should be absent, weak, or strong:

$h^2 + k^2 + l^2$	Reflection	$h^2 + k^2 + l^2$	Reflection
1		5	
2		6	
3		7	
4		8	

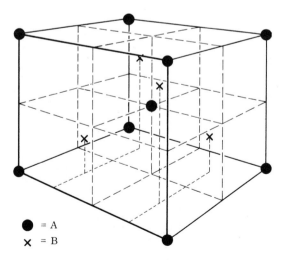

● = A
✕ = B

FIGURE 18-3

(d) Calculate the Cu–O distance, and explain what the coordination number of copper is in the crystal and the geometric figure formed by its oxygen neighbor.

14. (30 min) Zinc sulfide crystallizes in a cubic unit cell, with the zinc atoms in a fcc arrangement and the sulfur atoms in the centers of alternate small cubes at the positions: $(\frac{1}{4}, \frac{1}{4}, \frac{1}{4})$, $(\frac{1}{4}, \frac{3}{4}, \frac{3}{4})$, $(\frac{3}{4}, \frac{1}{4}, \frac{3}{4})$, and $(\frac{3}{4}, \frac{3}{4}, \frac{1}{4})$. The side of the unit cell is 6 Å.

(a) Locate the Zn and S atoms in the perspective drawing of the unit cell in Fig. 18-4.

(b) Calculate the density of ZnS, using the above information, plus the atomic weights of S and of Zn.

(c) Calculate the $\sin \theta$ values for first-order reflections from 200 and 110 planes (these are the actual planes), using x rays of 0.2 Å wavelength. Explain whether the intensity of the two reflections should be normal or reduced in intensity because of interference effects.

(d) Calculate the Zn–S distance, that is, the sum of the Zn and S radii; indicate such an interatomic distance on the drawing.

(e) What is the number and geometric arrangement of Zn nearest neighbors to each S atom?

15. (30 min) MnO has a cubic unit cell, of side 4.47 Å. Its density is 5.2 g/cc (atomic weight of Mn is 55).

(a) Calculate the number of MnO units per unit cell, and determine from this which type it is (simple, body- or face-centered, with the alkali halide crystals in mind). Give also the coordination number of Mn and of O.

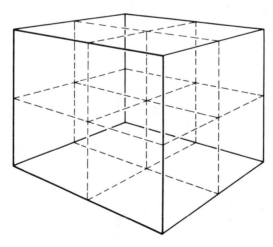

FIGURE 18-4

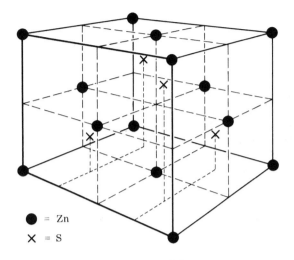

FIGURE 18-5

(b) Assuming the unit cell to be body-centered cubic (of side 4.47 Å), calculate sin θ for reinforcement of 0.5 Å x rays, when reflected from 210 planes (first-order reflection). From which of the following three planes should the first-order reflection be the most intense (assuming the correct respective θ values are employed): 100, 110, or 210? Why?

(c) Define the term double repulsion (as applied to MX-type crystals).

16. (30 min) The first-order Bragg reflections from 100-, 110-, and 111-type planes are at sin θ values of 0.0643, 0.0908, and 0.0557, in the case of NaI, using x rays of 0.4 Å wavelength. The reflections from the 100- and 110-type planes show the normal decrease in intensity with increasing order, but the first-order peak for the 111-type plane is less intense than is the second-order one. (Atomic weights: Na, 23; I, 127; radius of Na^+ is 0.95 Å.)

(a) Show whether NaI has the ZnS-, NaCl-, or CsCl-type of structure. You are asked to show that the above data conform to one of these types and fail to conform to the other two.

(b) Calculate the length in Å of the side of the unit cell.

(c) Calculate the density of NaI, and the radius of I^-. The ZnS structure is shown in Fig. 18-5.

17. (13 min) The intermetallic compound CuPd has a cubic unit cell of side 3.0 Å. From a powder x-ray pattern, first-order reflections are found at sin θ values of 0.0472, 0.0666, and 0.115; these are to be presumed to be due to reflections from 100-, 110-, and 111-type planes, but not necessarily in the order stated. The wavelength of the x rays is 0.2 Å. Show from this data whether the unit cell is simple cubic, body-centered, or face-centered, and sketch the cell, with probable location of the atoms. Assign sin θ values to the proper planes.

18. (30 min) A certain element (hypothetical) exists in three crystalline modifications: sc (simple cubic), bcc, and fcc.
(a) The diffraction of 0.5 Å x rays is studied for one of these modifications and it is found that strong reflections occur from 111 *actual* places at a sin θ value of 0.108, and from 110 *type* planes at a sin θ value of 0.175. Show whether this particular sample was sc, bcc, or fcc, and calculate a, the side of the unit cell. (b) If the radius of the atom of this element is 1.5 Å, calculate the ratio of the density of the fcc form to that of the sc form. Bear in mind that regardless of structure, nearest neighbor atoms are considered to be in contact. (c) Calculate the fraction void space for the sc form. (d) Calculate F (in terms of f) for 531 planes, in the case of the bcc form.

19. (16 min) A metal which crystallizes in the cubic system gives the following Bragg reflections. Explain whether the unit cell is sc, fcc, or bcc and calculate the value of a, if x rays of wavelength 1.54 Å are involved. Also, index the reflections.

θ	sin θ	(sin θ)^2	hkl
21.8	0.371	0.138	
25.4	0.429	0.185	
37.4	0.606	0.369	
45.4	0.712	0.510	

20. (24 min) The mineral perovskite has the formula $Ca_xTi_yO_z$, and has a cubic structure. The atoms of the unit cell are located as follows:
Ca: (000), (100), (110), (010), (001), (101), (111), and (011)
Ti: ($\frac{1}{2}\frac{1}{2}\frac{1}{2}$)
O: ($\frac{1}{2}\frac{1}{2}0$), ($0\frac{1}{2}\frac{1}{2}$), ($\frac{1}{2}0\frac{1}{2}$), ($1\frac{1}{2}\frac{1}{2}$), ($\frac{1}{2}\frac{1}{2}1$), and ($\frac{1}{2}1\frac{1}{2}$)
(a) Give the formula (i.e., the values of x, y, and z). (b) Calculate the density of perovskite if a is 3.8 Å. (c) State whether Bragg reflections should be strong, weak, or absent from the following Miller index (actual) planes (hkl): 100, 200, 400, 110, 220.
(d) Give the positions of the atoms in the repeating unit (as opposed to the unit cell locations above).

21. (18 min) (final examination question) Suppose that Cs metal crystallizes in a sc structure of the same unit cell side, a, as does CsCl (which is bcc). Calculate: (a) The ratio of the density of Cs metal to CsCl; (b) The ratio of the sin θ value for first-order reflection from Cs metal 111 *type* planes to the sin θ value for first-order reflection from CsCl 111 type planes; (c) The relative intensity of the reflections from Cs metal 210 actual planes versus the reflections from CsCl 210 actual planes.

ANSWERS

1. (a) The $\sin \theta$ values are in the ratio, $1 : 1.41 : 1.73$, that is, $1 : \sqrt{2} : \sqrt{3}$. This corresponds to interplanar distances in the ratio $1 : 1/\sqrt{2} : 1/\sqrt{3}$, which is characteristic of a **simple cubic** type.

(b) Using the value for 100-type planes, $d_{100} = a = \lambda/2 \sin \theta = 1.54/2 \times 0.55 = $ **3.42 Å.**

(c) Since the crystal is sc, there is one atom per unit cell; hence

$$\rho = \frac{210}{6 \times 10^{23}(3.42)^3 10^{-24}} = \textbf{8.75 g/cc}$$

(d) The number of atoms per unit volume may be expressed as (number of atoms per cm^2 in a given type plane)/(interplanar distance). For 110 planes, d is $a/\sqrt{2} = 2.42$ Å, and the number of atoms per cc is $1/(3.42)^3 \, 10^{-24} = 2.5 \times 10^{22}$. The surface density is then $2.5 \times 10^{22} \times 2.42 \times 10^{-8} = $ **6.05 $\times$ 10^{14} atoms/cm^2.**

2. The unit cell evidently consists of eight spheres touching and with centers at the corners of a cube. Each sphere is of radius $a/2$, and the unit cell contains eight-eighths or one sphere, so the volume of atoms it contains is $\frac{4}{3}\pi(a/2)^3 = 0.525a^3$. Then **47.5%** is void space.

3. The unit cell contains one atom, so the density is

$$\rho = \frac{69.7}{6 \times 10^{23}(4.51)^2(7.51)10^{-24}} = \textbf{0.75 g/cc}$$

(Actually, things are a little more complicated as there are eight atoms in this unit cell—something we ignore here.) The projection on the xy plane looks like a simple cubic array, so d_{110} is $4.51/\sqrt{2} = $ **3.2 Å.**

4. (a) See Fig. 18-6. Note that the additional Ca atoms shown by open circles must be added to complete the full unit cell.

(b) There are four Ca atoms and eight F atoms in the unit cell so

$$\rho = \frac{4 \times 40 + 8 \times 19}{6 \times 10^{23}(5.45)^3 10^{-24}} = \textbf{3.22 g/cc}$$

(c) As shown in Fig. 18-6, this is half of the body diagonal of a small cube, so $d = \frac{1}{2}(a/2)\sqrt{3} = 5.45 \times 1.73/4 = $ **2.35 Å.**

(d) Each fluorine has four calcium atoms at the alternate corners of a small cube, that is, at the corners of a **tetrahedron.** Each calcium atom has eight fluorines, at the corners of a **cube.**

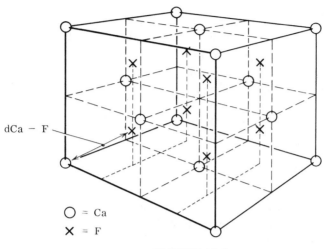

dCa – F

$\bigcirc$ = Ca
$\times$ = F

FIGURE 18-6

(e) The first four orders of 100 correspond to 100, 200, 300, and 400 reflections. The 100 reflections will be absent since there is an in-between set of identical planes. The 200 reflections will be weak, since there is an in-between set of planes, but occupied by the other kind of atom. 300 planes do not correspond to any crystal planes in this case, so there should be no reflection of this type. The 400 reflections should be strong.

(f) From the Bragg equation, $\sin \theta = n\lambda/2d = n(1.5/2 \times 5.45) = 0.137n$. The values are then 0.137, 0.275, 0.412, and 0.55.

5. (a) The reciprocals of the $\sin \theta$ values are in the ratio $1:1.41:0.58$, or $1:2/\sqrt{2}:1/\sqrt{3}$. This corresponds to a **body-centered** cubic structure.

(b) From the 100-type reflection, $d = 0.8/2 \times 0.133 = 3$; hence $a = 6\,\text{Å}$.

(c) There are two atoms in the unit cell, so

$$\rho = \frac{2 \times 133}{6 \times 10^{23}(6)^3 10^{-24}} = \textbf{2.0 g/cc}$$

(d) Each Cs has **eight** nearest neighbors, at the corners of the cube.

(e) Cs atoms are touching along the body diagonal, which is then equal to four radii. Then $r = 1.73a/4 = \textbf{2.6 Å}$.

(f) If we take the simple cubic system, $a' = 2r = a\sqrt{3}/2$. There is now one atom in the unit cell so the densities will be in the ratio

$$\frac{2/a^3}{1/a'^3} = 6\sqrt{3}/8$$

Cs would then be less dense in the simple cubic form. If we suppose the alternative form to be face-centered, the face-diagonal equals $4r$, so

$4r = a''\sqrt{2}$ and $a'' = a\sqrt{3}/\sqrt{2}$. The densities will now be in the ratio

$$\frac{2/a^3}{4/a''^3} = 3\sqrt{3}/4\sqrt{2}$$

The fcc form will thus be the denser.

6. KCl will have four atoms of each kind in its fcc structure, whereas CsF will have one of each kind in its body-centered type of structure. The densities will then be in the ratio

$$\frac{\rho_{KCl}}{\rho_{CsF}} = \frac{4M_{KCl}/a_{KCl}^3}{M_{CsF}/a_{CsF}^3} = 4\left(\frac{1}{2}\right)\left(\frac{1}{1.5}\right)^3 = \mathbf{0.59}$$

7. (a) In the NaCl-type of lattice, the side of the unit cell is made up of one M^+ and one X^- diameter, since oppositely charged ions are touching along the side. Then $a_{NaCl} = d_{Na^+} + d_{Cl^-} = 3d_{Na^+}$ and $a_{KCl} = d_{K^+} + d_{Cl^-} = 3.43d_{Na^+}$ so $a_{KCl}/a_{NaCl} = \mathbf{1.14}$.

(b) $\dfrac{\rho_{NaCl}}{\rho_{KCl}} = \dfrac{M_{NaCl}/(a_{NaCl})^3}{M_{KCl}/(a_{KCl})^3} = (1.14)^3\dfrac{58.5}{74.5} = \mathbf{1.16}$

(c) The projection on the xy plane is shown in Fig. 18-7.

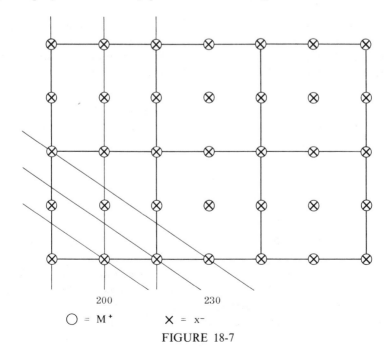

200   230

○ = M^+ ✗ = x^-

FIGURE 18-7

8. The ratio of densities will be given by

$$\frac{\rho_{Mn}}{\rho_{Cu}} = \frac{54.9/(1.05a)^3}{63.6/(a)^3} = 0.745$$

The density of Mn is then 0.745 that of Cu, or **6.6 g/cc.**

9. There will be four atoms per unit cell in the γ form, and two in the β form, so the ratio of densities will be

$$\frac{\rho_\gamma}{\rho_\beta} = \frac{4/(3.68)^3}{2/(2.9)^3} = \mathbf{0.98}$$

10. The atoms, assumed spherical, must be touching along the diagonal, so $4r = \sqrt{3}a$, where r is the radius of an atom. The unit cell contains 2 atoms, whose volume is then $2 \times 4 \times \pi \times r^3/3$ as compared to the volume a^3 of the cell. The fraction of occupied volume is then

$$\frac{2 \times 4\pi \times (\sqrt{3}/4)^3}{3} = 0.68$$

The fraction of void space is then **0.32.**

11. Inspection of the situation shows that planes containing all the atoms will be 200, 110, and 220. The interplanar spacing is the largest for the 100 planes, $a/\sqrt{2}$, so $a = 1.5\sqrt{2} = 2.1$ Å. There are two atoms per unit cell, and the density is then $2 \times 25/(2.1)^3 \times 10^{-24} \times 6 \times 10^{23} = \mathbf{0.90\,g/cc.}$

12. (a) The atoms are located in Fig. 18-8.

(b) There are eight corner A atoms and one central atom, or two net atoms. There are two face atoms of B; hence there is one net atom. The formula is then A_2B.

(c) The density is

$$\rho = \frac{2 \times 40 + 120}{6 \times 10^{23}(6)^3 10^{-24}} = \mathbf{1.54\,g/cc}$$

(d) For B atoms only, inspection shows $d_{100} = a$. Furthermore, the B atoms are evidently arranged in a simple cubic array of side a, so $d_{110} = a/\sqrt{2}$ and $d_{111} = a/\sqrt{3}$. Then $d_{100} = \mathbf{6\,Å}$, $d_{110} = \mathbf{4.25\,Å}$, and $d_{111} = \mathbf{3.47\,Å}$.

(e) 100 planes contain A and B atoms, with planes half-way between containing only A atoms. Partial interference is expected, so the intensity should be **weak.** 200 planes contain all atoms, so the intensity should be

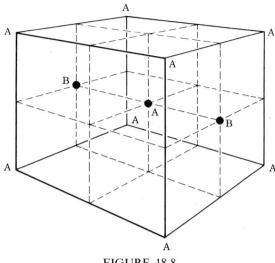

FIGURE 18-8

strong. For the next two cases, it is helpful to make the xy projection shown in Fig. 18-9. It is then evident that 110 planes contain only A atoms, with planes half-way between containing only B atoms. Partial interference will be present, so the intensity should be **weak.** Finally, 220 planes contain all atoms, so the intensity should be **strong.**

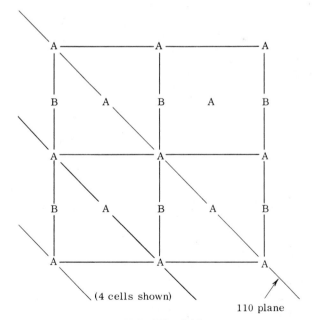

(4 cells shown)

110 plane

FIGURE 18-9

(f) A atoms are of two types: those having two B atoms $a/2$ distance away as nearest neighbors, and those having 4 B atoms $a/\sqrt{2}$ distance away. For A, the coordination numbers are **2** and **4**. Each B atom has two A atoms at $a/2$ and four A atoms at $a/\sqrt{2}$. The first two are the closer, so the coordination number of B is **two**. The figure is a distorted (compressed) octahedron.

13. By the rules for counting atoms, there are eight corner atoms of the A type, or one net atom, plus one interior atom for a total of two atoms. There are four interior B-type atoms, hence four net atoms. The formula could then be CuO_2 or Cu_2O. The latter is the more reasonable from a valence standpoint, so $O = A$ and $Cu = B$, referring to Fig. 18-3.

(b) The density is then

$$\rho = \frac{4 \times 63.5 + 2 \times 16}{6 \times 10^{23}(4.26)^3 10^{-24}} = \textbf{6.2 g/cc} \tag{6-2}$$

(c) It is helpful first to index the $(h^2 + k^2 + l^2)$ numbers and refer to Fig. 18-10. 100 planes of either Cu or O will have identically occupied planes

$h^2 + k^2 + l^2$	Index
1	100
2	110
3	111
4	200
5	210
6	211
7	
8	220

half-way between, so the intensity will be **zero**. 110 planes likewise: **zero**. 111 planes: again, looking at the body-centered oxygen lattice, there will be a set of nonidentical planes in between, so the intensity will be **weak**. 200 planes will have a second but not identical set in between, so the intensity will be **weak**. By looking at the xy projection we can see that 210 planes will not contain all the atoms and in-between planes will not be identical, so the intensity should be **weak**. 220 planes, however, do contain all the atoms and are identical, so the intensity should be **strong**. Finally, there is no index giving the integer 7 for $(h^2 + k^2 + l^2)$, so **no intensity.**

(d) The smallest Cu—O distance is evidently half the body diagonal of one of the small cubes; that is, $d = \frac{1}{2}(a/2)\sqrt{3} = 4.26 \times 1.73/4 = \textbf{1.83}$ Å. Each copper has two oxygen nearest neighbors (lying on a straight line).

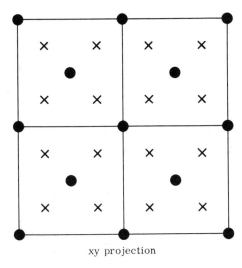

xy projection

FIGURE 18-10

14. (a) See Fig. 18-11.

(b) There are evidently four atoms of each kind in the unit cell, so the density is

$$\rho = \frac{4(65.4 + 32)}{6 \times 10^{23}(6)^3 \times 10^{-24}} = \mathbf{3.0\,g/cc}$$

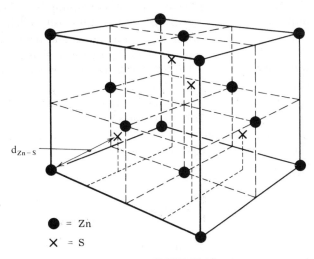

d_{Zn-S}

● = Zn

✕ = S

FIGURE 18-11

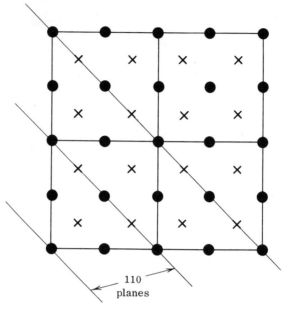

xy projection (4 cells)

FIGURE 18-12

(c) d_{200}^2 and d_{110}^2 will be $\frac{36}{4}$ and $\frac{36}{2}$, so d_{200} and d_{110} are 3 Å and 4.22 Å, respectively. The respective sin θ values are then $0.2/2 \times 3 = \mathbf{0.033}$ and 0.0237.

Inspection of the drawing shows that 200 planes contain only atoms of one kind and that there will be in-between planes containing the other kind of atom, so that partial interference will occur and the intensity will be **weak**. 110-type planes contain both kinds of atoms, but 110-actual planes will be separated by an additional equivalent set halfway between, as is clear from the *xy* projection. The intensity of 110 reflections will thus be **zero,** owing to complete interference.

(d) The Zn–S distance is half the body diagonal of a small cube, $d = \frac{1}{2}(a/2)\sqrt{3} = 6 \times 1.73/4 = \mathbf{2.58\,\mathring{A}}$. (See Fig. 18-12.)

(e) Each sulfur atom has four zinc nearest neighbors, at the alternate corners of a cube or, in other words, in tetrahedral geometry.

15. (a) We have

$$5.2 = \frac{n(55 + 16)}{6 \times 10^{23}(4.47)^3 10^{-24}} = 1.31n$$

from which $n = 4$. This suggests the NaCl or **face-centered** type. The coordination number of each kind of atom is then **6.**

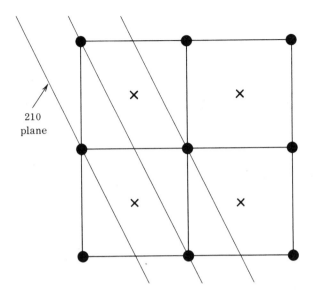

210
plane

FIGURE 18-13

(b) If the unit cell were body-centered, that is, of the CsCl type, the projection of the xy plane would be as shown in Fig. 18-13. The 210 planes are now seen to contain only one kind of atom, with in-between planes containing the other kind, so partial cancellation will occur. $d_{210}^2 = a^2/5$ or $d_{210} = 0.445 \times 4.47 = 2$ Å. Sin θ is then $\lambda/2 \times 2 = 0.5/4 = \mathbf{0.125}$. Returning to intensities, 100 reflections will be weak, for the same reason as with the 210 reflections. Since 110 planes contain all the atoms and are identical, these reflections will be strong.

(c) In ionic MX-type crystals, it is assumed that nearest oppositely charged ions touch (i.e., this is how the ionic radii are defined). Double repulsion is present if like-charged ions are also touching.

16. (a) The sin θ values are in the ratio $1:1.41:0.865$ and their reciprocals are in the ratio $1:0.71:1.15$, or $1:1/\sqrt{2}:2/\sqrt{3}$. Turning to the three possible types, for CsCl the three distances are $a/2$, $a/\sqrt{2}$, and $a/2\sqrt{3}$, which does not conform. For ZnS, we have $a/4$, $a/2\sqrt{2}$, and $a/2\sqrt{3}$, which does not conform. Finally, for the NaCl type, the distances are $a/2$, $a/2\sqrt{2}$, and $a/2\sqrt{3}$, which again does not conform. The information about intensities provides a clue, however. The 100 and 110 reflections must correspond to an interplanar distance between adjacent planes, that is, no in-between planes, while first-order 111 reflections must correspond to a set of planes having an in-between plane giving partial cancellation. We can, in fact, rule out the CsCl and ZnS structures immediately, since 100 reflections should give alternation in intensity, contrary to observation. In the NaCl structure, alternation would only occur with 111 reflections, and the weak 111 peak would then correspond to the interplanar distance for I^- containing planes, that is, $a/\sqrt{3}$. The 100, 110, and 111 interplanar distances that should be used are thus $a/2$, $a/2\sqrt{2}$, and $a/\sqrt{3}$, which do conform to the ratio obtained from the $\sin \theta$ values.

(b) Using the $\sin \theta$ value for 100 reflections, $d_{100} = 0.4/2 \times 0.0643 = 3.11\,\text{Å}$ and $a = \mathbf{6.2\,\text{Å}}$.

(c) The density is given by

$$\rho = \frac{4(23 + 127)}{6 \times 10^{23}(6.2)^3\,10^{-24}} = \mathbf{4.2\,g/cc}$$

The side is made up of adjacent Na^+ and I^- ions, so $a = 2r_{Na^+} + 2r_{I^-}$ or $r_{I^-} = 3.1 - 0.95 = \mathbf{2.15\,\text{Å}}$.

17. Since no intensity data is given, we shall have to assume that the atoms are to be treated as indistinguishable. The $\sin \theta$ values are in the ratio $1:1.41:2.43$ and their reciprocals, $1:0.71:0.412$. If simple cubic, $d_{100}:d_{110}:d_{111}$ would be $1:0.71:0.58$, and no arrangement of the d ratios from the $\sin \theta$ values can be made to fit. If the crystal were bcc we would have the ratio $1:1.41:0.58$, and if the set of d ratios from the data is put in the order $0.71:1:0.412$, it is seen that they do fit. Finally, a test calculation assuming a fcc structure would give the ratio $1:0.71:1.15$, which cannot be made to fit. The crystal is then **body-centered** cubic. A possible unit cell is shown in Fig. 18-14, and the $\sin \theta$ values are for d_{110}, d_{100}, and d_{111} in the order as originally given.

18. (a) Since 111 actual planes give a strong reflection, the form must be either sc or fcc, and by Eq. (18-4) $d_{111} = 0.5/2 \times 0.108 = 2.32\,\text{Å}$. Since $d_{111} = a/\sqrt{3}$, $a = 1.73 \times 2.32 = \mathbf{4.0\,\text{Å}}$. Next, for sc, d_{110} is the same for

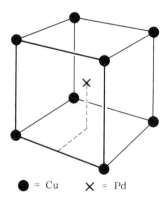

● = Cu ✗ = Pd

FIGURE 18-14

actual and type planes and is equal to $a/\sqrt{2}$, giving an expected sin θ value of $0.5/2 \times 2.82 = 0.09$ (which is wrong). If the crystal form is fcc, the d for 110 type planes must actually be d_{220}, which is $a/2\sqrt{2}$, and calculation now gives the observed 0.175 value for sin θ. The crystal was therefore **fcc.**

(b) The atoms across a face diagonal are touching in the fcc form, so $\sqrt{2}a = 4r$, and the unit cell volume of a^3 is then given by $(4/\sqrt{2})^3 r^3$ or $16\sqrt{2}r^3$. There are four atoms in the unit cell, so the volume per atom is $4\sqrt{2}r^3$. In the case of an fcc crystal, corner atoms are touching so $a = 2r$, and the volume per unit cell and per atom is therefore $8r^3$. The density ratio is the inverse of the above volumes per atom, so

$$\frac{\text{fcc}}{\text{sc}} = \frac{8r^3}{4\sqrt{2}r^3} = \textbf{1.41}$$

(c) The unit cell has one atom whose volume is $4\pi r^3/3$, while the volume of the unit cell itself is a^3 or $(2r)^3 = 8r^3$. The fraction of occupied space is then the ratio of the above, or $\pi/6 = 53\%$, and the void fraction is thus 47%.

(d) The phase function, P, of Eq. (18-2) is

$$2\pi(5 \times 0 + 3 \times 0 + 1 \times 0) + 2\pi(5 \times \tfrac{1}{2} + 3 \times \tfrac{1}{2} + 1 \times \tfrac{1}{2})$$

or

$$P = 0 + 9\pi$$

The structure factor is then

$$F = f(\cos 0 + \cos 9\pi) + fi(\sin 0 + \sin 9\pi)$$
$$= f(1 - 1) + f(0 + 0) = \textbf{0}$$

19. Since sin θ is proportional to $1/d$, Eq. (18-1) tells us to look for the smallest set of integers that are in the ratio of the $(\sin \theta)^2$ values. If each is divided by 0.138, the numbers 1, 1.34, 2.67, and 3.70 result; multiplying these by 3 gives the set 3, 4, 8, and 11. This sequence is characteristic of a **fcc** crystal.

20. (a) There are 8 corner and hence 1 whole Ca atom, 1 interior Ti atom, and 6 face or 3 whole O atoms. The formula is then **CaTiO$_3$.** (b) The density is given by $\rho = 1 \times 136/3.8 \times 10^{-24} \times 6 \times 10^{-23} = \textbf{4.1 g/cc.}$ (c) The projection down the z axis, shown in Fig. 18-15 shows that 100 planes should give a **weak** reflection as there is an intervening but not identical plane. 200 reflections should be **strong;** 400 **weak,** because they will be second order 200 reflections; 110 **weak,** because of an intervening but not identical

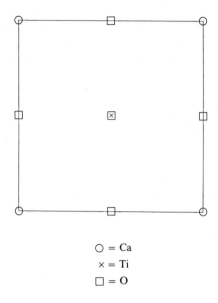

$\bigcirc$ = Ca
$\times$ = Ti
$\square$ = O

FIGURE 18-15

set of planes; and 220 reflections will be **strong**. (d) The smallest unit that will generate the crystal with unit translations along the x, y, and z directions consists of Ca: (000), Ti: $(\frac{1}{2}\frac{1}{2}\frac{1}{2})$, and O: $(\frac{1}{2}\frac{1}{2}0)$, $(0\frac{1}{2}\frac{1}{2})$, and $(\frac{1}{2}0\frac{1}{2})$.

21. (a) Since a is the same in both cases, and unit cells contain either one Cs or one CsCl unit, the ratio of the densities is simply that of the respective formula weights: $\rho_{Cs}/\rho_{CsCl} = 133/168.5 = $ **0.79**. (b) For Cs, 111 type planes have the spacing $a/\sqrt{3}$, while for CsCl, the spacing is $a/2\sqrt{3}$. Since $\sin\theta$ is proportional to $1/d$, the ratio is thus

$$\frac{(\sin\theta)_{Cs}}{(\sin\theta)_{CsCl}} = \frac{(a/2\sqrt{3})}{(a/\sqrt{3})} = \frac{1}{2}$$

(c) On making a projection down the z axis, one sees that 210 planes will contain all the Cs atoms in Cs or in CsCl, but not the Cl atoms of the latter. The reflection will then be **strong** for Cs, but **weak** for CsCl because of the intervening but not identical planes.

19
NUCLEAR CHEMISTRY

COMMENTS

We concern ourselves here with a few of the more general and straightforward principles of radio and nuclear chemistry. Nuclear transmutation reactions will be limited to those that can occur with incident energies of the bombarding particle of 5 to 10 million electron volts (mev). In effect these are limited to reactions of the $A(x, y)B$ type, where x and y represent a neutron, proton, alpha particle, or deuteron, and a neutron, proton, or alpha particle, respectively. A and B are the initial and product nuclei. Two additional processes are to be understood to be included in the above designation. They are the (n, γ) and the $(n, 2n)$ reactions, which can occur with relatively low-energy neutrons. You will be expected to write equations for transmutation processes and to balance them as to nuclear charge or atomic number, and mass number.

The calculation of energy changes will be required. Remember the Q of a transmutation reaction is given by the sum of the weights of the reactants minus the sum for the products. Since your equation should be balanced as to mass numbers, Q can equally well be obtained by similarly combining sums of mass defects.

A few of the more important aspects of nuclear systematics will come up. You should know the general appearance of the plot of mass defect versus mass or atomic number. If the stable isotopes are located on a plot of Z versus A (atomic number versus mass number), they are found to lie roughly along a line that has a slope of $\frac{1}{2}$ for light elements, but which then begins to decrease in slope, so that heavy elements tend to be neutron rich relative to light ones.

The principal modes of radioactive decay considered here will be beta emission, positron emission, and alpha particle emission, although you should know about electron capture as an alternative to positron emission.

The calculation of Q for decay processes presents no problem in the case of alpha decay, but remember that the mass of the emitted electron in beta decay is taken care of automatically, since the atomic weights or mass defects you use are for the neutral atoms. The related consequence in the case of positron decay is that the mass of two electrons must be added to that of the product isotope. Keep in mind, too, that alpha decay is largely confined to the heaviest elements, that isotopes which lie below the line of stable isotopes will beta decay by negative electron emission, whereas those that lie above the line will decay by positron emission or by electron capture. These rules can be thought of as manifestations of the tendency of an unstable isotope to decay in such a way as to bring it closer to the line of stable isotopes.

Radioactive decay is a quantum mechanical process, and all that can be said about an individual nucleus is that it has a certain probability λ of decaying, per unit time. If a large number of disintegration events are occurring per unit time, because the sample contains a large number of atoms, statistical fluctuations become small and radioactive decay can be treated as a first-order rate process with λ as the first-order specific-rate constant. A number of problems involve calculations on this basis. Included as well are cases of a radioactive daughter or product isotope.

Where physical constants are given for specific isotopes, such data will be approximately correct. Some minor liberties have been taken, especially in the case of half-life figures, in order to simplify calculations. Take special note of the short-cut to evaluating $e^{-\lambda t}$, where t happens to be a simple fraction or multiple of a half-life, that is given below.

EQUATIONS AND CONCEPTS

Definitions

Z	atomic number
A	mass number
N	number of atoms
D	disintegration rate, ordinarily in dis/sec
λ	decay constant
$T_{1/2}$	half-life
k	rate of production of an isotope
σ	cross section, usually in barns, where 1 barn $= 10^{-24}$ cm^2
nv	neutron flux, neutrons/cm^2-sec
$_{+}\beta, _{-}\beta$	electron or positron emitted in radioactive decay
β	v/c or velocity relative to light
Q	energy of a nuclear transmutation process
Δ	mass defect; atomic weight $= A + \Delta$
γ	gamma quantum
ev	electron volt

Units and Conversion Factors

Energy
  $1 \text{ ev} = 1.59 \times 10^{-12} \text{ erg}$
  $1 \text{ mev} = 10^6 \text{ ev} = 23 \times 10^6 \text{ kcal/mole}$
Also, the conversion of one atomic weight unit into energy gives 930 mev; that is, $\frac{1}{6} \times 10^{23} \text{ g} \approx 930 \text{ mev}$. Roughly a change in the sum of atomic weights of reactants versus products in a transmutation reaction of 0.001 mass unit means an energy release or uptake of 1 mev (0.93 mev).

Disintegration rate
  $1 \text{ curie} = 3.7 \times 10^{10} \text{ dis/sec}$
  $1 \text{ mc} = 0.001 \text{ c}$
  $1 \text{ rutherford (rd)} = 10^6 \text{ dis/sec}$

Frequently used atomic weights
  Neutron: 1.0089
  Proton (hydrogen atom, actually): 1.0081
  Electron: 0.00055
  Alpha particle (helium atom, actually): 4.0039
  Deuteron (deuterium atom): 2.0147

Rate of Production of an Isotope

$k = (nv)\sigma N$, where N is the number of atoms of target isotope, σ is the cross section, and (nv) is the neutron flux.

Decay Equations

$$D = \lambda N \qquad (19\text{-}1)$$

$$D = D^0 e^{-\lambda t} \qquad (19\text{-}2)$$

$$T_{1/2} = 0.69/\lambda \qquad (19\text{-}3)$$

Note that if $t = T_{1/2}$, $e^{-\lambda t} = \frac{1}{2}$; and, in general, if $t = nT_{1/2}$, $e^{-\lambda t} = (\frac{1}{2})^n$.

Case of Constant Rate of Production

$$D = k(1 - e^{-\lambda t}) \qquad (19\text{-}4)$$

At $t = \infty$, $D =$ saturation value; that is, $D^\infty = k$.

Parent–Daughter Equation

$$D_2 = \frac{\lambda_2}{\lambda_2 - \lambda_1} D_1^0 (e^{-\lambda_1 t} - e^{-\lambda_2 t}) \qquad (19\text{-}5)$$

where D_2 is the daughter activity after time t if none is present at $t = 0$. If $\lambda_2 > \lambda_1$, the limiting condition of *transient equilibrium* is reached at sufficiently large t:

$$D_2 = \frac{\lambda_2}{\lambda_2 - \lambda_1} D_1 \tag{19-6}$$

If $\lambda_2 \gg \lambda_1$, the limiting condition becomes $D_2 \cong D_1$ and is known as *secular equilibrium*. If the daughter is in turn radioactive, and so on, so that a decay chain is involved, these equilibria still apply to the ith species provided that the half-life of the parent is greater than that of any of the daughters.

Specific Activity

D/N, but often given in terms of disintegration rate per gram or similar basis.

Relativity Equation

$$m = \frac{m_0}{\sqrt{1 - \beta^2}} \tag{19-7}$$

where m_0 is the rest mass of the particle.

PROBLEMS

1. (12 min) As a result of nuclear reactions induced by cosmic radiation, there is a steady level of ^{14}C (5,800 y half-life, where y is a year) present in the atmosphere, in amounts corresponding to 14 dis/sec-g of total carbon. Calculate the fraction of carbon atoms that this figure represents, that is, the fraction of atmospheric carbon that is ^{14}C.

2. (9 min) Calculate the specific activity of tritium gas, in mc/cc (STP), if the half-life is 3 y. (It is sufficient to set up the expression containing only numbers that, when multiplied out, will give the desired specific activity.)

3. (21 min) Given

$$^{226}_{88}Ra \xrightarrow[\text{or } 5.06 \times 10^{10} \text{ s}]{1,600 \text{ y}} {}^{222}Rn \xrightarrow{4 \text{ d}} {}^{218}RaA$$

(a) Give the atomic number of ^{218}RaA. (b) Calculate the specific activity of Ra in dis/sec-g. (c) A sample initially consists of pure Ra, of activity

1,000 dis/sec. How many dis/sec of Rn will be present 8 d later and 1 y later?

---

4. (6 min) Calculate the velocity of an electron whose total mass (per mole) is 0.605 mg.

5. (12 min) $^{209}_{83}Bi$ undergoes fission with 50-mev protons to give two identical fragments (and no other products). (a) Write the balanced equation for the process. (b) Would you expect these two fission products to be radioactive, and, if so, by what mode should they decay? Explain. (c) If the mass defect of ^{209}Bi is 0.05, and that of the fission products is -0.04, calculate the Q for the fission process.

6. (19 min) (final examination question) ^{131}I decays with a 7-d half-life with emission of a 0.6 mev beta and a 0.3 mev gamma.
(a) Write a balanced equation for the decay of ^{131}I.
(b) If the exact atomic weight of ^{131}I is 130.985, calculate the exact atomic weight of the isotope formed by its decay.
(c) ^{131}I is one of the uranium-fission products. On placing a piece of uranium in the Oak Ridge reactor, fissions occur at a rate such that 10^8 atoms/sec of ^{131}I are formed. Calculate the dis/sec of ^{131}I that will be present in the uranium 7 d later and 70 d later.

7. (6 min) Q is -5.4 mev for the process $^{14}_{7}N + ^{1}_{0}n = ^{13}_{6}C + ^{2}_{1}H$. What is the atomic weight of ^{13}C if the values for ^{14}N, ^1n, and ^2H are 14.0075, 1.0089, and 2.0147, respectively?

8. (13 min) A sample of sodium undergoes the (p, n) reaction

$$^{23}_{11}Na + ^{1}_{1}H = ^{1}_{0}n + ^{23}_{12}Mg$$

(a) Explain whether the product ^{23}Mg, if radioactive, is more likely to be a positron or a negative electron emitter. (b) If the Q for the above reaction is -3.5 mev, calculate the atomic weight of the ^{23}Mg. (c) Assuming ^{23}Mg to be a positron emitter, write the balanced equation for its decay, and calculate the Q for the decay process, that is, the energy of the emitted positron, in mev.
Atomic weights: ^{23}Na: 22.9920; ^1H: 1.0078; n: 1.0086; e^-: 0.00055.

9. (9 min) Isotope X is so positioned that it finds it possible to decay either by β^- or by β^+ emission; that is,

$$_{Z}X^{A} \rightarrow ^{A'}_{Z'}Y + _{-}\beta + Q_1$$
$$_{Z}X^{A} \rightarrow ^{A''}_{Z''}W + _{+}\beta + Q_2$$

Suppose that Q_1 and Q_2 are both zero. Calculate $(Z' - Z'')$, $(A' - A'')$, and $(W' - W'')$ (W = atomic weight).

---

10. (12 min) Radioactive species A is the parent of radioactive species B. The half-lives of A and B are 2 d and 1 d, respectively. If a sample initially consists of equal dis/sec of A and B, what percent of the total dis/sec will the activity of A amount to 4 d later? Show your work.

11. (6 min) (final examination question) Radio sodium (^{24}Na, half-life, 15 h) is produced at a constant rate of 10^7 atoms/sec by irradiation of NaCl in a reactor. If the NaCl is irradiated for 30 h, then allowed to stand for 45 h, what will be the resulting dis/sec?

12. (30 min) One-half gram of molybdenum foil is irradiated in a nuclear reactor at a flux of 2×10^{12} neutrons/cm^2-sec. The cross section of Mo is 3 barns for the reaction ^{100}Mo$(n, \gamma)^{101}$Mo. Also, ^{101}Mo is radioactive and decays with a 15-min half-life to an isotope X of a neighboring element. X is also radioactive and decays with a 3-min half-life to a stable isotope Y of still another element.

(a) Calculate k, the rate of production of ^{101}Mo.

(b) Write balanced equations for the decay of ^{101}Mo and of its daughter.

(c) The foil is irradiated for $\frac{1}{2}$ h and then allowed to stand for 1 h. Assuming the answer to (a) is $k = 10^7$ sec^{-1}, calculate the activity, in dis/sec, of the ^{101}Mo at the end of the $\frac{1}{2}$-h irradiation. Calculate also the activities, in dis/sec, of the ^{101}Mo and of its daughter X at the end of the 1-h period of standing.

(d) Calculate Q, in mev, for the (n, γ) reaction, assuming that ^{100}Mo and ^{101}Mo have the same mass defect.

13. (15 min) The following fission-product chain occurs in uranium fission:

$$^{88}\text{Kr} \xrightarrow{2\,\text{h}} {}^{88}\text{Rb} \xrightarrow{20\,\text{min}} {}^{88}\text{Sr (stable)}$$

The fission yield of the chain is 1%. One milligram of uranium placed in a reactor undergoes fissions at the constant rate of 10^8 fissions/sec.

(a) Calculate the equilibrium dis/sec of ^{88}Kr and of ^{88}Rb in this uranium sample.

(b) After equilibrium has been reached, the uranium sample is removed from the reactor, dissolved, and the ^{88}Kr is quickly separated out in pure form. Now calculate the dis/sec of ^{88}Rb associated with the Kr 20 min after its separation. Calculate also the dis/sec of ^{88}Kr and of ^{88}Rb present in the Kr sample 6 h after its separation.

14. (9 min) Given the following parent–daughter relationship

$$^{140}Ba \longrightarrow {}^{140}La + beta \qquad (\text{half-life, 12 d})$$
$$^{140}La \longrightarrow {}^{140}Ce\,(\text{stable}) + beta \qquad (\text{half-life, 10 h})$$

Also, a sample initially contains 1,000 dis/sec of ^{140}Ba (and no ^{140}La). Calculate the dis/sec of ^{140}Ba after 24 d and the dis/sec of ^{140}La after 10 h and after 24 d.

15. (6 min) (final examination question) A sample of ^{24}Na (15 h half-life) is known to contain some ^{31}Si (3 h half-life) impurity. A sample having 1,000 counts/min initial activity had an activity of 200 counts/min after 30 h. What percent of the initial activity was due to ^{31}Si?

16. (18 min) On irradiation of 2 g of ^{154}Gd with high energy neutrons, it is found that radioactive Eu is produced at the rate of 10^6 atoms/sec; the total irradiation time is 1 d. The decay rate of the Eu is such that it takes 5 y for the disintegration rate of a sample to decrease by 30%. (Atomic numbers: Sm, 62; Eu, 63; Gd, 64; Tb, 65.)
   (a) Write a balanced equation for the transmutation reaction whereby the Eu is produced. (b) This Eu isotope decays by $_-\beta$ emission; write the balanced equation for its decay. Calculate (c) the half-life of Eu isotope and (d) the dis/sec produced by the above irradiation.

17. (15 min) Given the sequence

$$U \xrightarrow{10^9\,y} UX_1 \xrightarrow{20\,d} UX_2 \xrightarrow{1\,min}$$

Also, a sample of uranium in secular equilibrium with UX_1 and UX_2 has a total disintegration rate of 1,000 dis/sec. The uranium is then chemically separated from the UX_1 and UX_2. (a) Calculate the total activity of the uranium sample 10 d after the separation, and (b) calculate the number of atoms of UX_1 in the original sample.

18. (16 min) A sample of RaE (^{210}Bi, 5 d half-life) is known to be contaminated with some radium (1,800 y half-life). The initial activity of the sample is 20 mc (millicuries) and 10 d later the activity has dropped to 8 mc. Assume the detection equipment to have been 100% efficient so that the above figures represent total dis/sec. (a) How many mc of radium are present in the sample? Show your work. (b) Assuming that the sample contained 10 mc of radium, calculate the initial mole fraction of RaE, that is, the ratio

$$\frac{(\text{moles RaE})}{(\text{moles RaE}) + (\text{moles Ra})}$$

19.   (12 min) The parent–daughter sequence is given: Nb → Zr → stable, where the half-lives of Nb and Zr are 63 and 35 d, respectively. A sample initially having only Nb activity is allowed to stand for a year (actually 54 weeks) at the end of which time the disintegration rate of the Zr daughter is found to be 1,000 dis/min. Calculate the initial disintegration rate of the sample.

20.   (15 min) (final examination question) One mg of cobalt metal is irradiated in a nuclear reactor for 1 week, at the end of which time it was found that the cobalt had 5 curies of activity (as ^{60}Co). ^{60}Co has a 5-y half-life and decays to stable ^{60}Ni. Calculate the number of atoms of ^{60}Ni present immediately after the irradiation and after a long enough time that all of the ^{60}Co has decayed.

21.   (15 min) (final examination question) A well-known parent–daughter sequence is

$$^{140}\text{Ba} \xrightarrow{12\,\text{d}} {}^{140}\text{La} \xrightarrow{2\,\text{d}} {}^{140}\text{Ce} \longrightarrow \text{(stable)}$$

The half-lives are as indicated. A sample consists initially of pure ^{140}Ba, and 12 d later its *total* disintegration rate (Ba + La) is found to be 2,000 dis/sec. Calculate the initial disintegration rate and the dis/sec caused by La alone when the sample was 1 d old.

22.   (15 min) (final examination question) A sample consists initially of 1,000 dis/min of pure ^{238}U. Calculate the total disintegration rate of the sample 2 d later. The decay sequence is

$^{238}\text{U} \longrightarrow {}^{234}\text{Th} + \text{alpha}$ (half-life 4×10^9 y)

$^{234}\text{Th} \longrightarrow {}^{234}\text{Pa} + \text{beta}$ (half-life 24 h)

$^{234}\text{Pa} \longrightarrow {}^{234}\text{U} + \text{beta}$ (half-life 1.2 min)

---

23.   (18 min) A sample of chlorine is bombarded with protons and various (x, y)-type reactions occur, that is, $\text{Cl}(x, y)I$, where I denotes the product nucleus. As a result of this bombardment various radioactive products are obtained, one of which is a positron emitter.

(a) Using the stable isotope chart shown in Fig. 19-1 (X's locate stable isotopes), explain what this product probably is, and write a balanced reaction for its formation.

(b) Assuming that the mass defect of the product exceeds that of the Cl isotope it is formed from by 10 mg/mole, calculate Q for the reaction.

(c) Write a balanced equation for the positron decay of the product isotope.

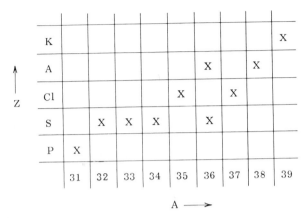

FIGURE 19-1

24. (15 min) (final examination question) ^{197}Au yields ^{198}Au (half-life, 2 h) on neutron irradiation. Some gold foil is irradiated in a medium-flux reactor, whose nv value is 10^{14} neutrons/cm^2-sec, for 24 h, during which time 0.1% of the ^{197}Au in the sample is transmuted. Calculate the cross section of ^{197}Au and the dis/sec of the ^{198}Au per gram of ^{197}Au.

25. (15 min) (final examination question) ^{95}Zr (which may be produced by uranium fission) decays with a 70 d half-life to give ^{95}Nb, which in turn is radioactive with a 35 d half-life. A sample consisting initially of 1 mc of pure ^{95}Zr is allowed to stand for 70 d. Calculate the dis/sec of ^{95}Nb present at the end of this period. Explain briefly why all fission products are expected to be radioactive and to decay by negative electron emission.

26. (9 min) Given the following sequence of nuclear processes

$$_{Z}X^{A} + n \longrightarrow {}_{Z'}Y^{A'} + p \qquad Q_1$$
$$_{Z'}Y^{A'} \longrightarrow {}_{Z}X^{A} \quad \text{plus electron plus gamma} \qquad Q_2 = 4.5 \text{ mev}$$

Is the electron in step 2 a negative electron or a positron? Calculate Q_1.

27. (12 min) On irradiation of 1 g of Ge metal in a reactor whose neutron flux is 10^{12} neutrons/cm^2-sec, the process ^{76}Ge$(n, \gamma)^{77}$Ge occurs. The half-life of ^{77}Ge is 1 min.

(a) Calculate the cross section for the above process if the dis/sec of ^{77}Ge present 1 min after a 1 h irradiation is 10^6.

(b) Calculate the dis/sec of ^{77}Ge that should be present had the Ge been irradiated for 1 min instead of 1 h. The measurement is again made 1 min after the end of the irradiation.

(A bonus of 10% extra credit): ^{77}Ge decays to a 40-h ^{77}As daughter. How many dis/sec of ^{77}As were present in case (a) above?

Z	16	17	18	19	20	21	22	23
11								X
10					X	X	X	
9				X				
8	X	X	X					

$A \longrightarrow$

(X denotes stable isotope)

FIGURE 19-2

28. (15 min) A portion of the stable isotope chart is reproduced in Fig. 19-2. It is found that slow neutron irradiation of fluorine gives radioactive fluorine isotope A, and that fast neutron irradiation of fluorine gives radioactive isotopes A and B. Also, alpha-particle bombardment of oxygen gives radioactive fluorine isotopes A and C. Considering only transmutation reactions of the $A(x, y)B$ type (see Comments), discuss what the mass numbers of isotopes A, B, and C must be. Explain the probable mode of decay of B, and C.

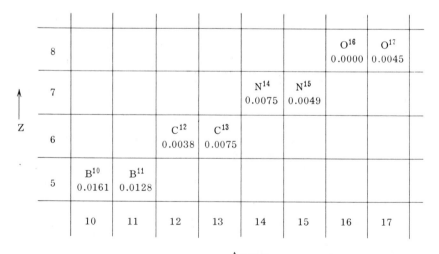

FIGURE 19-3

29. (6 min) Balance the following nuclear reactions, indicating what the missing particles are and inserting mass and atomic numbers:

$$^{31}P + neutron = {}^{30}P + ?$$
$$^{16}O + ? = {}^{14}N + alpha$$
$$^{12}_{5}B = ? + beta$$

30. (15 min) A portion of the stable isotope chart is given in Fig. 19-3, where the numbers are the respective mass defects. Additional mass defects are 0.0089, 0.0081, and 0.0039 for ^1_0n, ^1_1H, and ^4_2He, respectively. Give three plausible transmutation reactions for obtaining ^{14}C, and calculate the Q, in mev, for one of them. The mass defect for ^{14}C is 0.0077.

ANSWERS

1. The number of ^{14}C atoms that will average 14 dis/sec is given by $N = D/\lambda = DT_{1/2}/0.69 = 14 \times 5,800 \times 365 \times 24 \times 3,600/0.69 = 3.7 \times 10^{12}$. The desired ratio is then $3.7 \times 10^{12}/(6 \times 10^{23}/12) = \mathbf{7.3 \times 10^{-11}}$.

2. The basic equation is $D = \lambda N$, where N is now the number of atoms per cc (STP), or $N = 2 \times 6 \times 10^{23}/22,400$ (remember, 2 atoms/molecule of tritium gas). The desired specific activity is then

$$D = (0.69/3 \times 365 \times 24 \times 3,600)(2 \times 6 \times 10^{23}/22,400)$$
$$\times (1/3.7 \times 10^{10})$$
$$= \mathbf{10.6 \, c/cc \, (STP)}$$

3. (a) Since the mass numbers are changing by four units, the sequence is evidently one of alpha decay, and the daughter isotopes are therefore $^{222}_{86}Rn$ and $^{218}_{84}RaA$.

(b) We use the equation $D = \lambda N$, where N will be the number of atoms per gram.

$$D = \frac{0.69}{5.06 \times 10^{10}} \frac{1 \times 6 \times 10^{23}}{226}$$
$$= \mathbf{3.6 \times 10^{10} \, dis/sec} \qquad \text{(i.e., 1 curie)}$$

(c) D_2, the dis/sec for the Rn daughter, is given by

$$D_2 = \frac{\lambda_2}{\lambda_2 - \lambda_1} D^0_1 (e^{-\lambda_1 t} - e^{-\lambda_2 t})$$

where $e^{-\lambda_1 t}$ is essentially unity as the time is short compared to the half-life of radium, and $e^{-\lambda_2 t} = \frac{1}{4}$ since 8 d is two half-lives of radon. Also, in this case $\lambda_2/(\lambda_2 - \lambda_1)$ is practically unity. Then $D_2 = 1,000(1 - \frac{1}{4}) = $ **750 dis/sec.** After 1 y the radon will be in secular equilibrium with the radium, which still has decayed but slightly, so D will be **1,000 dis/sec.**

4. The atomic weight of the electron is 0.00055 or 0.55 mg/mole, so m/m^0 is $0.605/0.55 = 1.10$. Then $1.10 = 1/\sqrt{1 - \beta^2}$ from which $\beta^2 = 0.18$ and $\beta = 0.428$ and the velocity is **1.28×10^{10} cm/sec.**

5. (a) $^{209}_{83}\text{Bi} + ^1_1\text{H} = 2^{105}_{42}\text{Mo}$.

(b) Since heavy elements are neutron rich compared to light elements, the fission products should lie below the line of stable isotopes and should be beta radioactive.

(c) $Q = 0.05 + 0.0081 - 2(-0.04) = 0.138$ or **128 mev.**

6. (a) $^{131}_{53}\text{I} \rightarrow ^{131}_{54}\text{Xe} + _{-1}^{0}\beta + $ gamma.

(b) The Q for the reaction can be taken as $0.6 + 0.3 = 0.9$ mev or 0.00097 mass units. Then, $0.00097 = 130.985 - A$, or $A = $ **130.984.**

(c) After 7 d, half of saturation will be reaction, so $D = k/2 = 5 \times 10^7$ **dis/sec.** After 70 d, D will be at the saturation value or $D = 10^8$ **dis/sec.**

7. The Q of -5.4 mev corresponds to -0.0058 change in the sum of the mass defects: $-0.0058 = 0.0075 + 0.0089 - 0.0147 - \Delta_{13_C}$. Then $\Delta_{13_C} = 0.0075$ or $A = $ **13.0075.**

8. (a) The stable isotope of Mg is ^{24}Mg, so ^{23}Mg lies above the general line of stable species and should therefore be a **positron** emitter. (b) The Q value means that the products weigh $3.5/0.93$ or 3.8 milli mass units more than the reactants. Then

$$1.0086 + W_{^{23}\text{Mg}} = 22.9920 + 1.0078 + 0.0038$$

from which $W_{^{23}\text{Mg}} = $ **22.9950.** (c) As somewhat of a shortcut, combination of the following equations is useful:

$$
\begin{array}{ll}
^{23}_{11}\text{Na} + ^1_1\text{H} = ^1_0 n + ^{23}_{12}\text{Mg} & Q = -3.5 \text{ mev} \\
^{23}_{12}\text{Mg} = ^{23}_{11}\text{Na} + _{+}\beta & Q_1 = \\
\hline
^1_1 H = ^1_0 n + _{+}\beta & Q_2 = 1.0078 - 1.0086 - 0.0011 \\
& \quad = -0.0019 \text{ or } -1.8 \text{ mev}
\end{array}
$$

Then $Q_1 - 3.5 = -1.8$ or $Q_1 = $ **1.7 mev.**

9. Evidently, $Z' = Z + 1$ and $Z'' = Z - 1$, so $(Z' - Z'') = 2$. Also, $A = A' = A''$, so $(A' - A'') = 0$. Finally,

$$Q_1 = W_X - W_Y$$
$$Q_2 = W_X - W_W - 2W_e$$

Since Q_1 and Q_2 are equal, $(W' - W'')$ or $(W_Y - W_W) = 2W_e = $ **0.0011 mass units.**

10. It is convenient to make the calculation in three parts. Let D^0 be the initial disintegration rate. Then after 4 d $D_A = D^0/4$. Also, in 4 d the amount of B initially present will have decayed by four half-lives so $D_B = D^0/16$. Finally, new B will have grown in from the decay of A according to the equation,

$$D'_B = \frac{\lambda_B}{\lambda_A - \lambda_B} D^0 (e^{-\lambda_A t} - e^{-\lambda_B t})$$

$$= \frac{1/1}{1/1 - 1/2} D^0(\tfrac{1}{4} - \tfrac{1}{16}) = \tfrac{3}{8} D^0$$

The fraction of total activity resulting from A will thus be

$$\frac{D^0/4}{D^0/4 + D^0/16 + 3/8 D^0} = \frac{4}{11} \quad \text{or} \quad \textbf{36.5\%}$$

11. The basic equation is $D = k(1 - e^{-\lambda t}) = 10^7(1 - \tfrac{1}{4}) = 0.75 \times 10^7$ dis/sec at the end of the irradiation. Forty-five hours or 3 half-lives later, one-eighth this amount will be present or **9.4×10^5 dis/sec.**

12. (a) $k = (nv)$ number of atoms of ^{100}Mo $= 2 \times 10^{12} \times 3 \times 10^{-24} \times 0.5 \times 0.097 \times 6 \times 10^{23}/100 = $ **1.75×10^9 sec^{-1}** (the figure 0.097 comes from the natural abundance of ^{100}Mo).

(b) The isotope tables inform us that ^{101}Mo is a beta emitter, so

$$^{101}_{42}\text{Mo} \rightarrow {}^{101}_{43}\text{Tc} + {}_{-1}^{0}\beta$$

The isotope X is then ^{101}Tc and it too is a beta emitter, so the equation for its decay is

$$^{101}_{43}\text{Tc} \rightarrow {}^{101}_{44}\text{Ru} + {}_{-1}^{0}\beta$$

(c) The dis/sec of ^{101}Mo after $\tfrac{1}{2}$ h (or two half-lives) irradiation will be $D = 10^7(1 - \tfrac{1}{4}) = $ **7.5×10^6 dis/sec.** After standing for 1 h (or four half-lives),

1/16 of this will be left, or **4.7 × 10⁵ dis/sec.** The 1 h period is sufficient for the daughter to come into transient equilibrium, so its disintegration rate D_2 will be

$$D_2 = \frac{\lambda_2}{\lambda_2 - \lambda_1} D_1$$

$$= \frac{1/3}{1/3 - 1/15} 4.7 \times 10^5$$

$$= \mathbf{5.9 \times 10^5 \, dis/sec}$$

(d) If ^{100}Mo and ^{101}Mo have the same mass defect, Q is simply that for the neutron, or 0.0089. Or, $Q = \mathbf{8.3 \, mev}$.

13. (a) Since 1% of fissions lead to this chain, the effective k value is 10^6. At equilibrium there will then be $\mathbf{10^6 \, dis/sec}$ of the ^{88}Kr, and likewise of ^{88}Rb, which will have come into secular equilibrium.

(b) We assume the time of the separation procedure itself is short, so we want the dis/sec of ^{88}Rb(D_2) grown in after 20 min:

$$D_2 = \frac{\lambda_2}{\lambda_2 - \lambda_1} D_1^0 (e^{-\lambda_1 t} - e^{-\lambda_2 t})$$

Twenty minutes is $\frac{1}{6}$ of a half-life of ^{88}Kr, so $e^{-\lambda_1 t}$ is $(\frac{1}{2})^{1/6} = 0.9$; $e^{-\lambda_2 t} = \frac{1}{2}$. Then

$$D_2 = \frac{1/20}{1/20 - 1/120} 10^6 (0.9 - 0.5)$$

$$= \mathbf{4.8 \times 10^5 \, dis/sec} \text{ of } ^{88}\text{Rb}$$

Six hours after separation, the ^{88}Kr will have decayed to $\frac{1}{8}$ its initial value, or to 1.25×10^5 dis/sec, and the ^{88}Rb will have come into transient equilibrium, and its activity will be $(6/5)1.25 \times 10^5 = \mathbf{1.5 \times 10^5 \, dis/sec}$.

14. After 24 d or two half-lives, one-fourth the ^{140}Ba will be left, or 250 dis/sec. The amount of ^{140}La grown in after 10 h will be

$$D = \frac{1/10}{1/10 - 1/12 \times 24} 1{,}000 (e^{-\lambda_1 t} - e^{-\lambda_2 t})$$

In this case $e^{-\lambda_1 t}$ will be nearly unity [actually

$$e^{(-0.69)(10)/(12)(24)} = e^{-0.024} = 0.98]$$

and $e^{-\lambda_2 t}$ will be $\frac{1}{2}$. Then $D_{La} = 1.03 \times 1,000 \times 0\,5 = $ **515 dis/sec.** After 24 d, the ^{140}La will be essentially in transient equilibrium with the ^{140}Ba, so its dis/sec will be $1.03 \times 250 = $ **258 dis/sec.**

15. After 30 h negligible ^{31}Si is left, so the 200 counts/min are entirely due to the ^{24}Na. Since 30 h is two ^{24}Na half-lives, the initial ^{24}Na activity must have been $200/\frac{1}{4}$ or 800, and there was a 200 count/min or **20%** ^{31}Si impurity.

16. (a) $^{154}_{64}$Gd $+ ^{1}_{0}n = ^{154}_{63}$Eu $+ ^{1}_{1}$H. [Of the simple reaction types, only an (n, p) reaction will reduce the atomic number by one.]
  (b) $^{154}_{63}$Eu $\rightarrow ^{154}_{64}$Gd $+ _{-1}\beta^{0}$.
  (c) It helps if you notice that the fraction of Eu remaining after 5 y is 0.7 and that this is just $1/\sqrt{2}$ so that 5 y is half of a half-life. The half-life is then **10 y.**
  (d) One day's irradiation will produce $10^6 \times 24 \times 3,600$ or 8.65×10^{10} atoms. Then

$$D = \lambda N = \frac{0.69}{10 \times 365 \times 24 \times 3,600} 10^6 \times 24 \times 3,600$$
$$= \textbf{187 dis/sec}$$

17. (a) Initially, all three species have the same activity, or 333 dis/sec for each. Ten days after the separation, or half of a UX_1 half-life, the dis/sec of UX_1 will be $333(1 - \sqrt{1/2})$ or 100 dis/sec. Because of its short half-life, the UX_2 will be in secular equilibrium with the UX_1, so its activity will also be 100 dis/sec. The total activity of the sample will then be $333(U) + 100(UX_1) + 100(UX_2)$ or **533 dis/sec.**
  (b) $N = D/\lambda = 333 \times 20 \times 24 \times 3,600/0.69 = $ **8.35 $\times 10^8$ atoms.**

18. (a) Ten days corresponds to two RaE half-lives, so $\frac{3}{4}$ of the original RaE activity has disappeared, while the Ra half-life is so long its activity has remained essentially constant. The drop in disintegration rate of $20 - 8 = 12$ mc thus corresponds to $\frac{3}{4}$ of the original RaE activity, which was then **16 mc.** (b) By Eq. (19-1), the mole fraction of RaE is

$$N_{RaE} = \frac{10/\lambda_{RaE}}{10/\lambda_{RaE} + 10/\lambda_{Ra}}$$

Clearly, the $10/\lambda_{RaE}$ term in the denominator can be neglected, so the equation simplifies to

$$N_{RaE} = \frac{\lambda_{Ra}}{\lambda_{RaE}} = \frac{T_{RaE}}{T_{Ra}} = \frac{5}{1,800 \times 365} = \textbf{7.6} \times \textbf{10}^{-6}$$

19. For the purpose of a slide rule calculation, it is helpful to put Eq. (19-5) in the form

$$D_2 = \frac{T_1}{T_1 - T_2} D_1^0 e^{-\lambda_1 t}(1 - e^{-\Delta \lambda t})$$

where $\Delta \lambda$ is given by $0.69(1/63 - 1/35)$ and corresponds to an effective half-life of $1/(1/63 - 1/35)$ or $63 \times 35/(63 - 35)$ or about 80 d. The year then is about 5×80 d, so $\exp(-\Delta \lambda t)$ is about $\frac{1}{32}$ or about 0.03. The expression for D_2 then becomes

$$D_2 = \frac{63}{63 - 35} D_1^0 \left(\frac{1}{2}\right)^6 (1 - 0.03)$$

since the 54-week year corresponds to six 63-d half-lives. Then

$$D_1^0 = 1,000 \times \frac{28}{63} \times \frac{6.4}{0.97} = \textbf{29,000 dis/min}$$

20. The ^{60}Co half-life is so much longer than the irradiation time that the activity must build up essentially linearly with time. We can then take an average activity of 2.5 curies, and over a week $2.5 \times 3.7 \times 10^{10} \times 7 \times 24 \times 3,600$ or $\textbf{5.6} \times \textbf{10}^{16}$ disintegrations would have occurred; this is just the number of ^{60}Ni atoms produced. The total number of ^{60}Co atoms present is given by Eq. (19-1) as

$$N_{Co} = \frac{(5 \times 3.7 \times 10^{10})/0.69}{5 \times 365 \times 24 \times 3,600} = \textbf{4.2} \times \textbf{10}^{19}$$

Again, this is just the number of ^{60}Ni atoms present after complete decay of the ^{60}Co.

21. From Eq. (19-5)

$$D_{La} = \frac{\lambda_{La}}{\lambda_{La} - \lambda_{Ba}} D_{Ba}^0 \left(\frac{1}{2} - \frac{1}{64}\right)$$

since 12 d corresponds to one Ba half-life and six La half-lives. Further, since λ and $T_{1/2}$ are inversely related, the equation further reduces to

$$D_{La} = \frac{12}{12 - 2} D_{Ba}^0 (0.485)$$

or $D_{La} = 0.58 \, D_{Ba}^0$. Also, $D_{Ba} = 0.5 \, D_{Ba}^0$, so $D_{tot} = 1.08 \, D_{Ba}^0$, from which $D_{Ba}^0 = 2,000/1.08 = $ **1,850 dis/sec.** Next, Eq. (19-5) is to be evaluated for $t = 1$ d, which corresponds to $\frac{1}{12}$ of a Ba half-life and $\frac{1}{2}$ of a La one, so

$$D_{La} = \frac{12}{12 - 2} \times 1,850\left[\left(\frac{1}{2}\right)^{1/12} - \left(\frac{1}{2}\right)^{1/2}\right]$$
$$= 1.2 \times 1,850 \, (0.945 - 0.71) = \textbf{522 dis/sec}$$

22. The ^{238}U is producing ^{234}Th at essentially a constant rate, in view of the long half-life of the former. Eq. (19-5) reduces to Eq. (19-4) with $k = 1,000$; that is,

$$D_{^{234}Th} = 1,000(1 - (\tfrac{1}{2})^{1/2})$$

where the exponential has been replaced by $(\tfrac{1}{2})^{1/2}$ since 12 d is half of a half-life. The disintegration rate of ^{234}Th at the end of 12 d is then $1,000(1 - 0.71) = 290$ dis/sec. The ^{234}Pa will be in transient equilibrium with the ^{234}Th; in fact since 1.2 min is less than 1% of 24 h, Eq. (19-6) reduces to $D_2 = D_1$, so the disintegration rate of the ^{234}Pa will be essentially 290 dis/sec. The total disintegration rate is then $1,000 + 290 + 290$ or **1,580 dis/sec.**

23. (a) For it to be a positron emitter, the product must lie above a stable isotope. A (p, α) reaction would reduce the charge of the nucleus by one and the mass number by three, thus giving ^{32}S and ^{34}S, which are stable. A (p, n) reaction, however, would yield ^{35}A and ^{37}A. The latter is apt to decay by electron capture, whereas the former will probably have sufficient energy of decay to go by positron emission. The equation is

$$^{35}_{17}Cl + {}^{1}_{1}H = {}^{35}_{18}A + {}^{1}_{0}n$$

(b) Q will be given by the sum of the mass defects of the reactants minus that of the products

$$Q = \Delta_{Cl} + 8.1 - \Delta_{A} - 8.9 = -10 - 0.8 = -10.8 \text{ mg/mole}$$

or $-10.8 \times 0.93 = $ **−10 mev.**

(c) $^{35}_{18}A = {}^{35}_{17}Cl + {}^{0}_{+1}\beta$

24. If 0.1% of the ^{197}Au has reacted, then evidently $(nv)\sigma t = 0.001$ or $\sigma = 0.001/10^{14} \times 24 \times 3,600 = $ **1.16 $\times$ 10^{-22}.** The 24-h irradiation is sufficiently long so that the dis/sec of ^{198}Au is essentially at the saturation value. Thus D (per gram) $= k = (nv)\sigma \times$ (number of atoms/g) $= 10^{14} \times 1.16 \times 10^{-22} \times 6 \times 10^{23}/197 = $ **3.52 $\times$ 10^{13} dis/sec-g.**

25. Let species 1 and 2 refer to ^{95}Zr and ^{95}Nb, respectively. Then

$$D_2 = \frac{\lambda_2}{\lambda_2 - \lambda_1} D_1^0 (e^{-\lambda_1 t} - e^{-\lambda_2 t})$$

$$= \frac{1/35}{1/35 - 1/70} 1(\tfrac{1}{2} - \tfrac{1}{4})$$

$$= \mathbf{0.5\,mc}$$

Fission processes ordinarily refer to heavy elements and these are neutron rich relative to lighter elements. The fission products will then be neutron rich relative to the line of stable isotopes, and they will therefore tend to be beta radioactive.

26.   The balanced equations are

$$^A_Z X + n \rightarrow {}_{Z-1}^{A} Y + p$$
$$_{Z-1}^{A} Y \rightarrow {}^A_Z X + {}_{-1}^{0}\beta$$

If we add the two equations

$$n \rightarrow p + {}_{-}\beta \qquad Q = Q_1 + Q_2$$

Since $Q = 0.0089 - 0.0081 = 0.0008$, or 0.7 mev, $Q_1 = 0.7 - 4.5 = \mathbf{-3.8\,mev.}$

27.   (a) A 1-h irradiation would produce essentially the saturation level of activity, and the 10^6 dis/sec 1 min later would then correspond to half that value, that is, $k = 2 \times 10^6$. Then $2 \times 10^6 = (nv)\sigma$ (number of atoms of ^{76}Ge) or $2 \times 10^6 = 10^{12}\sigma$ $(0.078 \times 6 \times 10^{23}/76)$. (The isotopic abundance of ^{76}Ge is 7.8%.) The cross section is then $\sigma = \mathbf{3.28 \times 10^{-27}}$.

(b) An irradiation of 1 min, that is, one half-life, would produce half of the saturation level, so one would expect half of the previous value, or $\mathbf{5 \times 10^5}$ **dis/sec.**

Bonus: Since the ^{77}Ge builds up to essentially saturation in the first few minutes, practically all the ^{76}Ge transmuted during the 1-h irradiation must end up as ^{77}As, via the decay of ^{77}Ge. Very little ^{77}As will have decayed in 1 h, so that the number of atoms of ^{77}As present is nearly given by $N = (nv)\sigma t \times$ (number of atoms of ^{76}Ge) $= 2 \times 10^6 t = 2 \times 10^6 \times 3,600 = 7.2 \times 10^9$ atoms. Then $D = \lambda N = (0.69/40 \times 3,600)(2 \times 10^6 \times 3,600) = 0.69 \times 2 \times 10^6/40 = \mathbf{3.4 \times 10^4}$ **dis/sec.**

28.   Fluorine is, of course, element 9, and the chart shows ^{19}F as the only stable isotope. The most likely product of slow neutron irradiation would be

$^{19}F(n, \gamma)^{20}F$ so $A = {}^{20}F$. The additional fluorine isotope that fast neutrons could yield would be from $^{19}F(n, 2n)^{18}F$ [as noted under Comments at the beginning of the chapter, this type of reaction is included in the $A(x, y)B$ category], so B must be ^{18}F. To obtain fluorine isotopes by alpha bombardment of oxygen, we must be dealing with an (α, p) reaction. The three possibilities are

$$^{16}O(\alpha, p)^{19}F$$
$$^{17}O(\alpha, p)^{20}F$$
$$^{18}O(\alpha, p)^{21}F$$

Since ^{19}F is stable, isotope C must be ^{21}F. Finally, A and C will be beta emitters, decaying to stable Ne isotopes, whereas B will be either a positron emitter or will decay by electron capture, to stable ^{18}O.

29. The balanced equations are

$$^{31}_{15}P + {}^{1}_{0}n = {}^{30}_{15}P + 2^{1}_{0}n$$
$$^{16}_{8}O + {}^{2}_{1}H = {}^{14}_{7}N + {}^{4}_{2}He$$
$$^{12}_{5}B = {}^{12}_{6}C + {}_{-1}^{0}\beta$$

30. Some possibilities are

$$^{13}_{6}C + {}^{1}_{0}n = {}^{14}_{6}C + \text{gamma}$$
$$^{14}_{7}N + {}^{1}_{0}n = {}^{14}_{6}C + {}^{1}_{1}H$$
$$^{11}_{5}B + {}^{4}_{2}He = {}^{14}_{6}C + {}^{1}_{1}H$$
$$^{17}_{8}O + {}^{1}_{0}n = {}^{14}_{6}C + {}^{4}_{2}He$$

Q for the second reaction (which is the one by which ^{14}C is commonly produced) would be $0.0075 + 0.0089 - 0.0077 - 0.081 = 0.0006$ or **0.56 mev.**

20
QUANTUM THEORY
AND WAVE MECHANICS

COMMENTS

The subject of wave mechanics contains such vast forests of mathematical manipulations that it is very difficult to present it in a physical chemistry course without departing from what has been, up to this point, a certain philosophy of approach. This philosophy is that in physical chemistry the material should be presented sufficiently completely and with sufficient rigor that the student can appreciate in a critical way just what assumptions and approximations have gone into a theory and what its limitations are therefore apt to be. It is, after all, only by knowing the limits of something that one can define its extent; it is along this path that true understanding lies.

The problems of the preceding chapters have been kept simple in the sense that they have rested on a deliberately limited number of equations and concepts. On the other hand, you have been expected to use these equations and concepts fluently, to combine them, to vary them, to adapt them to specific situations: in other words, to use them not by rote, but in an understanding way.

When it comes to wave mechanics and quantum theory, the temptation generally yielded to in textbooks is to assemble a tremendous array of results of rather complex calculations. You are not given the details of these calculations, nor are you made aware of, much less made able to appreciate, the really monumental approximations that are often made. An example would be the wave-mechanical descriptions of bonding. The consequence is that you, the student of physical chemistry, are asked to memorize and to accept a number of pronunciementos without any real opportunity for criticizing or basis for appreciating them.

Another objection, whose subtleties require more space for discussion than is appropriate here, is that such an approach is essentially a dogmatic one. It instills uncritical acceptance—an attitude that negates the philosophy of the scientific method.

Nonetheless, and in spite of the above grumbling, quantum and wave mechanics *are* important. Their central role in providing expressions for translational, rotational, and vibrational energy states was illustrated in Chapters 5 and 7, in connection with statistical thermodynamics. The emphasis in this chapter is mainly on electronic energy states, and mainly therefore on the energetics and spatial arrangements of electrons in molecules. This chapter, then, serves also as a preamble to the following one on the group theory approach to chemical bonding. Four topics have been selected, each with a definite purpose, and each is approached in various ways so as to develop some appreciation of how the various equations and concepts behave and can be used.

The first topic is simply that of the Bohr atom. You are expected to be familiar with the quantization principle and how the energy and potential energy of a Bohr electron vary with different parameters. You gain, then, some experience with the meaning of the principal quantum number n for an atom.

The second topic brings in the Schrödinger wave equation in perhaps its simplest possible application—that of the particle in a box. It turns out that the one-, two-, and three-dimensional-box situations have some quite useful applications to chemistry. Long conjugated-double-bond systems can be likened to one-dimensional boxes in which one π electron from each carbon atom is placed. Aromatic rings can be thought of roughly as two-dimensional boxes, and an atom, very roughly, as a three-dimensional box. Working these problems gives you the encouragement of knowing that a solution to the wave equation, which *you* can verify, permits you to make complete although quite qualitative calculations about the energy states of real systems.

The wave-mechanical solutions for the hydrogen and hydrogen-like atom constitute the third topic. The three quantum numbers for the three-dimensional box now become n, l, and m, and the eigenfunction solutions are now somewhat messy spherical-harmonic-type functions. Obtaining them is straightforward but very complicated algebraically. There is not time in the course to go through them in detail, so you have to accept the results as mathematically correct.

It turns out that the great bulk of wave-mechanical language about atoms, molecules, and chemical bonding uses these hydrogen-like wave functions as its vocabulary. You need a fluent knowledge of the general behavior of these functions. They are formed by the product of three subsidiary functions, a radial one and two types of angular ones, and each of these varies with its quantum numbers in a systematic way. In due course, they will come to seem old friends.

Finally, a somewhat different type of wave-mechanical situation is taken up, namely, that of vibrating atoms. We restrict ourselves to the simplest case of a Hooke's law force between atoms and ask mainly that you become familiar with how the spacing of vibrational states depends on the force constant and masses. In addition, one observes empirically that changes in vibrational state are most probable if the quantum number changes by only one unit—we call this a selection rule, and one observes that, for light absorption to occur readily, the vibrational mode whose change in quantum number is involved must be one that produces a change in dipole moment. That is, the motion of atoms in a so-called infrared active mode must be one in which the dipole moment of the molecule varies as the atoms oscillate.

EQUATIONS AND CONCEPTS

Bohr Atom

Quantization: $mvr = nh/2\pi$ (20-1)

Force balance: $mv^2/r = Ze^2/r^2$ (20-2)

On elimination of v: $r = \dfrac{n^2h^2}{4\pi^2mZe^2}$ (20-3)

Energy statement: $\varepsilon_{tot} = \varepsilon_{kin} + \varepsilon_{pot}$

or $\varepsilon = mv^2/2 - Ze^2/r = Ze^2/2r - Ze^2/r = -Ze^2/2r$

or $\varepsilon = -\dfrac{2\pi^2mZ^2e^4}{h^2}\dfrac{1}{n^2}$ (20-4)

If Z and n are both unity, then insertion of numbers for the various constants gives $\varepsilon = -13.6$ electron volt (ev). The atomic energy unit (a.e.u.) is twice this, or 27.3 ev.

De Broglie equation:

for light: $h\nu = mc^2$ or $h/\lambda = mc = p$ (20-5)

where p is the momentum. By analogy, for particles:

$\lambda = h/p$ (20-6)

Schrödinger Wave Equation

$$-\frac{h^2}{8\pi^2m}\nabla\psi + U\psi = \varepsilon\psi$$ (20-7)

where U is the potential energy, in general a function of coordinates, and ε is the energy of the particle. In one dimension, the equation becomes

$$-\frac{h^2}{8\pi^2 m}\frac{d^2\psi}{dx^2} + U(x)\psi = \varepsilon\psi \tag{20-8}$$

Particle in a Box

In the case of a box, U is constant (and hence may be taken as zero) within the box and infinite at the wall of the box and thereafter. If a is the side of a one-dimensional box, the solution is

$$\psi = A\sin\frac{n\pi x}{a} \qquad \text{where } n = 1, 2, 3, \ldots, \text{etc.} \tag{20-9}$$

The integral values of n arise from the requirement that ψ be zero at $x = 0$ and $x = a$; $\psi^2(x)$ is interpreted as equal to the probability of the particle being present at that x value, and this probability must be zero at the walls. The solution may be verified by carrying out the differentiations and substituting back into the Schrödinger equation; the constant A is evaluated by requiring that $\int_0^a \psi^2\,dx = 1$, that is, that there be unit probability of the particle being somewhere within the box.

On making the above substitution, it is found that ψ is a solution if

$$\varepsilon = \frac{n^2 h^2}{8ma^2} \tag{20-10}$$

The energy is thus limited to the values given by $n = 1, 2, 3$, etc.

For a two- and a three-dimensional box, it turns out that the wave equation separates into two and three one-dimensional box-type equations. The allowed energies are then

$$\text{two-dimensional box:} \qquad \varepsilon = \frac{h^2}{8ma^2}(n_x^2 + n_y^2) \tag{20-11}$$

$$\text{three-dimensional box:} \qquad \varepsilon = \frac{h^2}{8ma^2}(n_x^2 + n_y^2 + n_z^2) \tag{20-12}$$

where the n's independently take on the values 1, 2, 3, and so on.

The following conversion relationships may be useful: $h^2/8ma^2 = 6 \times 10^{-11}$ erg or 37.8 ev if the particle is an electron and $a = 1$ Å. Energy differences are often expressed as wave numbers or as the reciprocal of the wavelength of light of that energy per quantum. Then $1/\lambda = v/c$ and the energy corresponding to 1,000 wave numbers is 1,000 cm^{-1} = 2×10^{-13} erg. Per mole of particles, light of wave number 1,000 cm^{-1} has 2.87 kcal of energy.

Pauli Exclusion Principle

No two electrons may have the same quantum numbers. If spin is disregarded, two electrons may occupy each quantum state.

Hydrogen-Like Atom

In abbreviated form, the wave function is $H\psi = \varepsilon\psi$, where the Hamiltonian operator H consists of a kinetic-energy and a potential-energy term. It is the assumption of wave mechanics that the kinetic-energy term is

$$-\frac{h^2}{8\pi^2 m}(\partial^2\psi/\partial x^2 + \partial^2\psi/\partial y^2 + \partial^2\psi/\partial z^2) \qquad (20\text{-}13)$$

and the potential-energy term is assembled by writing down all the various coulomb interactions between electrons and between electrons and nuclei. For a single electron and a nucleus of charge Z, the potential energy is $-Ze^2/r$, so the complete wave equation becomes

$$-\frac{h^2}{8\pi^2 m}(\partial^2\psi/\partial x^2 + \partial^2\psi/\partial y^2 + \partial^2\psi/\partial z^2) - \frac{Ze^2}{r}\psi = \varepsilon\psi \qquad (20\text{-}14)$$

It is found that solutions exist only for energy values

$$\varepsilon = -\frac{2\pi^2 m e^4 Z^2}{h^2 n^2} \qquad (20\text{-}15)$$

that is, the identical result as that from the simple Bohr theory.

The actual functions ψ are complicated, however, and are usually given in a polar coordinate system as shown in Fig. 20-1. They depend on the familiar three quantum numbers, n, l, and m, and the total can be written as a product of a function depending on radius alone (and on n and l), a function depending on θ alone (and on l and m), and a function depending on Φ alone (and on m). Thus

$$\psi(r, \theta, \Phi) = R_{nl}(r)\Theta_{lm}(\theta)\Phi_m(\varphi) \qquad (20\text{-}16)$$

In Table 20-1 $\rho = (2Z/na_0)r$ where a_0 is the radius of the first Bohr orbit for a hydrogen atom.

Notice that $R(r)$ consists of three types of terms. First, the term $(Z/a_0)e^{-\rho/2}$ appears each time (note that ρ depends on n; if we define $\sigma = Zr/a_0$, the exponential terms become $e^{-\sigma/n}$). The second term is a collection of constants whose value gets smaller as n increases. The third term consists of a

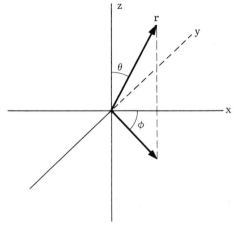

FIGURE 20-1

TABLE 20-1

Quantum numbers	$R(r)$
$n = 1, l = 0$	$R = [(Z/a_0)^{3/2}\, e^{-\rho/2}][2]$
$n = 2, l = 0$	$R = [(Z/a_0)^{3/2}\, e^{-\rho/2}]\left[\dfrac{1}{2\sqrt{2}}\right][(2 - \rho)]$
$l = 1$	$R = [(Z/a_0)^{3/2}\, e^{-\rho/2}]\left[\dfrac{1}{2\sqrt{6}}\right][\rho]$
$n = 3, l = 0$	$R = [(Z/a_0)^{3/2}\, e^{-\rho/2}]\left[\dfrac{1}{9\sqrt{3}}\right][(6 - 6\rho + \rho^2)]$
$l = 1$	$R = [(Z/a_0)^{3/2}\, e^{-\rho/2}]\left[\dfrac{1}{9\sqrt{6}}\right][(4 - \rho)\rho]$
$l = 2$	$R = [(Z/a_0)^{3/2}\, e^{-\rho/2}]\left[\dfrac{1}{9\sqrt{30}}\right][\rho^2]$
$n = 4, l = 0$	$R = [(Z/a_0)^{3/2}\, e^{-\rho/2}]\left[\dfrac{1}{96}\right][(24 - 36\rho + 12\rho^2 - \rho^3)]$
$l = 1$	$R = [(Z/a_0)^{3/2}\, e^{-\rho/2}]\left[\dfrac{1}{32\sqrt{15}}\right][(20 - 10\rho + \rho^2)\rho]$
$l = 2$	$R = [(Z/a_0)^{3/2}\, e^{-\rho/2}]\left[\dfrac{1}{96\sqrt{5}}\right][(6 - \rho)\rho^2]$
$l = 3$	$R = [(Z/a_0)^{3/2}\, e^{-\rho/2}]\left[\dfrac{1}{96\sqrt{35}}\right][\rho^3]$

polynomial in ρ, always of degree equal to $(n - 1)$; there are $n - l - 1$ roots or values such that $R(r) = 0$. As a result, the wave functions have $n - l - 1$ nodes as one goes out radially (see Table 20-1).

Notice that Θ consists essentially of a polynomial in $\cos \theta$ and $\sin \theta$, of degree equal to l, and shifting from an entirely $\cos \theta$ function for $m = 0$ to an entirely $\sin \theta$ function for $m = l$. (See Table 20-2.)

$$\Phi_m(\varphi)$$

$$\Phi = \frac{1}{\sqrt{2\pi}} e^{im\varphi} \quad \text{or} \quad \Phi|m| \quad \begin{cases} = \dfrac{1}{\sqrt{\pi}} \cos |m|\varphi \\[2ex] = \dfrac{1}{\sqrt{\pi}} \sin |m|\varphi \end{cases} \tag{20-17}$$

The Φ function is, as you see, very simple; it may be given in imaginary form or, by taking a linear combination of the Φ's for $+m$ and $-m$, in real form.

To summarize, $R(r)$ goes through several nodes depending on the n and l values, then tails off exponentially. The Θ functions show various lobes in a polar plot, and the Φ functions show various lobes in their polar plots. If a measure of electron density is wanted, one should use the square of the functions, and, usually $[R(r)]^2$ is multiplied by $4\pi r^2$ so as to give the chance of finding an electron in the spherical shell between r and $r + dr$.

You now have the information to assemble the complete ψ function for any n, l, and m value.

It has become conventional to display the Θ and Φ functions as polar plots. One objection is that such plots show merely the angular dependence of the functions yet are used as though they represent actual spatial "lobes" extending from an atom. A way of showing the spatial occupancy of various orbitals is in terms of domains [see A. W. Adamson, *J. Chem. Ed.* **42**, 140 (1965)]. A domain is defined as that region in space in which a given orbital has a higher electron density than does any other of the set under consideration.

Vibrational States

We assume a Hooke's law dependence of potential energy with separation between the two atoms

$$U = \tfrac{1}{2} k(r - r_e)^2 \tag{20-18}$$

where k is the force constant and r_e is the equilibrium distance. On inserting this function for U in the wave equation, one obtains solutions provided the

TABLE 20-2

Quantum numbers		$\Theta_{lm}(\theta)$
$l = 0$		$\Theta = \dfrac{\sqrt{2}}{2}$
$l = 1$	$m = 0$	$\Theta = \dfrac{\sqrt{6}}{2}\cos\theta$
	$m = \pm 1$	$\Theta = \dfrac{\sqrt{3}}{2}\sin\theta$
$l = 2$	$m = 0$	$\Theta = \dfrac{\sqrt{10}}{4}(3\cos^2\theta - 1)$
	$m = \pm 1$	$\Theta = \dfrac{\sqrt{15}}{2}\sin\theta\cos\theta$
	$m = \pm 2$	$\Theta = \dfrac{\sqrt{15}}{4}\sin^2\theta$
$l = 3$	$m = 0$	$\Theta = \dfrac{3\sqrt{14}}{4}(\tfrac{5}{3}\cos^3\theta - \cos\theta)$
	$m = \pm 1$	$\Theta = \dfrac{\sqrt{42}}{8}\sin\theta\,(5\cos^2\theta - 1)$
	$m = \pm 2$	$\Theta = \dfrac{\sqrt{105}}{4}\sin^2\theta\cos\theta$
	$m = \pm 3$	$\Theta = \dfrac{\sqrt{70}}{8}\sin^3\theta$

energy has the values

$$\varepsilon = (v + \tfrac{1}{2})\frac{h}{2\pi}(k/\mu)^{1/2} \qquad (20\text{-}19)$$

where v is the vibrational quantum number and has the values 1, 2, 3, and so on, and μ is the reduced mass, $\mu = (m_1 m_2)/(m_1 + m_2)$. It is observed that transitions between vibrational states occur most easily if the change in v is $+1$ or -1.

k values are generally around 3 to 20 ($\times\ 10^5$ dyne/cm). They tend to decrease somewhat as the size of the atoms increases, and they increase by 50 to 100% on going from a single to a double to a triple bond.

PROBLEMS

1. (10 min) The deBroglie wavelength of a particle is given by $\lambda = h/p$ where p denotes the momentum of the particle. Show that a Bohr orbit for a hydrogen-like atom has a circumference that is an integral number of deBroglie wavelengths of an electron of that momentum.

2. (12 min) It can be shown that the circumference of a Bohr orbit is an integral number of deBroglie wavelengths of the electron of that energy. Calculate the energy of an electron in a Bohr orbit of an atom of helium if the circumference of the orbit is three times the deBroglie wavelength of the electron. The answer may be given in atomic energy units, that is, in units of the energy of the $n = 1$ state of the hydrogen atom as in the Bohr model.

---

3. (10 min) Show by what factor the velocity of an electron in a Bohr orbit for a hydrogen atom will change if the value of the principal quantum number n is doubled.

4. (10 min) The potential (not the total energy of the electron) in a hydrogen atom is 27 ev according to the Bohr theory. In units of 27 ev, what should be the ionization potential for He^+? Show your reasoning.

5. (10 min) The first ionization potential of Na is about 2/5 that of a hydrogen atom. Assuming that the Bohr equations apply to the outer electron of Na, but with an effective nuclear charge Z_{eff}, to what value of Z_{eff} does the ionization potential of Na correspond? Explain your reasoning.

6. (7 min) The ionization potential of the outer electron of sodium is 0.2 atomic energy units. Treating the Na atom as a hydrogen-like kernel plus the outer electron, find the effective nuclear charge seen by that electron.

---

7. (12 min) (final examination question) Calculate the effective nuclear charge that the valence electron of Li "sees" assuming the simple Bohr theory and given that the ionization potential of this electron is 5.4 ev.

8. (15 min) From the wave equation, the energy of an electron in a one-dimensional box is given by

$$\varepsilon = n^2 h^2/8ma^2$$

where n is the quantum number and a is the length of the box. If a is the length of a carbon–carbon bond, insertion of the numerical constants gives the result that the energy to promote an electron from $n = 1$ to $n = 2$ corresponds to light of a wavelength about 170 Å.

If this same model is applied to a long conjugated system of double bonds, the chief modifications are that one electron from each carbon atom occupies the lower energy levels, and, of course, the length of the box is now xa where x is the number of carbon atoms. Calculate the wavelength for the absorption of light to give the first excited state of a 12-carbon conjugated chain.

9. (10 min) The energy of a particle in a one-dimensional box is given by $\varepsilon = h^2/8ma^2$ (a is the length of the box; n is taken to be unity here). Show that the product of the momentum p of the particle and a is of the order of magnitude of h. Neglect relativistic effects.

10. (10 min) The ground-state energy of an electron in a one-dimensional box of 3 Å length is about 4 ev. The radius of a hydrogen atom is about 1 Å, and, as a very rough analogy, suppose that the hydrogen electron is thought of as being in a three-dimensional cubic box of 1 Å on the side. Estimate the energy of the hydrogen electron on this basis.

11. (10 min) The ground-state energy for an electron in a 3-Å one-dimensional box is about 100 kcal/mole. For what size three-dimensional box would the ground-state energy be equal to the average kinetic energy of the electron from thermal motion at 25°C?

12. (15 min) An electron confined to a one-dimensional box 1.4 Å in length has a ground-state energy corresponding to light of wavelength about 700 Å. Benzene, as a rough approximation, may be considered to be a two-dimensional box that encompasses the regular hexagonal shape. The C—C bond length in benzene is 1.4 Å, so the side of the box would be about 2.8 Å.

Estimate wavelength for transition from ground state to first excited state of benzene, assuming it is just the π-bonding electrons that are involved.

13. (12 min) A cubic box 10 Å on the side contains 8 electrons. Applying the simple particle-in-a-box theory, calculate the value of $\Delta\varepsilon$ for the first excited state of this system.

14. (12 min) A linear conjugated system, that is —C=C—C=C—C=, contains p carbon atoms in the chain; the average C—C distance may be taken to be 1.5 Å (and end effects neglected). If $\Delta\varepsilon$ for the transition from the ground to the first excited state is 2 ev, show what the value of p must be, according to the particle-in-a-box treatment.

---

15. (10 min) The eigenfunction for a $1s$ electron of a hydrogen-like atom is given by $\psi = (const)e^{-Zr/a^0}$ where a^0 is the radius of the first Bohr orbit for hydrogen. Show that the radius at which there is a maximum probability of finding a $1s$ electron (in any direction) is just $r_{max} = a^0/Z$.

16. (10 min) One of the eigenfunctions for a hydrogen-like atom behaves as is shown graphically in Fig. 20-2. The first figure shows the variation of ψ with r; the second two show how ψ varies in the xy and xz planes. Explain which eigenfunction this must be, that is, to what value of n, l, and m it

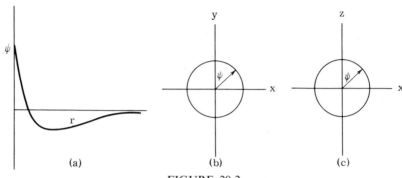

FIGURE 20-2

corresponds. [*Note:* Figs. 20-2b,c, are to be understood as giving the locus of points in the xy and xz planes for which ψ has a given constant value.]

17. (15 min) An excited carbon atom has the electronic structure $2p^6$. The electron density will then be proportional to $\psi^2_{p_x} + \psi^2_{p_y} + \psi^2_{p_z}$. Show by a trigonometric analysis what the geometric nature of the angular distribution of this electron density will be, assuming hydrogen-like wave functions.

18. (15 min) Consider a vector **r** whose direction in space is given by the angles θ and φ, according to the usual system (see Fig. 20-1). Derive the trigonometric relationships for the x, y, and z projections of this vector. Compare your result with the hydrogen-like wave functions for a p electron.

19. (20 min) The usual textbook illustration of p orbitals shows pairs of spheres on x, y, and z axes. The cross section through the xy plane is then

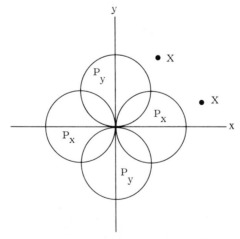

FIGURE 20-3

that shown in Fig. 20-3. Clearly, however, it cannot be true that there is no electron density at points such as those marked X. Construct a new representation in which you draw the boundaries of the region or regions in the xy plane in which the electron density of p_x electrons is *greater* than that of p_y electrons. [*Note:* The answer is obtainable with very little numerical calculation.]

20. (10 min) Assemble the Hamiltonian for the H_2 molecule. Explain which of the terms is likely to be the hardest to handle in evaluating the energy of this system of four particles.

21. (15 min) Consider a vector **r** oriented in space at angles θ and φ, as shown in Fig. 20-1. Show that the product of the projection on the x axis and that on the y axis gives a trigonometric function which corresponds to that for one of the d orbitals in the wave-mechanical solution for a hydrogen-like atom.

22. (10 min) A hydrogen-like wave function ψ_{nlm} is of the form $\psi_{nlm} = (\text{const})r^4 \sin^4 \theta \sin 4\varphi$. Explain what the values of n, l, and m are.

23. (15 min) Assemble the hydrogen-like wave function for $n = 4$, $l = 3$, $m = 3$. Sketch the polar plots for Θ^2 and Φ^2, and for their product.

24. (10 min) Give the angular part (excluding constants) of the hydrogen-like wave function for $n = 5$, $l = 4$, $m = 4$.

25. (15 min) The hydrogen-like wave functions for $n = 2$ are

$$\psi_1 = \frac{1}{4\sqrt{2\pi}}\left(\frac{Z}{a_0}\right)^{3/2} \rho e^{-\rho/2} \cos \theta$$

$$\psi_2 = \frac{1}{4\sqrt{2\pi}}\left(\frac{Z}{a_0}\right)^{3/2} \rho e^{-\rho/2} \sin \theta \cos \varphi$$

$$\psi_3 = \frac{1}{4\sqrt{2\pi}}\left(\frac{Z}{a_0}\right)^{3/2} (2 - \rho)e^{-\rho/2}$$

$$\psi_4 = \frac{1}{4\sqrt{2\pi}}\left(\frac{Z}{a_0}\right)^{3/2} \rho e^{-\rho/2} \sin \theta \sin \varphi$$

where a_0 is the radius of the first Bohr orbit for hydrogen, and $\rho = Zr/a_0$. The coordinate system is shown in Fig. 20-1.

Explain what the l and m quantum number values are for each of the above wave functions, or, alternatively, explain which is the ψ_{2s} function, which is the ψ_{2p_x} function, which is the ψ_{2p_y} function, and which is the ψ_{2p_z} function.

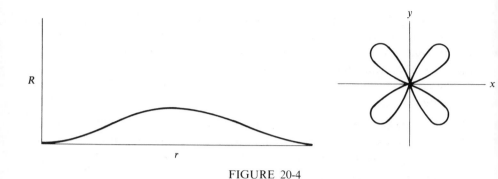

FIGURE 20-4

26. (15 min) (final examination question) Figure 20-4 shows a plot of the radial wave function R and a polar plot of the Φ wave function, in the xy plane, for a particular hydrogen-like orbital. Explain what the values of n, l, and m must be, or at least what can definitely be said about them.

27. (15 min) (final examination question) Figure 20-5 shows a sketch of the radial wave function R and of the polar plot of the Θ wave function, both in the xz plane, for a particular hydrogen-like orbital. Explain what the values of n, l, and m must be.

28. (11 min) A particular hydrogen-like orbital has its maximum extension in the xy plane, and in this plane, the polar plot is as shown in Fig. 20-6. The radial function has no nodes. Show what the values of n, l, and m must be.

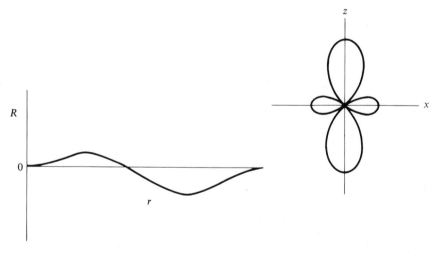

FIGURE 20-5

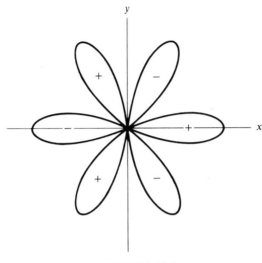

FIGURE 20-6

29. (9 min) A pair of hybrid orbitals can be formed by the combinations

$$\psi_1 = A(2s + 2p_x) \qquad \psi_2 = A(2s - 2p_x)$$

(a) Sketch the polar plot of the angular portion of ψ_1 and ψ_2 in the xy plane.
(b) Sketch the domains for ψ_1 and ψ_2 in the xy plane. A domain is that region in space (in the xy plane in this case) in which a given ψ function has a greater electron density than any other of the set being considered.

---

30. (12 min) The force constants for the HF, HCl, Cl—C, and C—C bonds are about 10, 5, 3.5, and 5, respectively (in units of 10^5 dyne/cm). Explain what the sequence would be if these bonds were arranged in order of increasing wave number of first vibrational absorption. You may assume that each of the above is a diatomic molecule, hypothetical in the case of Cl—C and C—C.

31. (8 min) The force constants for the H—Cl and D—Cl bonds are about the same, and the absorption frequency for the $v = 0$ to $v = 1$ transition for HCl occurs at $2,890\ \text{cm}^{-1}$. Calculate the absorption frequency for the $v = 0$ to $v = 1$ transition for DCl.

32. (8 min) Give, with a brief explanation, the number of fundamental vibrational absorption frequencies resulting from different modes of vibration that you would expect to observe in the infrared-absorption spectrum of Cl_2, HCl, H_2O, and NO_2 (assume the latter to be linear O—N—O).

33. (10 min) The normal modes of vibration are shown in Fig. 20-7 for H_2O, NO_2, CO_3^{2-} (a planar molecule), and ND_3 (a pyramidal molecule).

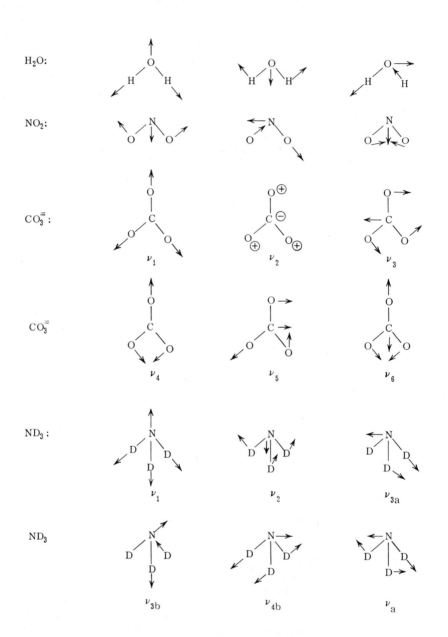

FIGURE 20-7

Plus and minus signs signify motion up and down, respectively. Explain which modes should be infrared active.

34. (10 min) A diatomic gas at high temperature shows a series of vibrational absorptions. By accident, some of the data are lost, but it is known that the absorption frequencies included 5,600, 11,200, and 14,000 wave numbers. Explain what the probable quantum-number assignments are for these three transitions. (The figures are hypothetical, since no deviation from the simple parabolic potential energy curve is assumed, and selection-rule restrictions are ignored.)

35. (15 min) The fundamental absorption frequency for HCl is at 2,890 cm^{-1}, which corresponds to a $\Delta\varepsilon$ value of 5.75×10^{-13} erg. Consider a series of homopolar molecules, $X-X$, the force constant for which is constant and equal to that for HCl. What should the atomic weight of X be in order that the fundamental absorption frequency for X_2 be equal to the average kinetic energy at 25°C? (This absorption would not occur, that is, the transition should not be infrared active, but in principle the $\Delta\varepsilon$ value could be obtained by other means.)

36. (20 min) The fundamental absorption frequency for $C\equiv N^-$ is about 2,000 cm^{-1}. Estimate the cyanide-stretching frequency that you might expect to observe in, say, $Fe(CN)_6^{3-}$. Base your estimate on two considerations. First, assume that a change in electron distribution occurs so that the cyanide group is more nearly $C=N$. Second, suppose that the carbon is so anchored by being bonded to the iron that the effective mass of the carbon is essentially infinite.

37. (10 min) The four fundamental vibrational modes for CO_2 are shown in Fig. 20-8. Only two fundamental absorptions are found in the infrared absorption spectrum, however. These are at 667 cm^{-1} and 2,350 cm^{-1}, with the former more intense. Explain how you would assign these frequencies.

38. (12 min) Assume that the first vibrational absorption ($v = 0$ to $v = 1$) occurs at 2,000 cm^{-1} for CO. Calculate k, the force constant for the CO bond. Calculate also the wave number for this same absorption in ^{14}CO.

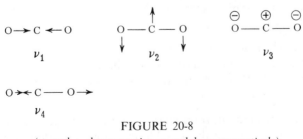

FIGURE 20-8

(*+ and − denote motion up and down, respectively*)

39. (15 min) (final examination question) Excitation to the first vibrational excited state of $H^{35}Cl$ occurs with infrared radiation of frequency $2,900\,cm^{-1}$. Calculate the expected position of this same absorption in the case of $D^{35}Cl$ (D is deuterium).

ANSWERS

1. Bohr's basic postulate is that angular momentum is quantized, that is, that $mvr = nh/2\pi$, where n is an integer. Then $2\pi r = nh/p$, since $p = mv$, and therefore $2\pi r = n\lambda$, since $\lambda = h/p$.

2. Since the energy of an electron in the field of a nucleus of charge Z is, by Eq. (20-4), proportional to Z^2/n^2, it can be written as $\varepsilon = (a.e.u)Z^2/2n^2$. By Eq. (20-1), $2\pi r = nh/mv$, or including Eq. (20-6), $2\pi r = n\lambda$. Thus $n = 3$ in the present case (and $Z = 2$). The energy of the electron in a.e.u. is then simply $2^2/3^2 = \mathbf{2/9}$!

3. The radius of a Bohr orbit is proportional to n^2 so r is quadrupled if n is doubled. From the basic postulate that $mvr = nh/2\pi$, we then conclude that v must be halved when n is doubled.

4. The potential energy is $-Ze^2/r$ and is thus twice the total energy, $-Ze^2/2r$. This last gives the ionization potential, which in units of e^2/r (i.e., units of 27 ev), is then $\frac{1}{2}$ for the hydrogen atom. Since the total energy is proportional to Z^2 (referring to the expression from which r has been eliminated, $E = -2\pi^2 me^4 Z^2/h^2 n^2$), it follows that the ionization potential for He^+, for which Z is two, should be four times larger or equal to 2 in units of 27 ev.

5. According to the Bohr equation,

$$E = -\frac{2\pi^2 me^4 Z^2}{h^2 n^2}$$

the ionization energy for Na should be $Z^2/9$ times that for hydrogen since $n = 3$ for the outer sodium electron. We take Z to be Z_{eff}, so $Z_{eff}^2/9 = \frac{2}{5}$, whence $Z_{eff} = \mathbf{1.9}$.

6. The outer electron of sodium is in an $n = 3$ level, so were the nuclear charge unity, the ionization energy would be $13.6/9$ or 1.5 ev. The actual energy is 0.2×27.2 or 5.44 ev. It then follows using Eq. (20-4), that $Z_{eff}^2 = 5.44/1.51$ or $Z_{eff} = \mathbf{1.9}$.

7. By Eq. (20-4) the ionization energy for the hydrogen atom is 13.6 ev. The last electron in Li is in the $n = 2$ level, whose ionization energy would be 13.6/4 or 3.4 ev if Z were unity. Evidently the effective Z is then given by $Z_{eff}^2 = 5.4/3.4$, whence $Z_{eff} = \mathbf{1.26}$.

8.   Let $\varepsilon = An^2$ for the case of the single bond or box of length a. Then $\Delta\varepsilon = A(4 - 1) = 3A$ for the transition from $n = 1$ to $n = 2$; hence $A = \Delta\varepsilon/3$. In the case of the 12-carbon chain, the 12 π electrons present will fill up the first six levels, so the transition to the first excited state will involve promoting an electron from $n = 6$ to $n = 7$ (by the Pauli principle, two electrons can go into each level, with spins opposed). In addition, for this larger box, $\varepsilon' = A'n^2$ where $A' = A/12^2$ since $a' = 12a$. $\Delta\varepsilon'$ for the desired transition is then $\Delta\varepsilon' = A'(7^2 - 6^2) = A'(49 - 36) = 13A' = 13A/144 = (13/144)(\Delta\varepsilon/3)$ $= 0.030\Delta\varepsilon$. Since for light, wavelength is inversely proportional to energy, the desired wavelength is $170/0.030 = \mathbf{5{,}700\ Å}$.

9.   For a nonrelativistic particle, we can write $\varepsilon = mv^2/2$, so $2m\varepsilon = (mv)^2$ $= p^2$. But $2m\varepsilon = h^2/4a^2$, so $p = h/2a$ and $pa = h/2$. If p is interpreted as the order of precision with which momentum can be known, and a, likewise for position, the conclusion that $pa = h/2$ amounts to a demonstration of the uncertainty principle.

10.   For a three-dimensional box the energy turns out to be the sum of ε_x, ε_y, and ε_z. Then for the ground state $\varepsilon_{box} = 3\varepsilon_{(1-D\,box)}$. Then for a box 3 Å on the side, we expect 3×4 or 12 ev energy. The hydrogen atom "box" is only 1 Å, however, and since ε is inversely proportional to a^2, the final expectation is $3 \times 4 \times 9 = 108$ ev.

11.   First, for a three-dimensional box we have 100 kcal for each direction $(\varepsilon_x + \varepsilon_y + \varepsilon_z)$ or 300 kcal. However, $E_{kin} = 3RT/2 = 0.9$ kcal/mole, which is a factor of about 330 smaller. Since the quantum-mechanical energy depends inversely on the square of the dimension, the size of the desired box must be $(330)^{1/2}$ or 18 times larger than the reference size of 3 Å, that is, 54 Å.

12.   As shown in Fig. 20-9, a regular hexagon can be constructed from six equilateral triangles, so the side of each is 1.4 Å, and a box enclosing the hexagon would have a side equal to 2.8 Å.

The formulas for the particle in a one- and two-dimensional box are

$$\varepsilon_1 = \frac{h^2}{8ma^2}n^2$$

and

$$\varepsilon_2 = \frac{h^2}{8ma^2}(n_1^2 + n_2^2)$$

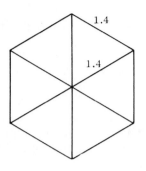

FIGURE 20-9

From the information given we know ε_1 for the case of $n = 1$ and $a = 1.4\,\text{Å}$; hence $\varepsilon_2 = (\varepsilon_1/4)(n_1^2 + n_2^2)$ for a two-dimensional box of side 2.8 Å. The succession of energy levels is that shown in Fig. 20-10, and it is seen that the six π electrons for benzene will fill the $(1, 1)$ and the $(2, 1)$ and $(1, 2)$ levels. The first transition will then be from a $(2, 1)$ or $(1, 2)$ level to a $(2, 2)$ level, and the change in $(n_1^2 + n_2^2)$ is then $8 - 5$ or 3. We therefore expect $\varepsilon_2 = 3\varepsilon_1/4$, and, since wavelength is inversely proportional to energy, the corresponding wavelength should be $700 \times \frac{4}{3}$ or 930 Å.

13. From Eq. (20-12), $\varepsilon = (37.8/a^2)(n_x^2 + n_y^2 + n_z^2)$ or, in this case, $\varepsilon = 0.378(n_x^2 + n_y^2 + n_z^2)$. The energy level scheme is as shown below, where the integers in parentheses give the various quantum number combinations. As indicated by Fig. 20-11, two electrons (with paired spins) go into each level, and eight electrons just fill the first and second (triply degenerate) levels. The first excited state then is obtained by promoting an electron from a $(1, 1, 2)$ to a $(1, 2, 2,)$ type level. $\Delta\varepsilon$ is thus $0.378(9 - 6) = \mathbf{1.13\,ev.}$

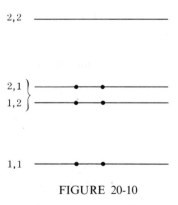

FIGURE 20-10

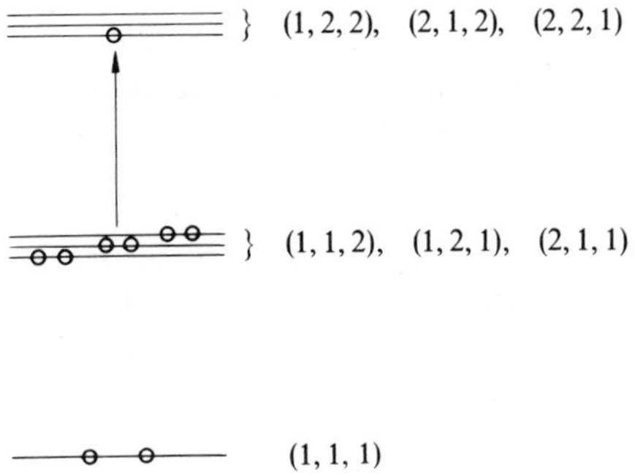

FIGURE 20-11

14. By Eq. (20-10), $\varepsilon = 37.8n^2/a^2$. Since end effects are to be neglected, $\overset{\circ}{a} = 1.5(p - 1)$. Further, the p carbon atoms supply p π electrons, which, when paired, just fill the first p levels. The transition must then be from the $n = p/2$ to the $n = p/2 + 1$ level. Thus

$$\Delta\varepsilon = 2 = \frac{37.8}{1.5^2(p - 1)^2}[(p + 1)^2 - p^2]$$

or

$$\frac{p + 1}{(p - 1)^2} = 0.12$$

On solving the above equation, the best integral value of p is found to be $p = 9.0$.

15. The probability of finding an electron in unit element of volume at distance r is given by ψ^2, but the chance of finding it at distance r, irrespective of direction, is given by $4\pi r^2\psi^2$. Then this last probability P is given by

$$P = (\text{const})r^2 e^{-2Zr/a^0}$$

and setting $dP/dr = 0$, we obtain

$$0 = (\text{const})[2re^{-2Zr/a^0} - r^2 2(Z/a^0)e^{-2Zr/a^0}]$$
$$= (\text{const})2re^{-2Zr/a^0}(1 - Zr/a^0)$$

The solution is then $Zr/a^0 = 1$ or $r = a^0/Z$.

16. The second two parts of Fig. 20-2 indicate that ψ does not depend in value on direction in space and therefore is spherically symmetric. It is thus an s function. There is one node, or crossing of the axis in the radial plot, which means that, with $l = 0$, n must be 2. The function is therefore that for a $2s$ electron.

17. The radial portions of the wave functions for ψ_{p_x}, ψ_{p_y}, and ψ_{p_z} are spherically symmetrical and need not be considered. Furthermore, the coefficients of the angular portions are the same, so that all we need to consider is the sum of the square of the trigonometric factors

$$
\begin{aligned}
\sum \psi^2 &= \cos^2 \theta + \sin^2 \theta \cos^2 \varphi + \sin^2 \theta \sin^2 \varphi \\
&= \cos^2 \theta + \sin^2 \theta \, (\cos^2 \varphi + \sin^2 \varphi) \\
&= \cos^2 \theta + \sin^2 \theta = 1
\end{aligned}
$$

Thus the sum of the squares turns out to be independent of direction, and the total electron density is spherically symmetric.

18. The z projection is independent of φ, and is simply $\mathbf{r} \cos \theta$. The projection on the xy plane is $\mathbf{r} \sin \theta$, and the projections of *this* on the x and y axis are simply $\mathbf{r} \sin \theta \cos \varphi$ and $\mathbf{r} \sin \theta \sin \varphi$, respectively. These trigonometric statements exactly correspond to the angular portions of ψ_{p_z}, ψ_{p_x}, and ψ_{p_y}, respectively.

19. First $\psi_{p_z}^2$ is zero, since in the xy plane $\theta = 90°$ and $\cos \theta$ is zero. The magnitudes of $\psi_{p_x}^2$ and $\psi_{p_y}^2$ are then proportional to $\cos^2 \varphi$ and $\sin^2 \varphi$. Now the cosine and sine of an angle are equal at 45°, so $\psi_{p_x}^2$ and $\psi_{p_y}^2$ must be equal at $\varphi = 45°$. It then follows that the boundary of equal electron density for the two functions is simply a line through the origin at 45° (or as shown in Fig. 20-12, two such lines).

20. As shown in Fig. 20-13, there are six interparticle distances. r_{AB} enters only in the internuclear-repulsion energy, which can be evaluated separately. The Hamiltonian is then that for electron 1 plus that for electron 2 plus the interelectron-repulsion term:

$$
\mathcal{H} = -\frac{h^2}{8\pi^2 m}\nabla_1^2 - \frac{h^2}{8\pi^2 m}\nabla_2^2 - \frac{e^2}{r_{1A}} - \frac{e^2}{r_{2A}} - \frac{e^2}{r_{1B}} - \frac{e^2}{r_{2B}} + \frac{e^2}{r_{12}}
$$

The last term, in r_{12}, is the most difficult to deal with in evaluating the energy, since it involves both electrons.

21. (See Fig. 20-14.) The projection of $\mathbf{r}$ on the xy plane is $\mathbf{r} \sin \theta$, and the projections of this on the x and y axes are $\mathbf{r} \sin \theta \cos \varphi$ and $\mathbf{r} \sin \theta \sin \varphi$.

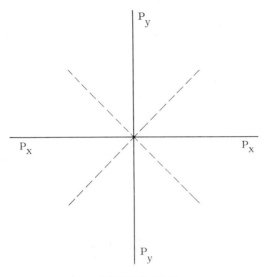

FIGURE 20-12

The product of these two projections is then $r^2 \sin^2 \theta \cos \varphi \sin \varphi$. But $\cos \varphi \sin \varphi = \sin 2\varphi$, so we have $r^2 \sin^2 \theta \sin 2\varphi$, which is the angular function for the d_{xy} orbital.

22. First, the radial portion of a hydrogen-like wave function is a simple power of r (rather than a polynomial in r) when l has the maximum possible value, and the power is then $(n - 1)$. Thus in this case **n = 5**, and **l = 4**. The θ function is a simple power of $\sin \theta$ if m has the maximum value for that l, and the power is then l. Thus in this case **m = 4**. This is further confirmed by the function in φ.

23. From the tables of functions (Tables 20-1 and 20-2) at the beginning

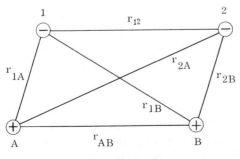

FIGURE 20-13

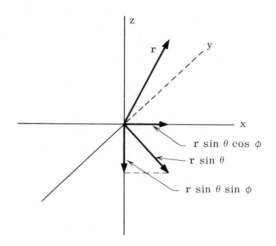

FIGURE 20-14

of this section, we obtain

$$\psi = [(Z/a_0)^{3/2}e^{-\rho/2}]\left[\frac{1}{96\sqrt{35}}\right][\rho^3]\left[\frac{\sqrt{70}}{8}\right][\sin^3\theta]$$

$$\times \left[\frac{1}{\sqrt{\pi}}\sin 3\varphi\right] \quad (\text{or } \cos 3\varphi)$$

The functions Θ and Φ, as far as their angular dependence is concerned, are then

$$\Theta^2 \cong \sin^6\theta \quad \text{and} \quad \Phi^2 \cong (\cos 3\varphi)^2$$

The Θ^2 function simply consists of two lobes, at $\theta = 90°$ and $180°$. These are narrow because of the high power to which $\sin\theta$ is raised. This is shown in Fig. 20-15a.

The Φ^2 function will show two lobes in the interval between 0 and $120°$; that is, it will be at a maximum at $\varphi = 0$, a minimum at $\varphi = \pi/6$ or $30°$, a maximum at $60°$, a minimum at $90°$, and so on, as shown in Fig. 20-15b.

The product $\Theta^2\Phi^2$ simply leads to six elongated lobes lying in the xy plane, as in Fig. 20-15c.

24. The Φ function will be proportional to the $e^{4i\varphi}$ or, alternatively, either to $\cos 4\varphi$ or $\sin 4\varphi$. Examination of the table of θ functions shows that, for $m = l$, Θ is proportional to $(\sin\theta)^l$. In this case, then, Θ is proportional to $\sin^4\theta$. The angular part of ψ is $\sin^4\theta \times e^{4i\varphi}$.

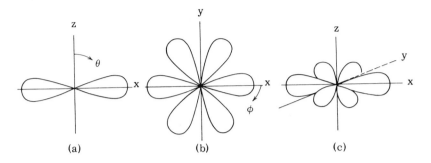

FIGURE 20-15

25. First, ψ_3 must be ψ_{2s} or the one with $l = m = 0$, since it is the only spherically symmetric function. Next, ψ_1 must be ψ_{2p_z}, since it is symmetric about the z axis; it also follows that for it, $l = 1$ and $m = 0$. ψ_2 and ψ_4 must then be for $l = 1$ and $m = \pm 1$. Further, ψ_2 is at a maximum at $\varphi = 0$ and ψ_4 is at a maximum when $\varphi = 90°$, so ψ_2 must be ψ_{2p_x} and ψ_4 must be ψ_{2p_y}.

26. Since the R function shows no nodes, $n - l - 1 = 0$ or $n = l + 1$. The Φ function must be one in sin ϕ since there is no maximum at $\phi = 0$; further, the presence of a maximum at $\phi = 45°$ means that $\boldsymbol{m} = \mathbf{2}$. If the maximum extension is in the xy plane, then the l value is the maximum possible for the m value, or $\boldsymbol{l} = \mathbf{2}$. Then $\boldsymbol{n} = \mathbf{3}$.

27. First, since there is one node in the R function, $n - l - 1 = 1$, or $n - l = 2$. Second, the Θ function is evidently symmetric about the z axis, and so corresponds to a function in cos ϕ only, and hence one for which $\boldsymbol{m} = \mathbf{0}$. There is only one polar node, which corresponds to the case of $l = 2$. It then follows that $\boldsymbol{n} = \mathbf{4}$.

28. Since there are no nodes to the radial function, $n - l - 1 = 0$, or $n - l = 1$; since the maximum extension is in the xy plane, the Θ function must only contain sin θ, and the m value must therefore be a maximum for that l. The lobes shown are entirely due to the Φ function, and since there is a polar node at 30°, or $\frac{1}{3}$ of 90°, it follows that $\boldsymbol{m} = \mathbf{3}$. The l value for which this m is the maximum possible is $\boldsymbol{l} = \mathbf{3}$. Finally, n is therefore **4**.

29. (a) As indicated in Fig. 20-16a, the positive regions combine to reinforce, and the negative ones, to partially cancel, to give ψ functions with principal lobes either in the plus or the minus x direction. (b) Clearly, ψ_1 dominates in all regions to the right of the y axis, and ψ_2, in all regions to the left of it. See Fig. 20-16b.

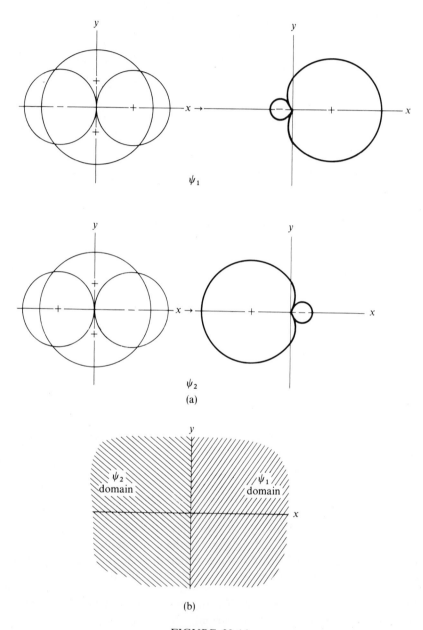

ψ_1

ψ_2

(a)

(b)

FIGURE 20-16

30. The energy change and hence the wave number corresponding to the first vibrational transition is proportional to $(k/\mu)^{1/2}$ where μ is the reduced mass. Per mole, the reduced masses for HF, HCl, Cl—C, and C—C will be roughly 1, 1, $12 \times 36/48$ or 9, and $12 \times 12/(2 \times 12)$ or 6. The ratios k/μ are then roughly 10, 5, 0.4, and 0.8. The order of increasing wave number is Cl—C, C—C, HCl, and HF.

31. Since the absorption frequency is proportional to $(k/\mu)^{1/2}$, it follows that

$$\frac{\Delta\varepsilon_{DCl}}{\Delta\varepsilon_{HCl}} = \left(\frac{k_{DCl}}{\mu_{DCl}}\frac{\mu_{HCl}}{k_{HCl}}\right)^{1/2} = \left(\frac{\mu_{HCl}}{\mu_{DCl}}\right)^{1/2}$$

The two reduced masses are 35.5/36.5 and 71/37.5 or 0.975 and 1.89. Their ratio is then 0.515, and the square root of the ratio is about 0.72. The absorption frequency for DCl will be $2,890 \times 0.72$ or about **2,080 cm^{-1}**.

32. Recalling Chapter 5, we can determine the number of vibrational degrees of freedom by the equation $f = 3n - 5$ for a linear molecule and $f = 3n - 6$ for a nonlinear one. There is thus only one mode for Cl_2 and for HCl. However, Cl_2 will not show any absorption, as no change in dipole moment is expected during vibration.

In the case of H_2O, $9 - 6$ or 3 modes are expected. These are all asymmetric, so all three should be infrared active. Finally, $9 - 5$, or 4 modes are expected for a linear O—N—O (actually it is bent). Of these, one is symmetric ($\leftarrow$O—N—O$\rightarrow$) so three infrared ones are expected. To summarize, the distinct absorption frequencies expected are Cl_2, none; HCl, one; H_2O, three; NO_2, three.

33. Vibrational modes will not be infrared active if they do not change the dipole moment of the molecule. On this basis, we see that all the H_2O modes are active. Those for NO_2, although arranged differently, are the same as for H_2O and all are active. The v_1 mode for CO_3^{2-} is a "breathing" mode that does not change the geometry of the molecule and hence should not be infrared active. The other five should be. In the case of ND_3, all should involve dipole-moment change and hence all should be infrared active.

34. The energy of a vibrational state is given by $\varepsilon = (v + \frac{1}{2})hv_0$, where $v = 0, 1, 2$, and so on. The energy difference between two states must, then, be some integer times hv_0, and, looking at the wave numbers given, the first two differ by 5,600, whereas the last two differ by 2,800. Evidently 2,800 is the largest possible value of hv_0, and the simplest assumption is that $hv_0 = 2,800$. The absorption at 5,600 cm^{-1} is then from $v = 0$ to $v = 2$; that at 11,200 cm^{-1} is from $v = 0$ to $v = 4$, and that at 14,000 cm^{-1} is from $v = 0$ to $v = 5$.

35. The fundamental absorption energy (i.e., from $v = 0$ to $v = 1$) is $\Delta\varepsilon = hv_0 = (1/2\pi c)(k/\mu)^{1/2}$; that is, $\Delta\varepsilon$ is proportional to $(k/\mu)^{1/2}$. Now the kinetic energy per mole at 25°C is $3RT/2$ or 890 cal/mole or 6.1×10^{-14} erg/molecule. Then

$$\left(\frac{k_{X_2}}{\mu X_{X_2}}\right)^{1/2} = \frac{6.1 \times 10^{-14}}{5.75 \times 10^{-13}}\left(\frac{k_{HCl}}{\mu_{HCl}}\right)^{1/2}$$

Since the k's are equal, the above reduces to $\mu_{X_2} = (9.5)^2 \mu_{HCl}$; μ_{HCl} is 35.5/36.5 or 0.97, so $\mu_{X_2} = 86$ and the atomic weight of X should then be $2 \times 86 = $ **172.**

36. The fundamental absorption frequency or corresponding $\Delta\varepsilon$ is proportional to $(k/\mu)^{1/2}$, so we look for what changes there should be in k and in μ. First, by analogy to the frequencies for stretching of C—C, C=C, and C≡C, we guess that k for C=N will be about $\frac{2}{3}$ of k for C≡N. Second, the reduced mass for a free CN molecule or ion is 6.5, $(12 \times 14/26)$, whereas that for a CN group anchored at the carbon would be $\infty \times 14/\infty = 14$. The new $\Delta\varepsilon$ and hence new frequency then is given by $(2/3)^{1/2}/(14/6.5)^{1/2}$ time the old one or, for the coordinated cyanide, we expect absorption at $2,000 \times 0.56$ or $1,120$ cm^{-1}.

37. v_1 should not be observed, as no dipole-moment change occurs with this symmetric vibration. v_2 and v_3 are actually identical, but in planes at right angles; they are thus degenerate. Since bending is easier than stretching, we would expect the 667-cm^{-1} absorption to be due to v_2 plus v_3; this conclusion is reinforced by the information that the 667-cm^{-1} absorption is more intense than the 2,350-cm^{-1} one. This would be expected, other things being equal, because of the twofold degeneracy. The 2,350-cm^{-1} absorption is then assigned to the v_4 asymmetric-stretching vibration.

38. By Eq. (20-19), $\Delta\varepsilon = (h/2\pi)(k/\mu)^{1/2}$, so $(k/\mu)^{1/2} = 4 \times 10^{-13} \times 2\pi/6.6 \times 10^{-27} = 3.8 \times 10^{14}$ sec^{-1}. μ in atomic weight units is $12 \times 16/28$ or 6.85; therefore

$$k = \frac{(3.8 \times 10^{14})^2 \times 6.85}{6 \times 10^{23}} = \textbf{16.5} \times \textbf{10}^{-5}\textbf{ dyne/cm}$$

If $\Delta\varepsilon'$ corresponds to the energy for the same transition in ^{14}CO, then $\Delta\varepsilon'/\Delta\varepsilon = (\mu/\mu')^{1/2}$. μ' is $14 \times 16/30$ or 7.47, and μ/μ' is therefore 0.915. $\Delta\varepsilon'$ is then $2,000 \times (0.915)^{1/2} = $ **1,900 cm^{-1}.**

39. The force constant is assumed to remain the same. The energy of the transition will therefore be proportional to $(1/\mu)^{1/2}$. μ for H^{35}Cl is $1 \times 35/36$ or 0.97, and μ of D^{35}Cl is $2 \times 35/37$ or 1.98. The new transition energy is then $\Delta\varepsilon = 2,900(0.97/1.89)^{1/2} = $ **2,100 cm^{-1}.**

21

ELEMENTARY GROUP THEORY AND CHEMICAL BONDING

COMMENTS

This last chapter constitutes an experiment in that the subject matter contained here is not normally a part of the undergraduate course in physical chemistry. I feel, however, that material of this type should be included in such a course, and we are now doing so at the University of Southern California. The reason is that, although the wave-mechanical model for atomic structure has indeed become quite important to current chemical thinking, its actual use for the most part consists in providing a language with which to discuss chemical bonding. The most recurring words in this vocabulary are ones such as sigma bonds, pi bonds, orbital overlap, hybrid-bond orbitals, and so on. The term *orbital*, moreover, is used almost exclusively to refer to the hydrogen-like orbitals given in the preceding chapter, hence their emphasis there. It further turns out that most of the qualitative conclusions about bonding can be deduced by determining which hydrogen-like orbitals have a spatial distribution that so fits in with the known molecular geometry as to allow the proper kind of overlapping of orbitals of adjacent atoms. Expressed in severely simple form, if the overlap involves regions of the orbitals of like sign, electrons available to occupy such orbitals are called bonding; if essentially no such sign correlation is present, the orbitals are called nonbonding.

The practice (one might almost say the art) of carrying out the detailed but still necessarily largely approximate calculations on the quantitative degree of bonding is one that belongs more to the field of chemical physics. I can see no particular justification for converting the undergraduate physical chemistry course into one on chemical physics. Physical chemistry lies at the heart of all chemistry; its roots are deeply grounded in the principles of the scientific method. It should not become a course in specialized mathematics.

The symmetry properties of molecules have always been of fundamental interest to chemistry, however, from the early days when the tetrahedral nature of the bonding around a carbon atom was first deduced from the occurrence of geometric and optical isomerism. It is thus really no accident that knowledge of molecular geometry still is crucial in applying today's sophisticated vocabulary of wave mechanics. We do owe a great debt to the mathematicians, to the mathematically inclined scientists, to the crystallographers, in that they have developed a logic of treating symmetry properties which permits great generality and rigor in drawing important conclusions about molecular properties.

This logic is that of the group-theoretical treatment of symmetry elements. As a mathematical discipline, it can be rather formidable, but if we extract only those portions of major importance to our purpose, and further confine these to the more straightforward situations, the topic becomes manageable. Of great help is the work of some chemists to provide reference books in this area. Several such are cited at the end of this section.

As has been the practice throughout this collection of problems, certain aspects of the general topic have been selected for major emphasis on the grounds that it is better to become reasonably familiar with a few important points than to be confused about many. Accordingly, the problems of this chapter fall into four groups. The first consists of a set of questions involving the nature of a group and how to determine the point group of an object or molecule. The second deals more specifically with the use of symmetry operations. The third and fourth groups undertake the application of group theory to the very important matter of determining the characters of the reducible representations of a set of sigma or pi bonds and of ascertaining from this the equivalent irreducible representations and the types of orbital functions that therefore form a basis for bonding.

In the section that follows, some of the important concepts and relationships that you will need are developed. Since the material is not to be found in the usual physical chemistry text, it is outlined in more detail than has been the practice in the corresponding sections of preceding chapters. This book is not intended to function as its own textbook, however, and it will still be necessary for you to do some studying among the references cited below. Good luck!

SUGGESTED REFERENCES

Chemical Applications of Group Theory, F. A. Cotton, Wiley–Interscience, New York, 1963.

Quantum Chemistry, H. Eyring, J. Walter, and G. E. Kimball, Wiley, New York, 1944.

Quantum Chemistry, K. S. Pitzer, Prentice-Hall, New York, 1953.

EQUATIONS AND CONCEPTS

Symmetry Elements and Operations

The objects that we shall be dealing with, namely, simple molecules, generally have a high degree of symmetry. In fact, they can usually be reduced to a collection of points representing centers of nuclei and arranged in some simple geometric figure. Thus $PtCl_4^{2-}$ takes the form shown in Fig. 21-1a, and for symmetry purposes, Fig. 21-1a may be further reduced to a stereotyped one as shown in Fig. 21-1b. In this latter drawing, plus signs and small circles denote otherwise equivalent points lying above and below the plane of the circle, respectively. If, while our backs are turned, the circle were rotated a quarter turn, on looking at it again we could not tell whether anything had been done to it. The points whose interchange could not be so detected are considered equivalent. If the points had each been labeled, say by writing a number beside each, however, the quarter rotation would have been evident. The points are then not identical. A full rotation, or none at all, would have left the figure in an identical configuration.

A symmetry element may be defined as a geometry operation that takes an object into an equivalent configuration, and one special symmetry operation, called E, is that which takes it into an identical one.

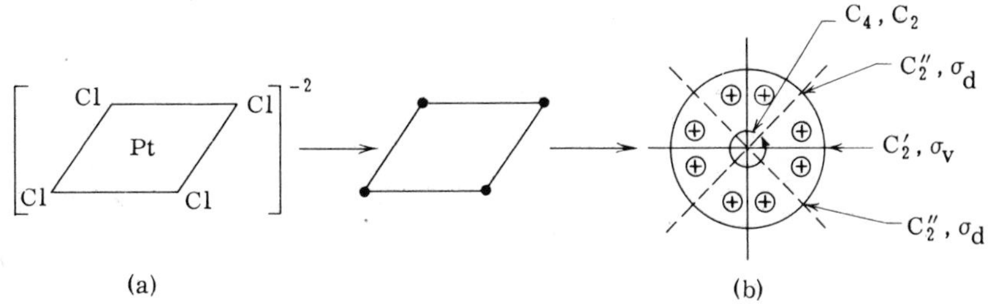

(a) (b)

FIGURE 21-1

There is a certain limited number of types of symmetry operations, and one kind of assignment you will have is that of determining which types are present in a given case. The first thing you seek is an axis of rotation, denoted C_n, where n is the number of positions in a full circle which produce equivalent configurations. There may be more than one, and the principal axis is that for which n is the largest. Except for the one or two cases of tetrahedral molecules, there will always be a single principal axis in the examples used here.

In the case of $PtCl_4^{2-}$, the principal axis is C_4. There are, in addition, some C_2 axes. First, there are two which go through diagonally opposite chlorines (shown as full lines in the stereotype), and then two more which go through the middle of opposite sides of the square (shown as dotted lines in the stereotype). These are different, and so are called C_2' and C_2''. The symmetry elements so far consist of E, C_4, C_2', and C_2''. Note that the C_2' and C_2'' axes are perpendicular to the C_4 axis.

Next, look for planes of symmetry. There are two main types. A plane perpendicular to the principal axis is called σ_h. Alternatively (or as well) there may be one or more planes of symmetry containing the principal axis. If there are no secondary axes, such a plane is designated σ_v. If there are secondary axes, there will be n of them (this is implied in the C_n operation), and one must also look for vertical planes of symmetry that bisect the angle between two secondary axes. Such planes are labeled σ_d. $PtCl_4^{2-}$ has a σ_h, two σ_v's, and two σ_d's.

A remaining type of symmetry is the center of inversion, denoted by i. Also σ_h may enter in combination with rotation on the principal axis; thus in Fig. 21-2 there is a C_2 axis, but a quarter rotation followed by reflection gives an equivalent configuration. Such a combined rotation–reflection operation is called S_n (S_4 in this example of Fig. 21-2).

For later purposes it is important to count up every distinct symmetry operation. A C_4 axis affords three: a quarter rotation or C_4^1, a half rotation

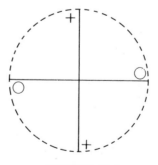

FIGURE 21-2

or C_2^4 (but designated as C_2^1 or just C_2), and C_4^3, or a three-quarter rotation. The S_4 axis similarly is broken down into its distinct stages: S_4^1, S_4^2 (but this is the same as C_2, so is omitted), and S_4^3. Thus for $PtCl_4^{2-}$ the entire set of distinct symmetry operations is

$$E, 2C_4, C_2, 2C_2', 2C_2'', i, 2S_4, \sigma_h, 2\sigma_v, 2\sigma_d \qquad \text{(16 in all)}$$

It will turn out that, if you carry out two symmetry operations in succession, the result will be identical to carrying out some one operation. That is, if you number the points that define the geometry of an object (as in a stereo-typed diagram) the result will be to have the points *as numbered* in a configuration identical to that which some single symmetry operation would give. Thus in the case of $PtCl_4^{2-}$, some trivial examples would be C_4^1 followed by $E = C_4^1$, or E followed by $\sigma_h = \sigma_h$. Other examples are C_4^1 followed by $C_4^1 = C_2$; S_4^1 followed by $i = C_4^1$; C_2' followed by $\sigma_h = \sigma_v$ (see if you can verify this one).

The order of a group is the number of symmetry operations that are involved. Thus the order for the C_3 group is 3, that for C_{2v} is 4.

Symmetry Groups

If one takes the collection of distinct symmetry operations that go with an object, it turns out that the product of any two of them will yield a third which is in the same set, and never a completely new one. Because of this, and other characteristics, such a set of symmetry operations is to be called a group. We speak of a group multiplication table as the table which gives the product of all possible pairs of symmetry operations.

As a second example, and one for which it will be easier to display the entire multiplication table, consider the molecule NH_3. This is pyramidal

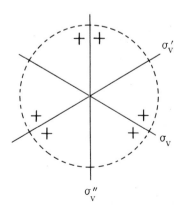

FIGURE 21-3

TABLE 21-1

C_{3v}	E	C_3^1	C_3^2	σ_v	σ_v'	σ_v''
E	E	C_3^1	C_3^2	σ_v	σ_v'	σ_v''
C_3^1	C_3^1	C_3^2	E	σ_v'	σ_v''	σ_v
C_3^2	C_3^2	E	C_3^1	σ_v''	σ_v	σ_v'
σ_v	σ_v	σ_v''	σ_v'	E	C_3^2	C_3^1
σ_v'	σ_v'	σ_v	σ_v''	C_3^1	E	C_3^2
σ_v''	σ_v''	σ_v'	σ_v	C_3^2	C_3^1	E

in shape, and the principal axis is C_3. We shall take rotation to be clockwise, and C_3^1 denotes a 120° degree rotation and C_3^2 a 240° rotation. There are no secondary axes; there is no σ_h but there are three σ_v planes. The stereotyped figure having this symmetry is shown in Fig. 21-3, and the multiplication table (which you should verify) is as given in Table 21-1. Notice, as will always be true, that each symmetry element occurs once in each row and that E either occurs along the diagonal or is symmetrically disposed to it.

There are a number of important types of symmetry in objects, and the groups that their sets of symmetry elements form have been organized into a classification scheme. First, there are groups of high symmetry, as represented by diatomic molecules ($C_{\infty v}$ and $D_{\infty h}$), tetrahedral molecules (T_d), regular octahedral molecules (O_h). We shall be concerned here mainly with groups for which there is a principal axis (called *rotational groups*), and the naming scheme is as follows:

1. C_n. The molecule possesses an axis of symmetry only.

2. C_{nv}. The molecule has also n σ_v planes. (If it has one it must have n as a consequence of the C_n axis.)

3. C_{nh}. The molecule has a σ_h plane.

If the molecule or object possesses a secondary twofold axis (and it must then have n of them), it must belong to one of the D groups:

4. D_n. There is no σ_h and no σ_d's.

5. D_{nh}. There is a σ_h.

6. D_{nd}. There is no σ_h but there is a σ_d (necessarily n σ_d's).

7. If in addition to the C_n principal axis there is an S_{2n} axis collinear with it, but no other symmetry elements, the group is S_{2n}.

8. If there is only a plane of symmetry, the group is C_s, and if no symmetry elements at all are present, it is called C_1.

Representations of Groups

An aspect of group theory that we shall not go into detail on is that, within a group, there can be a set or smaller group of symmetry elements that form what is called a *class*. In the case of the group C_{3v} for which the multiplication table was given above, the symmetry operations C_3^1 and C_3^2 form a class, designated as $2C_3$. Also the element σ_v, σ_v', σ_v'' form another class, designated as $3\sigma_v$. The number in front is the order of the class, and it is a rule that the order must be an integral factor of the order of the whole group.

Now it is generally possible to find a set of numbers whose multiplication table will be consistent with that of the given group. It will always be true, for example, that one can assign the number unity to each symmetry operation. Often there will be other simple choices in which 1 or -1 is assigned. Thus for the C_{3v} group two such assignments are

E	$2C_3$	$3\sigma_v$
1	1	1
1	1	-1

You can verify that use of these numbers is consistent with the multiplication table. In addition, however, there will be many possible ways in which a matrix can be assigned to each operation such that the multiplication table is obeyed. Such sets of numbers or matrices are called representations of the group, and there are in principle an infinite number of possibilities. There will, however, be some one set of representations that is simpler than all the others and cannot be expressed in terms of any other. Such representations are called the irreducible representations of the group. In the case of the C_{3v} group, there is a third irreducible representation in addition to the two given above, and this one is a two-by-two matrix.

It turns out, however, that for our purposes it is not necessary to have the detailed display of these matrix representations, but only to know the sum of their diagonal elements. This sum is called the *trace*[1] of the matrix. If, of course, the representation consists of simple numbers only, the trace is just the set of numbers. The collection of diagonal sums or trace that a given representation shows for the set of symmetry operations of the group is called the *character*[1] of the representation. Thus, in the case of the C_{3v} group, the complete character table is

C_{3v}	E	$2C_3$	$3\sigma_v$
A_1	1	1	1
A_2	1	1	-1
E	2	-1	0

[1] An alternative nomenclature designates the sum of the diagonal elements as the character rather than the trace of a matrix.

You can verify the correctness of the A_1 and A_2 representations, as noted above, by replacing the entries in the multiplication tables by their assigned number. The E representation, however, consists of two-dimensional matrices, and it is the actual matrices that would have to be multiplied to verify the multiplication table. Essentially, however, the traces serve to identify the representation for our purposes.

A partial compilation of character tables for various symmetry groups is given at the end of this introduction. As in the C_{3v} example above, each column is headed by the name of the class of symmetry operation, multiplied by the number of operations in that class. Each row corresponds to an irreducible representation, identified at the left by its conventional symmetry symbol; the set of traces of each row constitutes the character of that representation.

Reducible Representations

As is discussed further below, one can generate additional representations of a group by locating unit vectors along or perpendicular to the bonds in the molecule belonging to that symmetry group. If such a set of vectors is put through each of the symmetry operations, the changes in their x-, y-, and z-axis components form a set of matrices which constitute a representation of the group. This will in general be a reducible representation, however, and it becomes important to see into what irreducible representations it can be reduced. Fortunately, it is only necessary to know the character of the reducible representation in order to do this.

We have a reducible representation P and know its trace for each of the symmetry classes of operations S. P can then be expressed as a sum of irreducible representations I_i, each multiplied by a coefficient which gives the number of times I_i enters, that is, by a_i. If h denotes the order of the group, Eq. (21-1) holds.

$$a_i = \frac{1}{h} \sum_S \chi_P(S)\chi_{I_i}(S) \tag{21-1}$$

Here $\chi_P(S)$ denotes the trace of the reducible representation P with respect to the symmetry operation S, and $\chi_{I_i}(S)$ denotes the trace of the ith irreducible representation for that same symmetry operation. The sum is over all the symmetry operations. Each entry in the sum should be weighted by the order of the class of the particular symmetry operation.

An illustration is doubtless in order! Referring to the C_{3v} group again, the character of a reducible representation P is given below.

C_{3v}	E	$2C_3$	$3\sigma_v$
A_1	1	1	1
A_2	1	1	-1
E	2	-1	0
P	3	0	1

(21-2)

The desired coefficients are then

$$a_{A_1} = \tfrac{1}{6}[1 \times 3 + (2) \times 1 \times 0 + (3) \times 1 \times 1] = 1$$

$$a_{A_2} = \tfrac{1}{6}[1 \times 3 + (2) \times 1 \times 0 + (3) \times (-1) \times 1] = 0$$

$$a_E = \tfrac{1}{6}[2 \times 3 + (2) \times (-1) \times 0 + (3) \times 0 \times 1] = 1$$

Then $P \approx A_1 + E$.

Construction of Hybrid Orbitals

Recalling Chapter 20, we gave a fairly complete listing of hydrogen-like wave functions. With the exceptions of s functions, these all contained some type of trigonometric function, that is, were angularly dependent. These trigonometric functions can be expressed as some combination of projections of a unit vector on the x, y, and z axes as, for example, was asked for in Problem 20–21. For p orbitals, the trigonometric functions are $\cos \theta$, $\sin \theta \cos \varphi$, and $\sin \theta \sin \varphi$; these correspond to the z, x, and y projections of a vector, hence the designation p_z, p_x, and p_y. Similarly, the five d orbitals can be designated d_{z^2}, $d_{x^2-y^2}$, d_{xy}, d_{xz}, and d_{yz}. A hybrid orbital is simply a linear or additive combination of such individual orbitals, and the usual purpose of making such a combination is to obtain a set of hybrid functions whose angular shapes are identical but which are pointed in different directions in space. Thus a suitable set of combinations of an s and the three p orbitals gives a set of functions whose polar plots are identical but are oriented toward the four corners of a tetrahedron.

It is possible to apply our group theory to determine what types of orbitals can in principle be combined to give hybrid orbitals of a geometry that matches that of a given molecule. If each of the various p and d orbital functions is put through the symmetry operations of some group, they are found to transform in the same way as their algebraic subscript indicates. That is, p_x transforms as does the x projection of a vector, and so on. For example, the z projection of a vector is unchanged by any symmetry operation of the C_{3v} group, so the matrix giving its transformation is unity for each of the symmetry operations. The x and y projections get mixed up when carried through some symmetry operation of this group, and their transformation must be given by a matrix of order 2. The traces for these matrices turn out to be the same as for the E representation of the C_{3v} group, and hence the combination (x, y); therefore (p_x, p_y) is said to form the basis for the E irreducible representation.

These transformation properties have been worked out for the p and d orbital functions for all the symmetry groups, and if you look at tables of the characters of groups you will observe that along with each irreducible representation are given the simpler algebraic functions that transform in the same way. Thus the complete character table for C_{3v} is shown in Table 21-2.

The s orbital always transforms as does the A_1 representation, but in addition you see that p_z and d_{z^2} also do. There are no simple functions that transform as does the A_2 representation, but the combination (p_x, p_y) goes with the E representation, as do $(d_{x^2-y^2}, d_{xy})$ and (d_{xz}, d_{yz}). E representations are always two-dimensional (the characters are those of two-dimensional matrices), so there will always be a pair of orbitals that, together, go with this representation.

Note that E as a symbol for a representation has nothing to do with E as a symbol for the identity operation. Note also that the trace of the E operation gives the dimensionality of the representation. Thus A and B representations are always of dimension one, E representations of dimension two, and T representations of dimension three.

We have now reached a point where a fairly straightforward procedure can be described whereby the potentially available types of hybrid orbitals for sigma and for π bonding in a given molecule can be determined. By sigma bonding we mean that bonding occurs through electron occupancy of orbitals of adjacent atoms whose lobes are symmetric around the line joining the two nuclear centers. This is the ordinary single bond or electron-pair bond. The symmetry properties of such a bond can be represented by an arrow pointing from one atom to the other in the case of bonds between unlike atoms. Thus for ammonia, we would draw three arrows, pointing to each of the three hydrogens.

TABLE 21-2

$$C_{3v}$$

	E	$2C_3$	$3\sigma_v$	*Algebraic functions*	
A_1	1	1	1	z	$x^2 + y^2, z^2$
A_2	1	1	-1		
E	2	-1	0	(x, y)	$(x^2 - y^2, xy), (xz, yz)$

Such a set of arrows or vectors forms the basis for a reducible representation of the symmetry group (C_{3v} in this case); that is, the matrices formed by the new set of x, y, and z components after carrying out each of the symmetry operations would obey the group multiplication table. The representation is a reducible one, however, and it is necessary to express it in terms of the irreducible representations of the group. To do this, we only need to know the character of the reducible representation, and this can be obtained very simply. The procedure is to apply each of the symmetry operations of the group, in turn, to the set of vectors representing the sigma bonds, and determine which vectors remain unchanged in each case. That number is the trace of the reducible representation for that symmetry operation. Thus, in the case of NH_3, applying the symmetry operation E leaves all three vectors unchanged, so the trace is 3. The operation C_3 changes all of them, so the trace is zero. The reflection σ_v leaves one unchanged, so the trace is 1. These are just the traces of the representation P above [Eq. (21-2)]. We can then apply Eq. (21-1) to conclude that $P = A_1 + E$, and from the complete character table for C_{3v}, it follows that possible hybrid orbital combinations are $sp_x p_y$, $p_z p_x p_y$, d_{z^2}, $d_{xz} d_{yz}$, and so forth. In the case of nitrogen, we do not expect d orbitals to be energetically very available, so conclude that the hybridization is probably sp^2 or possibly p^3.

The example of ammonia illustrates what is, in a sense, a weakness or at least a complexity in the above approach. The use of bond directions to obtain the traces of the reducible representation can lead to difficulty if not all of the potentially bonding electrons of the atom are involved. In the case of NH_3, there is a lone pair of electrons which, while not used in $N-H$ bonds, should nonetheless be included. The arrangement of *electrons* around the nitrogen is thus nearly tetrahedral and not that of a triangular pyramid. The bonding electrons thus have a symmetry which is contained in the symmetry of the whole set of electrons. As a practical recognition of this situation, one must expect that the best over-all combination of orbitals will be some combination of the various possibilities derived from the C_{3v} group. In the case of NH_3, this means that some mixture of $(sp_x p_y)$ and $(p_x p_y p_z)$ is actually involved (still excluding d orbitals on energetic grounds). In terms of hybridization, linear combinations of s, p_x, p_y, and p_z can be made to give orbitals having tetrahedral symmetry, while linear combinations of s, p_x, p_y can give a set having triangular symmetry. Ammonia lies between these extremes of geometry. However, this type of complication will not be considered further since your introduction to group theory and bonding is intended to be just that—an introduction.

The treatment of π bonding is slightly more involved. As illustrated in Fig. 21-4, a π bond is one formed through the overlap of lobes above and below the line of nuclear centers, so that the bond does not have circular symmetry around this line. In most cases, and in all considered here, there will

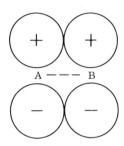

FIGURE 21-4

be a central atom that forms two or more bonds to other atoms. Typically, these secondary or coordinated atoms have p (rather than d or higher-type) orbitals available for bonding. Recalling that the p orbital (or actually its polar plot) consists of plus- and minus-sign lobes, the set of three p orbitals then form a set of three pairs of such lobes. One of these pairs will be collinear with the line of nuclear centers and hence potentially available for sigma bonding, but the other two will be at right angles to this line and to each other. For symmetry purposes, they can be represented by arrows at right angles to the line of nuclear centers.

Each set of arrows again forms the basis for a reducible representation. Again, the procedure is to carry out each symmetry operation of the group and to determine for each how many arrows are unchanged. That number is the trace of the reducible representation for that symmetry operation. There is now the additional possibility that a symmetry operation will invert the direction of an arrow but leave it otherwise unchanged. Each such instance contributes -1 to the trace. In this manner the sets of traces for the two types of π-bonding reducible representations are obtained, and by means of Eq. (21-1) they are then expressed in terms of reducible representations. The algebraic functions and hence orbitals that correspond to these tell you what are the possible types of hybrid-bond combinations.

As an example, a nonplanar molecule of the type AX_3 could possibly form π bonds with either of the two sets of p orbitals of the X groups. These are shown in Fig. 21-5, and the equivalent picture is the second one, which shows two sets of arrows all perpendicular to the AX bonds, but one set pointing upward (set 1) and the other set pointing sideways (set 2). The symmetry or point group is that of NH_3 or C_{3v}, and the symmetry operations are E, C_3, and σ_v. The E operation leaves all arrows unchanged, so the trace for each set is 3. The C_3 operation changes all arrows, so the trace for each set of arrows is zero. Finally, the σ_v reflection leaves one arrow unchanged in the case of set 1 and inverts the direction of one arrow in the case of set 2.

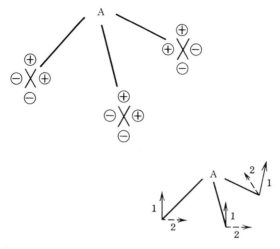

FIGURE 21-5

The characters are then

	E	$2C_3$	$3\sigma_v$
P_{π_1}	3	0	1^a
P_{π_2}	3	0	-1

a Same as for sigma bonds.

Then $P_{\pi_1} = A_1 + E$ and so would require the same type of hybrid orbital as would sigma bonding. It is usually assumed that sigma bonding takes precedence, so set 1 of orbitals might not be able to do much π bonding. On applying Equation 1 we find $P_{\pi_2} = A_2 + E$. Here again some use of the same orbitals would be needed as are involved in the sigma bonds; the A_2 representation does not correspond to any of the simple orbital types, so it is of no help. We conclude that our AX_3 molecule is able to get along mainly on its sigma bonds.

PROBLEMS

1. (10 min) Give the point group of (a) a book whose covers and pages are completely blank, (b) a normal book, and (c) the benzene molecule, as shown in Fig. 21-6. Explain your answers.

2. (20 min) Explain what the point group is for each member of the series

TABLE 21-3 CHARACTER TABLES

C_{2v}	E	C_2	$\sigma_v(xz)$	$\sigma_v'(yz)$		
A_1	1	1	1	1	z	x^2, y^2, z^2
A_2	1	1	-1	-1		xy
B_1	1	-1	1	-1	x	xz
B_2	1	-1	-1	1	y	yz

C_{3v}	E	$2C_3$	$3\sigma_v$		
A_1	1	1	1	z	$x^2 + y^2, z^2$
A_2	1	1	-1		
E	2	-1	0	(x, y)	$(x^2 - y^2, xy)(xz, yz)$

C_{4v}	E	$2C_4$	C_2	$2\sigma_v$	$2\sigma_d$		
A_1	1	1	1	1	1	z	$x^2 + y^2, z^2$
A_2	1	1	1	-1	-1		
B_1	1	-1	1	1	-1		$x^2 - y^2$
B_2	1	-1	1	-1	1		xy
E	2	0	-2	0	0	(x, y)	(xz, yz)

C_2	E	C_2		
A	1	1	z	x^2, y^2, z^2, xy
B	1	-1	x, y	yz, xz

C_{2h}	E	C_2	i	σ_h		
A_g	1	1	1	1		x^2, y^2, z^2, xy
B_g	1	-1	1	-1		xz, yz
A_u	1	1	-1	-1	z	
B_u	1	-1	-1	1	x, y	

D_2	E	$C_2(z)$	$C_2(y)$	$C_2(x)$		
A	1	1	1	1		x^2, y^2, z^2
B_1	1	1	-1	-1	z	xy
B_2	1	-1	1	-1	y	xz
B_3	1	-1	-1	1	x	yz

D_3	E	$2C_3$	$3C_2$		
A_1	1	1	1		$x^2 + y^2, z^2$
A_2	1	1	-1	z	
E	2	-1	0	(x, y)	$(x^2 - y^2, xy)(xz, yz)$

TABLE 21-3 CHARACTER TABLES (continued)

D_4	E	$2C_4$	C_2	$2C_2'$	$2C_2''$		
A_1	1	1	1	1	1		$x^2 + y^2, z^2$
A_2	1	1	1	-1	-1	z	
B_1	1	-1	1	1	-1		$x^2 - y^2$
B_2	1	-1	1	-1	1		xy
E	2	0	-2	0	0	(x, y)	(xz, yz)

D_{2h}	E	$C_2(z)$	$C_2(y)$	$C_2(x)$	i	$\sigma(xy)$	$\sigma(xz)$	$\sigma(yz)$		
A_g	1	1	1	1	1	1	1	1		x^2, y^2, z^2
B_{1g}	1	1	-1	-1	1	1	-1	-1		xy
B_{2g}	1	-1	1	-1	1	-1	1	-1		xz
B_{3g}	1	-1	-1	1	1	-1	-1	1		yz
A_u	1	1	1	1	-1	-1	-1	-1		
B_{1u}	1	1	-1	-1	-1	-1	1	1	z	
B_{2u}	1	-1	1	-1	-1	1	-1	1	y	
B_{3u}	1	-1	-1	1	-1	1	1	-1	x	

D_{3h}	E	$2C_3$	$3C_2$	σ_h	$2S_3$	$3\sigma_v$		
A_1'	1	1	1	1	1	1		$x^2 + y^2, z^2$
A_2'	1	1	-1	1	1	-1		
E'	2	-1	0	2	-1	0	(x, y)	$(x^2 - y^2, xy)$
A_1''	1	1	1	-1	-1	-1		
A_2''	1	1	-1	-1	-1	1	z	
E''	2	-1	0	-2	1	0		(xz, yz)

D_{4h}	E	$2C_4$	C_2	$2C_2'$	$2C_2''$	i	$2S_4$	σ_h	$2\sigma_v$	$2\sigma_d$		
A_{1g}	1	1	1	1	1	1	1	1	1	1		$x^2 + y^2, z^2$
A_{2g}	1	1	1	-1	-1	1	1	1	-1	-1		
B_{1g}	1	-1	1	1	-1	1	-1	1	1	-1		$x^2 - y^2$
B_{2g}	1	-1	1	-1	1	1	-1	1	-1	1		xy
E_g	2	0	-2	0	0	2	0	-2	0	0		(xz, yz)
A_{1u}	1	1	1	1	1	-1	-1	-1	-1	-1		
A_{2u}	1	1	1	-1	-1	-1	-1	-1	1	1	z	
B_{1u}	1	-1	1	1	-1	-1	1	-1	-1	1		
B_{2u}	1	-1	1	-1	1	-1	1	-1	1	-1		
E_u	2	0	-2	0	0	-2	0	2	0	0	(x, y)	

TABLE 21-3 CHARACTER TABLES (continued)

T_d	E	$8C_3$	$3C_2$	$6S_4$	$6\sigma_d$		
A_1	1	1	1	1	1		$x^2 + y^2 + z^2$
A_2	1	1	1	-1	-1		
E	2	-1	2	0	0		$(2z^2 - x^2 - y^2,$ $x^2 - y^2)$
T_1	3	0	-1	1	-1		
T_2	3	0	-1	-1	1	(x, y, z)	(xy, xz, yz)

O_h	E	$8C_3$	$6C_2$	$6C_4$	$3C_2$	i	$6S_4$	$8S_6$	$3\sigma_h$	$6\sigma_d$		
A_{1g}	1	1	1	1	1	1	1	1	1	1		$x^2 + y^2 + z^2$
A_{2g}	1	1	-1	-1	1	1	-1	1	1	-1		
E_g	2	-1	0	0	2	2	0	-1	2	0		$(2z^2 - x^2 - y^2,$ $x^2 - y^2)$
T_{1g}	3	0	-1	1	-1	3	1	0	-1	-1		
T_{2g}	3	0	1	-1	-1	3	-1	0	-1	1		(xz, yz, xy)
A_{1u}	1	1	1	1	1	-1	-1	-1	-1	-1		
A_{2u}	1	1	-1	-1	1	-1	1	-1	-1	1		
E_u	2	-1	0	0	2	-2	0	1	-2	0		
T_{1u}	3	0	-1	1	-1	-3	-1	0	1	1	(x, y, z)	
T_{2u}	3	0	1	-1	-1	-3	1	0	1	-1		

CH_4, CH_3Cl, CH_2Cl_2, $CHCl_3$, and CCl_4. Assume all like atoms equivalent in a given molecule.

3. (10 min) Explain what the symmetry point group is for water

O
/ \
H   H, and for half-deuterated water H

O
/ \
H   D.

4. (15 min) Explain what the point group must be for a molecule such as PCl_5 (this is a triangular bipyramid). To what lower symmetry group does one go if the two apical chlorides are not equivalent (e.g., if one is ^{36}Cl and the other ^{35}Cl)? Again, explain.

5. (10 min) A "lazy susan" has three equivalent salt shakers on it, positioned 120° apart. Although the salt shakers are themselves asymmetric in shape, they are placed on the lazy susan turntable in equivalent orientations. Show what the symmetry group of the lazy susan salt shakers is. The arrangement is shown in Fig. 21-7.

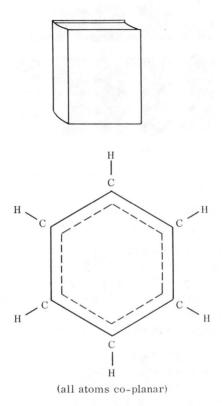

(all atoms co-planar)

FIGURE 21-6

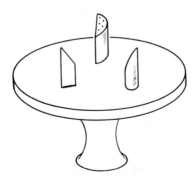

FIGURE 21-7

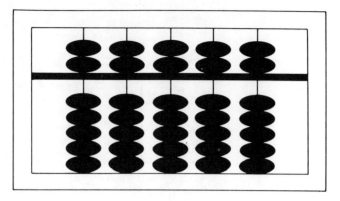

FIGURE 21-8

6. (10 min) Show what the point group is for a Chinese abacus whose counters are all in their lowest position, as shown in Fig. 21-8.

7. (10 min) Figure 21-9 shows a stereotype figure whose symmetry fits one of the point groups. The plus signs and circles represent points which are equivalent, except that those denoted by the plus sign lie above the plane of the circle and those denoted by the circles lie an equal distance below the plane. Show to which point group the figure belongs. In the process of doing this, enumerate the various symmetry elements present.

8. (10 min) The stereotype figure for one of the point groups is shown in Fig. 21-10. Determine what symmetry elements are present and explain to what point group they belong. The convention is that plus signs and small circles denote otherwise equivalent points which lie above and below the plane of the circle, respectively.

9. (15 min) List as many symmetry elements as you can for a regular tetrahedron.

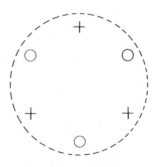

FIGURE 21-9

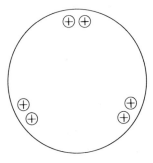

FIGURE 21-10

10. (10 min) The stereotype figure for one of the point groups is shown in Fig. 21-11. Determine what symmetry elements are present, and explain to what point group they belong. The plus signs denote equivalent points lying above the plane of the circle.

11. (10 min) To what symmetry group does *cis*-dichloroethylene belong? Explain. Assume the molecule to be planar.

12. (11 min) List all symmetry elements, and from this determine the point group to which the hypothetical complex ion species $CoCl_4^{2-}$ belongs. The ion has the shape of a square pyramid, as shown in Fig. 21-12.

13. (15 min) (final examination question) Assume that benzene has the structure shown in Fig. 21-13. The carbon atoms lie on a plane, but the double bonds do not "resonate" and are fixed in the positions shown. Alternate hydrogen atoms lie above the carbon plane, and the others, below. List all the symmetry elements and operations, and give the point group for this molecule. Although the hydrogens lie above or below the carbon plane, the

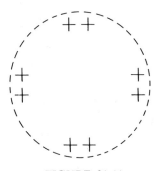

FIGURE 21-11

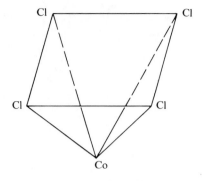

FIGURE 21-12

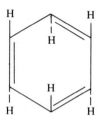

FIGURE 21-13

C—H bonds are in planes perpendicular to the carbon ring; the carbon ring may be assumed to form a regular hexagon.

14.   (18 min) Explain to what point group the following belong. (a) A triangular antiprism, that is, an equilateral triangular prism, the top and bottom of which have been rotated 60° relative to each other (Fig. 21-14a). (b) The square planar Pt(II) complex shown in Fig. 21-14b (ignore the hydrogens on the nitrogens). (c) CH_2ClBr. Assume the geometry to remain that of a regular tetrahedron.

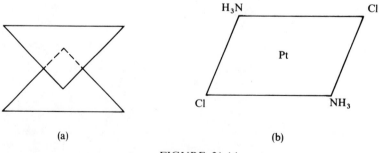

(a)   (b)

FIGURE 21-14

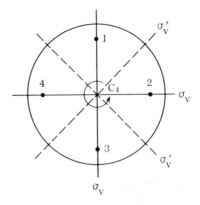

FIGURE 21-15

15. (15 min) (final examination question) Show what symmetry elements and operations are present and give the point group for (a) ortho-dichlorobenzene and (b) para-dichlorobenzene.

---

16. (15 min) The circle shown in Fig. 21-15 has four equivalent points, distinguished for the purpose of symmetry operations by the numbers 1 through 4. There is evidently a C_4 axis and there are the two σ_v planes shown by the full lines. Show that the presence of the above symmetry elements implies the presence of two additional σ_v' planes (shown as dashed lines).

17. (15 min) The stereotype figure for the C_{2v} group is shown in Fig. 21-16. The plus signs are distinguished for symmetry purposes by their numbers.

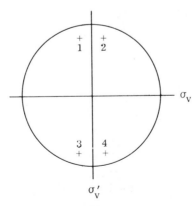

FIGURE 21-16

Work out the multiplication table for this group, using the form provided below.

	E	C_2	σ_v	σ_v'
E				
C_2				
σ_v				
σ_v'				

18.   (20 min) The molecule H_2O_2 has the symmetry C_{2h} when it is in the *trans*-planar configuration. Alternatively, a generalized figure is that of a circle having the two sets of points located in its plane, as shown in Fig. 21-17.

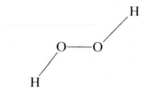

	E	C_2	i	σ_h
E				
C_2				
i				
σ_h				

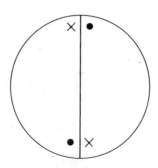

FIGURE 21-17

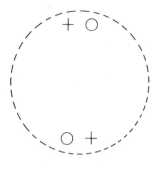

FIGURE 21-18

The points here represent objects symmetric above and below the plane of the circle so there is a σ_h. Work out the multiplication table for this group.

19. (15 min) Show that, if an object has a C_2 axis and one σ_v plane, there must be a second σ_v plane at a right angle to the first one.

20. (15 min) The stereotype figure for the D_2 group is shown in Fig. 21-18. The circles and plus signs denote otherwise equivalent points which lie below and above the plane of the circle, respectively. Determine the symmetry elements present and construct their multiplication table.

21. (10 min) The multiplication table for a hypothetical symmetry group is given below. Give the characters of two representations of this group.

	E	A	B	C
E	E	A	B	C
A	A	E	C	B
B	B	C	E	A
C	C	B	A	E

22. (10 min) The object in Fig. 21-19 consists of a circle and two points which are equivalent except that the one denoted by the plus sign lies above the plane of the circle and the other lies below the plane. Carry out the symmetry operations of the D_4 group to generate the stereotype figure for this symmetry.

If now the figure is to have D_{4h} symmetry, show what additional points must be generated.

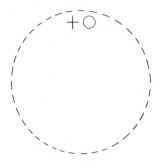

FIGURE 21-19

23. (9 min) The point group D_3 has the symmetry operations E, C_3^1, C_3^2, C_2, C_2', and C_2''. The stereotype figure is shown in Fig. 21-20. Draw lines to show your location of the C_2 axes, and show what the product $C_3^1 C_2$ is. In the figure, $+$ and $\bigcirc$ denote points lying above and below the circle, but otherwise the same.

24. (15 min) Given the stereotype drawing for the space group D_3 shown in Fig. 21-20, complete the multiplication table for the group. The classes are E, $2C_3$, and $3C_2$. Show whether the set of traces $E = 1$; C_3, $C_3^2 = -1$; and C_2, C_2', $C_2'' = 1$ is or is not the character of a representation of this group.

	E	C_3	C_3^2	C_2	C_2'	C_2''
E	E	C_3	C_3^2	C_2	C_2'	C_2''
C_3	C_3					
C_3^2	C_3^2					
C_2	C_2					
C_2'	C_2'					
C_2''	C_2''					

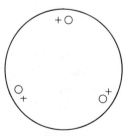

FIGURE 21-20

25. (15 min) The character table for the group D_3 is given together with the characters for a reducible representation P, except that one entry is missing. Explain what can be said about the makeup of P in terms of the irreducible representations, and what a possible value for the missing character might be.

	E	$2C_3$	$3C_2$
A_1	1	1	1
A_2	1	1	-1
E	2	-1	0
P	5	-1	?

26. (20 min) An example of a molecule having D_{3h} symmetry would be the species $Co(NH_3)_6^{3+}$, which has been postulated as a reaction intermediate, provided we ignore the hydrogens and just look at the geometry of nitrogens around the cobalt, as shown in Fig. 21-21.

The angular portion of the hydrogen-like wave function ψ_{p_z} is simply $\cos \theta$, and $\cos \theta$ is just the projection on the z axis of a unit vector. When put through the various symmetry operations, this z projection behaves in the same way as does one of the irreducible representations of the character table. Show which representation this is.

27. (15 min) The multiplication table for the group C_{4v} is shown below.

	E	C_4^1	C_4^2	C_4^3	σ_v	σ_v'	σ_d	σ_d'
E	E	C_4^1	C_4^2	C_4^3	σ_v	σ_v'	σ_d	σ_d'
C_4^1	C_4^1	C_4^2	C_4^3	E	σ_d	σ_d'	σ_v'	σ_v
C_4^2	C_4^2	C_4^3	E	C_4^1	σ_v'	σ_v	σ_d'	σ_d
C_4^3	C_4^3	E	C_4^1	C_4^2	σ_d'	σ_d	σ_v	σ_v'
σ_v	σ_v	σ_d'	σ_v'	σ_d	E	C_4^2	C_4^3	C_4^1
σ_v'	σ_v'	σ_d	σ_v	σ_d'	C_4^2	E	C_4^1	C_4^3
σ_d	σ_d	σ_v	σ'	σ_v'	C_4^1	C_4^3	E	C_4^2
σ_d'	σ_d'	σ_v'	σ_d	σ_v	C_4^3	C_4^1	C_4^2	E

(A molecule of this symmetry would be octahedral $Co(NH_3)_5Cl^{3+}$, if the hydrogens are ignored and only the geometry of the five nitrogens and the chlorine are considered.)

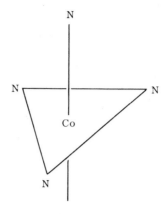

FIGURE 21-21

The following are two proposed representations:

	E	C_4	C_4^2	C_4^3	σ_v	σ_v'	σ_d	σ_d'
X	1	-1	1	-1	-1	-1	1	1
Y	1	1	1	-1	1	1	-1	-1

One is actually a representation and the other is not. Show which is not a representation. (This is not as frightful as it looks; reflect a moment on what is the minimum demonstration required.)

28. (20 min) A reducible representation in the T_d point group has the traces $E = ?$, $C_3 = 1$, $C_2 = -1$, $S_4 = -3$, and $\sigma_d = 1$. The trace under the symmetry operation E is missing. Explain what is the simplest choice for this missing number, and what the makeup of P would then be in terms of irreducible representations.

29. (15 min) The water molecule belongs to the point group C_{2v}. Determine the traces for a representation P consisting of the two $O-H$ sigma bonds, and from this, what types of orbitals could be used in forming the two sigma bonds.

30. (15 min) (final examination question) P denotes a reducible representation in the C_{3v} symmetry group. The traces for the E and σ_v operations are 3 and -1, respectively, but that for the C_3 operation is not given. Give, with explanation, a possible value for the missing trace, and the corresponding breakdown of P into irreducible representations. Describe a macroscopic object (i.e., not a molecule or its model) that would have C_{3v} symmetry.

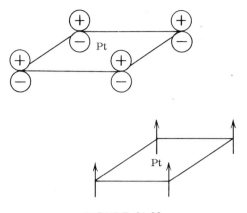

FIGURE 21-22

31. (15 min) As a pyramidal molecule, the point group for NH_3 is C_{3v}. Determine the traces for a representation P, consisting of three N—H bonds, and from this the irreducible representations to which it is equivalent and thus the possible types of orbitals that might be used in the bonding.

32. (20 min) The four sigma bonds formed by the central atoms of a square planar molecule such as $PtCl_4^{2-}$ (whose symmetry group is D_{4h}) can be constructed by the hybridization scheme dsp^2. Explain which d and p orbitals are likely to be the ones involved, from qualitative reasoning. Determine the traces for the reducible representation of the four sigma bonds, and determine by means of the character table which are the irreducible representations into which it goes. Show that your guess as to orbitals used is confirmed.

33. (20 min) Referring to Problem 32, there may also be π type bonds between the chlorines and the Pt atom. The chlorine p orbitals that would be involved come in two sets, one of which is shown in Fig. 21-22. For symmetry purposes, each p orbital can be replaced by an arrow, and by taking the set of four arrows through the D_{4h} symmetry operations, the set of traces for the reducible representation P_π is obtained.

Obtain this set of characters and determine to what set of reducible representations P_π is equal, and from this, decide which Pt orbitals might be used in π bonding.

34. (30 min) Consider a hypothetical transition metal M, which forms a planar trichloride in which the chlorines are all identical, as illustrated in Fig. 21-23. The point group is then D_{3h}.

(a) Determine the traces for the reducible representation consisting of the three sigma bonds that M could form, and then determine the equivalent

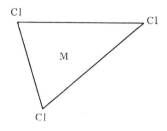

FIGURE 21-23

irreducible representations and from these the types of hybrid bonding that M could use.

(b) M might also form π bonds with chlorine p orbitals, and there are two sets which could be involved, as shown in Fig. 21-24. Set 1 has the axes of the p orbital lobes parallel to the C_3 axis, and set 2 has them perpendicular to it. Determine the representations P_1 and P_2 for these two sets of bonds (remember that if a symmetry operation leaves a p orbital unchanged, except that the signs are inverted, a contribution of -1 to the trace is made). Determine the irreducible representations equal to P_1 and to P_2, and from these, the types of hybrid bonds that M could make.

35. (20 min) You have been informed that, for a tetrahedral molecule such as methane, the bond hybridization of the carbon atom is sp^3. Show that the representation of four tetrahedrally oriented sigma bonds is indeed equal to a set of irreducible representations for which these particular orbital functions form a basis.

What alternative hybridization schemes are also possible?

It will help, in figuring out the traces for the sigma-bond representation, to draw a tetrahedron with apices at opposite corners of a cube.

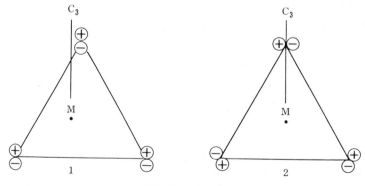

FIGURE 21-24

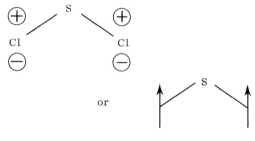

FIGURE 21-25

36. (15 min) Water belongs to the group C_{2v}, as does SCl_2. In the latter case there is some potentiality for π bonding. One set of chlorine π orbitals would be those lying in the plane perpendicular to that of the paper, as shown in Fig. 21-25. Determine the representation for the two π bonds of this type, and from this explain what possible type or types of hybridization of sulfur orbitals could form such bonds.

37. (12 min) Atoms A and B have the following orbitals available for bonding:

$$A: d_{xy} \quad \text{and} \quad d_{x^2-y^2} \qquad B: s, p_y$$

Which of the four combinations could produce sigma bonding along the bond axis? Which could produce π bonding? Which could produce neither? Use sketches to explain your answers.

ANSWERS

1. (a) The blank book has a twofold axis, as shown in Fig. 21-26 (i.e., the symmetry element C_2). There are no other axes, so it belongs to one of the C groups. There are two planes of symmetry containing C_2, so the point group is C_{2v}.

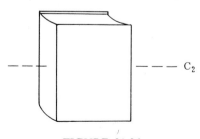

FIGURE 21-26

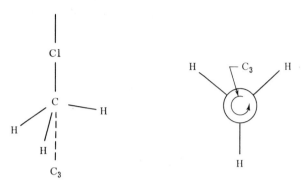

FIGURE 21-27

(b) The ordinary book has no elements of symmetry, so it belongs to the trivial group C_1.

(c) Benzene has a sixfold axis perpendicular to the plane of the molecule, and three twofold axes in the plane of the molecule going through the carbons, and three more going through the midpoints of opposite carbon–carbon bonds. The molecule then belongs to one of the D groups. There is a σ_h, so the point group is D_{6h}.

2. First, CH_4 and CCl_4 are easy. We know they are tetrahedral, and so belong to the group T_d. Next, CH_3Cl and $CHCl_3$ will belong to the same group. Examination reveals a threefold axis (in Fig. 21-27 we are looking down the C—Cl bond in CH_3Cl). There is no σ_h plane, but there are three σ_v planes. The point group is thus C_{3v}.

In considering CH_2Cl_2, it is helpful to place the molecule in a rectangular box, as shown in Fig. 21-28. We can find a C_2 axis, but none perpendicular to it and no σ_h planes. There are two σ_v planes, however, one containing the two hydrogens and the other containing the two chlorines. The group is then CH_2Cl_2 C_{2v}.

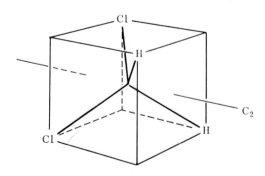

FIGURE 21-28

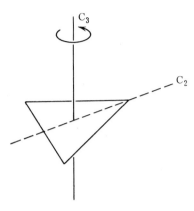

FIGURE 21-29

3. Water has a C_2 axis and two σ_v planes. There are no other symmetry elements, so it belongs to the group C_{2v}. HDO has only a plane of symmetry and thus belongs to the group C_s.

4. First, the principal axis is evidently C_3. Since there is a σ_h, there is also an S_3 axis coincident with the C_3. Next, we observe $3\sigma_v$, and three C_2 axes, as shown in Fig. 21-29. The molecule then belongs to the group D_{3h}. If the two apical chlorines are not equivalent, we still have the C_3 axis, but no σ_h and similarly no S_3. Further, the C_2 axes are now missing but the σ_v planes are still there. The group is then C_{3v}. Such a big change for such a small alteration in the chemistry!

5. From the description there is evidently a threefold axis C_3. Since the salt shakers are themselves asymmetric, there can be no σ_h, no σ_v, and no other axes of symmetry. The group must then be C_3.

6. There is one C_2 axis (coincident with the middle pin) and two σ_v planes. No other symmetry elements are present, so the group is C_{2v}.

7. First, there is clearly a C_3 axis (considering, for example, just the plus signs). However, rotation by $2\pi/6$ followed by reflection in the plane of the circle is also a symmetry operation; hence there is an S_6 axis. Finally, there is a center of inversion i. No other symmetry elements are present, so the point group is S_6.

8. First there is evidently the major axis C_3. There are in addition three C_2 axes at right angles, so we must be involved with one of the D groups. There is a σ_h, so the group is D_{3h}.

9. The answer to this one can be found in standard texts. Briefly, it helps greatly to draw the tetrahedron with its apices at alternate corners of a cube,

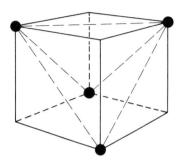

FIGURE 21-30

as shown in Fig. 21-30. You can see immediately that there are three C_2 axes, along the x, y, and z directions; next you might notice four C_3 axes, each through a cube diagonal.

Turning to symmetry planes, there is clearly one through each face diagonal, containing two apices of the tetrahedron and another through the other diagonal, or six in all. Finally, and perhaps less evidently, is the presence of three S_4 axes, parallel to the x, y, and z directions (and coincident with the C_2 axes).

10. The major axis is evidently C_4. There are no C_2 axes at right angles so we do not have one of the D groups. There is no S axis, and no σ_h. There are four σ_v planes, however; so the group must be C_{4v}.

11. There is a C_2 axis and $2\sigma_v$ but no σ_h (see Fig. 21-31). The group is then C_{2v}.

12. First, there is a C_4 axis, but no σ_h plane and no secondary axes. There are $2\sigma_v$ and $2\sigma_v'$ and the point group is evidently C_{4v}. The symmetry elements are then E, $2C_4$ (i.e., C_4^1 and C_4^3), $2\sigma_v$, and $2\sigma_v'$.

13. There is evidently a C_3 axis, but no σ_h, σ_v, or i. There are, however, three secondary C_2 axes, each passing through the midpoints of opposite pairs of carbon atoms. The point group is then D_3.

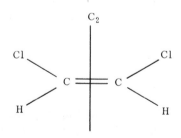

FIGURE 21-31

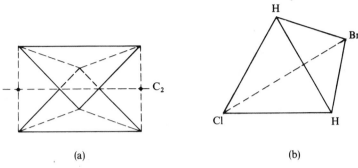

FIGURE 21-32

14. (a) There is a C_3 axis (and also an S_6 one), a center of inversion, $3C_2$, and $3\sigma_v$. The C_2 axes are perpendicular bisectors of lines drawn from a point of one triangle to a nearest point of the other triangle, as illustrated in Fig. 21-32a. There is no σ_h, and the point group is then $\boldsymbol{D}_{3d}$. The symmetry elements are E, C_3^1, C_3^2, C_2, C_2', C_2'', i, S_6^1, and S_6^5.

(b) There is one unique C_2 axis (perpendicular to the squares), and two secondary C_2 axes (through opposite Cl's and N's), as well as two σ_v planes and an inversion center. The group must be $\boldsymbol{D}_{2h}$, and the symmetry elements are E, C_2, C_2', C_2'', i, $\sigma(xy)\sigma(xz)\sigma(yz)$.

(c) Two alternative illustrations are given in Figs. 21-32b and 21-30. Evidently most of the symmetry elements of a regular tetrahedron are absent. There are no C_2 or C_3 axes, no σ_h, and only a plane of symmetry containing Cl and Br. The group is then $\boldsymbol{C}_s$.

15. (a) Most of the symmetry elements of benzene are gone; there is a C_2 axis in the plane of the ring and bisecting the C–C bond of the carbons to which Cl's are attached, and two σ_v's. There are no secondary axes, so the group is $\boldsymbol{C}_{2v}$. (b) There is a C_2 axis perpendicular to the ring, a σ_h, and two σ_v's. There are also two secondary C_2 axes, so the group must be $\boldsymbol{D}_{2h}$.

16. The formal procedure for the required type of demonstration is to show that the product of two symmetry operations gives the third, in this case, that $\sigma_v C_4 = \sigma_v'$. The sequence in Fig. 21-33 shows that this requirement is met.

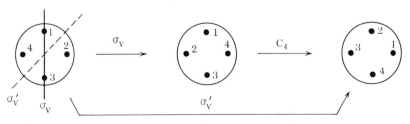

FIGURE 21-33

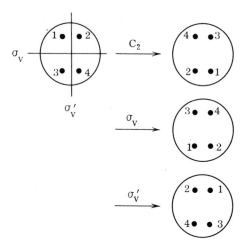

FIGURE 21-34

17. It is helpful to diagram the change that each symmetry operation brings about (Fig. 21-34). The resultant of two operations in succession is then easy to see. The table can then be filled out as shown below.

	E	C_2	σ_v	σ_v'
E	E	C_2	σ_v	σ_v'
C_2	C_2	E	σ_v'	σ_v
σ_v	σ_v	σ_v'	E	C_2
σ_v'	σ_v'	σ_v	C_2	E

18. It is helpful to diagram the change that each symmetry operation brings about (Fig. 21-35). The product of two can then be followed easily. The resulting table is that shown below.

	E	C_2	i	σ_h
E	E	C_2	i	σ_h
C_2	C_2	E	σ_h	i
i	i	σ_h	E	C_2
σ_h	σ_h	i	C_2	E

19. There are two ways of doing this that would be acceptable. First, a figure having C_{2v} symmetry would be a circle with two pairs of nonequivalent

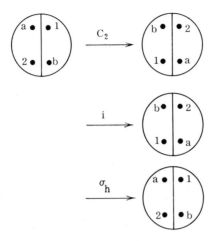

FIGURE 21-35

points (see Fig. 21-36). As shown in Fig. 21-37, the effect of carrying out C_2 and then σ_v is the same as carrying out σ_v'; hence a σ_v' exists.

Alternatively, and more elegantly, the operation C_2 on a general point (x, y, z) is (where we take the z axis as coincident with C_2)

$$(x, y, z) \xrightarrow{C_2} (-x, -y, z)$$

Reflection through the xz plane gives $(-x, y, z)$. This sequence of two steps is evidently equivalent to reflection on (x, y, z) through the yz plane; hence if one σ_v exists, the presence of C_2 implies the existence of a second σ_v' at right angles to the first plane.

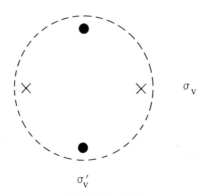

FIGURE 21-36

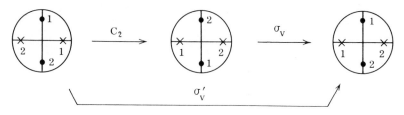

FIGURE 21-37

20. As illustrated in Fig. 21-38, the symmetry elements consist of a C_2 axis, a C_2' and a C_2'' axis, and, of course, E. As an aid to working out the multiplication table, the effect of each of the first three is illustrated. We then obtain the following:

	E	C_2	C_2'	C_2''
E	E	C_2	C_2'	C_2''
C_2	C_2	E	C_2''	C_2'
C_2'	C_2'	C_2''	E	C_2
C_2''	C_2''	C_2'	C_2	E

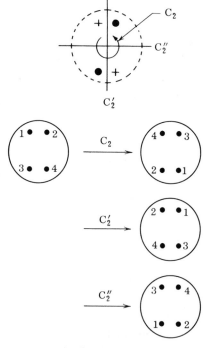

FIGURE 21-38

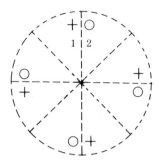

FIGURE 21-39

21. The multiplication table actually corresponds to the D_2 group. One representation is obviously obtained by assigning unity to each operation. A few trials will produce another one in which 1 or -1 is assigned to each operation. Three such are shown in the table below.

E	A	B	C
1	1	1	1
1	-1	-1	1
1	-1	1	-1
1	1	-1	-1

22. First, label the initial points as 1 and 2. Application of the C_4 symmetry operation generates the addition points shown in Fig. 21-39. Inspection shows that the figure now has the four C_2 axes indicated by the dashed lines, and required by the D_4 point group. No other symmetry elements are required. Inclusion of a σ_h plane simply adds a matching point above or below the existing ones.

23. The C_2 axes are shown in Fig. 21-40, and also the result of C_3^1 followed by C_2. This result is the same as C_2'.

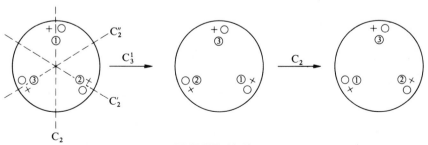

FIGURE 21-40

24.   The completed multiplication table is given below.

	E	C_3	C_3^2	C_2	C_2'	C_2''
E	E	C_3	C_3^2	C_2	C_2'	C_2''
C_3	C_3	C_3^2	E	C_2'	C_2''	C_2
C_3^2	C_3^2	E	C_3	C_2''	C_2	C_2'
C_2	C_2	C_2''	C_2'	E	C_3^2	C_3
C_2'	C_2'	C_2	C_2''	C_3	E	C_3^2
C_2''	C_2''	C_2'	C_2	C_3^2	C_3	E

By way of practical hints, first do the various C_3 and C_3^2 products, and C_2C_2, $C_2'C_2'$, and $C_2''C_2''$ since for these the answer follows almost by inspection. A great deal of time can be saved in filling out the remaining blanks by remembering that each symmetry operation must appear once in each row and each column.

The set of traces are **not** those of a representation of the group. It is only necessary to find one failure to establish this. For example, $C_3C_3 = C_3^2$, but $(-1)(-1) \neq -1$.

25.   The number of times the ith irreducible representation enters into the reducible representation P is given by $a_i = \frac{1}{6} \sum_S \chi_P(S)\chi_i(S)$ (the order of the group is six in this case). Let c denote the missing trace in the representation for P, then

$$a_{A_1} = \tfrac{1}{6}[1 \times 5 + (2) \times 1 \times (-1) + (3) \times 1 \times c]$$
$$= \tfrac{1}{6}(3 + 3c) = \tfrac{1}{2}(1 + c)$$

c can then have the values 1, 3, and so forth, and -1 (i.e., values such that $(1 + c)$ is divisible by two) to give a positive integer or zero.

$$a_{A_2} = \tfrac{1}{6}[1 \times 5 + (2) \times 1 \times (-1) + (3) \times (-1) \times c]$$
$$= \tfrac{1}{6}(3 - 3c)$$
$$= \tfrac{1}{2}(1 - c)$$

c can then have the values 1, -1, -3, and so forth.

$$a_E = \tfrac{1}{6}[2 \times 5 + (2) \times (-1) \times (-1) + (3) \times 0 \times c]$$
$$= \tfrac{1}{6}(12)$$
$$= 2$$

We can then say for sure that P contains $2E$. If we take $c = 1$ as a simple choice, then $P = 2E + A_1$. If $C = -1$, then $P = 2E + A_2$.

26. None of the symmetry operations converts a z component into an x or y component, and the only question is whether the z component is unchanged or reversed in sign. The traces of the z representation (or the representation for which z forms a basis) are then either 1 or -1. On going through the various operations, we have

$E:$ z unchanged, so write 1 $\sigma_h:$ z inverted, so write -1

$C_3:$ z unchanged, so write 1 $S_3:$ z inverted, so write -1

$C_2:$ z inverted, so write -1 $\sigma_v:$ z unchanged, so write 1

This sequence corresponds to the representation A_2''.

27. All that is needed is to find one instance where replacement of the symmetry operations by numbers fails to conform to the multiplication table. The X and Y representations differ first in the group of C_4 operations, so one can start there.

	representation X	representation Y
$(C_4)(C_4) = C_4^2$	$(-1)(-1) = 1$ OK	$(1)(1) = (1)$ OK
$(C_4)(C_4^2) = C_4^3$	$(-1)(1)$ $= -1$ OK	$(1)(1) = (1)$ wrong, since C_4^3 is supposed to be replaceable by -1.

Under the terms of the problem, the finding of this one mismatch is sufficient to determine that representation Y is incorrect. This might have been guessed immediately since C_4 and C_4^3 are in the same class and should have the same trace.

28. The number of times the ith irreducible representation enters into the reducible representation P is given by $a_i = \frac{1}{24} \sum_S \chi_P(S)\chi_i(S)$. (The order of the group is 24 in this case.)

On applying this equation to the A_1 representation we obtain

$$a_{A_1} = \tfrac{1}{24}[1 \times (p) + (8) \times 1 \times 1 + (3) \times 1 \times (-1) + (6)$$
$$\times 1 \times (-3) + (6) \times 1 \times 1]$$
$$= \tfrac{1}{24}(p + 8 - 3 - 18 + 6)$$
$$= \tfrac{1}{24}(p - 7)$$

Since the coefficient should be a small integer, i.e., 0, 1, 2, etc., possible values of p are 7, 31, 55. A good guess at this point would be 7.

Proceeding to the A_2 representation,

$$a_{A_2} = \tfrac{1}{24}[1 \times (p) + (8) \times 1 \times 1 + (3) \times 1 \times (-1) + (6)$$
$$\times (-1) \times (-3) + (6) \times (-1) \times 1]$$
$$= \tfrac{1}{24}(p + 8 - 3 + 18 - 6)$$
$$= \tfrac{1}{24}(p + 17)$$

Possible values of p are 7, 31, 55, for $a = 1, 2,$ or 3.

For the E representation we find $a_E = \tfrac{1}{24}(2p - 14)$, from which p could be 7, 19, 31 for $a = 0, 1,$ or 2. Next $a_{T_1} = \tfrac{1}{24}(3p - 21)$, from which p could be 7, 15, 23 for $a = 0, 1,$ or 2. And $a_{T_2} = \tfrac{1}{24}(3p + 27)$, from which p could be 7, 15, 23, 31 for $a = 2, 3, 4,$ or 5. The simplest choice for p is evidently 7, in which case the representation P is given by $A_2 + 2T_2$.

29.   The two sigma bonds, shown as arrows in Fig. 21-41, are unchanged by the E operation, are both changed by C_2, are unchanged by σ_v, and are both changed by σ_v'. The traces for P are then

	E	C_2	σ_v	σ_v'
P	2	0	2	0

The coefficients of the irreducible representations that make up this reducible representation are

$$a_{A_1} = \tfrac{1}{4}(1 \times 2 + 1 \times 0 + 1 \times 2 + 1 \times 0) = 1$$
$$a_{A_2} = \tfrac{1}{4}(1 \times 2 + 1 \times 0 - 1 \times 2 + 1 \times 0) = 0$$

Similarly $a_{B_1} = 1$ and $a_{B_2} = 0$. Then $P \approx A_1 + B_1$. Turning to the lists of algebraic functions, one could use a p_z and a p_x orbital (i.e., p^2 hybridiza-

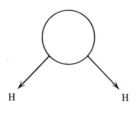

FIGURE 21-41

tion). Other possible combinations would be $p_z d_{xz}$, $d_{z^2} p_x$, and $d_{x^2} d_{xz}$; these are improbable in the case of water, since the oxygen atom has no d orbitals of low enough energy to be useful in bonding.

30. The order of the C_{3v} group is 6, so application of Eq. (21-1) to P must lead to coefficients a_i which are integral multiples of 6. For the A_1 representation, the missing trace could then be 0, 3, 6, and so forth. For the A_2 representation, it could again have the above values, and likewise for the E representation. A possible object would be a triangular pyramid whose base was an equilateral triangle.

31. The identity operation leaves the three bonds unchanged, so the E trace of P is 3. The C_3 operation changes the identity of all the bonds, so its trace is zero, and the σ_v operation leaves one bond unchanged, so its trace is 1. Thus for P

	E	$2C_3$	$3\sigma_v$
P	3	0	1

Applying the formula $a_i = \frac{1}{6} \sum_S \chi_P(S)\chi_i(S)$ (since the order of the group is six), we obtain

$$a_{A_1} = \tfrac{1}{6}[1 \times 3 + (2) \times 1 \times 0 + (3) \times 1 \times 1] = 1$$
$$a_{A_2} = \tfrac{1}{6}[1 \times 3 + (2) \times 1 \times 0 + (3) \times (-1) \times 1] = 0$$
$$a_E = \tfrac{1}{6}[2 \times 3 + (2) \times (-1) \times 0 + (3) \times 0 \times 1] = 1$$

Then $P = A_1 + E$. From the listing of algebraic functions which transform as do A_1 and E, possible orbital combinations would be $sp_x p_y$, $p_z p_x p_y$, $sd_{x^2-y^2} d_{xy}$, and so on.

32. If four sigma bonds, as shown in Fig. 21-42 by the arrows, are put through the various symmetry operations, we get the following traces for P.

E: all unchanged, so 4 i: all changed, so 0
C_4^1, C_4^3: all changed, so 0 S_4: all changed, so 0
C_2:   all changed, so 0 σ_h: none changed, so 4
C_2':   two unchanged, so 2 σ_v: two unchanged, so 2
C_2'':   all changed, so 0 σ_d: all changed, so 0

The answer to the first part of the question is that we look qualitatively for orbitals that might point toward the four chlorines. The p orbitals should then be p_x and p_y, and the d orbital should be $d_{x^2-y^2}$.

It is not really necessary to go down the entire list of irreducible representations to find each coefficient. From our guess as to which orbitals are involved, we would expect that the representation of the four sigma bonds P be expressible as $P = A_{1g} + B_{1g} + E_u$, since these three are the ones to which the expected orbitals correspond. As a check,

$$a_{A_{1g}} = \tfrac{1}{16}[1 \times 4 + (2) \times 1 \times 0 + 1 \times 0 + (2) \times 1 \times 2$$
$$+ (2) \times 1 \times 0 + 1 \times 0 + 2 \times 1 \times 0 + 1 \times 4$$
$$+ (2) \times 1 \times 2 + (2) \times 1 \times 0]$$
$$= \tfrac{1}{16}(4 + 4 + 4 + 4) = 1$$
$$a_{B_{1g}} = \tfrac{1}{16}(4 + 0 + 0 + 4 + 0 + 0 + 0 + 4 + 4 + 0) = 1$$
$$a_{E_u} = \tfrac{1}{16}(8 + 0 + 0 + 0 + 0 + 0 + 0 + 8 + 0 + 0) = 1$$

The guess is then confirmed and $P \approx A_{1g} + B_{1g} + E_u$. (All the other coefficients come out zero.)

33. The traces are as follows:

E: all arrows unchanged, so 4 σ_h: all arrows inverted, so -4
C_4^1 and C_2: all changed, so 0 σ_v: two arrows unchanged, so 2
C_2: two arrows inverted, so -2 σ_d: all changed, so 0
C_2: i, and S_4: all changed, so 0

Then

$$a_{A_{1g}} = \tfrac{1}{16}(4 - 4 - 4 + 4) = 0 \qquad a_{A_{2g}} = \tfrac{1}{16}(4 + 4 - 4 - 4) = 0$$
$$a_{B_{1g}} = \tfrac{1}{16}(4 - 4 - 4 + 4) = 0 \qquad a_{B_{2g}} = \tfrac{1}{16}(4 + 4 - 4 - 4) = 0$$
$$a_{E_g} = \tfrac{1}{16}(8 + 8) = 1 \qquad\qquad a_{A_{1u}} = \tfrac{1}{16}(4 + 4 - 4 - 4) = 0$$
$$a_{A_{2u}} = \tfrac{1}{16}(4 + 4 + 4 + 4) = 1 \qquad a_{B_{1u}} = \tfrac{1}{16}(4 - 4 + 4 - 4) = 0$$
$$a_{B_{2u}} = \tfrac{1}{16}(4 + 4 + 4 + 4) = 1 \qquad a_{E_u} = \tfrac{1}{16}(8 - 8) = 0$$

Therefore $P = E_g + B_{2u} + A_{2u}$. From the character table, we see that d_{xz}, d_{yz} together form a basis for the E_g representation and p_z forms the basis for the A_{2u} one; no simple functions correspond to B_{2u}. We conclude that the platinum could form π bonds using two of its d orbitals and its p_z orbital.

34. (a) For the set of three sigma bonds,

E: all unchanged, so $\chi = 3$ σ_h: all unchanged, so 3
C_3: all changed, so 0 S_3: all changed, so 0
C_2: one unchanged, so 1 σ_v: one unchanged, so 1

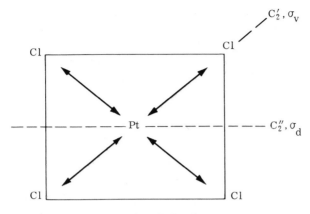

FIGURE 21-42

Then

$$a_{A_1'} = \tfrac{1}{12}[1 \times 3 + (2) \times 1 \times 0 + (3) \times 1 \times 1 + 1 \times 3$$
$$+ (2) \times 1 \times 0 + (3) \times 1 \times 1] = 1$$
$$a_{A_2'} = \tfrac{1}{12}(3 - 3 + 3 - 3) = 0$$
$$a_{E'} = \tfrac{1}{12}(6 + 6) = 1 \qquad \text{(the rest of the coefficients will be zero)}$$

Thus, $P = A_1' + E'$, and from the listing of functions that serve as bases for the representations, we see that a possible hybrid bond would be sp_xp_y. Other possibilities would be $d_{z^2}p_xp_y$, $sd_{x^2-y^2}d_{xy}$.

(b) For the first set of p orbitals,

E: all unchanged, so $\chi = 3$ $\qquad$ σ_h: all inverted, so -3
C_3: all changed, so 0 $\qquad$ S_3: all changed, so 0
C_2: one inverted, so -1 $\qquad$ σ_v: one unchanged, so 1

Then

$$a_{A_1'} = \tfrac{1}{12}(3 - 3 - 3 + 3) = 0 \qquad a_{A_2'} = \tfrac{1}{12}(3 + 3 - 3 - 3) = 0$$
$$a_{E'} = \tfrac{1}{12}(6 - 6) = 0 \qquad\qquad a_{A_1''} = \tfrac{1}{12}(3 - 3 + 3 - 3) = 0$$
$$a_{A_2''} = \tfrac{1}{12}(3 + 3 + 3 + 3) = 1 \qquad a_{E''} = \tfrac{1}{12}(6 + 6) = 1$$

Thus $P_1 = A_2'' + E''$, and the possible hybrid bond combination would be $p_z d_{xz} d_{yz}$.

Finally, for the second set of p orbitals,

E: all unchanged, so $\chi = 3$ $\qquad$ σ_h: all unchanged, so 3
C_3: all changed, so 0 $\qquad$ S_3: all changed, so 0
C_2: one inverted, so -1 $\qquad$ σ_v: one inverted, so -1

On evaluating the various coefficients we find $P_2 = A'_2 + E'$. No orbital functions of A'_2 symmetry are listed, and those of E' symmetry will have to compete with the sigma-bond hybrid. No good set of hybrid bonds seem possible in this case, although some bonding could occur with the use of the two orbitals $(p_x, p_y$ or $d_{x^2-y^2}, d_{xy})$ not used for sigma bonding.

Note: Although this problem has been quite long to work out in detail, the various operations are simple enough that you should have been able to write down P, P_1, and P_2 almost directly and have evaluated the coefficients almost by merely scanning the character table.

35. It is perfectly fair, in the interests of time, to solve this problem backwards. From the functions listed in the character table, we see that an sp^3 combination must correspond to $A_1 + T_2$. The characters for the sigma-bond representation must be such as to give coefficients of unity for the A_1 and T_2 representations and zero for the others. The trace, χ_E, for the E operation on four sigma bonds must be four, of course, and we can then write

for T_2 $a_{T_2} = 1 = \frac{1}{24}(4 \times 3 - 3\chi_{C_2} - 6\chi_{S_4} + 6\chi_{\sigma_d})$

Since the sum of terms inside the parentheses must be equal to 24, a good guess is that χ_{C_2} and χ_{S_4} are zero and that χ_{σ_d} is two. Then for the A_1 representation,

$a_{A_1} = \frac{1}{24}(4 + 8\chi_{C_3} + 0 + 0 + 12)$

This now requires that χ_{C_3} be unity. With these answers in mind, it is now easy to turn your figure of the tetrahedron and to verify the characters.

You can answer the second question directly by noting that (xy, xz, yz) form a basis for the T_2 representation. This means that an $sd_{xy}d_{xz}d_{yz}$, that is, an sd^3, hybridization would work equally well (if such d orbitals were low lying in energy).

36. The trace for the E operation is obviously 2 and that for C_2 is zero. If we take σ_v to be in the plane of the paper, the effect of this reflection is to exchange the plus and minus lobes of a π orbital or to invert the direction of the arrow; hence the trace is -2. The one for σ'_v will then be zero. Applying the usual formula,

$a_{A_1} = \frac{1}{4}(2 + 0 - 2 + 0) = 0$

$a_{A_2} = \frac{1}{4}(2 + 0 + 2 - 0) = 1$

$a_{B_1} = \frac{1}{4}(2 - 0 - 2 - 0) = 0$

$a_{B_2} = \frac{1}{4}(2 - 0 + 2 - 0) = 1$

Then $P = A_2 + B_2$, and $d_{xy}d_{yz}$ would be a possible basis for these representations. Another possibility would be $p_y d_{xy}$.

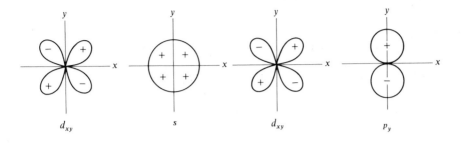

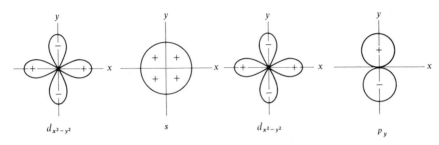

FIGURE 21-43

37. As illustrated in Fig. 21-43, the combination d_{xy-s} gives no bonding because the overlaps are symmetric but opposite in sign product. d_{xy-py} will give π bonding, $d_{(x^2-y^2)-s}$ will give sigma bonding, and $d_{(x^2-y^2)-py}$ gives no bonding.

APPENDIX
PARTIAL DERIVATIVES

COMMENTS

This short collection of problems on partial differentiation is added as an afterthought but, I hope, as a useful one. Quite commonly, students entering physical chemistry find that the calculus course they took or are taking leaves them unprepared to handle the types of partial differential relationships that occur, for example, in chemical thermodynamics.

The mathematics department is apt to regard the subject of partial differentiation as part of a more advanced course in partial differential equations and may therefore minimize the subject in the standard introductory calculus course. What is involved in beginning physical chemistry, however, is not the formidable matter of partial differential equations, but rather the fairly straightforward notion of a partial derivative, plus the relationships stemming from the concept of a total (or perfect, or exact) differential.

On first encounter with these relationships, it is easy to be a little suspicious of whether they really work; it is certainly easy to feel insecure about using them! The problems that follow are designed to help relieve this feeling. A simple system of related variables is set up, consisting of the length, width, perimeter, diagonal, and area of a rectangle. The problems then consist of finding a desired partial derivative in terms of known other ones. The desired partial can always be found directly, so you have the opportunity to check the correctness of your manipulations. These problems, incidentally, are not original; their source is lost in the antiquity of my student days!

EQUATIONS AND CONCEPTS

Partial Derivatives

If a quantity y depends on the variables u, v, and w, then

$$y = f(u, v, w)$$

then the following partial derivatives exist

$$\left(\frac{\partial y}{\partial u}\right)_{v,w} \quad \left(\frac{\partial y}{\partial v}\right)_{u,w} \quad \text{and} \quad \left(\frac{\partial y}{\partial w}\right)_{u,v}$$

The subscripts signify that these variables are held constant while the function is differentiated with respect to the third variable. Thus $V = nRT/P$ for an ideal gas, and

$$\left(\frac{\partial V}{\partial T}\right)_{n,p} = \frac{nR}{P} \quad \left(\frac{\partial V}{\partial n}\right)_{T,P} = \frac{RT}{P}$$

and

$$\left(\frac{\partial V}{\partial P}\right)_{n,T} = -\frac{nRT}{P^2}$$

Total Differentials

If u, v, and w change simultaneously by small increments, then, in the limit, the resulting change in $y = f(u, v, w)$ is

$$dy = \left(\frac{\partial y}{\partial u}\right)_{v,w} du + \left(\frac{\partial y}{\partial v}\right)_{u,w} dv + \left(\frac{\partial y}{\partial w}\right)_{u,v} dw \qquad (1)$$

Thus for an ideal gas

$$dV = \left(\frac{nR}{P}\right) dT + \left(\frac{RT}{P}\right) dn - \left(\frac{nRT}{P^2}\right) dP$$

Relationships between Partial Derivatives

For simplicity, we take y to be a function of two variables, u and v. We assume that the functions and their derivatives are continuous and single-valued, as is ordinarily true of physical–chemical relationships. Then

$$y = f(u, v) \quad \text{and} \quad dy = \left(\frac{\partial y}{\partial u}\right)_v du + \left(\frac{\partial y}{\partial v}\right)_u dv \qquad (2)$$

The condition may be imposed that y be held constant:

$$0 = \left(\frac{\partial y}{\partial u}\right)_v du + \left(\frac{\partial y}{\partial v}\right)_u dv \qquad (3)$$

Or, expressing the constancy of y in the equation itself,

$$\left(\frac{\partial y}{\partial u}\right)_v = -\left(\frac{\partial y}{\partial v}\right)_u \left(\frac{\partial v}{\partial u}\right)_y \tag{4}$$

or

$$\left(\frac{\partial v}{\partial u}\right)_y = -\frac{(\partial y/\partial u)_v}{(\partial y/\partial v)_u} \tag{5}$$

Thus for 1 mole of an ideal gas (i.e., $n = 1$),

$$\left(\frac{\partial P}{\partial T}\right)_V = -\left(\frac{\partial P}{\partial V}\right)_T \left(\frac{\partial V}{\partial T}\right)_P \tag{6}$$

Equation (6) may be checked by computing each partial derivative separately and substituting into the equation.

The condition may be imposed that some other dependent variable z is held constant:

$$\left(\frac{\partial y}{\partial u}\right)_z = \left(\frac{\partial y}{\partial u}\right)_v + \left(\frac{\partial y}{\partial v}\right)_u \left(\frac{\partial v}{\partial u}\right)_z \tag{7}$$

Thus energy is also a function of T and P, so

$$\left(\frac{\partial V}{\partial T}\right)_E = \left(\frac{\partial V}{\partial T}\right)_P + \left(\frac{\partial V}{\partial P}\right)_T \left(\frac{\partial P}{\partial T}\right)_E \tag{8}$$

Finally, if $y = f(u, v)$ as before, and u is some function of w, that is, $u = g(w)$, then

$$\left(\frac{\partial y}{\partial w}\right)_v = \left(\frac{\partial y}{\partial u}\right)_v \left(\frac{\partial u}{\partial w}\right)_v \tag{9}$$

Higher Derivatives

For functions of the type encountered in physical chemistry, the order of successive differentiation is immaterial. Thus

$$\frac{\partial}{\partial u}\left(\frac{\partial y}{\partial u}\right)_v = \left(\frac{\partial^2 y}{\partial u^2}\right)_v \tag{10}$$

and

$$\frac{\partial}{\partial v}\left(\frac{\partial y}{\partial u}\right)_v = \frac{\partial}{\partial u}\left(\frac{\partial y}{\partial v}\right)_u = \frac{\partial^2 y}{\partial u \, \partial v} = \frac{\partial^2 y}{\partial v \, \partial u} \tag{11}$$

Test for a Total Differential

Equation (2) may be written in the form

$$dy = A \, du + B \, dv \tag{12}$$

where $A = (\partial y/\partial u)_v$ and $B = (\partial y/\partial v)_u$. It is not necessarily true, however, that any equation of the form of (12) is the total differential of some function. Thus the expression

$$M \, du + N \, dv \tag{13}$$

will be a total of perfect differential of a function y only if

$$M = \left(\frac{\partial y}{\partial u}\right)_v \quad \text{and} \quad N = \left(\frac{\partial y}{\partial v}\right)_u \tag{14}$$

If relationships (14) hold, the derivative of M with respect to v must be equal to the derivative of N with respect to u:

$$\left(\frac{\partial M}{\partial v}\right)_u = \frac{\partial^2 y}{\partial v \, \partial u} = \left(\frac{\partial N}{\partial u}\right)_v \tag{15}$$

The test, then, of whether an expression such as (13) is a total differential is that the cross differentiation (15) gives the same result. Conversely, where we have a known total differential, the cross differential equation provides a useful additional equation. Such equations are known as *Euler relationships*. Thus

$$dE = T \, dS - P \, dV$$

is a known total differential from the combined first and second laws of thermodynamics. Then it must be true that

$$\left(\frac{\partial T}{\partial V}\right)_S = -\left(\frac{\partial P}{\partial S}\right)_V$$

PROBLEMS

Let a represent the area, p the perimeter, d the diagonal, b the breadth, and l the length of a rectangle. One can easily write down from analytical geometry all the various relationships between the above variables, and from these obtain directly a variety of partial differential quantities.

Thus,

$a = bl$

hence $(\partial a/\partial b)_l = l$, $(\partial a/\partial l)_b = b$, and so on. In this way a complete family of partial differentials could be evaluated. Alternatively, however, one partial differential can be obtained from another through the use of the various rules given above.

The following table of partial differentials is given

$$\left(\frac{\partial a}{\partial l}\right)_b = b \qquad \left(\frac{\partial l}{\partial b}\right)_d = -\frac{b}{l} \qquad \left(\frac{\partial p}{\partial b}\right)_l = 2$$

$$\left(\frac{\partial a}{\partial b}\right)_l = l \qquad \left(\frac{\partial d}{\partial b}\right)_l = \frac{b}{d} \qquad \left(\frac{\partial l}{\partial b}\right)_p = -1$$

In the following problems you are asked to find the stated partial derivative first by the direct method, and then by using only the partial derivatives given in the table above, plus the various rules for interconverting derivatives.

1. Find $(\partial a/\partial l)_d$.
2. Find $(\partial a/\partial b)_p$.
3. Find $(\partial a/\partial p)_l$.
4. Find $(\partial a/\partial b)_d$.
5. Find $(\partial a/\partial d)_l$.
6. Find $(\partial b/\partial a)_l$.
7. Find $(\partial l/\partial b)_a$.

Find, using appropriate derivatives from any of those given above, including Problems 1 through 7,

8. $(\partial d/\partial l)_p$.
9. $(\partial d/\partial b)_p$.
10. $(\partial a/\partial d)_p$.

11. Find $(\partial b/\partial p)_a$, and by appropriate change of variables derive $(\partial a/\partial p)_a$ from it.

12. Show that $(\partial P/\partial V)_T = -P/V$ for 1 mole of an ideal gas. By interchange of T and V find $(\partial P/\partial T)_V$. Confirm your result by a direct method.

ANSWERS

1. *Direct method*: $a = bl = (d^2 - l^2)^{1/2}l$ then

$$\left(\frac{\partial a}{\partial l}\right)_d = \frac{1}{2}\frac{1}{(d^2 - l^2)^{1/2}}(-2l)(l) + (d^2 - l^2)^{1/2}$$

$$= -l^2/b + b$$

Indirect method: We want to express $(\partial a/\partial l)_d$ in terms of the differentials in the table. We write

$$da = \left(\frac{\partial a}{\partial b}\right)_l db + \left(\frac{\partial a}{\partial l}\right)_b dl$$

By Eq. (7),

$$\left(\frac{\partial a}{\partial l}\right)_d = \left(\frac{\partial a}{\partial b}\right)_l \left(\frac{\partial b}{\partial l}\right)_d + \left(\frac{\partial a}{\partial l}\right)_b$$

and, using the table of differentials,

$$\left(\frac{\partial a}{\partial l}\right)_d = l\frac{-l}{b} + b = -\frac{l^2}{b} + b$$

2. *Direct method*: $a = lb$ and $p = 2b + 2l$ so $l = (p - 2b)/2$ and $a = (p - 2b)/2b$. Then

$$\left(\frac{\partial a}{\partial b}\right)_p = \tfrac{1}{2}(-2)b + \frac{p - 2b}{2} = -b + \frac{p - 2b}{2} = l - b$$

Indirect method: Consider a as a function of b and l. Then

$$da = \left(\frac{\partial a}{\partial b}\right)_l db + \left(\frac{\partial a}{\partial l}\right)_b dl$$

Divide by db, holding p constant:

$$\left(\frac{\partial a}{\partial b}\right)_p = \left(\frac{\partial a}{\partial b}\right)_l + \left(\frac{\partial a}{\partial l}\right)_b\left(\frac{\partial l}{\partial b}\right)_p$$

From the table of differentials,

$$\left(\frac{\partial a}{\partial b}\right)_p = l + b(-1) = \boldsymbol{l - b}$$

3. *Direct method:* $a = lb$ and $b = (p - 2l)/2$, so $a = l[(p - 2l)/2]$. Then $(\partial a/\partial p)_l = \boldsymbol{l/2}$.

Indirect method: By Eq. (9), $(\partial a/\partial p)_l = (\partial a/\partial b)_l(\partial b/\partial p)_l$, and from the table of values, $(\partial a/\partial p)_l = (l)(1/2) = \boldsymbol{l/2}$.

4. *Direct method:* We want to obtain $a = f(b, d)$. First, $a = bl$, and second, $l = (d^2 - b^2)^{1/2}$; so $a = b(d^2 - b^2)^{1/2}$. Then

$$\left(\frac{\partial a}{\partial b}\right)_d = (d^2 - b^2)^{1/2} + \frac{b}{2}\frac{(-2b)}{(d^2 - b^2)^{1/2}} = \frac{\boldsymbol{l - b^2}}{\boldsymbol{l}}$$

Indirect method: Try considering a as a function of b and l. Then

$$da = \left(\frac{\partial a}{\partial b}\right)_l db + \left(\frac{\partial a}{\partial l}\right)_b dl$$

On dividing by db at constant d,

$$\left(\frac{\partial a}{\partial b}\right)_d = \left(\frac{\partial a}{\partial b}\right)_l + \left(\frac{\partial a}{\partial l}\right)_b\left(\frac{\partial l}{\partial b}\right)_d = l + b\frac{-b}{l}$$
$$= \boldsymbol{l - b^2/l}$$

5. *Direct method:* We want $a = f(d, l)$, that is, $a = l(d^2 - l^2)^{1/2}$. Then

$$\left(\frac{\partial a}{\partial d}\right)_l = \tfrac{1}{2}l\frac{2d}{(d^2 - l^2)^{1/2}} = \frac{ld}{(d^2 - l^2)^{1/2}} = \frac{\boldsymbol{ld}}{\boldsymbol{b}}$$

Indirect method: By Eq. (9), we can write

$$\left(\frac{\partial a}{\partial d}\right)_l = \left(\frac{\partial a}{\partial b}\right)_l\left(\frac{\partial b}{\partial d}\right)_l = l\frac{d}{b} = \frac{\boldsymbol{ld}}{\boldsymbol{b}}$$

6. *Direct method:* We want $b = f(a, l)$, that is, $b = a/l$. Then

$$\left(\frac{\partial b}{\partial a}\right)_l = \frac{1}{l}$$

Indirect method: This is really trivial: $(\partial b/\partial a)_l = (\partial a/\partial b)_l^{-1} = 1/l$.

7. *Direct method:* We write $l = a/b$, so $(\partial l/\partial b)_a = -a/b^2 = -l/b$.
Indirect method: Consider $l = f(b, a)$; then

$$dl = \left(\frac{\partial l}{\partial b}\right)_a db + \left(\frac{\partial l}{\partial a}\right)_b da$$

By Eq. (3)

$$0 = \left(\frac{\partial l}{\partial b}\right)_a \left(\frac{\partial b}{\partial a}\right)_l + \left(\frac{\partial l}{\partial a}\right)_b$$

from which $(\partial l/\partial b)_a = -l/b$.

8. To find $(\partial d/\partial l)_p$:
Direct method: We want $d = f(l, p)$. Since $d = (l^2 + b^2)^{1/2}$ and $b = (p - 2l)/2$, we have

$$d = \left[l^2 + \frac{(p - 2l)^2}{4}\right]^{1/2} = \left(2l^2 - pl + \frac{p^2}{4}\right)^{1/2}$$

Then

$$\left(\frac{\partial d}{\partial l}\right)_p = \frac{1}{2} \frac{4l - p}{(2l^2 - pl + p^2/4)^{1/2}} = \frac{4l - p}{2d} = \frac{l - b}{d}$$

Indirect method: This one is a little more difficult. We have the variables d, l, and p, but none of the previous differentials involves these three. We must then use Eq. (7) to permute in a fourth variable so as to give us combinations, taken three at a time, which we do have. A little reflection gives

$$\left(\frac{\partial d}{\partial l}\right)_p = \left(\frac{\partial d}{\partial l}\right)_b + \left(\frac{\partial d}{\partial b}\right)_l \left(\frac{\partial b}{\partial l}\right)_p$$

But

$$\left(\frac{\partial d}{\partial l}\right)_b = -\frac{(\partial b/\partial l)_d}{(\partial b/\partial d)_l} = -\frac{-l/b}{d/b} = \frac{l}{d}$$

The other derivatives have already been obtained, so

$$\left(\frac{\partial d}{\partial l}\right)_p = \frac{l}{d} + \frac{b}{d}(-1) = \frac{l-b}{d}$$

9.   To find $(\partial d/\partial b)_p$:

Direct method: We want $d = f(b, p)$. We have $d = (l^2 + b^2)^{1/2}$ and $l = (p - 2b)/2$ so

$$d = \left[\frac{(p - 2b)^2}{4} + b^2\right]^{1/2}$$

Then

$$\left(\frac{\partial d}{\partial b}\right)_p = \frac{1}{2}\frac{-p + 4b}{(p^2/4 - pb + 2b^2)^{1/2}} = \frac{b-l}{d}$$

Indirect method: The procedure is much the same as for Problem 8. We write

$$\left(\frac{\partial d}{\partial b}\right)_p = \left(\frac{\partial d}{\partial b}\right)_l + \left(\frac{\partial d}{\partial l}\right)_b\left(\frac{\partial l}{\partial b}\right)_p$$

But

$$\left(\frac{\partial d}{\partial l}\right)_b = -\frac{(\partial b/\partial l)_d}{(\partial b/\partial d)_l} = -\frac{-l/b}{d/b} = \frac{l}{d}$$

Then

$$\left(\frac{\partial d}{\partial b}\right)_p = \frac{b}{d} + \frac{l}{d}(-1) = \frac{b-l}{d}$$

10.   To find $(\partial a/\partial d)_p$:

Direct method: First,

$$p^2 = [2(b + l)]^2 = 4d^2 + 8a$$

or

$$a = \frac{p^2}{8} - \frac{d^2}{2}$$

whence

$$\left(\frac{\partial a}{\partial d}\right)_p = -\boldsymbol{d}$$

Indirect method: We first write

$$\left(\frac{\partial a}{\partial d}\right)_p = \left(\frac{\partial a}{\partial d}\right)_l + \left(\frac{\partial a}{\partial l}\right)_d\left(\frac{\partial l}{\partial d}\right)_p$$

From the results of Problems 5, 1, and 8, respectively, we can evaluate the three quantities on the right to obtain

$$\left(\frac{\partial d}{\partial d}\right)_p = \frac{ld}{b} + \left(-\frac{l^2}{b} + b\right)\frac{d}{l-b} = -\boldsymbol{d}$$

11. To find $(\partial b/\partial p)_d$, and from it $(\partial a/\partial p)_d$, we first set

$$\left(\frac{\partial b}{\partial p}\right)_d = \left(\frac{\partial b}{\partial p}\right)_l + \left(\frac{\partial b}{\partial l}\right)_p\left(\frac{\partial l}{\partial p}\right)_d = \frac{1}{2} + (-1)\left(\frac{\partial l}{\partial p}\right)_d$$

Then

$$\left(\frac{\partial p}{\partial l}\right)_d = \left(\frac{\partial p}{\partial l}\right)_b + \left(\frac{\partial p}{\partial b}\right)_l\left(\frac{\partial b}{\partial l}\right)_d = \left(\frac{\partial p}{\partial l}\right)_b + 2\left(\frac{-l}{b}\right)$$

and

$$\left(\frac{\partial p}{\partial l}\right)_b = -\left(\frac{\partial p}{\partial b}\right)_l\left(\frac{\partial b}{\partial l}\right)_p = -(2)(-1) = 2$$

Then $(\partial p/\partial l)_d = 2 - 2l/b = (2b - 2l)/b$, and

$$\left(\frac{\partial b}{\partial p}\right)_d = \frac{1}{2} - \frac{b}{2b-2l} = -\frac{l}{2b-2l}$$

To find $(\partial a/\partial p)_d$, first write

$$\left(\frac{\partial a}{\partial p}\right)_d = \left(\frac{\partial b}{\partial p}\right)_d\left(\frac{\partial a}{\partial b}\right)_d = \left(-\frac{l}{2b-2l}\right)\left(\frac{\partial a}{\partial b}\right)_d$$

Then

$$\left(\frac{\partial a}{\partial b}\right)_d = \left(\frac{\partial a}{\partial b}\right)_l + \left(\frac{\partial a}{\partial l}\right)_b\left(\frac{\partial l}{\partial b}\right)_d = l + b\left(\frac{-b}{l}\right) = l - \frac{b^2}{l}$$

Finally,

$$\left(\frac{\partial a}{\partial p}\right)_d = -\frac{l}{2b - 2l}\frac{l - b^2}{l} = -\frac{l^2 - b^2}{2b - 2l} = \frac{b + l}{2} = \frac{p}{4}$$

12. For an ideal gas, $PV = RT$ or $P = RT/V$. Then

$$\left(\frac{\partial P}{\partial V}\right)_T = -\frac{RT}{V^2} = -\frac{P}{V}$$

For the second part, we write

$$\left(\frac{\partial P}{\partial T}\right)_V = -\left(\frac{\partial P}{\partial V}\right)_T\left(\frac{\partial V}{\partial T}\right)_P = \frac{P}{V}\left(\frac{\partial V}{\partial T}\right)_P$$

To confirm by the direct method, we note that $(\partial P/\partial T)_V = R/V$ and that $(\partial V/\partial T)_P = R/P$. Then $R/V \overset{?}{=} (P/V)(R/P) = R/V$.   Q.E.D.

INDEX